Publishing Sacrobosco's *De sphaera* in Early Modern Europe

Matteo Valleriani · Andrea Ottone
Editors

Publishing Sacrobosco's *De sphaera* in Early Modern Europe

Modes of Material and Scientific Exchange

 Springer

Editors
Matteo Valleriani
Max Planck Institute for the History
of Science
Berlin, Germany

Technische Universität Berlin
Berlin, Germany

Tel Aviv University
Tel Aviv, Israel

Andrea Ottone
Department of Economics, Management,
Quantitative
University of Milan
Milan, Italy

Max Planck Institute for the History
of Science
Berlin, Germany

ISBN 978-3-030-86602-0   ISBN 978-3-030-86600-6 (eBook)
https://doi.org/10.1007/978-3-030-86600-6

This Springer imprint is published by the registered company Springer Nature Switzerland AG
The registered company address is: Gewerbestrasse 11, 6330 Cham, Switzerland

Preface

The combination of the history of science and book history is not very common, despite the proximity of the two disciplines. But it becomes particularly urgent in research projects focused on a precise corpus of historical printed sources.

Such is the case for the project *The Sphere: Knowledge System Evolution and the Shared Scientific Identity of Europe*, for which a corpus of treatises and textbooks on cosmology and astronomy has been built to trace the process of homogenization of scientific knowledge that took place during the early modern period. After several studies concerned with the evolution of geocentric astronomy, the necessity emerged to investigate more closely all the historical actors involved in the production and circulation of such knowledge. Since the first phase of the research project, these actors were identified with the authors of the texts and those texts' producers, namely printers and publishers.

A previous international working group investigated the intellectual profile of the authors of the commentaries; its results were published in 2020 by Springer Nature in M. Valleriani (ed.). *De sphaera of Johannes de Sacrobosco in the Early Modern Period: The Authors of the Commentaries*. The present volume is the second part of the same research endeavor.

Until the end of 2018, it was still highly unclear how to approach the investigation of the producers of university textbooks in the early modern period. On the one hand, it had become evident that such actors played a decisive role because they were the last level of the decisional process: it was the printer or the publisher who ultimately decided which scientific works and which illustrations would constitute a new textbook in astronomy. Printers and publishers were also responsible for textbooks as products—and therefore for their costs and end prices. In this respect, they held influence over one of the most relevant parameters that could determine the success (or not) of a specific textbook and, with that, the spread (or not) of a specific scientific aspect.

On the other hand, there was almost no literature concerning the history of early modern textbook production and even less about its marketing and distribution.

More general investigations and studies were therefore needed in order to begin, as a historian of science, my first incursion into the history of the book. Fortunately, I soon encountered a monograph that offered me the fundamentals to start with and, as it turned out soon thereafter, much more. This monograph is the seminal work of Angela Nuovo, *The Book Trade in the Italian Renaissance*, republished by Brill in 2013. Angela Nuovo's work clearly showed me that I was missing an important piece of the puzzle concerning the process of evolution of knowledge, namely the understanding not just of printed book production in the early modern period but, even more relevantly, of the rules of the academic book market, and of the business model within whose framework printers and publishers acted while operating in that market. I needed to look at the editions of the *Sphaera* corpus from the perspective of the early modern manager who produced and sold them. When the first opportunity occurred, I traveled to Milan and met Angela in the fabulous spaces of Braidense National Library.

Angela helped me understand fundamental aspects of the early modern book market, especially in relation to the emergence of the *privilegium* as a means of protecting a product rather than the knowledge within. Yet both of us recognized that the academic book market was still largely an unexplored field; for the most part, we could only guess about the real business opportunities and difficulties offered by that

Fig. 1 Back row from left: Olga Nicolaeva, Stefano Gulizia, Angela Nuovo, Oliver Duntze, Saskia Limbach, Victoria Beyer. Back row from right: Jochen Büttner, Falk Eisermann. Middle row from left: Alissar Levy, Manuela Bragagnolo, Insa Christiane Hennen, Isabelle Pantin, Teresa Hollerbach, Catherine Rideau-Kikuchi, Christoph Sander, Ian Maclean, Paul Grendler, Alejandra Ulla Lorenzo. Front from left: Andrea Ottone, Matteo Valleriani, Leo Corry. Richard L. Kremer was absent at the moment of the photograph

market. Finally, Angela added a decisive point to the discussion: she told me that I could easily continue this exploration by requesting the support of a connoisseur of the early modern book world, someone who worked with her but was living in Berlin like me. It is here that Andrea Ottone came into the game. The present volume is the result of what we discussed the very first time we met, one week after the Braidense rendezvous. Andrea took up the challenge.

First of all, we defined the problem and recognized that we needed help. We described our research question and prepared what we called the Mission Statement. By means of this document, we contacted the contributors to this volume and asked them to write a chapter of the book, focusing on one or more of the aspects mentioned in the Statement. They did it, and once all the chapters were ready, we invited them and a number of experts to Berlin to act in the capacity of discussants.

In February 2020, we met for three days and discussed all the individual chapters (Fig. 1). After this meeting, we edited new versions of the chapters on the basis of the feedback we had mutually provided to one another. What followed is an editorial history comprised of a continuous and assiduous correspondence among the contributors and especially with the editors, reviewers, librarians, and so forth, on matters of content, style, formalities, image copyrights (because illustrations matter when a volume pertains to book production!), and the *cosmic* work of achieving a standard for bibliographic references—when investigating subject matter at the intersection of the history of science and book history, antique books become both the means and the object of investigation.

Berlin, Wilmersdorf   Matteo Valleriani
March 19, 2021

Acknowledgments

First of all, we would like to thank the contributors to this volume and the discussants who attended the February 2020 meeting in Berlin. The list of people without whom we would never have been able to realize this project is very long. The first place certainly goes to Victoria Beyer, who assiduously supported the project from the very beginning by preparing the first reader of the chapters to discuss at the meeting, helping to organize the meeting itself, formatting the review and production copy of the entire volume, preparing the indexes, and finally controlling the proofs during each stage of production. During the phase of production, we also enjoyed the invaluable support of Nana Citron, whom we would also like to thank. We would also like to express our gratitude to Megan Drinkwater for the relevant help she provided in the preparatory stage of the workshop.

To retrieve all the electronic copies of the numerous images of the volume, as well as the permissions to publish them, we were in the experienced hands of Urte Brauckmann, who works at the library of the Max Planck Institute for the History of Science (MPIWG). To Urte we would like to express our profound gratitude. Urte was occasionally supported by Sabine Bertram, whom we would also like to thank. The meeting in Berlin was made possible thanks to the financial support of Department 1 of the MPIWG, and for this, we would like to thank its Director, Jürgen Renn. The publication is available in Open Access thanks to the generous financial support of the library of the MPIWG, to whose head, Esther Chen, we would like to offer thanks at least as generous as her support has been. About two-thirds of the chapters were linguistically edited by Zachary Gresham, who found the elegance in our prose with patience and precision. Last but not least, we would like to thank Lindy Divarci, the publication manager of Dept. 1 of the MPIWG, who supervised several of the labors described here.

Matteo Valleriani
Andrea Ottone

Contents

1   Printers, Publishers, and Sellers: Actors in the Process
    of Consolidation of Epistemic Communities in the Early
    Modern Academic World   1
    Matteo Valleriani and Andrea Ottone

2   Printerly Ingenuity and Mathematical Books in the Early
    Estienne Workshop 25
    Richard J. Oosterhoff

3   Erhard Ratdolt's Edition of Sacrobosco's *Tractatus de
    sphaera*: A New Editorial Model in Venice? 61
    Catherine Rideau-Kikuchi

4   Printers, Booksellers, and Bookbinders in Wittenberg
    in the Sixteenth Century: Real Estate, Vicinity, Political,
    and Cultural Activities 99
    Insa Christiane Hennen

5   Scholars, Printers, and the Sphere: New Evidence
    for the Challenging Production of Academic Books
    in Wittenberg, 1531–1550 147
    Saskia Limbach

6   Sacrobosco at the Book Fairs, 1576–1624: The Pedagogical
    Marketplace 187
    Ian Maclean

7   The Iberian and New World Circulation of Sacrobosco's
    Sphaera in the Early Modern Period 225
    Alejandra Ulla Lorenzo

8   The Giunta's Publishing and Distributing Network and Their
    Supply to the European Academic Market 255
    Andrea Ottone

9   **Mathematical Books in Paris (1531–1563): The Development
     of Publishing Strategies in a Competitive International Market** 289
     Isabelle Pantin

10   **Paratexts, Printers, and Publishers: Book Production in Social
     Context** 337
     Matteo Valleriani and Christoph Sander

11   **The *Sphaera* in Jesuit Education** 369
     Paul F. Grendler

12   **Printing Sacrobosco in Leipzig, 1488–ca. 1521: Local Markets
     and University Publishing** 409
     Richard L. Kremer

13   **Publishing Mathematical Books of Parisian *Calculatores*
     (1508–1515)** 459
     Alissar Levy

Index 485

Editors and Contributors

About the Editors

Matteo Valleriani is research group leader at the Department I at the Max Planck Institute for the History of Science in Berlin, Honorary Professor at the Technische Universität of Berlin, and Professor by Special Appointment at the University of Tel Aviv. He investigates the relation between diffusion processes of scientific, practical, and technological knowledge and their economic and political preconditions. His research focuses on the Hellenistic period, the late Middle Ages, and the early modern period. Among his principal research endeavors, he leads the project "The Sphere: Knowledge System Evolution and the Shared Scientific Identity of Europe" (https://sphaera.mpiwg-berlin.mpg.de), which investigates the formation and evolution of a shared scientific identity in Europe between the thirteenth and seventeenth centuries. In the context of this project, Matteo Valleriani implemented the development of machine learning technology and of the physics of complex systems in the humanities. The project is also part of the investigations led by Matteo Valleriani in the context of BIFOLD (https://bifold.berlin). Among his publications, he has authored the book *Galileo Engineer* (Springer 2010), is editor of *The Structures of Practical Knowledge* (Springer Nature 2017), and published *De sphaera of Johannes de Sacrobosco in the Early Modern Period. The Authors of the Commentaries* (Springer Nature 2020).

Andrea Ottone earned a doctorate in history at the University of Naples. He is currently a postdoctoral research fellow at the University of Milan's Department of economics, management, and quantitative methods in the context of the ERC founded EMoBookTrade project (Grant Agreement n° 694476). He is also a fellow at Berlin's Max Planck Institute for the History of Science. He teaches classes in European history at the Technische Universität Berlin.

Contributors

Paul F. Grendler University of Toronto, Toronto, ON, Canada

Insa Christiane Hennen LEUCOREA, Wittenberg, Germany

Richard L. Kremer Department of History, Dartmouth College, Hanover, New Hampshire, USA

Alissar Levy École nationale des chartes, Paris, France

Saskia Limbach Georg-August-Universität Göttingen, Göttingen, Germany

Ian Maclean All Souls College, University of Oxford, Oxford, England

Richard J. Oosterhoff University of Edinburgh, Edinburgh, Scotland

Andrea Ottone Department of Economics, Management, Quantitative, University of Milan, Milan, Italy;
Max Planck Institute for the History of Science, Berlin, Germany

Isabelle Pantin Institut d'Histoire Moderne et Contemporaine (IHMC), Ecole Normale Superieure, Paris, France

Catherine Rideau-Kikuchi Laboratoire Dynamiques patrimoniales et culturelles, Université de Versailles Saint-Quentin, Versailles, France

Christoph Sander Bibliotheca Hertziana, Max Planck Institute for Art History, Rome, Italy

Alejandra Ulla Lorenzo Universidad de Santiago de Compostela, Santiago, Spain

Matteo Valleriani Max Planck Institute for the History of Science, Berlin, Germany;
Technische Universität Berlin, Berlin, Germany;
Tel Aviv University, Tel Aviv, Israel

Chapter 1
Printers, Publishers, and Sellers: Actors in the Process of Consolidation of Epistemic Communities in the Early Modern Academic World

Matteo Valleriani and Andrea Ottone

Abstract This chapter proposes a global view of the set of dynamics of interplay that were generated in the early modern publishing sector around a single astronomical work, the *Tractatus de sphaera* by Johannes de Sacrobosco. The *Sphaera*, a thirteenth-century tract of geocentric cosmology, rather than remaining a static text, became over the centuries a multiauthored dynamic textual tradition. This essay argues that publishers, printers, and booksellers had a fair share of agency not only in perpetuating but also in shaping the evolution of this long-lasting textual tradition. The present essay traces the ways this agency was configured.

Keywords *Sphaera* · Johannes de Sacrobosco · Cosmology · History of science · Book history · Network theory · Digital Humanities

1 Introduction

Early modern astronomy is a constellation of great discoveries by scientists such as Nicolaus Copernicus (1473–1543), Galileo Galilei (1564–1642), and Johannes Kepler (1571–1630). Such great discoveries, proposing and striving to promote a new heliocentric worldview, went down in history associated with great events: the publication of the respective works by which the new cosmology was disclosed to an educated audience. While the emphatic perspectives of the great scientists have been used and re-used to reconstruct the early modern history of astronomy and cosmology, little has been done to understand the nature of the scientific knowledge possessed by their aforementioned educated audience.

M. Valleriani (✉) · A. Ottone
Max Planck Institute for the History of Science, Berlin, Germany
e-mail: valleriani@mpiwg-berlin.mpg.de

M. Valleriani
Technische Universität Berlin, Berlin, Germany

Tel Aviv University, Tel Aviv, Israel

A. Ottone
Department of Economics, Management, Quantitative, University of Milan, Milan, Italy

© The Author(s) 2022
M. Valleriani and A. Ottone (eds.), *Publishing Sacrobosco's* De sphaera *in Early Modern Europe*, https://doi.org/10.1007/978-3-030-86600-6_1

In the context of the research project *The Sphere: Knowledge System Evolution and the Shared Scientific Identity of Europe*, a corpus of 359 early modern editions has been collected with the focus on one particular text: Johannes de Sacrobosco's (d. 1256) *Tractatus de sphaera.* Collected editions span from 1472, the year that saw the first two printed editions of Sacrobosco's text, to 1650, which is approximately when the *Tractatus* loses scholarly relevance.[1] The *Sphaera* was used at virtually all European universities, gymnasia, and other institutions of higher education during the early modern period. Originally compiled in Paris during the thirteenth century, this text comfortably transitioned from manuscript to early print culture, gaining an outstanding visibility as the standard text for introductory classes of astronomy at the Faculties of Liberal Arts, conceptually incapsulated within the framework of the *quadrivium.*

Sacrobosco's text was however not a standalone source that European scholars and students used to learn cosmological rudiments. On the contrary, Sacrobosco's tract was often published along with commentaries and textual apparatuses, thus making the *Sphaera* a common space for scholarly engagement. In fact, the term *Sphaera* designated not only Sacrobosco's own treatise but, more generally, was used as a label for specific collections of texts used in astronomical teaching (Valleriani 2017). Just as there was an *articella* for medicine, there was a *corpus astronomicum* called *Sphaera.* Such a *corpus astronomicum*, a proper introduction to geocentric cosmology, was first shaped during the late Middle Ages (Pedersen 1975), but it continued its evolution until the first half of seventeenth century, long after the outbreak of the Copernican revolution. While Sacrobosco's text remained at the pivot of this corpus for over three centuries, the corpus of texts surrounding it became increasingly elaborated. Regional trends developed over time and many updated commentaries were appended, especially during the sixteenth century.

Previous studies have emphasized how a general tendency toward the homogenization of astronomical knowledge emerged, especially from the 1530s and the 1540s. In particular, the impulses of the Reformation transformed the curriculum of the Faculty of Liberal Arts of Wittenberg into a model that was imitated all over Europe until the end of the century and beyond (Valleriani et al. 2019; Zamani et al. 2020).

In order to understand the mechanics of this evolutionary process, the whole corpus of *Sphaera* treatises has been gathered and dissected into "text-parts." In the context of this methodology, single text-parts are defined as textual passages not smaller than a paragraph, and that cover a defined subject matter with relative completeness. One text-part in the corpus of Sacrobosco's *De sphaera*, for instance, is the *Theoricae novae planetarum* by Georg von Peuerbach (1423–1461). This text was first included in the *Sphaera* treatises as early as 1482, and by 1537, it had been reprinted seventeen times in as many known editions of the *Sphaera.* If we include literary addenda such as epigrams, sonnets, and other types of composition that are usually considered "literary paratexts"—often printed in scientific books beginning in the sixteenth century—a text-part could be much more modest in length.

---

[1] The database is accessible through the project website: https://sphaera.mpiwg-berlin.mpg.de. Accessed 08 June 2021.

A representative example might be the short *carmen* written by Donato Villalta (1510–1560) and dedicated to the scholar Pierio Valeriano (1477–1560), printed for the first time in 1537 and reprinted a further thirty-two times.

The corpus in its entirety contains 540 defined text-parts. These have been not only identified by publication dates, printers, and publishers, but are also accompanied by in-depth investigations of their authors. Most of the editions credit only a certain number of authors (usually two or three) on the title pages. By dissecting the works into text-parts, however, a number of uncredited texts were revealed, meaning that the text-parts' authors' names are not declared on the title pages, thus making them unretrievable at a metadata level.

Text-parts tend to recur among editions in the *Sphaera* corpus. By singling out text-parts that were published at least twice, with the second instance released at least one year after the first, the number drops to 241 text-parts, meaning that 299 text-parts were published either only once or more than once but only in the same year. The remaining 241 text-parts recur 1,394 times. Recurrences range from just one instance to a maximum of eighty-seven.

On the basis of the analysis described above, the geo-temporal manifestation of recurrences allows for the identification of editions as either imitated or imitating models on the basis of the combinations of text-parts they contain. In particular, it emerged that editions produced in Wittenberg gained a hegemonic position in the European production of introductory textbooks for cosmology and astronomy. Much of the process of general homogenization of knowledge, therefore, took place through this general tendency to mutual imitation in premodern scholarship.

On an abstract level, text-parts—intended as semiotic signs of knowledge—can be conceived as atoms that migrated and re-aggregated in different constellations of content over time and space. A specific edition corresponds therefore to a definite combination of text-parts. When repeated via imitation, these can be singled out as redactional formulas used by publishers to lure specific audiences. This dynamic of circulation and re-aggregation of text-parts is at the basis not only of the process of homogenization of scientific knowledge in the West but also of its progressive mathematization and its practical turn (Valleriani 2017).

On a more pragmatic level, instructors performing their teaching duties were compelled to choose textbooks that best suited their pedagogical purposes and scientific inclinations, whereas publishers handled multiple redactional formulas, for instance, by determining a constellation of text-parts enriched and adorned with variants of illustrations, diagrams, and tables to best suit consumers and gain slices of a crowded market. From a historical perspective, therefore, the question emerges as to what lies behind the choice made by an instructor over a specific cluster of text-parts, or the choice made by a publisher to offer the market a specific redactional formula. How were these choices made? Which were the typologies of the actors involved? Where did the inputs come from? In other words, understanding the abstract process of the circulation of knowledge—here described as a mechanism of appearance, reappearance, and mixing up of text-parts—still requires the human factor. Behind the assembling and reassembling of texts, there were whole communities of people interacting with one another over a short or a very long distance.

To understand this dynamic at work, one initial analytical stage involved authors alone (Valleriani 2020a). A first round of investigation regarded all scholars explicitly cited on title pages in their capacity as authors, commentators, and editors. When an edition was authored by at least two scholars who were also alive at the time of publication, a potential relationship, via printer or publisher, among the two was assumed. The results clearly indicated that such communities of authors, though extant, were not sufficient to explain the wide circulation of their texts. The recurring text-parts were therefore matched with a corresponding author. Hence, a longer list of 222 commentators emerged, comprising both credited authors and uncredited authors, identified by way of the atomized text-part analysis described above.

By applying the same formal conditions, a network of potential communities of authors emerges in a more encompassing picture (Fig. 1).

The number of contemporary authors who were potentially in contact with one another is 130 but the components of the network are distributed over a span of 172 years, starting in 1472. The general network therefore is highly disconnected. While there clearly is a big component that groups four distinct sub-regions, the rest of the network is constituted of a high number of smaller components of various sizes. The big component comprises for the most part the authors included in the several editions of the *Sphaera* produced in Wittenberg, the printing center that gave birth to the hegemonic redactional model of the *Tractatus* in Europe.[2] However, the graph in its entirety is not structured to enable the circulation of a great number of text-parts at a European level, as determined by previous studies.

The studies hitherto accomplished, however, completely neglected other relevant components of this thread, namely publishers and printers who worked hard to bring these clusters of texts into a material form. While studies concerned with early modern book producers and distributors are abundant, rarely has the focus been on studying book traders as a collective body operating around a single intellectual piece. Even more rarely has this task been attempted in the frame of a *longue-durée* research. By going back two centuries before the emergence of modern science, the goal is to retrieve the collective *modus operandi* of the European printing community while engaging in the production of one of the most widely used scientific textbooks of the time.

---

[2] In this respect, it is worth mentioning that some anonymous text-parts which were included in the Wittenberg textbooks and which, therefore, became greatly influential all over Europe were compiled and edited by Georg Joachim Rheticus (Valleriani et al. 2022).

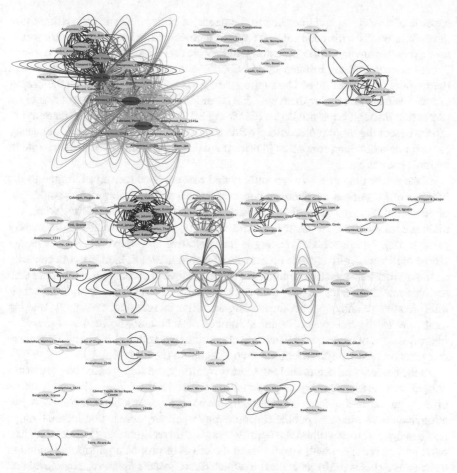

Fig. 1 Network of authors of text-parts constituting the treatises collected in the *Sphaera* corpus, which contain, among others, Johannes de Sacrobosco's *Tractatus de sphaera*. Authors are pairwise connected to each other, when their texts appeared in the same editions and where both alive at time of publication. Network data and network visualization by Beate Federau

2 Printers as a Collective Body of Actors

In the context of knowledge communication networks, publishers, printers, and book-sellers played a significant part. Scholarship has long dismissed the idea that the book industry was merely a gear in the factory of written culture. Publishers, printers, and bookdealers at large have been increasingly recognized as holding a fair share of agency in shaping and influencing the textual and visual outlook of literature while processing it for printed circulation (Darnton 1982). Their role became particularly determinant in the process of assigning intellectual products an intrinsic commercial value. By working at the intersection between authors and users, bookdealers were capable of absorbing and interpreting the needs of both poles and translating those needs into books with good sales records. At times, publishers and printers would take excessive agency while replicating literary works (Chap. 9). This utterly enraged contemporary authors, irritated by the liberty with which the former reinterpreted their works, interpolating and corrupting their texts and thus endangering their reputation—aside from causing them financial harm. Authors growingly sought copyright protection through book privileges (Ginsburg 2013; Squassina 2017). The authors' efforts in seeking those protections provide a vivid testimony of publishers' tendency toward intellectual appropriation in print publishing, revealing their primary role in textual production.

If the active dialogue between authors and readers is conceived as limited to the lifespan of the author, the dialogue between readers and an authored text could, and often would, survive the author. Depending on the impact of a literary work, the dialogue may endure for centuries. And virtually, no reader would hold a merely passive role. The process of reading is interpretative and transformative. A reader with a quill would already take up the role of a commentator, though not necessarily an impactful one. A restricted number of readers, however, would do much more than scribble marginal notes on their own copies (Grafton 2021). They would make their interpretative readings public, thus taking an authorial role and, eventually, making their way to the title page. So did a number of scholars who read and provided interpretations of a given text. This type of continuative relationship between readers and texts revived the life of a literary work, keeping the ball rolling.

In the context of a centuries-long literary tradition, the actors of the book industry gained even greater agency in perpetuating the fortune of a text that outlived its primary author or multitude of authors. In this continuing process of textual perpetuation and transformation, publishers could play a primary role in commercializing a text disengaged from authorial paternity. They could assemble and reassemble text-parts and merge them with visual aids in the effort of proposing a formula suited to the market. This is the exact context in which Sacrobosco's *Sphaera*, a medieval text with a plethora of living and mostly non-living commentators, endured for almost two centuries of print culture. Voided of the active role of its original author, the *Sphaera* became a standard on which editors, commentators, publishers, printers, and correctors performed before a participative audience. In this configuration, publishers and printers (when not the same person) were those who usually had the last word on how to fuse textual and metatextual elements—the intellectual, material, and visual features that made a given edition appealing to users and competitive in the marketplace of textbooks for higher education. Printers and publishers of the *Sphaera*, moreover, were mostly active and experienced in academic book production and distribution (Chap. 7); not rarely they were "accredited university booksellers" (Chap. 2). Publishers and printers knew better how to turn a book into a bestseller, and the *Sphaera*, a work that long survived its primary author, was no exception.

Terms such as publisher, printer, bookseller, and to some extent even *consumer* will be extensively used in the pages that follow; but they are open to several levels of critique and accusations of reductionism. First of all, the very configuration of the early modern publishing industry hardly allows historians to sharply distinguish one professional figure from another. These roles may at times be distinct on the title page of a single edition, but it still holds true that in the everyday life these men and women would not distinguish their professions as sharply as we sometimes do, and in the heat of the book trade, they would deploy the expertise of each.

Furthermore, the very use of terms like publisher, printer, and seller as single individuals does not do justice to the complexity of a publishing house, a print shop, a bookshop, or a network of sellers. A single publishing house could use consultants and accountants, informants whose role in the planning and delivery of a single edition gets no mention on a title page (Rück and Boghardt 1994; Giesecke 1998).[3] The print shop was a collective body of artisans, more or less literate, who had the ultimate responsibility of translating the intellectual efforts of scholars and entrepreneurs in a tangible commodity.[4] A single bookshop functioned through the collaboration of masters and apprentices. With regard to consumers, although the same terminological awareness used for printers, publishers, and sellers is not necessary, it is nevertheless worth mentioning that the act of distinguishing consumers, producers, and dealers of printed books may be useful at an explicative level, but it again carries an element of reductionism. Publishers, printers, and sellers, when not the same person, could be themselves eager consumers and might therefore place themselves in the position of their own customers.

To chase the complex, unfolding mechanism that for almost two centuries brought the *Sphaera* corpus to a large circulation in Europe, the decision here has been to gather exemplificative stories into three sections covering respectively the levels of production, distribution, and consumption. The aim has been to recreate both the sequence of motion of an edition from the press to the shelves, and to follow the *Sphaera* corpus through the three main knots of the book industry network: publisher, dealer, and collector. The sequence follows a commonsense-based view of the market, but the circular motion of this process should not be overlooked, because of the mutual influence of each level on the others.

3   Production

Producing the *Sphaera*, like many mathematical and astronomical works, presented several graphic challenges. Works of geometry, astronomy, and the natural sciences employed visual aids to communicate content. A greater number and a better quality of diagrams, images, and tables made the difference between one edition and another. In the artisanal world of early modern printing, quantity and quality were two parameters that affected costs. Publishing houses and printing shops would make no secret

---

[3] We are grateful to Falk Eisermann for bringing this methodological aspect to the attention of the working group.

[4] For an example of the lively activity and craftmanship of a Renaissance print shop, see (Gerrotsen 1991).

of enhanced visual aids in promoting their editions; they announced the augmented features of their editions with a rhetoric of mastery and ingenuity.

An example of this comes from the context of Paris printing community, with particular reference to Simon de Colines (1480–1546) and Henri Estienne (1460–1520) (Chap. 2). Part of this rhetoric was plain advertisement strategy and self-promotion. However, it was an aspect of the printing craft that integrated users and producers, and it opens up our understanding of a factor that needs stress: printing astronomical works was no amateur business. Whoever adventured in production of this kind needed specific skills at hand and the ability to handle augmented costs with adequate commercial strategies. In the case of the *Sphaera*, this was an even more critical point: for a work of large-scale consumption, the transnational competition could be fierce, thus making adequate revenues critical.[5] If on one hand a fiercer competition encouraged innovation, and innovation primarily involved more visual aids and explicatory apparatuses such as tables, indexes, and diagrams, on the other hand these quality-enhancing elements were taken at a greater risk of market failure. Furthermore, quality-enhancing innovations and their consequent augmented costs required the consideration that there was only so much a publisher could ask the target audience to spend on a product (Milazzo 2020). With reference to the *Sphaera*, the wallets of primer consumers could be quite thin, as students of the quadrivium were not necessarily the wealthiest consumers on the book market (Chap. 8).

In the planning phase, the craft of publishers was to conceive a formula that the market would welcome, gather enough funding to finance it, maintain the channels of transmission (eventually build new ones for the purpose of a single project), and guess the right print run for the market to absorb. Much of this work required financial and logistical know-how along with a practical sense of the market merged with empirical means of assessment (Chap. 8). But, aside from this operative skill set, publishers were those who would best interpret the appetite of consumers; ultimately, they would bet money (most often not their own) and their reputations (which would later influence their access to credit lines) on an editorial formula that merged content and outlook and satisfied the expectations of the audience. At times, they would try to shock the market with innovations. In this way, they hoped to penetrate a rather conservative environment in which the preservation of a past model was a virtue and innovation could be perceived as a form of corruption. When successful, publishers would create a new niche demand, profitably go around their competitors still bound to an old formula and succeed out of their commercial intuition.

Once a specific formula proved successful, the market readjusted around it. The new formula could imply a novel outlook, refashioned content, or a newly translated text. A new redactional formula would eventually gain momentum and become a

---

[5] The term *transnational* is here used to capture the ongoing process of modern state building. The term is being favored to *international* to signify the fact that the process of state formation was not yet complete. In the context of commercial networks, *transnational* poses an emphasis on the role of political and normative structures as an element of the governance of a supernational integrated reality, such as the book market. In the context of epistemic networks and groups of cultural correlation, the term *transregional* is being used instead to place emphasis on elements of cultural, linguistic, or geomorphological assimilation.

model for others to follow. The *Sphaera* saw the juxtaposition of different redactional formulas that experienced a period of hegemony over the market, only to be later replaced by new redactional formulas (Pantin 2020) (Chap. 3). For a literary tradition like that of Sacrobosco's *Sphaera*, variations in redactional formulas may have implied more than just a shift in the format; most often they involved the aggregation or re-aggregation of text-parts, the addition of new or clearer visual aids, the introduction of short manuals for building and applying mathematical instruments, and the increasing enrichment of the text—originally a qualitative introduction to cosmology—with computational tables to support mathematical workflows. The abovementioned studies on the formation of text-part clusters have shown that when such combinations proved appealing to the market, they were rapaciously imitated throughout Europe, either with or without the consent of those who initiated the new redactional formulas.

When it comes to the interaction between the *Sphaera* and wider print culture, Wittenberg plays a key role on both a quantitative and qualitative level (Chaps. 4, 5 and 10). The vibrant town in Saxony presents most of the common characteristics of the printing centers that produced editions of the *Sphaera*: a lively university community with a laborious print industry, mutually supporting each other in the interplay of supply and demand.

Wittenberg, a modest town in its own terms, came under the spotlight of Europe in the times of the Reformation. Its university became pivotal for central Europe and its scholars earned international resonance. With enrollment growing and local theologians rising in fame, the local print industry experienced a dramatic burst. From the 1530s on, the town participated in the production and dissemination of the *Sphaera* corpus with a redactional model that soon became dominant on the European market.

The explanation for such great success needs stress. Wittenberg being one of the centers of the cultural-religious debate of the time, its book production also received considerable attention, at least from areas sympathetic to the Reformation. The ongoing religious controversy, however, contributed in general to rise attention towards Reformed scholars, even in Catholic lands, at least until Catholic censorship developed into a firm structure (in a process that began in the 1560s) and brought these names from fame to infamy in certain regions of Europe (Sander 2018). This is clearly the case of Philip Melanchthon's (1497–1560) initiative to promote the 1531 and 1538 editions of the *Sphaera* (Chap. 5). These editions, which soon became a standard in German lands, gained ground in Catholic lands as well, with Venice quickly using it as a templet for local editions. When a single edition sold well, transnational attention rose over its redactional formula. This would eventually justify cross-confessional cultural transfers in Catholic lands equally interested in participating in the transregional commercialization of new editorial formula of the *Sphaera*.

Wittenberg more than other printing centers simplifies and magnifies the interlocking of the intellectual atmosphere of a college town and that of a busy printing industry. With growing attention toward the small town as a cultural epicenter, the local university also experienced rapid growth in the student population. The local print industry followed. However, unlike older universities (like Paris), which could count on a long tradition of ruling the book trade even in manuscript culture (Chap. 2), Wittenberg's university took quite a while before establishing an official university press and a structured regulation of the local industry. This left much of the dealing to the private initiatives of scholars, investors, and craftsmen. The laissez-faire system adopted in Wittenberg generated a heated dialectic between the professionals involved in the making of books: authors, editors, publishers, printers, and binders (Chap. 5). Observing the unfolding of this tension is to observe part of the inner mechanisms of the infrastructure that produced and distributed the *Sphaera*. The unfolding of these interactions highlights the pressing priorities of the various parties involved in the production chain of books, the *Sphaera* among them: the desire for quality and accuracy on the part of authors and investors; the necessity of earning profits that publishers needed to keep themselves afloat in a difficult market; the struggle of editors, printers, and binders, who tried to make a living while operating at the bottom of the food chain.

Hence, tracing the history of the most fortunate rendition of the *Sphaera* does not necessarily mean following a history of success. Such is the case for Joseph Klug (1490–1552), the printer behind one of the most influential editions of Sacrobosco, whose business sank, along with his reputation, leading to financial misery despite the visible legacy he left in the propagation of the *Sphaera* corpus. Behind Klug's financial ruin lays, evidence shows, the strangulating tug of war between quality and the necessities of competitive pricing.

At the intersection of all these demands were printers who were left with the dilemma of accepting ill-paid contract jobs or handing those opportunities to competitors, only to be cut out of future initiatives (Chap. 5). The ecosystem of the print industry seen from the microcosm of Wittenberg proves even more profitable to historians due to the wealth of information on the urban fabric of the university town (Chap. 4). A planimetric view of this community of scholars, entrepreneurs, and artisans reveals the compartmentalization or the alliance between the professions involved in the book industry. The respective extent of the estates owned by any of the characters involved in the production of the Wittenberg *Sphaera*, and their placement in administrative positions of the town become symbolic elements in reconstructing power relations and structures of the fairly pyramidal system that was the early modern book industry.

4 Distribution

Distribution dynamics may be as transformative for a text as the printing process. It is through wide circulation that editorial formulas gain momentum, earn popularity, and eventually become dominant (Chap. 10). Market frictions are determinant in putting different formulas to the test, and it is through the spinning of several coexisting editions in the book market that different redactional formulas and graphic outlooks merge to create new editorial models.

The intellectual market of astronomical and mathematical texts seemed to follow precise trends and patterns that infused academic centers with a particular dynamism in the discipline and established their leadership in longevity. Such is clearly the case for Paris in mid-sixteenth century. In the field of mathematical and astronomical texts, Paris often set a standard for other publishing centers in terms of both layout and content (Chaps. 9 and 13). Paris, however, was not alone. As to how redactional and visual models migrated from print center to print center, the most intuitive answer would be that this happened with the circulation of the commodities themselves. Books circulated virally in the transnational market, and publishers—ever aware of one another—possessed enough sensibility to figure a good editorial idea from a less fortunate one. They then decided which models to follow, imitate, or reinterpret.

However, ideas could also follow the migratory trajectories of people. With regard to the *Sphaera*, the German printer Erhard Ratdolt (fl. 1477–ca. 1528), active in Venice in the late fifteenth century, provides an example of how a single editorial model could propagate as a consequence of the relocation of a single printer who carried his know-how and professional idiosyncrasies from city to city (in this particular case from Augsburg to Venice) (Chap. 3).

The human factor in the migration of ideas is surely an element to bear in mind, but in investigating the proto-industrial world of early modern printed books, the market-driven dynamics of the circulation, filtration, and optimization of ideas is an element difficult to resist. Following the idea that better-selling books earn superior commercial value, thus raising the attention of other publishers and triggering the imitative mechanism, an adequate knowledge of the transformative potential of the market is called for.

Nothing epitomizes the challenges of the transnational book market better than book fairs, and nothing represents the phenomenon of Renaissance book fairs better than the Frankfurt fair.[6] To investigate the representation of Sacrobosco's editions at the Frankfurt fair is to measure the transnational aspirations of the several editions that entered the market between the sixteenth and seventeenth centuries. With at least twenty instances of the *Tractatus* being officially declared at the fair, early modern editions of the *Sphaera* seem to have been conceived as literary products aimed at a transnational rather than at a localized market. Furthermore, official declarations at the fair (as shown from surviving catalogues) do not capture the complete picture of what was actually traded at the venue (Chap. 6). Thus, if the absence of an official mention of the *Sphaera* at regular exhibitions would have been a significant indication of a primarily local circulation of the *Tractatus*, its episodic yet substantial presence in official documents of the Frankfurt fair is evidence of its transnational circulation, which was likely even larger than evidenced. In fact, the non-regular mention of the *Sphaera* at the Frankfurt fair, in light of its mass production throughout Europe, opens up other relevant issues. To be officially declared at the fair, products had to meet criteria of novelty (Maclean 2021, 12). Hence, the recurrence of official declarations of the *Sphaera* at Frankfurt is an indicator of alleged or true instances

---

[6] For an overview of Renaissance book fairs see (Nuovo 2013, 281–314). For more information on the Frankfurt Fair, see (Maclean 2021) and (Chap. 6).

of redactional innovations. Ultimately, considering the *Tractatus* in the scope of transnational commercial venues (such as fairs) clarifies the market drive behind instances of innovation that justified the migration of paratexts, text-parts, and other visual and textual apparatuses. Furthermore, chasing its several appearances at the Frankfurt fair helps detail the geographic trajectories that the editions of the *Sphaera* followed on the transnational market. The presence of Catholic printing centers like Rome in the listing of *Sphaera* editions at Frankfurt (a largely Protestant commercial trading center from the first phase of the Protestant Reformation) confirms the cross-confessional vocation of the product. Instead, the absence of *Sphaera* editions stemming from relevant print centers such as Paris and Wittenberg—both especially influential in setting the editorial standard of the overall corpus—complicates the view of the ways in which these editions found their way through the transnational market.

Another way for publishers to reach out to a transnational audience was by building an independent distribution infrastructure framed by existing channels of the European book trade and trade at large. An example that stands out is that provided by the Giunta publishing firm. Florentine in origin, cosmopolite by vocation, the Giunta built a commercial empire with trading posts in some of the most relevant printing centers of Catholic Europe (Chap. 8). Given the large scope of their commercial network, the magnitude of their output, and the sophistication of their publishing choices, the question is raised as to where the *Sphaera* fit in their global portfolio. The answer that emerges is that to a large-scale publisher with a muscular position in the continental market, the *Tractatus* looked like a less-than-impressive deal. As intellectual merchandise, the *Tractatus* was aimed at an audience that the Giunta regarded with only moderate interest. Students of the quadrivium, as a social group and commercial target, were large in number but had fairly modest means. Publishers such as Giunta were accustomed to moving large, multivolume works of high-class scholarship for consumers in the high professions. These were generally people of good financial standing who had a legitimate need for quality imprints. Hence, they represented a far more appealing group of customers. They were medical practitioners, lawyers, clergymen, or institutions, both secular and ecclesiastical, such as courts, monasteries, convents, and whole administrative or ecclesiastical districts. In comparison, students halfway through their education were much less significant consumers.

A further demotivating factor was the fierce competition to serve quadrivium students. The over three hundred editions of the *Sphaera* and the war of pirate reprints show that the commercial race was brutal (Chap. 6). Furthermore, the technical skills deployed to make an old text like the *Sphaera* look like a new and attractive one (new visual aids and a refreshing alchemy of old and new text-parts that could also battle the second-hand market) made the engagement time consuming, costly, and risky. Placing the *Sphaera*, or any other early modern textbook of this kind, in the midst of the free market proves relevant to understanding it not only as an intellectual piece, but also as a commercial artifact.

In the context of the integrated book market of Renaissance Europe, there were commercial ecosystems that stood out for a few peculiarities. This is the case of the Iberian Peninsula, a commercial area that, as far as the circulation of the *Sphaera* in Latin was concerned (Latin being the standard language of higher education), was overly dependent on foreign imports, leaving most local production to vernacular versions (Chap. 7). The imbalance between vernacular and Latin editions in the publishing portfolio of local publishers mirrored the general structure of the Iberian print industry, which mostly catered to the local market rather than engaging in risky exports. But the predominance of vernacular editions of the *Sphaera* finds its explanation in the particular use that Iberian consumers made of the *Tractatus* and the different social and professional typologies that Iberian publishers targeted. While the archetypical user of Sacrobosco for most of the continental market remained Latin-reading students of the quadrivium, Iberian publishers aimed rather at more mundane groups, such as explorers and traders involved in maritime travel (Crowther 2020; Lanuza-Navarro 2020; Leitão 2008, 2013). In light of this, the Iberian tradition of learning from Sacrobosco's legacy appears to be more linked to the empire-building effort than to the formation of national elites, functionaries, and scholars to be employed in the efforts of modern state building. If on the one hand the Iberian Peninsula was an eager recipient of the trans- and sub-alpine production of the *Sphaera* corpus, on the other hand, due to the far-reaching radius of their commercial routes, Spain and Portugal were also responsible for expanding Sacrobosco's tradition from continental Europe to the New World.

5   Consumption

The consumption level has a twofold relationship with production and distribution dynamics, in that it functions equally as trigger and recipient of both. For the *Sphaera* corpus, the natural landing environment was the world of education.

The *Tractatus* was indeed handled in the book fair catalogues, such as that of Frankfurt, under the category *scholastica* (Chap. 6). Its wide circulation found a reason in the interconnection of two mutually dependent processes: on one side the increasing demand for a mathematical education, and on the other the evolution of the knowledge displayed in the *Sphaera* corpus from a qualitative introduction to geocentric cosmology to an introduction to mathematical astronomy. Christopher Clavius (1538–1612), for instance, the architect of the Jesuits' *Ratio studiorum*, considered mathematics the means to understand precepts of natural philosophy (Chap. 11) (Feldhay 1999, 2021; Price 2014). The layout of these textbooks, moreover, and in particular the design of their frontispieces and title pages clearly display the increased relevance of mathematical astronomy (Chap. 2); they therefore hint at a profound change in the role and function of *Sphaera* knowledge.

Coming back to the field of education, Jesuits, occupy a distinct space. Therefore, they provide a valuable viewpoint whence to observe the trajectory of astronomical studies and the *Sphaera* corpus in particular in the curricula of higher education. Moreover, the Jesuit movement sits almost halfway in the chronology of the history of the *Sphaera* in print culture (1570–1650). This is an invaluable feature if one considers that with their placement in the chronology of the early modern period, Jesuits were structuring their pedagogy by filtering much of the Renaissance tradition and stretching their vision toward the cultural and social challenges of the Baroque era, to which they contributed considerably by setting a competitive educative standard. Further, the setting of the Jesuit school curriculum—*Ratio studiorum*—was a process that animated a lively internal debate in the Society. Much of this debate was put on record for historians to assess the inner logic that guided them in establishing their educative paradigm. The inner debate over mathematical education in the Jesuit curriculum reveals tensions, disagreements, and reconciliations helpful in unpacking the black box of the Renaissance and Baroque pedagogy with regard to applied mathematics (Chap. 11).

The picture that emerges from the debate internal to the Society of Jesus is quite demotivating for mathematics enthusiasts. An increasing interest in mathematical learning is indisputable, especially if compared to previous centuries. Nonetheless, the period of transition between the Renaissance and the Baroque eras saw a resistance in pedagogical circles of the full-scale mathematization of the sciences, as was called for by some innovators like Clavius—also a prominent commentator of Sacrobosco—whose passionate defense of mathematical knowledge contributes to the understanding of his own cultural agenda as a user, teacher, and commentator of the *Tractatus*. The debate triggered in the Society of Jesus, however, reveals that to Jesuit hierarchies, mathematics was perceived as inapt to respond to the challenges of post-Tridentine society and inadequate to fit the cultural model that Jesuits aimed to pursue through their schools. If not isolated, Clavius's ideas concerning the role of mathematics in the Jesuit curriculum were clearly regarded as secondary, a factor that over time created a distinction between the general scientific tendencies and the curricular developments inside the order. Nonetheless, this distinction is extremely helpful in ranking mathematics and astronomy in the realm of late Renaissance and early Baroque education, thus allowing a tentative social and cultural profiling of the consumers interested in works of applied mathematics like the *Sphaera*.

The example of a consumption dynamic provided by Paris (Chap. 13) highlights an aspect of the early modern book market that is too often neglected: the tight relationship that existed between supply and demand. If large-scale distribution was an option for publishers and printers embedded in a proto-industrial market, the still largely artisanal production of the pre-mechanized printing press also required the careful handling of print runs in response to primarily local demands.

For instance, this type of producer–consumer interaction is clearly exemplified by the short, yet meaningful adventure of a group of Iberian scholars, the *calculatores*, who, for a limited span of years (1508–1515) established themselves in Paris, likely in the attempt to implant a foreign tradition of mathematical studies. This experiment seems to have in fact faded away soon after that community of mathematicians departed the city. Their short Parisian adventure however opened a small but fresh niche in the already vibrant market of mathematical works in Paris. This episode in the history of Parisian mathematical books should provide an example of how nuanced the pre-mechanized book market was in comprising both large- and short-scale modes of book production and consumption, more explicable in terms of an induced attempt at cultural promotion rather than independent streams of market demand.

Another outstanding example of how production could be tightly linked to demand comes from Leipzig (Chap. 12). Being a university town, Leipzig hosted a considerable number of consumers of the quadrivium curriculum—readers thus also interested in the *Sphaera*. Leipzig however was also the site of a relevant book fair. The town was therefore fully integrated in the commercial channels of the transnational book trade that pivoted around the Frankfurt fair. Admittedly, one was scheduled soon after the other to allow attendants to visit both (Maclean 2021, 24). Surprisingly, however, when it came to producing a large-consumption product such as the *Sphaera*, the Leipzig print industry used a thoroughly independent redactional model fully rooted on a local manuscript tradition, thus showing no interest in participating in the imitation war at play between other relevant printing centers. Evidence would then suggest that both consumers and producers were following their own self-determined agenda based on continuity. Likewise, the redactional formula of Leipzig did not inspire other European printers; the circulation of *Sphaera* imprints produced in Leipzig was primarily local. Most likely, copies served the nearby university of Wittenberg (at least until the latter initiated its own local tradition in the 1530s to set itself apart as a dominant transregional standard). This illustrates how the texts of the *Sphaera* corpus could either reach a global radius or remain largely relegated to serving the learning purposes of a restricted community. This fact alone may nuance any overly enthusiastic claims of automated scientific information sharing linked to new printing technology. In fact, large-scale production was an available option—but so was a reduced-scale production and distribution mode. A single scholarly tradition could be doctored to stay quiet and local.

6   Modes of Production of Early Modern Scientific Textbooks

As mentioned, the early modern European system of production and dissemination of written knowledge in print was a very complex one, and yet this was only one part of a much more complex system of production, innovation, and transmission of scientific knowledge. It has been highlighted how each part of this system was bound by a relationship of reciprocity. The purpose of this section is to settle these complexities and to break down the integrated system into smaller and more comprehensible parts. The focus will be solely on the multiple dynamics that pertained to the production of textbooks, which, as in the case of the *Sphaera*, were mainly intended to serve the purpose of the higher education.

Meaningful historical conclusions concerning the early modern academic book market can only be reached after acknowledging that dealing with textbooks from this era means dealing with sources that often remained in the same state as their printers conceived them for the market.[7] It is not completely clear why textbooks were handled differently than other texts. This feature certainly relates with their normative-pedagogical function, as these were instruments for teaching in the context of highly regulated educational institutions. But an overarching study concerning the normative features of early modern textbooks in reference to the evolution of their content, format, and market is still largely missing.

With regard to a tentative model of the workflow that brought a textbook to press, a standard way to begin the unfolding of any literary project (including textbooks) would be its authorial textual conception. In the case of the early modern editions belonging to the *Sphaera* corpus, authorship does not refer to the original text—which constituted the nucleus of the corpus. This was compiled in the thirteenth century, long before the printing press came to be. Rather, for a book like the *Sphaera*, the so-called authorial conception was mainly linked to the selection, philologic refinement, and eventual novel integration of the numerous commentaries and text-parts that deepened specific subjects touched on by the main text. Another form of semi-authorial intervention involved in the production of the *Sphaera* concerns translators, who gave birth to new vernacular renditions of both the main text and the commentaries that accompanied it. Such works were printed together with the *Tractatus* of Sacrobosco. Their authors were almost always scholars involved in quadrivial teaching (Valleriani 2020a). Scholars directly linked to the world of teaching also had direct insight into the chosen commercial target. This allowed them to link their intellectual initiative to specific teaching needs for the academic years to follow, thus assuring a publishing project with a minimum number of sales.

Publishers, for their part, were the professional figures tasked with translating the intellectual and pedagogic impulses of authors into feasible products. They were also the ones who would make a project financially viable by putting their reputation, their commercial networking capacity, and their financial credibility on the line (Burkart 2019, 42–50).

Wary of the niche market and of the redactional formulas in circulation with variable market acclaim, publishers worked with authors in the conceptualization of a piece. Publishers, however, were also up to the much more mundane task of drafting a functional plan of action. Consideration over the adoption of a specific redactional formula had to be weighed with consideration of the materiality of the commodity that was being planned (paper, format, types, iconographic apparatus, and so forth).[8] All of the above would require a set of costs that had to be balanced with an adequate retail price suitable to the pockets of targeted users. Even more detailed considerations over costs were on the way: storage, shipment, insurance, and copyright fees, to name a few. All considerations on costs and possible revenues had to be measured against the capacity of the market to absorb the product. Publishers whose know-how included

---

[7] According to Sarah Werner, early modern textbooks were sold stitched or paper wrapped (Werner 2019, 23). This feature might be related to the fact that such works are often preserved in their original state and not bound to other works, as this is often the case for other literary genres.

[8] For an example, concerning the decision-making process and its inter-links with considerations over the intended audience in the context of a large-scale printer-publisher such as Christophe Plantin (1520–1589), see (Renaud 2020).

skills of market predictability (Chap. 8) were responsible for proposing a feasible figure for print runs. Here is where the know-how of publishers merged with the exact knowledge of the scholars they collaborated with in regard to how ample or tight the most proximate market of reference would be. In the case of textbooks like the *Sphaera*, it is fair to hypothesize that the figure coincided with the number of students enrolled in quadrivium classes for the current year and prospectively for the years to come. This was perhaps the easiest variable to forecast, and the foreknowledge was plausibly capable of covering a good part of the initial costs. Anything beyond that number could translate into direct or indirect revenue, one may hypothesize.[9]

Conversely, a small print run, although it minimized risks, also made the project less profitable. However, a shallow-radius distribution network and small storage capacity were all considerable limits to large print runs and thus to larger profits. The task of publishers then was that of building a sufficient distribution network to make their initiatives sustainable and, even better, profitable. A big name in the printing community had a bigger reputation based on a larger network of local and transnational alliances. This allowed them a more ambitious plan, a greater capacity for cutting costs per copy by producing larger print runs, and easier access to lines of credit (based on the expectations that creditors had for the financial viability of the planned publishing initiative). The economy of scale was fully at work in the process of turning an intellectual effort into a salable commodity.

Economic considerations concerning the size of print runs, moreover, did not solely regard the book market as observed from the perspective of an individual printer and publisher as described above. The textbook market had its own characteristics, and these were valid all over Europe, though with more or less efficacy depending on specific territorial regulations. Following the argument developed by Paul Gehl for schoolbooks in sixteenth-century Italy (Gehl 2013), all textbooks were first and foremost designed, produced, and distributed for a local market. In other words, they were the result of a trade-off between the teachers and lecturers on one side and the printers and the publishers on the other.[10] In this trade-off, teachers and lecturers represented the educational institutions present on the local markets. This kind of trade-off could take place for a variety of reasons. The most relevant in the case of the *Sphaera* corpus was the fact that, as mentioned above, the same teachers and lecturers were also the authors of the commentaries or of other texts that, in the redaction of an edition, were added or appended to the original (Valleriani 2020a).

---

[9] The issue of revenue in the field of book trade is a nuanced one. Bookdealers did not solely base their trade on the exchange between commodities and cash. Bartering was also common practice. This could involve books in exchange for books (Maclean 2021, 50–51, 247–278), which could be traded for cash or used as currency to tighten commercial or political advantages or to maintain patronage-based liaisons. Booksellers, however, would also exchange books for ordinary commodities (Dondi and Harris 2013).

[10] For a focus on the commentator and lecturer Jacques Lefévre d'Étaples, see (Chap. 2); for the relation between Wittenberg printers and Philipp Melanchthon, (Chap. 5); for the trade-off between Paris printers and the group of the *calculatores*, (Chap. 13). For another example, concerned with the Parisian publisher and bookseller Guillaume Cavellat, see (Pantin 1998).

Some of these textbooks were then able to enter a transnational market. Gehl analyzes only the case when the production of a specific textbook (or of a specific text-part thereof) was taken over by a printer or publisher who had established a transnational market for their business. On the basis of his empirical analysis, access to the transnational market seems to have opened when a sufficient number of reprints or reissues had already taken place at a local level. In other terms, it is possible to hypothesize that a specific threshold of (re-)production had to be met in order for a textbook to access a wider distribution network. This hypothesis can be expanded by cases derived from Paris printers and publishers, who were working on the local market while active, at the same time, on a transnational one. The opportunity therefore existed for scholars to enter both markets at once by means of a single publication agreement. A known example is the relationship between the famous reformer of the mathematical curriculum of the university of Paris, Élie Vinet (1509–1587), and Guillaume Cavellat (1500–1576) (Chaps. 2 and 9) (Pantin and Renouard 1986).

Large-scale publishers guaranteed access to a wide transnational market by means of established channels of distribution, transnational alliances, sound marketing strategies, and regular attendance at fairs such as Frankfurt's (Chap. 6). But alongside good sales performances, there was another relevant way in which redactional models might have circulated and inspired imitative reprints; this involved the awareness that actors of the publishing industry had of alternative redactional models. As mentioned, Wittenberg's editions of the *Sphaera* soon became a dominant model in Europe (Valleriani et al. 2019; Zamani et al. 2020). These were however primarily conceived to cater to the local academic market: their absence from the Frankfurt fair's catalogue may be evidence that advertising them to a transnational audience was not a priority. Their emergence as dominant redactional models may then find an explanation in the interest they garnered among European authors and publishers regardless of their transnational visibility (Chap. 10). Sometimes, such awareness was made explicit by publishers, as in the case of the 1562, 1569, 1574, and 1586 editions by Girolamo Scoto and his heirs, who presented them as reprints of the previous Paris edition of Cavellat (*ex postrema impressione Lutetiae*).

All these distribution considerations had to jibe with publishers' knowledge of their own distribution capacity; publishers were well aware of the franchising structure they had built over the years, the alliances they held with colleagues around Europe, and their influence on the market. In sum, publishers knew the capacity and extent of their distribution network and planned print runs according to this factor, alongside estimations of market saturation.

A powerful weapon publishers and authors could consider deploying were book privileges. These were costly legal instruments granted to either authors or publishers (or, at times, to the former by the way of the latter). Privileges not only shielded grantees against pirated copies but also granted them a monopolistic position within their book market (most of the privileges had a limited geographic span). Privileges were among the itemized expenses that publishers took into account when planning a publishing project. Book privileges were granted only to editions that introduced true innovations to the content (mainly texts, images, or apparatuses). It is fair to say that the objective of gaining even a local monopoly worked as an incentive for innovation. Thus, textual or metatextual innovations in frequently republished works like the *Sphaera* were also market-driven elements.[11]

Most material production costs were negotiated in a dialogue that, at least in the case of Wittenberg, saw printers in a position of great disadvantage (Chap. 5). With publishers interested in getting away with the most convenient price for a single print run and willing to use local competition among printers as valid leverage, printers could be forced to make the most of a contracted job by downgrading the quality of their work to the minimum standard agreed upon with the publishers.

The complexity of the pre-production process was partially mirrored by the microcosm of the print shop, where diverse skills brought by diverse characters could meet and benefit from mutual cooperation. Mathematical texts such as the *Sphaera* required special expertise (Chap. 2), and the production of innovative diagrams and images required an astronomer to work with an engraver and for the two to agree upon the accuracy of the visual outcome. This necessity occasioned episodes of intellectual collaboration between professionals who would otherwise have little reason to work together. The act of correcting proof sheets could have been the mechanical practice of an ordinary corrector whose task was collating imprints with a rubber-stamped manuscript. However, clues suggest that quality editions made use of expert scholars to confirm that complex mathematical material would hold together (Pantin 2013). In certain cases, authors and printers could even be the same person, creating a fine short circuit between theoretical knowledge and mechanical know-how (Chap. 2) (Axworthy 2020).

Summing up, grasping the academic book market requires an understanding of the inherent mechanisms of both the local and global markets and their reciprocal interaction. On the local market, the dominant factor was represented by the close relationship between book producers and instructors, as well as the educational institutions in which they were active. On the global market, the dominant factors were twofold: from a material and economic perspective, the dominant factor was the distribution network of book producers, and from a more abstract perspective, the dominant factor was the mutual awareness among book producers in addition to the authors' networks. The European success of the Wittenberg *Sphaera* was due mostly to the latter. However, Wittenberg models were first imitated by great transnational printers and publishers in Venice and Paris, who in turn were echoed by other distribution networks.

---

[11] Another feature of book privileges worth mentioning is that they occasionally provide indirect clues on print runs. For example, it is known that in Venice, it was customary from the 1540s onward to grant book privileges only for editions exceeding four hundred copies (Nuovo 2013, 110).

7 Continuities and Further Research

This volume has been conceived as a continuation of the work published in 2020 concerning the authors of the commentaries of the *Sphaera* (Valleriani 2020b). The goal was to complete investigations of the actors, networks, and modes of transmission of knowledge involved in the perpetuation of the epistemic tradition linked to the *Tractatus de sphaera*.[12]

This volume is exclusively concerned with the circulation of the *Sphaera* in print, although it is fairly obvious that the printing press was not the exclusive circuit of dissemination and consumption of the *Tractatus*. Print culture and manuscript culture largely coexisted in the period represented by the *Sphaera* corpus, and manuscript redactions of Sacrobosco likely played a significant role in shaping the modes of transmission of astronomical and mathematical knowledge, as well as the dynamics of consolidation of epistemic communities (Dicke and Grabmüller 2003; Richardson 2009; Richardson and de Vivo 2011).

Secondly, the study mainly covers continental Europe, with the exception of brief coverage of the Iberian trans-Atlantic territories. If the tradition of Sacrobosco's scholarship has been pursued in the areas in which it flourished, the volume does not touch upon English-speaking regions and northern Europe. This is justified by the fact that such areas did not have a relevant role in producing printed editions of the *Sphaera*, with the exception of a few nautical manuals translated from Spanish into English in Britain, mainly based on excerpts or brief paraphrases of the text.

Finally, in compiling adequate case studies, one relevant center of book production, Antwerp, was not included. In the context of the print history of the *Sphaera*, Antwerp was in fact a late comer, and not an outstanding contributor in terms either quantitative or qualitative, with none of the local editions becoming a dominant model.

In spite of these limits, however, the volume covers forty-three percent of the sources of the corpus.[13] By means of these studies, it will now be possible to interpret data concerned with the social, economic, and institutional relationships among authors, printers, and publishers, and thus to determine whether the emergence of an epistemic family of treatises, characterized by their similarity to the Wittenberg model, is structurally related to the emergence of a social group. This is the direction of future research.

---

[12] To pursue the investigation presented in the first volume of the series, forty-three percent of the corpus was taken into consideration (https://sphaera.mpiwg-berlin.mpg.de/doi-visualisation-authors-volume).

[13] For a visualization of the sources of the *Sphaera* corpus that are mentioned in each chapter of the present book, see the "Visualizations" page on the *Sphaera* project website: https://sphaera.mpiwg-berlin.mpg.de/sphaera-printers-volume/. Accessed 16 June 2021.

Abbreviations

Digital Repositories

Sphaera Corpus*Tracer*   Max Planck Institute for the History of Science. https://db.
sphaera.mpiwg-berlin.mpg.de/resource/Start.   Accessed
07 June 2021.

References

Secondary Literature

Axworthy, Angela. 2020. Oronce Fine and Sacrobosco: From the edition of the *Tractatus de sphaera* (1516) to the *Cosmographia* (1532). In *De sphaera of Johannes de Sacrobosco in the early modern period: The authors of the commentaries*, ed. Matteo Valleriani, 185–264. Cham: Springer Nature. https://doi.org/10.1007/978-3-030-30833-9_8.

Burkart, Lucas. 2019. Early book printing and venture capital in the age of debt: The case of Michel Wenssler's Basel printing shop (1472–1491). In *Buying and selling. The business of books in early modern Europe,* ed. Shanti Graheli, 23–54. Leiden/Boston: Brill.

Crowther, Kathleen M. 2020. Sacrobosco's *Sphaera* in Spain and Portugal. In *De sphaera of Johannes de Sacrobosco in the early modern period: The authors of the commentaries*, ed. Matteo Valleriani, 161–184. Cham: Springer Nature. https://doi.org/10.1007/978-3-030-30833-9_7.

Darnton, Robert. 1982. What is the history of books? *Daedalus* 111: 65–83.

Dicke, Gerd, and Klaus Grubmüller, eds. 2003. *Die Gleichzeitigkeit von Handschrift und Buchdruck*. Wiesbaden: Harrassowitz Verlag.

Dondi, Cristina, and Neil Harris. 2013. Oil and green ginger. The *Zornale* of the Venetian bookseller Francesco de Madiis, 1484–1488. In *Documenting the early modern book world: Inventories and catalogues in manuscript and print*, eds. Malcolm Walsby and Natasha Constantinidou, 341–406. Brill: Leiden/Boston.

Feldhay, Rivka. 1999. The cultural field of Jesuit science. In *The Jesuits. Cultures, sciences, and the arts. 1540–1773*, eds. John W. O'Malley, Gauvin Alexander Bailey, Steve J. Harris, and T. Frank Kennedy, 107–130. Toronto/Buffalo/London: University of Toronto Press.

Feldhay, Rivka. 2021. Catholic Europe and sixteenth-century science: A path to modernity? In *Religious responses to modernity*, ed. Yohanan Friedmann and Christoph Markschies, 49–63. Berlin: DeGruyter.

Gehl, Paul F. 2013. Advertising or Fama? Local markets for schoolbooks in sixteenth-century Italy. In *Print culture and peripheries in early modern Europe. A contribution to the history of printing and the book trade in small European and Spanish cities*, ed. Benito Rial Costas, 69–100. Leiden: Brill.

Gerritsen, Johan. 1991. Printing at Froben's: An eye-witness account. *Studies in Bibliography* 44: 144–163.

Giesecke, Michael. 1998. *Der Buchdruck in der frühen Neuzeit*. Frankfurt am Main: Suhrkamp.

Ginsburg, Jane C. 2013. Proto-property in literary and artistic works: sixteenth-century papal printing privileges. *The Columbia Journal of Law and the Art* 36: 345–458.

Grafton, Anthony. 2021. The margin as canvas: A forgotten function of the early printed page. In *Impagination—Layout and materiality of writing and publication*, eds. Ku-ming Kevin Chang, Anthony Grafton, and Glenn W. Most, 185–207. Berlin: De Gruyter.

Lanuza-Navarro, Tayra M.C. 2020. Pedro Sánchez Ciruelo. A commentary on Sacrobosco's *Tractatus de sphaera* with a defense of astrology. In *De sphaera of Johannes de Sacrobosco in the early modern period: The authors of the commentaries*, ed. Matteo Valleriani, 53–89. Cham: Springer Nature. https://doi.org/10.1007/978-3-030-30833-9_3.

Leitão, Henrique (ed.). 2008. *Sphaera Mundi: A Ciência na Aula de Esfera. Manuscriptos científicos do Colégio de Santo Antão nas colecções da BNP*. Lisboa: Biblioteca Nacional de Portugal.

Leitão, Henrique. 2013. *Um Mundo Novo e una Nova Ciência. In 360º. Ciência descoberta, Catálogo da exposição*, ed. Henrique Leitão, 16–39. Lisboa: Fundação Calouste Gulbenkian.

Maclean, Ian. 2021. *Episodes in the life of the early modern learned book*. Leiden: Brill.

Milazzo, Renaud. 2020. In the mind of a publisher: Establishing the price of emblem books in Antwerp in the sixteenth century. *De Gulden Passer/the Golden Compasses* 98: 183–201.

Nuovo, Angela. 2013. *The book trade in the Italian Renaissance*. Leiden/Boston: Brill.

Pantin, Isabelle, and Philippe Renouard, eds. 1986. *Imprimeurs et libraires parisiens du XVIe siècle: Cavellat-Marnef et Cavellat*. Paris: Bibliothèque nationale.

Pantin, Isabelle. 1998. Les problèmes de l'édition des livres scientifiques: l'exemple de Guillaume Cavellat. In *Le livre dans l'Europe de la Renaissance: Actes du XXVIIIe Colloque international d'Etudes humanistes de Tours*, ed. Bibliothèque Nationale, 240–252. Paris: Promodis, Editions du Cercle de la Librairie.

Pantin, Isabelle. 2013. Oronce Finé mathématiien et homme du livre: la pratique éditoriale comme moteur d'évolution. In *Mise en forme des savoirs à la Renaissance. À la croisée des idées, des techniques et des public*, eds. Isabelle Pantin and Gérard Péoux, 19–40. Paris: Armand Colin.

Pantin, Isabelle. 2020. Borrowers and innovators in the printing history of Sacrobosco: The case of the "in-octavo" tradition. In *De Sphaera of Johannes de Sacrobosco in the early modern period. The authors of the commentaries*, ed. Matteo Valleriani, 265–312. Cham: Springer Nature. https://doi.org/10.1007/978-3-030-30833-9_9.

Pedersen, Olaf. 1975. The *Corpus astronomicum* and the traditions of mediaeval Latin Astronomy. *Copernicana* 13: 57–96.

Price, Audrey. 2014. Mathematics and mission: Deciding the role of mathematics in the Jesuit curriculum. *Jefferson Journal of Science and Culture* 1: 29–40.

Richardson, Brian. 2009. *Manuscript culture in Renaissance Italy*. Cambridge/New York: Cambridge University Press.

Richardson, Brian, and Filippo de Vivo. 2011. *Scribal culture in Italy 1450–1700*. Leeds: Society for Italian Studies

Rück, Peter and Martin Boghardt (eds.). 1994. *Rationalisierung der Buchherstellung im Mittelalter und Frühneuzeit*. Wetter: Druckerei Schröder.

Sander, Christoph. 2018. Johannes de Sacrobosco und die Sphaera-Tradition in der katholischen Zensur der Frühen Neuzeit. *NTM Zeitschrift für Geschichte der Wissenschaften, Technik und Medizin* 26: 437–474. https://doi.org/10.1007/s00048-018-0199-6.

Squassina, Erika. 2017. La protezione del *Furioso*: Ariosto e il sistema dei privilegi in Italia. *Bibliothecae.it* 6: 9–38.

Valleriani, Matteo. 2017. The tracts on the sphere. Knowledge restructured over a network. In *Structures of practical knowledge*, ed. Matteo Valleriani, 421–473. Dordrecht: Springer.

Valleriani, Matteo. 2020a. Prolegomena to the study of early modern commentators on Johannes de Sacrobosco's *Tractatus de sphaera*. In *De sphaera of Johannes de Sacrobosco in the early modern period: The authors of the commentaries*, ed. Matteo Valleriani: 1–23. Cham: Springer. https://doi.org/10.1007/978-3-030-30833-9_1.

Valleriani, Matteo (ed). 2020b. *De sphaera of Johannes de Sacrobosco in the early modern period: The authors of the commentaries*. Cham: Springer. https://doi.org/10.1007/978-3-030-30833-9.

Valleriani, Matteo, Beate Federau, and Olga Nicolaeva. 2022. The hidden praeceptor: How Georg Rheticus taught geocentric cosmology to Europe. *Perspectives on Science* 30(3).

Valleriani, Matteo, Florian Kräutli, Maryam Zamani, Alejandro Tejedor, Christoph Sander, Malte
    Vogl, Sabine Bertram, Gesa Funke, and Holger Kantz. 2019. The emergence of epistemic commu-
    nities in the *Sphaera* Corpus: Mechanisms of knowledge evolution. *Journal of Historical Network
    Research* 3: 50–91. https://doi.org/10.25517/jhnr.v3i1.63.
Werner, Sarah. 2019. *Studying early printed books. 1450–1800. A practical guide.* Hoboken, NJ:
    Wiley Blackwell.
Zamani, Maryam, Alejandro Tejedor, Malte Vogl, Florian Kräutli, Matteo Valleriani, and Holger
    Kantz. 2020. Evolution and transformation of early modern cosmological knowledge: A Network
    Study. *Scientific Reports–Nature* 10. https://doi.org/10.1038/s41598-020-76916-3.

Matteo Valleriani is research group leader at the Department I of the Max Planck Institute for
the History of Science in Berlin, Honorary Professor at the Technische Universität of Berlin,
and Professor by Special Appointment at the University of Tel Aviv. He investigates the relation between diffusion processes of scientific, practical, and technological knowledge and their
economic and political preconditions. His research focuses on the Hellenistic period, the late
Middle Ages, and the early modern period. Among his principal research endeavors, he leads the
project "The Sphere: Knowledge System Evolution and the Shared Scientific Identity of Europe"
(https://sphaera.mpiwg-berlin.mpg.de), which investigates the formation and evolution of a shared
scientific identity in Europe between the thirteenth and seventeenth centuries. In the context of this
project, Matteo Valleriani implemented the development of Machine Learning technology and of
the physics of complex system in the humanities. The project is also part of the investigations led
by Matteo Valleriani in the context of BIFOLD (https://bifold.berlin). Among his publications, he
has authored the book *Galileo Engineer* (Springer 2010), is editor of *The Structures of Practical
Knowledge* (Springer Nature 2017), and published *De sphaera of Johannes de Sacrobosco in the
Early Modern Period. The Authors of the Commentaries* (Springer Nature 2020).

Andera Ottone earned a doctorate in history at the University of Naples. He is currently a postdoctoral research fellow at the University of Milan's Department of economics, management, and
quantitative methods in the context of the ERC founded EMoBookTrade project (Grant Agreement no 694476). He is also a fellow at Berlin's Max Planck Institute for the History of Science.
He teaches classes in European history at the Technische Universität Berlin.

Chapter 2
Printerly Ingenuity and Mathematical Books in the Early Estienne Workshop

Richard J. Oosterhoff

Abstract Even though the first press in Paris was set up in 1469, in rooms owned by the Collège de la Sorbonne, it took some time before the University's *cursus ordinarius* was regularly set in print. One of the first concerted efforts to reconfigure textbooks using print was carried out by Wolfgang Hopyl and Johann Higman, beginning in the late 1480s. Their press—and their collaboration with the circle of Jacques Lefèvre d'Étaples—was taken up by the elder Henri Estienne and then Simon de Colines, who transformed the press into one of the most illustrious cases of the printing art in Europe, alongside Manutius and the later Estiennes. This chapter focuses on the routine claims these printers made about publishing the *Sphaera* of Sacrobosco and their own artful labors. It similarly examines their remarkable frontispieces within the context of a nascent tradition of astronomical frontispieces. Printerly claims to ingenuity and the relation of books and observation in these frontispieces undermine old historiographical dichotomies that oppose craft knowledge and book knowledge.

Keywords Parisian printers · Henri Estienne · Simon de Colines · Artisanal skills · Mathematical books · Book history

1 Introduction

On the twelfth of February of 1494,[1] the Netherlandish printer of Paris, Wolfgang Hopyl (fl. 1489–1523), added the final, lengthy colophon to an edition of Sacrobosco's *Sphaera*. The book was a remarkable accomplishment, bringing together a wide range of new elements of the printer's art, so Hopyl quite rightly styled himself an "ingenious printer" (*ingeniosus impressor*): "Printed at Paris in the neighborhood of St. Jacques, near the sign of St. George, in the year of Christ, creator of the stars,

---

[1] Or 1495 to us in the "new style," after the late sixteenth-century calendar reforms moved the year's beginning from March 25 to January 1.

R. J. Oosterhoff (✉)
University of Edinburgh, Edinburgh, Scotland
e-mail: richard.oosterhoff@ed.ac.uk

M. Valleriani and A. Ottone (eds.), *Publishing Sacrobosco's* De sphaera *in Early Modern Europe*, https://doi.org/10.1007/978-3-030-86600-6_2

25

February 12, 1494. Done by the ingenious printer Wolfgang Hopyl, who always keeps this adage firmly in mind: 'Great things are not done by strength or speed or bodily swiftness, but by planning, judgment, and authority.' With the aid of the most diligent correctors, Lucca Walter Conitiensis, Guillaume Gontier, Jean Griettan, and Pierre Griselle—lovers of mathematics" (Sacrobosco 1494).[2]

Before 1495, Johannes de Sacrobosco's (1195–1256) treatise had been printed in quarto format, without commentary—Hopyl himself had been the first to print Sacrobosco in Paris in 1489 (Sacrobosco et al. 1489). But in 1495 Hopyl spread the *Sphaera* out over large folio pages, set off with the extensive commentary of Jacques Lefèvre d'Étaples (1455–1536), then at the beginning of his reimagination of university teaching and especially mathematical works (Chaps. 9 and 13). As he did for Lefèvre's many other commentaries and textbooks, Hopyl carefully indexed the paragraphs of Lefèvre's comments with numbers printed in the margins, allowing the reader to flip back to the analytical index of the work, printed at the outset. Likely with Lefèvre's input, Hopyl and his shop had devised new woodcuts, rearranging the entire visual program of the *Sphaera*. Perhaps the most difficult innovation, from the point of view of both printer and reader, was the suite of tables throughout the book, which transformed the *Sphaera* from a largely qualitative work of description into a primer in calculation, beginning with the tutorial on sexagesimal arithmetic that opened the book.

All of these elements made the *Textus de sphaera* a challenging book to print, labor captured in the colophon in two ways. Hopyl offered his own account of the task with that line from Cicero (106–43 BCE): "Great things are not done by strength or speed or swiftness of body, but by planning, judgment, and authority."[3] Hopyl defined his ingenuity not as the quick flash of insight, but as methodical labor. Moreover, the labor had been shared, for in the next line he listed several of Lefèvre's students as "correctors" (*recognitores*), defined also by their love of mathematics (*matheseos amatoribus*). As Lefèvre himself indicated in the prefatory epistle, young associates such as Jean Griettan were "skilled in abacus and calculation" and had contributed significant work to the book, possibly to its tables and calculations (Sacrobosco 1494, sig. [i]v.).[4]

Elsewhere I have commented on the distinctive features of Lefèvre's commentary on the *Sphaera*, on the significance of this work's tables for fostering quantitative skills among Renaissance readers, and on its place within the larger typographical program of Lefèvre's circle (Oosterhoff 2018, 133–150; 2020, *forthcoming*). In this chapter, I wish to focus instead on the printer's relationship to Sacrobosco.

---

[2] For this and later editions of the work, the Appendix gives titlepage and colophon transcriptions, with further bibliographical references.

[3] Cicero, *De senectute*, §17. Thanks to Anthony Ossa-Richardson for first spotting this one. Hopyl or whichever of his associates composed this colophon reused the dictum for other books, e.g., see the many colophons transcribed by (Stein 1891). For more on Hopyl's remarkable programme, see (Delft 2010).

[4] The original text reads: "Affuit levamini domesticus noster Iohannes Griettanus, abaci, numerandique peritie, et relique Matheseos non inscite studiosus—scripsit opus, et quasi fesso humerum subiecit Atlanti." For a transcription, see (Rice 1972, esp. 8).

Beyond Paris, this unusual version of the *sphere* was soon copied in omnibus editions of the *Sphaera* printed in Venice (Sacrobosco et al. 1499, 1508, 1531a; Ptolemy et al. 1518a, 1518b).[5] Within Paris, however, the book remained connected to Hopyl (Sacrobosco et al. 1500), and then to the printers who took over his press, first Henri Estienne I (Sacrobosco et al. 1507, 1511, 1516) and then Simon de Colines (1480–1546) (Sacrobosco et al. 1521, 1527, 1531b, 1534, 1538). Beyond small corrections, Lefèvre himself did not substantially modify the book in its later imprints, except to include two smaller works with the edition of 1500: the medieval propositions of Boethius' (ca. 480–ca. 525) translation of Euclid (323 BCE–285 BCE), and a treatise on an astrolabe ring by the Jewish papal physician Bonet de Lattes (ca. 1450–ca. 1515).[6] The later editions of Colines add only small ornaments and some marginal annotations, likely at the hand of Oronce Fine (1494–1555), and a new frontispiece, also by Fine. Until its last edition in 1538, the *Textus de sphera* remained a visually distinctive and regularly reprinted item in the Estienne press catalog.

I shall argue that following Sacrobosco through the early Estienne press will illuminate a claim often made regarding print and the early modern sciences, namely that the print shop was a space in which handworkers and headworkers shared knowledge, creating space for artisans to claim intellectual prestige (Eisenstein 1980, Chap. 6).[7] The following section will focus on what these editions of the *sphere* reveal about the printer's claims to own distinctive abilities, what I shall call "printer's ingenuity." In the period approximately between 1490 and 1520, print was a much more fluid phenomenon, both socially and technically, than it would be later, so Estienne (and, as we shall see, his predecessor and partner Johann Higman (d. 1500)) represents an effort to defend the publisher's status as a craftsman. In the third section, I will especially depend upon the shifting roles of colophons and title pages. The next section will return to another experimental element in this edition of Sacrobosco: frontispieces. Using this evidence, I shall suggest that even these objects—reimagined wholly within the print shop—help us to detect subtly changing attitudes towards the very practice of astronomy.

2 Printer's Ingenuity and Mathematical Books

When Hopyl called himself *ingeniosus*, he was invoking complex Renaissance debates over the intellectual value of artisanal skill (Marr et al. 2018, 19–52). The root word *ingenium* was often associated with swift, powerful invention, drawing on Cicero's influential characterization of the innate abilities of outstanding orators. But

---

[5] The Appendix below fully lists titles and colophons of the various editions of the *Textus de sphaera*.

[6] Following editions of this Boethian translation are listed by (Folkerts 1970, 41–49). The translation usually only lists a selection of enunciations (generally without proofs) from the first four books (Folkerts 1970, 80–82). The text on the astrolabe ring was first published in (Lattes 1493).

[7] Recent studies have deepened the point, for instance see (Grafton 2018). A related insight has been often formulated within the theoretical framework of "trading zones" (Long 2015).

painters, sculptors, and other artisans also argued that their work also displayed the creative qualities of ingenuity. This ambivalence of *ingenium* as both intellectual and embodied was deepened through the question of speed: was it quick or slow? Innate or acquired? The common doublets *ars et ingenium* or *industria et ingenium* could be seen as opposing—or they could function in pleonasm, shading from one to the other, from quick wit to plodding diligence (Marr et al. 2018, 9, 46–50, 88; Baxandall 1963, 304–326, 1986, 15–16). Wielded by literary elites, the plasticity of these terms could swiftly turn *ingeniosus* from a term of praise into a demeaning association with grubby manual labor.[8] As Eisenstein has recently documented, throughout the early modern period, from Trithemius to Moxon, printers could be dismissed as "mere mechanics" (Eisenstein 2012, esp. 15–19).

The negative associations of craft may have kept Hopyl, his partner Johann Higman, and the elder Henri Estienne somewhat more darkly in the shadows of print history than they might otherwise have been. Instead, the seventeenth- and eighteenth-century historians of this dynasty focused on those accomplishments that were easily recognized by early Enlightenment *belles lettres*: the Herculean *Thesaurus linguae latinae* (first ed. 1531) of Robert Estienne (1503–1559) (Estienne 1531), or the many Greek and Latin editions edited and published by Henri Estienne [II] (Almeloveen 1683; Maittaire 1709).[9] Even that incisive Victorian Mark Pattison, setting straight the earlier bibliographers by rereading "the great printers Stephens" in their sixteenth-century culture, had few words for the elder Henri Estienne—father to Robert and grandfather to the more famous Henri; Pattison had little use for mathematical works such as Sacrobosco (Pattison 1865).

But the first generation of Hopyl, Higman, and Estienne vaunted their craft status, as I shall consider in a moment (This even though they could claim some credit in learning—Higman, at least, had studied for the BA at Paris). The business aspects of print mattered. In the 1490s Hopyl and Higman reveal an energetic entrepreneurial campaign to engage the widest possible range of markets: devotional works, classical standards like Virgil (70–19 BCE) and Seneca (d. 65), collections of medieval letters, scholastic *quaestiones* on Aristotle (385–322), university textbooks, Lorenzo Valla's (1407–1457) *Elegantiae* (Valla 1490–1491), and theological dialogues. Hopyl even published a little in Dutch and French. As Andrew Pettegree has observed, the late fifteenth-century print was no place for idealistic dreams, requiring a strong stomach for risk and a keen eye for saleable products (Pettegree 2010, 53–55). As scholars and printers experimented and negotiated over new ways of producing, distributing, and selling books, the famously selective and erudite Aldus Manutius (1449–1515) offered only one model of success (in the 1490s it was hardly evident that his press would find the public needed for success).

---

[8] For instance: the relationship between Robert Boyle and Robert Hooke turned on charges of Hooke's "ingenuity," noted by (Bennett 2006).

[9] Representative modern studies of the Estiennes include (Armstrong 1954; Considine 2008, Chaps. 1 and 2; Kecskeméti et al. 2003). Very different but equally illuminating pathways into this world of Northern humanist print include, beyond the works of Rice cited already (Bietenholz 1971; White 2013).

One place to gain a market was in the universities. The library that Beatus Rhenanus (1485–1547) assembled during his studies at Paris from 1503 to 1507 shows Lefèvre's Sacrobosco commentary a key ingredient in his studies in Lefèvre's circle (Oosterhoff 2018).[10] In the first instance, it appears that the 1495 edition of Sacrobosco was a bespoke product for a specific circle of teachers—its reprinting in Venice and its presence throughout European university libraries suggests the book found a much wider university market within a decade. It is important to stress, however, that this was not necessarily obvious or easy in the late 1480s, when Hopyl started to produce university textbooks. Severin Corsten has observed how the most widely used university texts—the standard manuals of logic to be mastered by arts bachelors—took decades to be widely published in print (Corsten 1987). Fifteenth-century manuscript production of university texts was highly regulated, with jobs flowing through the hands of *libraire jurés*, the four official booksellers who were legally sworn by the university to regulate the city's book trade—such booksellers had long subcontracted to shops of manuscript copyists, illuminators, binders, as well as presiding over bookstalls, and they retained all of these functions near the end of the fifteenth century.[11] Did *libraire jurés* spot an opportunity and therefore become printers? Or did printers, having already entered the book market, then compete to gain the university's support as *libraires jurés*? Either way, by the late 1480s—nearly twenty years after the first press was set up in Paris—some *libraires jurés* were also printers, such as Antoine Vérard (fl. 1485–1512), who specialized in French courtly texts, books of hours, histories, and liturgical books, which could be illuminated and bound for particular clients (Winn 1997). Larger establishments such as Vérard joined forces with smaller presses to complete larger jobs, combining their access to specific markets. It seems likely that Hopyl and Higman began their partnership for this reason; Hopyl was a *libraire juré*, and although both men retained their own premises, they partnered on academic texts from the 1480s until Higman died in 1500. By Higman's death, university books formed a large part of their shared and separate catalogs.

Another way to capture a market was to attach one's press to a famous author, as Lukas Cranach (1472–1553) (Chap. 5) did with Martin Luther (1483–1556) and Johann Amerbach (1440–1513) and Johann Froben (1460–1527) did with Erasmus (1466–1536). Jacques Lefèvre d'Étaples and close collaborators such as Josse Clichtove (1472–1543) and Charles de Bovelles (1479–1566) eventually would perform this function in Paris. Already, Hopyl and Higman had published several of their first works. Shortly after Higman's death, Henri Estienne I married Higman's widow, Guyone Viart (fl. ca. 1500–*post* 1520).[12] Possibly having apprenticed with Higman or Hopyl, he was now proprietor of Higman's press, and also inherited the close relationship with Lefèvre and his circle. For nearly sixty of almost 130 editions that Henri Estienne produced, Lefèvre was author, editor, translator, or contributor in

---

[10] On a wider range of reading practices, see (Oosterhoff 2015).

[11] For the production and distribution of university textbooks in the earlier period, see (Rouse et al. 1988).

[12] Court documents from 1517 indicate this, as edited in (Stein 1895).

some other way. Many more of the remainder of Henri's corpus was linked to one or another person in Lefèvre's network.[13]

As seen at the outset of this chapter, Hopyl presented his ingenuity as a matter of care, thoughtful judgment—and diligent craft. In fact, Hopyl, Higman, and Estienne stressed this language for themselves in many colophons, a manuscript object which persisted in printed works well into the first few decades of the sixteenth century, as title pages gradually filled out. What had been a protective sheet intended to protect a text block before consumers had their purchases bound, was becoming a site for advertisement, including publication details of printer, time, and place. Lefèvre's 1497 edition of the *Nicomachean Ethics* (Lefèvre d'Étaples 1497) advertised both the expense and *diligentia* of Hopyl and Higman; after Higman's press was taken over by Henri Estienne, the same formula advertised the partnership of Hopyl and Estienne for a Windesheim breviary (*Breviarium Canonicorum* 1502).[14] Of course, printers were not always the only ones with money at stake in producing careful copies. Nevertheless, formulas of this kind draw attention to the link between financial value and craft.

That link emerges only more strongly when we compare printers' privileges elsewhere. A word search of the EMoBookPrivileges database of such privileges in Venice turns up *ingenium* (or Italian variants) in a couple of dozen examples. Three early cases mention Johann von Speyer (fl. 1468–1477), one of the brothers who first set up a press in Venice in 1469: "The art of printing books was brought into this glorious city of ours, and day by day it grows more famous and populous through the work, study, and ingenuity of Master Johann von Speyer, who chose our city out of all the others...."[15] In the 1490s, another Venetian legal formulation could draw on the associations of cunning deceit attached to the word: "If anyone wants to print, or have printed the aforesaid volumes, he may not do so himself or through another, using any mode or connivance, for the next ten years...."[16] Occasionally, a formulation used *ingenium* to refer to authorial invention; but the main goal of such privileges was to protect the labor and livelihoods of *printers*. Indeed, the use of *ingenium* in these privileges often reinforces the idea that the work of printing was closely intertwined with the artisan's powers of invention. In one more occasional phrase, *ingenium* suggests the printer's capacity to make the text anew, as when the privilege indicates that the book (*volumen*) can only be printed by the named artisan or those selected by him, "in that style and arrangement which he intends

---

[13] For a bibliography, see (Rice 1972).

[14] The formula reads: "sumptibus impensis, ac diligentia" (Stein 1891, 17, 19).

[15] "MCCCCLXVIIII, die xviii septembris. Inducta est in hanc nostram inclytam civitatem ars imprimendi libros, in diesque magis celebrior et frequentior fiet per operam, studium et **ingenium** magistri Ioannis de Spira, qui caeteris aliis urbibus hanc nostram praeelegit..." (EMoBookPrivileges 11. Venezia, ASV, Collegio Notatorio 11, 56v, privilege resolution). Author's emphasis.

[16] "MCCCCLXXXXii, die xviiii augusti....Et sit qui velit imprimere, vel imprimi facere, non possit per se vel alium, aliquo modo, vel **ingenio**, per annos decem proximos futuros dicta volumina...." (EMoBookPrivileges 48. See also, ASV, Collegio Notatorio 14, 71r, privilege resolution). Author's emphasis. On ingenium as a legal device or trick, see (Marr et al. 2018, 46–47).

to be done according to his ingenuity and new invention."[17] These examples from Venice bestow creative powers on the printer, while also, within the legal frame of the privilege, protecting labor and financial investment.

Some of the most intriguing evocations of printerly ingenuity come from colophons of mathematical works such as the *Textus de sphaera* that I have already quoted from. In 1496, Higman and Hopyl together also published Lefèvre's most heroic mathematical work, a multimodal study of medieval numbers and music theory (Lefèvre d'Étaples 1496).[18] This was a book of daunting complexity, including Lefèvre's reworked demonstrations set off from enunciations, a treatise on music theory, an introduction to Boethian arithmetic, and a medieval number game—each treatise with unusual tables and the margins of most pages bearing diagrams. In the end, the printers recorded their efforts: "these two parts of the quadrivium, the best and leading parts of the liberal arts [i.e., arithmetic and music], alongside certain aids, Johann Higman and Wolfgang Hopyl took care to supply for the use of students, with diagrams and the weightiest of labors and cost. At Paris, in the year of salvation of our Lord, who set out all things in number and harmony, they put an end to this task in that year, on July 22, devoting their labors to studious men, to farewell every-where, forever. So also does David Laux, a Briton from Edinburgh, who diligently corrected the whole work from the exemplar."[19] The cumbersome Latin reinforces the cause of these men's difficulties: a complex layout of "little helps" (*amminicula*) and "diagrams" (*formulas*). The idea of *labores* is repeated, even as the printers avow their worth to *studiosi*. In 1514, reprinting this same work, Henri Estienne adopts these sentiments for his own.[20] In other works, Estienne sets his specific form of printerly ingenuity as diligence and labor combined with good judgment. Josse Clichtove's work *De mystica numerorum* (1513) includes the colophon "By Henri Estienne, careful and industrious craftsman of the art of printing books."[21] An edition of Clichtove's commentary on Lefèvre's astronomical *Theoricae* from

---

[17] "MCCCCLXXXXII, die xxi augusti. Et volumina impressa sint ipsius Francisci, excepto duntaxat illo impressore, quem praefatus Franciscus duxerit eligendum, cui soli liceat videlicet supradictum eius imprimere, seu imprimi facere, et vendere Statuta et volumina ipsa, eo modo et forma quibus pro **ingenio** suo et nova inventione facere intendit." (EMoBookPrivileges 109). See also, (ASV, Collegio Notatorio 14, 69v, petition). Author's emphasis.

[18] Judging from the date of another dedicatory letter from 1493, the book was probably published first in manuscript before that date.

[19] "Has duas Quadriuum partes et artium liberalium precipuas atque duces cum quibusdam amminic-ulariis adiectis: curarunt una formulis emendatissime mandari ad studiorum utilitatem Joannes Higmanus, et Wolgangus Hopilius suis grauissimis laboribus et impensis Parisii, anno salutis domini: qui omnia in numero atque harmonia formauit 1496, absolutumque reddiderunt eodem anno, die vicesima secunda Iullii suos labores ubicunque valebunt semper studiosis deuoventes. Et idem quoque facit David Lauxius Brytannus Edinburgensis, ubique ex archetypo diligens operis recognitor." (Lefèvre d'Étaples 1496, colophon).

[20] "curavit ex secunda recognitione una formulis emendatissime mandari ad studiorum utilitatem Henricus Stephanus suo grauissimo labore et sumptu Parhisiis, anno salutis domini: qui omnia in numero atque harmonia formavit 1514" (Lefèvre d'Étaples 1514).

[21] "Per Henricum Stephanum, artis excusoriae librorum sedulum & industrium opificem, e regione scholae Decretorum habitantem" (Clichtove 1513).

1517 closely follows the formula: "Henri Estienne, careful and industrious artisan of bookmaking…."[22] In each of these cases, the nature of the ideal printer is one of industrious care and craft. The vocabulary of diligence and industry overlaps with the well-known language of Herculean labors that Erasmus would project on the basis of editing Jerome's letters, and which Robert Estienne would again claim for himself in his exhaustive (and exhausting, as he claimed in his preface) *Thesaurus linguae latinae* (1531).[23]

The formative influence of this period on the nature of the book becomes clear when we look at colophons together with the title pages that gradually replaced them as the Estienne press grew into a dynasty (Smith 2000). The first impression of the *Textus de sphera* in 1495 bore only the simple sheet with title that early printers used to protect the block of text. In 1507, Henri Estienne the Elder added a floriated frame to the title, which he updated for the 1511 and 1516 editions (Fig. 1). Colophons, meanwhile, shrank. Even as he added title pages, Estienne streamlined his colophons: "Impressum Parisii in officina Henrici stephani e regione Schole decretorum sita. | Anno Christi siderum conditoris 1507. Decimo die Nouembris."

Changes in the next generation were subtle, but with dramatic effect. When Henri Estienne died around 1520, his widow Guyone Viart married once again, and Simon de Colines took over the press (quite likely one of Estienne's employees or colleagues) since Henri's son Robert was not old enough to acquire the rights of a master printer until 1526 (Armstrong 1954). Meanwhile, Colines continued to work closely with Lefèvre's circle, and in 1521 demonstrated both his inheritance of Estienne's press with an edition of Lefèvre's *Textus de sphaera*—now spelled with a classicizing ligature, "Æ."

Up-to-date orthography was the least of the changes Colines made. Indeed, his attention to page design and space can be seen as a significant step towards the "style of Paris" analyzed by Isabelle Pantin elsewhere in this volume (Chap. 9). Such design choices reflected Colines' own distinctive ingenuity. Perhaps even more than Higman—who had studied for the BA at the University of Paris—or Estienne the Elder, Colines represented the most material, messy labor of the printer's many tasks, for he became well known as a type-cutter. His edition of the *Sphaera*, therefore, was no longer in the gothic type still used by Estienne, but now was in a spaciously formed roman type. His reimpressions of works once printed by Higman and Estienne, retained the claim of diligence and labor advanced by his predecessors.[24] But when he devised his own branding, he preferred the formulation

---

[22] "Excudit hoc opus & impressit Henricus Stephanus, efformandorum librorum sedulus & industrius artifex: Parisiis in sua officina libraria e regione scholae Decretorum" (Lefèvre d'Étaples and Clichtove 1517).

[23] Robert Estienne claimed he had neglected his household and health for two years to complete the work. More broadly on such tropes, see (Pabel 2008; Considine 2008, *passim* but esp. 19–30; Marr et al. 2018, 9–15).

[24] For example, an edition of Lefèvre's edition of the *Nicomachean Ethics* first published in 1497 simply replaces his own name in the formula "absoluta sunt impensis, sumptibus & diligentia Simonis Colinæi" (Renouard 1894, 78).

Fig. 1 Left: Title pages of Lefèvre, *Textus de sphera* (Paris, 1495, 1507, 1511 [repeated 1516]). Universitätsbibliothek Basel, shelfmark CC II 7:3, https://doi.org/10.3931/e-rara-49305 / Public Domain Mark; Center: Bayerische Staatsbibliothek, Res/2 Astr.u. 45 f, https://nbn-resolving.org/urn:nbn:de:bvb:12-bsb101 96242-9; Right: Universitäts- und Landesbibliothek Düsseldorf, https://nbn-resolving.org/urn:nbn:de:hbz:061:1-18933 / CC BY-NC-SA 4.0

"printed in his own most splendid of types" (*pressit suis typis nitidissimis*). One poet distinguished Colines' skill in punching type (Visagier 1537, 56):

> Printers are three, who must be held the best,
> Beside them pale and meagre all the rest
> Stephanus for correctness; Colines for the art
> Of cutting type; and Gryphius, for his part,
> Dexterity alike of hand and mind
> Being his, a master of them both we find.[25]

In 1521, Colines displayed not only his ownership of Estienne's enterprise but also his prowess on a beautiful new title page, which included all the information once reserved for the colophon: title, printer's name, place, date (Fig. 2).

I have not found explicit links between mathematical texts like the *Sphaera* and printerly ingenuity. Nevertheless, such links are not hard to infer. As we have seen, mathematical works required particular care and skill to set beautifully (Chap. 13). Tables and diagrams were easy to set badly, and their sheer complexity required special attention.[26] Furthermore, getting mathematical details right demanded special ability from correctors of such books—a problem we can sense in the numbers of correctors listed in the first edition of the *Textus de sphaera*. The significance of this labor directly impacted costs, one tenuous line of evidence suggests. The nineteenth-century bibliographer of the Estiennes had access to book catalogs of Robert Estienne from the 1520s and 1530s, which include copies of Lefèvre's edition of Euclid and the *Textus de sphaera*, both published 1516. The massive Euclid (522 folio pages) is listed at twenty-five *sous*;[27] the *Textus de sphaera* (64 folio pages) was twelve *sous*. Meanwhile, large volumes of Lefèvre's commentaries on Aristotle's political works—294 large-format pages, but devoid of woodcuts—went for four to ten *sous* (Renouard 1843, 1–23).[28] Mathematical works like Lefèvre's commentary

---

[25] "Inter tot, norunt qui cudere, tres sunt | Insignes: languet caetera turba fame. | Castigat Stephanus, sculpit Colinaeus; utrunque | Gryphius edocta mente, manuque facit" (Armstrong 1954, 9).

[26] Some of the difficulty can be sensed from efforts to understand Erhard Ratdolt's diagrams for the *editio princeps* of Euclid (Baldasso 2013).

[27] That means more than one *livre tournois* (= twenty *sous* = 240 *deniers*), which was equivalent to the standard *florin*.

[28] According to Renouard's report, the only other work that commanded prices above ten *sous* was the *Libri logicorum*, at fifteen *sous*—it too was full of diagrams, illustrating logical forms. As contributors of this volume have rightly pointed out, Renouard's report of these earlier catalogues may be unreliable, so I advance my interpretation cautiously. Certainly, Renouard's evidence is

Fig 2 Frontispiece of Lefèvre, *Textus de sphaera*, 1521. First printed as a frontispiece to (Fine 1515). Universidad de Deusto, Biblioteca Bilbao Sótano, 2 Fondo Antiguo 871–96"1" G 33 a, https://hdl.handle.net/11656/4868 / CC BY-NC-ND 3.0

on Sacrobosco offered an ambitious printer a chance to boast, and perhaps also to charge accordingly.

There is another possible way mathematics was tied especially close to printerly skill. Colines worked closely with two outstanding print designers who aligned their craft with mathematical mastery. Colines collaborated with Geofroy Tory (ca. 1480–ante 1533) on one *Aediloquium* and several Books of Hours (Aediloquium 1530; Renouard 1894, 65).[29] Tory is most widely recognized now for his *Champfleury* (Tory 1529), which used the earlier works of Lefèvre's student Charles de Bovelles—published by Henri Estienne—to develop quasi-mathematical figures such as a Vitruvian man, and who also used the writings of Luca Pacioli (1447–1517) and Albrecht Dürer (1471–1528) to suggest mathematical principles of font design (Bowen 1979, 13–27). Tory's fertile profusion of mythology and Pythagorean allusion is aimed to destabilize hierarchies: certainly to offer French as a learned, literary language, but also to trumpet the significance of visual design—and implicitly the printer's craft—as an intellectual project. He alluded to Bovelles' vernacular mathematics in particular to exemplify the kind of theorizing he wanted more of.[30]

But perhaps the more audacious bid for joining mathematics to the printer's craft as a shared intellectual project is found on Colines' new title page for Lefèvre's *Sphaera* commentary. This woodcut (Fig. 2) had been designed by another of Colines' associates, Oronce Fine.[31] This was not Fine's first creative effort to use the printer's mode of self-advertisement to set himself before a public. As Isabelle Pantin has shown, Fine first used this engraving in 1515 as the frontispiece to a collection of *Theoricae* (Pantin 1993, 90, 2009). There, Fine advertised himself where one might expect, at the end where the colophon was preceded by an acrostic poem on Fine's name, concluding in the distich "if you seek the one who corrected this and artfully decorated it with diagrams, the first elements [i.e., first letters of the lines] will give it."[32] Fine invites the reader to play, to puzzle out his role in assembling the

---

slightly out of line with Robert Estienne's bookseller's catalogues from the 1540s (Proot 2018), where *average* prices appear to correspond chiefly to number of paper sheets used, rather than complexity of typesetting or illustrations. Certainly that confirms other evidence from the later sixteenth century, for examples see (Limbach 2019; Proot 2019) and the findings ERC project EMoBookTrade led by Angela Nuovo. On reflection, however, I think Renouard's evidence still fits. The contrasting evidence is from the later sixteenth century and even then, when comparisons can be drawn, includes examples that show technical books with many figures, like those in Hebrew and Greek, *were* sometimes priced more highly than books of similar size but simpler formatting. Paper may have been the overriding economic concern, but surely was not always the only one (Bruni 2018, 273–278).

[29] For more on the typographical design details behind this relationship, see (Amert 2012, 69–70).

[30] "…en remettant le bon estudiant a Euclides, & a la Geometrie en francois de messire Charles Bouille, en la quelle il me semble auoir autant fructifie & acquis dimmortalite de son nom, quil a en tous ses autres Liures & oeuures latins quil a faicts studieusement. Nous nauons point encores veu de tel Autheur en langage Francois, Pleust a Dieu que beaucop daultres feissent ainsi…" (Tory 1529, 12r).

[31] On Fine's mathematical program, see (Axworthy 2016). On his attention to paratexts, see (Oosterhoff 2016).

[32] "Si petis hoc mendis quis terserit, arte figuris | Hinc decorarit opus, prima elementa dabunt" (Fine 1515). The colophon also highlights the printer's ingenuity: "solertia et caracteribus Michaelis

work, together with the long title that indicated how "all the works here recently were corrected with the greatest of diligence, together with figures and the most suitable engravings added in their place, far more accurately than before." Diligence becomes here not merely a matter of plodding industry, but of outstanding wit and sharp judgment, cutting into the printer's form. Fine's most daring move then is also his most playful: the seated figure, set at a table of mathematical books and instruments between the Muse Urania and Ptolemy, is Fine himself.[33] The engraver *is* the astronomer.

3 Astronomical Practice in Frontispieces

Now I would like to contribute to a project that Isabelle Pantin has engaged in at various points, namely considering these early, experimental frontispieces and what they say about astronomy. Efforts to understand the programmatic role of frontispieces in making early modern astronomy have been—apart from Pantin's work—focused chiefly on later periods (Burnett 1998; Remmert 2006, 2011; Söderlund 2010; Kaoukji and Jardine 2010). My argument about the contents of these books, which I also make elsewhere, is that they represent a shift towards the actual practice of astronomical mathematics, particularly the skills of calculation (Oosterhoff *forthcoming*). In them, astronomy was not only about conceptualizing the movement of the heavenly spheres, but even for novices was increasingly about calculating those motions. I will not cover that ground again but will suggest that these frontispieces mark a trend in raising the status of mathematical work by setting calculations alongside bookwork. Practitioners such as Oronce Fine used the print shop as the locus from which to reshape their personas in relation to books, practice, and experience, and they projected that reshaping in frontispieces.

Images prefacing early printed editions give a sense of the approach these books taught. The first printed editions of these books had no images, only leaving blank spaces in those pages where manuscripts usually bore illustrations. A copy of the first Venice edition from 1472 in the Cambridge University Library shows what printers likely expected: a later reader drew in their own diagrams.[34] Like many books published in the first decades of print, the book bore no title page and no frontispiece. We find the first astronomical frontispiece in a Venetian edition published in 1482 by Erhard Ratdolt (1442–1528) (Sacrobosco et al. 1482), who had substantial experience with illustrated texts and had developed a particular interest in technical

---

Lesclencher, artis formularie industrij opificis. Sumptibus vero honestorum Bibliopolarum Ioannis Parui et Reginaldi Chauderon (apud quos venales habentur)." See also (Pantin 2012, 10–13). The 1525 edition printed by Reginald Chaulderon also names Fine in the title.

[33] For confirmation, see (Pantin 1993) citing (Boy 1971 [here as Linet, Hillard and Poulle], no. 35), which is the Clouet portrait of Fine included by (Thevet 1584, vol. 3, fol. 564r). The similarity is uncanny.

[34] Cambridge University Library, Inc.4.B.2.3.8. I am grateful to Roger Gaskell for this observation, and for supplying images.

books (Chap. 3). Therefore, Ratdolt's introduction of images reflects not only the experience of a printer but specifically a printer seeking to associate himself with mathematical practice. The book includes the *Sphaera*, Johannes Regiomontanus' (1436–1476) diatribe against Gerardus Cremonensis (1114–1187), and Georg von Peuerbach's (1423–1461) new *Theoricae* of the planetary motions (Pedersen 1978; Pantin 2012). The image comprised a fist holding an armillary sphere, labeled with the arctic and Antarctic poles, and the signs of the zodiac (Fig. 3).

On its own, the armillary sphere was short-lived as a frontispiece; later editions of these texts moved it a few pages deeper into the text, using it to accompany a discussion of the tropics, equinoxes, and the zodiac (Fig. 4, left). Instead, beginning with the Venetian edition of Johannes L. Santritter (fl. 1480–1492) from 1488, we find a woodcut that was later appropriated by Ottaviano Scoto (fl. 1479–1498) and copied in several other places (Fig. 4, right). The image depicts the three figures of Astronomia, Urania, and Ptolemy, with the heavens above and the earth below. The contrast of heavens and earth emphasizes the terrestrial profusion of life, with rabbits, a lizard, and a stag resting and foraging between sprouting grass and flowers. The bottom of the image is demarcated by craggy earth. In contrast, the regularly positioned stars shine through a regular, diaphanous arc that demarcates the heavenly spheres from earth's atmosphere. Heaven's intelligent order beams from the faces given to the sun on the left and the moon on the right, who benevolently gaze on earth. The eyes of the sun and moon underscore the visual nature of astronomy.

Such images organized a much older iconography of astronomy that framed astronomers with their instruments. A fifteenth-century manuscript of Nicole Oresme's (1320–1382) *Traitié de l'espere* opens with a miniature of the scholar working on a text, overshadowed by an enormous armillary sphere.[35] An Italian manuscript of the Greek text of Ptolemy's (b. 100) *Cosmographia* similarly presents Ptolemy as a king standing beside a table piled high with instruments and books, holding an astrolabe up to his eye.[36] These in turn pick up motifs traceable through the medieval iconography of the liberal arts. I will not extend this iconographical study, though that would be revealing; rather, I will focus on the formal elements of these images that frame the contents of these books, because these subtle formal shifts bring changing attitudes to the fore.[37]

This early frontispiece (Fig. 4, right) suggests a deep ambivalence about observation in astronomical practice. Certainly, it reflects on vision and the significance of images in late medieval astronomy, as Barker and Crowther have argued. But these are bookish practices of visualization (Crowther and Barker 2013, 439–441). The figures focus our attention on books for mediating the knowledge gained by observation and instruments. The only figure to look directly upward is the muse Urania, focusing her gaze with a shading hand. Her feet stand firmly on the earth;

---

[35] BNF, fonds français 565, fol. 1r, as cited by (Murdoch 1984, no. 162).

[36] BNM, MS graeca Z.388, frontispiece, cited by (Murdoch 1984, no. 162).

[37] An effort to summarize this iconography might start with the early works of Fritz Saxl and Verdier (1969), and would have to process the massive enterprise led by Blume et al. (2012, 2016), evocatively summarised in (Blume 2014).

Fig. 3 Frontispiece of Sacrobosco, *Sphera*, printed by Erhard Ratdolt in Venice with Regiomontanus' *Contra Cremonensia* and Peurbach's *Theorica nova* (Sacrobosco et al. 1482), frontispiece. Bayerische Staatsbibliothek, Ink I.502, https://nbn-resolving.org/urn:nbn:de:bvb:12-bsb000546 05-7 / CC BY-NC-SA 4.0

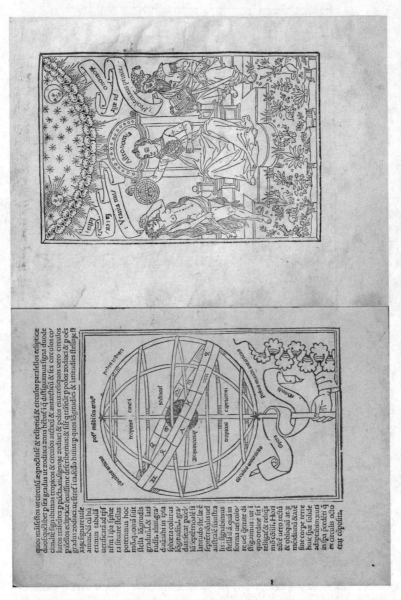

Fig. 4 Left: Sacrobosco with Peuerbach and Regiomontanus printed in Venice by Ottaviano Scoto I (Sacrobosco et al. 1490, a3v). This standardized image is precisely modeled on the 1482 Ratdolt frontispiece (Fig. 3), but now printed as a visualization tool later in the text. Courtesy of the Library of the Max Planck Institute for the History of Science, Rara J655sm. Right: Sacrobosco with Peuerbach and Regiomontanus from the same work (Sacrobosco et al. 1490, a1v (frontispiece)). This woodblock seems to have first been used as the frontispiece for the 1488 Santritter edition in Venice. Image courtesy History of Science Collections, University of Oklahoma Libraries; copyright the Board of Regents of the University of Oklahoma

with freely flowing hair, a naked body, and only enough cloth to cover her groin, she represents a corporeal vision that has not been chastened through learning. The art itself, Astronomia, sits at the center in a chaste, flowing robe, enthroned on what might be the lecturer's *cathedra*. Her feet, unlike Urania's, are firmly planted on the dais that bears her throne. She appears to be associated with observation, for in her right hand she holds an astrolabe by its ring, gazing across its sights—but in fact, she looks into Urania's face. On her other hand, Astronomia holds the reduction of the universe to its most abstract mathematical essentials: a small armillary sphere. The figure of Ptolemy, who sits with one foot on the dais and the other on the earth, reminds readers of this singular authority as the "prince of astronomers." This figure also underscores the centrality of books. The bearded sage—wearing the fur-lined robes and bejeweled headwear appropriate to the King of Egypt that Renaissance readers thought he was—reads from a book held open to the viewer. Three diagrams are just visible: a T-O map of the world, a dot in a circle commonly associated with the rotundity of the earth, and the circles of the tropics, with the ecliptic drawn across them. Ptolemy reveals to the reader the knowledge acquired from Urania and Astronomia through the written book.

In the next decades, this frontispiece combined with the armillary sphere to inspire an iconographical sub-tradition for astronomical books. But it is possible to detect two subtle shifts within this tradition: first, an increased emphasis on the schematic, diagrammatic, formal structure of mathematical astronomy; and second, an increased emphasis on the astronomer as an observer of the skies.

The first shift, emphasizing the schematic form of the heavens, is seen in the first edition of Lefèvre's *Textus de sphaera* of 1495 (Sacrobosco 1494), which thickened Sacrobosco with Lefèvre's commentaries as well as a new suite of illustrations.[38] These images tend to be angular, and devoid of text, characteristics already visible in the frontispiece (Fig. 5). In the rough, the figures of Urania, Astronomia, and Ptolemy indicate direct influence of the earlier Venice frontispiece: naked and long-haired Urania raises her arm, in precisely the same gesture to the heavens; a draped Astronomia gazes through the alidade in the center, with an armillary sphere in her left hand; Ptolemy expounds from a T-O map in a book. But the frontispieces also differ in important ways. The heavens have changed dramatically. The sun and moon smile down, but they and the stars have been brought down to earth, with no border separating them from the figures. They have been forced into mere decoration by a large armillary sphere reintroduced into the top half of the image. Urania no longer gazes at stars, but at the vast instrument that displays the polar circles and the heavenly tropics, with the zodiac wrapped around them. A second change is that all words are removed from the image, instead placed outside the border. Below the woodcut is a series of descriptions, keyed to letters in the image. The evacuation of words from the woodcut itself and the new central focus on the armillary sphere underscore the text's emphasis on schema, the stripped-down diagram. As we shall see further on, this text aims to lead the reader to mastery of information through calculation.

---

[38] Some of what follows builds on (Oosterhoff 2018, 133–150).

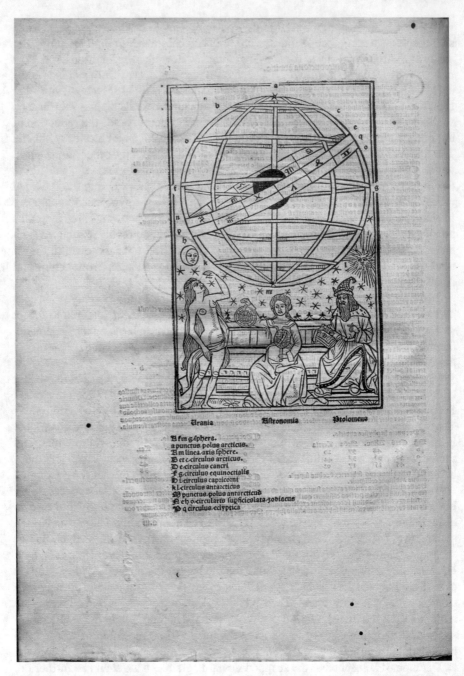

Fig. 5 Frontispiece of Lefèvre, *Textus de sphera* printed in Paris in 1494 (new style 1495) (Sacrobosco 1494). Universitätsbibliothek Basel, CC II 7:3, https://doi.org/10.3931/e-rara-49305 / Public Domain Mark

The second shift, towards observation, can be seen in Ottaviano Scoto's similar combination of armillary sphere and depiction of astronomy's teachers in 1496 posthumously published *Epytoma* (Regiomontanus 1496) of Ptolemy's *Almagest* by Johannes Regiomontanus (Fig. 6). Here the focus is again on an armillary sphere at the center of the frontispiece. The frill from the Santritter frontispiece which separates heavens, planets, and stars from the earth below has been reintroduced above the armillary sphere. Around the frontispiece a banner bears a now-familiar phrase, *Altior incubuit animus sub imagine mundi.*[39] But the greatest innovation is in the figures below the instrument. Ptolemy is now on the left, peering down into an open book that is invisible to us. The real source of knowledge is visible in the figure of Regiomontanus himself, who sits across from Ptolemy, with a closed book on his lap: he points a finger to the armillary sphere, which sits on the table between the astronomers. The modern astronomer has read the book and now turns to the instrument. The central image of Astronomia lecturing is eclipsed by the mathematically comprehended heavens themselves. The pedagogy of the arts disciplining the muses has been replaced by the discussion of experts, debating over a desk of books and instruments.

The iconography suggests Regiomontanus as an *alter Ptolemeus*, an astronomer who can be compared with the discipline's greatest ancient authority. Regiomontanus here stands at the beginning of Renaissance mathematicians' efforts to fashion themselves as authors, authorities, and indeed "famous men."[40] But the reason for Regiomontanus' fame is found in Regiomontanus' closed book and the finger he aims at the heavenly system. Regiomontanus persistently endeavored to reconcile the predictions yielded by the Ptolemaic theories of the planets and their motions with the results of his experience. He had grave concerns about the adequacy of the standard model, based on his own observations.[41] Regiomontanus vied with Ptolemy, using Ptolemy's tables as the starting point for further, sharper observation.

As an observer, Regiomontanus was an outlier among mathematical practitioners before Tycho Brahe's (1546–1601) systematic observations of the late sixteenth century. But Regiomontanus' frontispiece influenced someone rather more typical: Oronce Fine, whose frontispiece from 1515 we have seen (Fig. 2). By 1515 Oronce Fine had probably been in Paris for about five years. He was teaching at his *alma mater*, the Collège de Navarre, but also acquired a reputation as an illustrator of books. In the context of the earlier frontispieces, it is clear that Fine was self-consciously setting himself within a tradition. He partly draws on Lefèvre's frontispiece but also adapts many of the elements in the frontispiece to Regiomontanus' *Epytoma* of 1496, including its motto. The sun and moon are properly placed at the top, and the armillary sphere is labeled, as in the *Epytoma*. The biggest difference is in the figures below:

---

[39] Isabelle Pantin has found this image in frontispieces to (Grannollachs 1485), reprinted in 1489–1490 and 1492. See (Pantin 2009, 86–89).

[40] For the tradition of Urbino, see (Marr 2011, 48–56). By the 1490s Regiomontanus was also an authority on northern lists of *viri illustres*, notably Georg Tanstetter's account of "Viri mathematici" in (Peuerbach 1514, aa3v–aa6v). Much later genealogies are traced through title pages by (Remmert 2006, 259–262).

[41] These concerns are raised in a letter from 1463. See (Swerdlow 1990).

Fig. 6 Frontispiece of Regiomontanus, *Epytoma* (Venice: [Johannes Hamman], 1496). ETH-Bibliothek Zürich, Rar 4361, https://doi.org/10.3931/e-rara-528 / Public Domain Mark

Fine himself, whose table bears instruments new to these frontispieces; a square and a compass; and the pointed finger, which is directed towards Ptolemy, who gazes at the heavens through a quadrant. Fine thus seems to recognize Ptolemy as the source of observation, while he directs his own instruments to calculate from those given values.

The ambiguous nature of observation between books and firsthand experience is found in two more frontispieces Fine designed, one added to an edition of Lefèvre's *Textus de sphaera* from 1527 (Sacrobosco et al. 1527) (Fig. 7), and another to his own mathematical compendium of 1532 (Fine 1532; Axworthy 2020) (Fig. 8). Isabelle Pantin has explored these images more fully than I can do here (Pantin 2009). In the earlier illustration (Fig. 7), Fine has simplified the iconography. The frontispiece self-consciously advertises Fine's expertise—this was a precarious point in his career (Oosterhoff 2016, 556)—as Fine lays himself at full length below the heavenly model. Thus he presents himself as the noble soul in the banner borrowed from the frontispiece to Regiomontanus' *Epytoma* and repeated in his own earlier woodcut: "The lofty soul reclined [*incubuit*] below the image of the world" (*altior incubuit animus sub imagine mundi*). The reclining astronomer carelessly leaves his book open on the grass, unattended, as his mind inclines itself to the contemplation of heaven. Head on hand, Fine draws on the well-known iconography of melancholy, his contemplation preparing him to receive heavenly insight (Klibansky et al. 1964).

But the iconography is paradoxical, for while Fine seems to denote receptive repose, he also connotes the labor of intellectual industry. The verb form *incubuit* could be the perfect tense of either *incubare* or *incumbere*. The former could describe merely going to bed, but the most widely used dictionary of the Renaissance, Calepino, first noted that *incubare* implies a sustaining, internal activity, in preparation for something.[42] Likewise, the verb *incumbere* can mean "to recline," but also includes the more active sense of "to press, incline downward." For this sense, Calepino's first example was *incumbere studiis*, "to set down to one's studies." The word, therefore, encompasses both inspiration and labor, and the motto's dominant meaning seems to be one of diligent watchfulness. The high-minded must learn to see intellectually through lower means, through physical lines, illustrations, and diagrams drawn with the compass and square found in the earliest frontispiece. This industry and physical vigor of astronomical vision could not have escaped Fine himself, who was a skilled craftsman, renowned for his design of the visual arrangement of his many mathematical books, as well as frontispieces such as this one.

Indeed, this seems to be the point of the studious practitioner in Fine's frontispiece from the *Protomathesis* of 1532 (Fig. 8). The astronomer wields four instruments of mathematical practice: a book, an astrolabe, a quadrant, and a sundial. In fact, all of them are instruments Fine discusses in the book. The *Protomathesis* is a compendium of four books, the first two dedicated to arithmetic and geometry, the last two on cosmography and sundials. The organization of the cosmography and sundials is carefully orchestrated. The book ends with techniques for designing and using a

---

[42] (Calepino 1522, s.v.).

Fig. 7 Frontispiece of Lefèvre, *Textus de sphaera*, printed in Paris by Simon de Colines in 1527, and designed by Oronce Fine, and first included as the title page of the *Textus de sphaera* for this edition (Sacrobosco et al. 1527). Bibliothèque municipale de Bordeaux, shelfmark S 161(3), http://uranie.huma-num.fr/idurl/1/1511 / CC

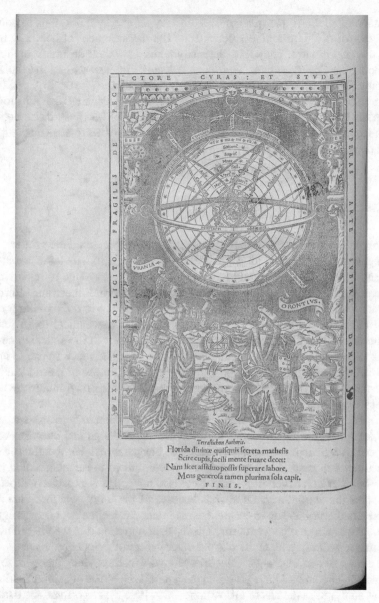

Fig. 8 Frontispiece of Oronce Fine, *Protomathesis* printed in Paris by Gerard Morrhius & Jean Pierre in 1532 (Fine 1532, AA8v, sig. O1v). On the title page of this work, Fine noted that "The author drew this figure by his own skill" (*Hanc Author proprio pingebat marte figuram*). ETH-Bibliothek Zürich, Rar 9724 GF, https://doi.org/10.3931/e-rara-9142 / Public Domain Mark

wide range of sundials and quadrants, from very simple cylinder dials to complex astrolabes. All of this machinery depends on the measurement of heavenly bodies and earthly coordinates. Therefore, the instruments draw their design from the foregoing book: the *Cosmographia, sive mundi sphaera*, Fine's contribution to spherical astronomy. The astronomer, surrounded by his instruments, looks to Urania, who points a telling finger at the model of the heavens. The pair marks a shift towards instruments and towards observation. But the parameters for those instruments are set by a sequence of tables that Fine presents at the center of his treatise on the sphere. The work of observing is inextricable from his making of books.

4 Conclusion

I have argued that the early Estienne press (like others) presented the printer's work in a language that evaluated diligence and craft, which included claims about good judgment. This is a familiar story in some respects: the language of ingenuity has long been part of the story of the rising status of the artist in Renaissance Europe.[43] This framework has even been applied, though less directly, to the familiar problem of explaining the rising status of mathematics from a mechanical to an intellectual enterprise. But that account usually is limited to later in the sixteenth century, perhaps with the exception of Oronce Fine.[44] The time and place I wish to engage here is earlier, and more central; the Estiennes offer a glimpse of these ideals early in their formation. These printers insisted on claiming their own intellectual—and laborious—contributions to the products they sold.

The nascent tradition of astronomical frontispieces explored here offers a parallel line of argument. Old dichotomies between hand knowledge and book knowledge simply do not explain these frontispieces. These images reveal, I have argued, a rising appreciation of the astronomer as an observer, committed to calculation and instruments—but the book never disappears. Bookish knowledge is expected to document and support the astronomer. Of course, artisans of the book, whether Fine or the Estiennes, had every reason to ensure that the intellectual value of craft was as much about books as about other practices. This is an assumption that we historians need to remember not only with textbooks but other genres of practical manuals.

But it may be that mathematical works offered a more obvious place for relating craft, insight, and labor. I have proposed the hypothesis—to be considered alongside the cases analyzed in the rest of this volume—that the *Sphaera* offered the Estiennes a particularly useful opportunity to make clear just how necessary printerly care and diligence were in order to produce reliable work. Recognizing the significance of

---

[43] Some classic bibliography is found in the appendix to (Emison 2004). One can find an extended bibliography on this question in relation to science in (Long 2011).

[44] One classic study relevant for Italy is (Biagioli 1989); for recent examples, see (Marr 2011). The tendency to focus on a later timeframe is evident in the recent collection of (Cormack et al. 2017). On Oronce Fine as an early exemplar, see the various studies in (Marr 2009).

this labor allows us to chart a middle course between Eisenstein's early claim that technical books benefitted from easy replication, and Adrian Johns' counterclaim that early modern books were defined by their unreliability (Eisenstein 1980; Johns 1998). Replication was difficult work; printers could claim ingenuity in privileges and colophons precisely because replication was possible but not easy.

Acknowledgements My thanks to Matteo Valleriani and Andrea Ottone for so kindly including me, even when logistics were impossible. I also owe gratitude to the entire working group for helpful comments, and especially to Alissar Levy and Saskia Limbach for correction on several key points; my obstinacy is to blame for the remaining errors. Thanks also to Alissar Levy for sharing a relevant section of her research underway at the *École nationale des chartes*.

Abbreviations

Digital Repositories

EMoBookPrivileges	Early Modern Book Privileges in Venice. Università degli Studi di Milano. https://emobooktrade.unimi.it. Accessed 07 June 2021
Sphaera Corpus*Tracer*	Max Planck Institute for the History of Science. https://db.sphaera.mpiwg-berlin.mpg.de/resource/Start. Accessed 07 June 2021

Archives and Special Collections

ASV	Archivio di Stato di Venezia
BnF	Bibliothèque Nationale de France
BNM	Biblioteca Nazionale Marciana

Appendix

1. ***Titles and colophons from the Paris editions of Jacques Lefèvre d'Étaples, Textus de sphera Johannis de Sacrobosco (published 1495–1538)***

Paris: [Johann Higman for[45]] Wolfgang Hopyl, 12 February 1494 [new style 1495]

---

[45] This judged from the fact that Higman's type was used to print this edition. See ISTC and USTC.

Title page: *Textus | De Sphera Johannis de Sacrobosco | Cum Additione (quantum necessa=| rium est) adiecta: Nouo commen=| tario nuper editio Ad vtilita\ tem studentium Philosophice Parisiensis. Aca=| demie: illustra | tus.*

Colophon: ¶ Impressum Parisij in pago diui Jacobi ad insigne sancti Georgij Anno Christi siderum conditoris | 1494 duodecima februarij Per ingeniosum impressorem Wolfgangum hopyl. Cui hec sententia semper fir=| ma mente sedet: Non viribus aut velocitatibus aut celeritate corporum res magne geruntur: sed Consilio, Sen| tentia, et Auctoritate ¶ Recognitoribus diligentissimis: Luca Uualtero Conitiensi, Guillermo Gonterio, | Johanne Griettano, et Petro Grisele: Matheseos amatoribus.

Bibliography: Rice LXXI; ISTC No.:ij00414000: Pr 8131.7; Hain-Cop. 14119; Klebs 874.18; Goff J414; Pell Ms 6715 (6680); CIBN J-275; Frasson-Cochet 174; Hillard 1150; Buffévent 313; Péligry 477; IBE 3279; IBPort 1026; IGI 5348 = 3779 (I); Sajó-Soltész 1947; Voull(Trier) 2302; Günt(L) 2238; Sack(Freiburg) 2124; Walsh 3669; BMC VIII 136; GW M14602.

Paris, [Johann Higman, for] Wolfgang Hopyl, 1 September 1500

Title page: *Text | us De Sphe | ra Johannis de Sa= | crobosco cum Additione | (quantum necessarium est) ad= | iecta: Nouo commentario nuper | edito Ad vtilitatem studentium Philosophi= | ce Parisiensis. Academie: illustratus | cum Compositione Anuli Astro= | nomici Boni Latensis. | Et Geometria Eu | clidis Mega | rensis.*

Colophon: ¶ Impressum Parisii in pago diui Jacobi ad insigne sancti Georgij Anno Christi siderum | conditoris 1494 duodecima februarij Per ingeniosum impressorem Wolfgangum hopyl. Cui hec senten-| tia semper firma mente sedet: Non viribus aut velocitatibus aut celeritate corporum res | magne geruntur: sed Consilio, Sententia, et Auctoritate.

Bibliography: Rice LXXIII; ISTC No. ij00423000; Copinger, 5207; Pr 8140.3; Cop.-Reich 5207-2396; Klebs 874,29.

Paris, 1503

Probably a ghost. Rice (LXXIIIa) reports this book as cited by Philippe Renouard, Bibliographie des éditions de Simon de Colines, 1520-1546 (*Paris: E. Paul, L. Huard et Guillemin, 1894), 435. But Rice had not seen the book itself, and it appears nowhere else; perhaps the collection of mathematical introductions from 1503 was mistaken as the* Textus de sphera.

Bibliography: Rice LXXIIIa.

Paris: Henri Estienne, 10 November 1507

Title page: *TEX | tus De Sphera Johannis de Sa- crobosco cum additione (quantum necessarium est) ad- | iecta: Nouo commen tario nuper | edito Ad vtilitatem studentium Philosophi- | ce Parisiensis. Academie: illustratus | cum Compositione Anuli Astro- | nomici Boni Latensis. | Et Geometria Eu- | clidis Megarensis.*

Colophon: ¶ Impressum Parisii in officina Henrici stephani e regione Schole decretorum sita. | Anno Christi siderum conditoris 1507. Decimo die Nouembris.

Bibliography: Rice LXXIV; USTC 143312; FB [=Pettegree et al. eds., *French Books*] 75639.

Paris: Henri Estienne, 10 November 1511

Title page: *TEX | tus De Sphera Johannis de Sa- crobosco Cum additione (quantum necessarium est) ad- | iecta: Nouo commentario nuper | edito Ad vtilitatem studentium Philosophi- | ce Parisiensis. Academie: illustratus | Cum Compositione Anuli Astro- | nomici Boni Latensis. | Et Geometria Eu- | clidis Megarensis.* [Within a floriated frame]

Colophon: Impressum Parisiis in officina Henrici stephani e regione Schole decretorum sita. Anno Christi siderum conditoris 1511. Decimo die Nouembris.

Bibliography: Rice LXXVI; USTC 143880; FB 75641.

Paris: Henri Estienne, 10 May 1516

Title page: *Tex-| tus De Sphe- |ra Joannis de Sa- | crobosco. Cum additione | (quantum neces-sarium est) ad- | iecta: Nouo commentario nuper | edito. A utilitatem studentium Philosophi- | ce Parisiensis. Academie illustratus | cum Compositione Anuli Astro- | nomici Boneti Latensis. Et Geometria Euclidis Megarensis.* [Within a floriated frame.]

Colophon: ¶ Geometrie Euclidis a Boetio translate Finis. | ¶ Impressum Parisijs in officina Henrici Stephani e regione Schole decretorum | sita Anno Christi siderum conditoris 1516. Decimo die Maij.

Bibliography: Rice LXXVII; USTC 203490; FB 75645.

Paris; Simon de Colines, 6 May 1521

Title page: *TEXTVS DE SPHÆ- | RA IOANNIS DE SACROBOSCO: INTRVCTORIA AD- | ditione (quantum necessarium est) commentarioque, ad vtilitatem studentium | philosophiæ Parisiensis Academiæ illustratus. Cum compositione Annu- | li astronomici Boneti Latensis: Et Geometria Euclidis Megarensis.* [woodcut of spheres, with Oronce Fine reclining below] *PARISIIS. | ¶ Vænit apud Simonem Colinærum, e regione scholæ Secretorum.*

Colophon: ¶ PARISIIS, EX AEDIBVS SIMO- | nis Colinaei, e regione scholæ Decretorum po- | sitis. Anno a Christo nato, primo & vi- | gesimo supra sesquimillesimum. | Sexto Calendas | maias. | [errata]

Bibliography: Rice LXXX; Renouard, *Colines*, 22-24; USTC 145470; FB 75648.

Paris: Simon de Colines, 1527

Title page: ☙ *TEXTVS DE SPHAERA IOAN-|NIS DE SACROBOSCO: INTRODVCTORIA ADDITIONE | (quantum necessarium est) commentarióque, ad vtilitatem studentium phi-| losophiæ Parisiensis Academiæ illustratus. Cum compositione Annuli | astronomici Boneti Latensis: Et Geometria Euclidis Megarensis. | [woodcut of spheres, with Oronce Fine below] | PARISIIS | Vænit apud Simonem Colinaeum. | 1527.*

Colophon: ¶ PARISIIS, EX AEDIBVS SIMONIS COLI=| næi, Anno à Christo nato, vigesimoseptimo supra | sesquimillesimum. XII Calendas | Septembres.

Bibliography: Rice LXXXI; Renouard, *Colines*, 100; USTC 145882; FB 75651.

Paris: Simon de Colines, 1531/2

Title page: ☙ *TEXTVS DE SPHAERA IOANNIS DE | SACROBOSCO: INTRODVCTORIA ADDITIONE | (quantum necessarium est) commentarióque, ad vtilitatem studentium Phi-| losophiæ Parisiensis Academiæ illustratus. Cum compositione Annuli | astronomici Boneti Latensis: Et Geometria Euclidis Megarensis. | [woodcut of spheres, with Oronce Fine below] | PARISIIS | Vænit apud Simonem Colinaeum. | 1531*

Colophon: PARISIIS, EX AEDIBVS SIMONIS COLINAEI, | Anno à Christo nato, tricesimoprimo supra sesquimillesimum, | pridie Nonas Februarij.

Bibliography: Rice LXXXIII; Renouard, *Colines*, 184; USTC 138085; FB 75652.

Paris; Simon de Colines, 1534

Title page: ☙ *TEXTUS DE SPHAERA IOANNIS DE* | *SACROBOSCO: INTRODVCTORIA ADDITIONE* | *(quantum necessarium est) commentarióque, ad vtilitatem studentium Phi-*| *losophiae Parisiensis Academiae illustratus. Cum compositione Annuli astro-*| *nomici Boneti Latensis: Et Geometria Euclidis Megarensis.* | *[plate 120x194mm]* | *PARISIIS* | *Vaenit apud Simonem Colinaeum.* | 1534.

Colophon: PARISIIS, EX AEDIBVS SIMONIS COLINÆI | Anno à Christo nato, tricesimoquarto supra sesquimillesimum, | septimo Idus Nouembris.

Bibliography: Rice LXXXIV; Renouard, *Colines*, 236; [not in USTC, FB].

Paris: Simon de Colines, 1538

Title page: ☙ *TEXTVS DE SPHAERA IOAN-*|*NIS DE SACROBOSCO: INTRODVCTORIA AD-* | *ditione (quantum necessarium est) commentarióque, ad vtilitatem* | *studentium Philosophiæ Parisiensis Academiæ illustratus. Cum* | *compositione Annuli astronomici Boneti Latensis: Et Geome-* | *tria Euclidis Megarensis.* [woodcut of spheres, with Oronce Fine below] | *PARISIIS* | *Vænit apud Simonem Colinaeum.* | *1538.*

Colophon: ¶ PARISIIS, EX AEDIBVS SIMONIS COLI=| næi, Anno à Christo nato, tricesimooctauo su- | pra sesquimillesimum, tertio | Idus Martias.

Bibliography: Rice LXXXV; Renouard, *Colines*, 295; USTC 147525; FB 75657.

2.   Compendia of Astronomical Works that include the *Textus de sphera*

Venice: Simon Bevilaqua, 1499

Title page: *Sphera Mundi cum* | *tribus Commentis* | *nuper editus videlicet* | *Cicchi Esculani* | *Francisci Capuani* | *de Manfredonia* | *Jacobi Fabri Stapulensis.*

Colophon: ¶ Impressum Venetiis per Simonem Paiensem dictum Biuilaquam | & summa diligentia correctum: ut legentibus patebit. Anno Cristi Side| rum conditoris. MCDXCIX. Decimo Calendas Nouembres.

Bibliography: Rice LXXII; ISTC No.: ij00419000; Goff J-419 Hain *14,125; BM 15th cent. V, 524; Klebs 874.26/7; USTC 993974; GW M14633; GW M14635.

Venice: Rubeo and Bernardino de Vercello for Giunta, 1508

Title page: *Nota eorum quæ in hoc libro continentur.* | *Oratio de laudibus astrologiae habita a Bartholomeo Vespucio flo-* | *rentino in almo Patavio Gymnasio anno. M.C. vi.* | *TEXTVS SPHAERAE IOANNIS DE SACRO BVSTO.* | *Expositio sphaerae Eximii artium & medicinæ doctoris Domini Fran-*| *cisci Capuani de mandfredonia.* | *Annotationes nonnullæ eiudem Bartholomei Vespucii hic idem intersertæ* | *Iacobi fabri stapulensis Commentarii in eandem sphæram.* | *Reuerendissimi Domini Petri de aliaco Cardinalis & piscopi Came-* | *racensis in eandem quæstiones subtilissimae numero xiiii.* | *REuerendissimi episcopi Domini Roberti linconiensis sphæræ compendium.* | *Disputationes IOannis de regio monte contra cremonensia deliramenta.* | *Theoricarum novarum textus cum expositione eiusdem Francisci Ca-*| *puani omnia nuper diligentia summa emendata.*

Colophon: Impressio Veneta per Ioannem Rubeum & Ber- | nardinum fratres Vercellenses ad instantiam iunctæ | de iunctis florenti Anno Domini .M.ccccc.viii. | die.vi.mensis maii.

Bibliography: Rice LXXV; USTC 800123.

Venice: Luc'Antonio de Giunta, 1518

Title page: *Sphera mundi [Johannis de Sacro Bosco] recogni* | *ta, cum commentariis et authoribus* | *in hoc volumine contentis, videlicet.*| *Cichi Esculani cum textu* | *Joannis Baptiste*

Capuani | Jacobi Fabri Stapulensis | Theodosii de Spheris cum textu | Michaelis Scoti Questiones | Petri de Aliaco, cardinalis, Questiones | Roberti Linconiensis Compendium | Theodosii iterum de Spheris cum textu | Tractatus de sphera solida | Theorice planetarum conclusiones [G. Purbachii], cum expositione | Campani Tractatus de sphera | ejusdem Tractatus de computo majori | Joannis de Monte Regio in Cremonensem disputatio | Theorice textus, cum Joannis Baptiste Capuani expositione | Ptolomeus de Speculis | Theorica planetarum Joannis Cremonensis pluri-| mum faciens ad disputationem Joannis de Monte Re-| gio, quam in aliis impressis non reperies.

Colophon: Venetiis, impensis nobilis viri domini Luce Antonii de Giunta, Florentini | die ultimo junii | 1518. *NB: The* Textus de sphera *is printed in double columns here; parts of the diagrams are missing, and the overall quality has deteriorated.*

Bibliography: Rice LXXIX; Panzer VIII no. 944; USTC 854129.

Venice: [Heirs of] Ottaviano Scoto, 1519

Title page: *Sphera | cum commentis in hoc volumine contentis. videlicet. | Cichi Esculani cum textu | Expositio Ioannis Baptiste Capuani in eandem | Jacobi Fabri Stapulensis | Theodosii de Speris | Michalis Scoti | Questiones Reverendissimi domini Petri de Aliaco, etc. | Roberti Linchoniensis Compendium | Tractatus de Sphera solida | Tractatus de Sphera Campani | Tractatus de computo maiori eiusdem | Disputatio Joannis de monte regio | Textus Theorice cum expositione Joannis Baptiste Capuani | Ptolomeus de Speculis.*

Colophon: Venetiis impensa heredum quondam Do-| mini octaviani Scoti Modoe-| tiensis, ac sociorum. | 19 Januari. | 1518.

Bibliography: Rice LXXVIII; Panzer VIII, no. 944; USTC 854128.

Venice: Luc'Antonio Giunta, 1531

Title page: ❧ *SPHERAE TRACTATVS* ❦ | ¶*IOANNIS DE SACRO BVSTO ANGLICI VIRI CLARISDS>* | ¶ *GERARDI CREMONENSIS THEORICAE PLANETARVM VETERES* | ¶ *GEORGII PVRBACHII THEORICAE PLANETARVM NOVAE.* | ¶ *Prosdocimi de beldomando patauini super tractatu sphærico commentaria, | nuper in lucem diducta per .L.GA. nunquam amplius impressa. | ¶ Ioannis baptistæ capuani sipontini expositio in sphæra & theoricis. | ¶ Ioannis de monte regio disputationes contra theoricas gerardi. | ¶ Michaelis scoti expositio breuis & quæstiones in sphæra. ¶ Iacobi fabri stapulensis paraphrases & annotationes. | ¶ Campani compendium super tractatu de sphera. | Eiusdem tractatuus de modo fabricandi spheram solidam. | ¶ Petri cardin. de aliaco episcopi cameracensis. 14. Quæstiones. | ¶ Roberti linconiensis episcopi tractatulus de sphæra. | ¶ Bartholomei uesputii glossulæ in plerisque locis sphæræ.| ¶ Eiusdem oratio. De laudibus astrologiæ | ¶ Lucæ Gaurici castigationes & figura etoto opere diligentissime reformatæ. | ¶ Eiusdem quæstio Nunquis sub æquatore sit habitatio. | ¶ Eiusdem Oratio de inventoribus & laudibus Astrologiæ. | Reuerendissimo cardin. episcopo.D.Bernardo Tridentinorum princip dicata. || ¶ ALPETRAGII ARABI THEORICA PLANETARVM NVPRERIME LA-| tinis mandata literis a calo calonymos hebreo neapolitano, ubi nititur saluare | apparentias in motibus Planetarum absque eccentricis & epicyclis. || MD XXI*

Colophon: Venetiis in ædibus Luceantonii Iunte Florentini anno | Domini.M.D.XXXI.Mense Ianuario.

Bibliography: Rice LXXXII; Panzer VIII, no. 1581; USTC 854145.

References

Primary Literature

Aediloquium. 1530. Paris: Simon de Colines.

Almeloveen, Theodore Janssen van. 1683. *De vitis Stephanorum, celebrium typographorum dissertatio epistolica*. Amsterdam: Janson/Waasberg.

Breviarium Canonicorum regularium Windechimensis ordinis sancti Augustini. 1502. Paris: Wolfgang Hopyl and Henri Estienne.

Calepino, Ambrogio. 1522. *Vocabularium, thesaurus copiosissimus*. Toscolano.

Clichtove, Josse. 1513. *De mystica numerorum*. Paris: Henri Estienne.

Estienne, Robert. 1531. *Dictionarium seu Latinae linguae Thesaurus. Cum Gallica fere interpretatione*. Paris: Robert Estienne.

Fine, Oronce, ed. 1515. *Theoricarum nouarum Textus Georgii Purbachii, cum expositione Capuani, Item de Prierio; Insuper Jacobi Fabri Stapulensis astronomicon: Omnia nuper summa diligentia emendata cum figuris ac commodatissimis longe castigatius insculptis quam prius suis in locis adiectis*. Paris: Michael Lesclencher for Jean Petit and Reginald Chauderon.

Fine, Oronce, ed. 1532. *Protomathesis*. Paris: Gérard Morrhy and Jean Pierre de Tour. https://hdl.handle.net/21.11103/sphaera.101190.

Grannollachs, Bernard de. 1485. *Sumario dela nobilissima arte e scientia de astrologia*. Naples/Venice/Rome.

Lattes, Bonetus de. 1493. *Boneti delatis hebrei medici Provenzalis, Anuli per eum compositi super astrologiam utilitates*. Rome.

Lefèvre d'Étaples, Jacques. 1496. *Arithmetica elementa; Musica elementa; Epitome in libros arithmeticos divi Severini Boetii; Rithmimachie ludus que et pugna numerorum appellatur*. Paris: Johannes Higman and Wolfgang Hopyl.

Lefèvre d'Étaples, Jacques, ed. 1497. *Decem librorum Moralium Aristotelis tres conversiones: prima Argyropili Byzantii; secunda Leonardi Aretini; tertia vero Antiqua per capita et numeros conciliate: communi familiarique commentario ad Argyropilum ad lectio*. Paris: Johann Higman and Wofgang Hopyl.

Lefèvre d'Étaples, Jacques. 1514. *Arithmetica elementa; Musica elementa; Epitome in libros arithmeticos divi Severini Boetii; Rithmimachie ludus que et pugna numerorum appellatur*. Paris: Henri Estienne.

Lefèvre d'Étaples, Jacques, and Josse Clichtove [comm.]. 1517. *Introductorium astronomicum theorias corporum coelestium duobus libris complectens, adiecto commentario declaratum*. Paris: Henri Estienne.

Maittaire, Michael. 1709. *Stephanorum historia: vitas ipsorum ac libros complectens*, 2 vols. London: Benjamin Motte for Christopher Bateman.

Peuerbach, Georg. 1514. *Tabulae eclypsium*. Vienna: Johannes Winterberger.

Ps. Ptolemy, Campanus da Novara, Pierre d'Ailly, Cecco d'Ascoli, Gerard of Cremona, Robert Grosseteste, Theodosius of Bithynia, Michael Scot, Joannes Regiomontanus, Jacques Lefèvre d'Étaples, and Francesco Capuano. 1518a. *Sphera mundi nouiter recognita cum commentarijs et authoribus in hoc volumine contentis*. Venice: Lucantonio Giunta. https://hdl.handle.net/21.11103/sphaera.100047.

Ps. Ptolemy, Campanus da Novara, Pierre d'Ailly, Cecco d'Ascoli, Robert Grosseteste, Theodosius of Bithynia, Michael Scot, Joannes Regiomontanus, Jacques Lefèvre d'Étaples, and Francesco Capuano. 1518b. *Sphera cum commentis in hoc volumine contentis*. Venice: Heirs of Ottaviano Scoto I. https://hdl.handle.net/21.11103/sphaera.101057.

Regiomontanus, Johannes. 1496. *Epytoma*. Venice: Johannes Hamman.

Sacrobosco, Johannes de. 1494 [*new style* 1495]. *Textus de Sphera Iohannis de Sacrobosco cum additione (quantum necessarium est) adiecta [Stapulensis]*. Paris: Wolfgang Hopyl. https://hdl.handle.net/21.11103/sphaera.101126.

Sacrobosco, Johannes de, Pierre d'Ailly, Francesco Capuano, Robert Grosseteste, Jacques Lefèvre d'Etaples, Joannes Regiomontanus, Georg von Peurbach, and Bartolomeo Vespucci. 1508. *Oratio de laudibus astrologiae habita a Bartholomeo Vespucio; Textus Sphaerae Ioannis de Sacro Busto; Expositio sphaerae Francisci Capuani de Manfredonia; Annotationes nonnullae eiusdem Vespucii; Iacobi Fabri Stapulensis Commentarii in eandem sphaeram; Petri de Aliaco in eandem quaestiones subtilissimae numero xiiii; Roberti Linconiensis sphaerae Compendium; Disputationes Ioannis de regio monte contra Cremonensia; Theoricarum novarum textus cum expositione eiusdem Francisci Capuani*. Venice: Giovanni and Bernardino Rosso for Giuntino Giunta. https://hdl.handle.net/21.11103/sphaera.100915.

Sacrobosco, Johannes de, Cecco d'Ascoli, Francesco Capuano, and Jacques Lefèvre d'Etaples. 1499. *Sphera Mundi cum tribus commentis nuper editis*. Venice: Simone Bevilacqua. https://hdl.handle.net/21.11103/sphaera.100273.

Sacrobosco, Johannes de, Gerardus Cremonensis, Georg von Peurbach, Nūr al-Dīn Ishāq al-Bitrūgi, Prosdocimus de Beldomandis, Luca Gaurico, Francesco Capuano, Joannes Regiomontanus, Michael Scot, Jacques Lefèvre d'Etaples, Pierre d'Ailly, Robert Grosseteste, and Bartolomeo Vespucci. 1531a. *Spherae tractatus Ioannis de Sacro Busto; Gerardi Cremonensis theoricae planetarum veteres; Georgii Purbachii theoricae planetarum novae; Prosdocimi de Beldomando super tractatu sphaerico commentaria; Ioannis Baptistae Capuani expositio in sphaera & theoricis; Ioannis de Monteregio disputationes contra Theoricas Gerardi; Michaelis Scoti expositio brevis & quaestiones in sphaera; Iacobi Fabri Stapulensis paraphrases & annotationes; Campani compendium super tractatu de sphera, eiusdem tractatulus de modo fabricandi spheram solidam; Petri de Aliaco 14. Quaestiones; Roberti Linconiensis tractatulus de sphaera; Bartholomei Vesputii glossulae in plerisque locis sphaerae; eiusdem oratio; De laudibus astrologiae Lucae Gaurici; eiusdem quaestio Nunquid sub aequatore sit habitatio; eiusdem Oratio de inventoribus & laudibus Astrologiae; D. Bernardo Tridentinor principi dicata; Alpetragii Arabi theorica planetarum a hebreo neapolitano*. Venice: Lucantonio Giunta. https://hdl.handle.net/21.11103/sphaera.100999.

Sacrobosco, Johannes de, Bonet de Lattes, and Euclid. 1500. *Textus de Sphaera cum additione adiecta [Stapulensis]; Compositione Anuli Astronomici Boni Latensis; Geometria Euclidis Megarensis*. Paris: Wolfgang Hopyl. https://hdl.handle.net/21.11103/sphaera.100889.

Sacrobosco, Johannes de, Bonet de Lattes, and Euclid. 1507. *Textus de Sphaera cum additione adiecta [Stapulensis]; Compositione Anuli Astronomici Boni Latensis; Geometria Euclidis Megarensis*. Paris: Henri Estienne I. https://hdl.handle.net/21.11103/sphaera.100029.

Sacrobosco, Johannes de, Bonet de Lattes, and Euclid. 1511. *Textus de Sphaera cum additione adiecta [Stapulensis]; Compositione Anuli Astronomici Boni Latensis; Geometria Euclidis Megarensis*. Paris: Henri Estienne I. https://hdl.handle.net/21.11103/sphaera.100919.

Sacrobosco, Johannes de, Bonet de Lattes, and Euclid. 1516. *Textus de Sphaera cum additione adiecta [Stapulensis]; Compositione Anuli Astronomici Boni Latensis; Geometria Euclidis Megarensis*. Paris: Henri Estienne I. https://hdl.handle.net/21.11103/sphaera.100990.

Sacrobosco, Johannes de, Bonet de Lattes, and Euclid. 1521. *Textus de Sphaera cum additione adiecta [Stapulensis]; Compositione Anuli Astronomici Boni Latensis; Geometria Euclidis Megarensis*. Paris: Simon de Colines. https://hdl.handle.net/21.11103/sphaera.100995.

Sacrobosco, Johannes de, Bonet de Lattes, and Euclid. 1527. *Textus de Sphaera cum additione adiecta [Stapulensis]; Compositione Anuli Astronomici Boni Latensis; Geometria Euclidis Megarensis*. Paris: Simon de Colines. https://hdl.handle.net/21.11103/sphaera.100998.

Sacrobosco, Johannes de, Bonet de Lattes, and Euclid. 1531b. *Textus de Sphaera cum additione adiecta [Stapulensis]; Compositione Anuli Astronomici Boni Latensis; Geometria Euclidis Megarensis*. Paris: Simon de Colines. https://hdl.handle.net/21.11103/sphaera.101043.

Sacrobosco, Johannes de, Bonet de Lattes, and Euclid. 1534. *Textus de Sphaera cum additione adiecta [Stapulensis]; Compositione Anuli Astronomici Boni Latensis; Geometria Euclidis Megarensis*. Paris: Simon de Colines. https://hdl.handle.net/21.11103/sphaera.100099.

Sacrobosco, Johannes de, Bonet de Lattes, and Euclid. 1538. *Textus de Sphaera cum additione adiecta [Stapulensis]; Compositione Anuli Astronomici Boni Latensis; Geometria Euclidis Megarensis*. Paris: Simon de Colines. https://hdl.handle.net/21.11103/sphaera.101011.

Sacrobosco, Johannes de, Johannes Regiomontanus and Georg Peuerbach. 1482. *Novitiis adolescentibus...Iohannis de sacro busto Sphericum opusculum; Georgiique Purbachii Theoricae; Contra Cremonensia Iohannis de Monte Regio disputationes*. Venice: Erhard Ratdolt. https://hdl.handle.net/21.11103/sphaera.100692.

Sacrobosco, Johannes de, Johannes Regiomontanus and Georg Peuerbach. 1489. *Novitiis adolescentibus...Iohannis de sacro busto Sphericum opusculum; Georgiique Purbachii Theoricae; Contra Cremonensia Iohannis de Monte Regio disputationes*. Paris: Wolfgang Hopyl. https://hdl.handle.net/21.11103/sphaera.100821.

Sacrobosco, Johannes de, Johannes Regiomontanus and Georg Peuerbach. 1490. *Spaerae mundi compendium foeliciter inchoat. Noviciis adolescentibus....* Venice: Ottaviano Scoto I. https://hdl.handle.net/21.11103/sphaera.100885.

Thevet, André. 1584. *Les vrais pourtraits et vies des hommes illustres*, 3 vols. Paris: Kerver and Chaudière.

Tory, Geoffroy. 1529. *Champfleury: au quel est contenu lart & science de la deue & vraye proportio[n] des lettres attiques, quo[n] dit autreme[n]t lettres antiques, & vulgairement lettres romaines proportionnees selon le corps & visage humain*. A vendre a Paris sus Petit Pont: A Lenseigne du Pot Casse par Maistre Geofroy Tory di Bourges, libraire & autheur du dict liure: Et par Giles Gourmont aussi libraire demourant en la rue Sainct Iaques a Lenseigne des Trois Coronnes.

Valla, Lorenzo. 1490–1491. *Elegantiae linguae Latinae. De pronomine sui*. Paris: [Johannes Higman, for] Wolfgang Hopyl.

Visagier, Jean, cit. and trans. 1537. *Epigrammata*. Lyons: Michel Parmentier.

Secondary Literature

Amert, Kay. 2012. *The scythe and the rabbit: Simon de Colines and the culture of the book in Renaissance Paris*, ed. Robert Bringhurst. Rochester, NY: RIT Cary Graphic Arts Press.

Armstrong, Elizabeth. 1954. *Robert Estienne: Royal printer; an historical study of the elder Stephanus*. Cambridge: Cambridge University Press.

Axworthy, Angela. 2016. *Le Mathématicien renaissant et son savoir. Le statut des mathématiques selon Oronce Fine*. Paris: Classiques Garnier.

Axworthy, Angela. 2020. Oronce Fine and Sacrobosco: From the edition of the Tractatus de Sphaera (1516) to the Cosmographia (1532). In *De Sphaera of Johannes de Sacrobosco in the early modern period: The authors of the commentaries*, ed. Matteo Valleriani, 185–264. Cham: Springer. https://doi.org/10.1007/978-3-030-30833-9_8.

Baldasso, Renzo. 2013. Printing for the doge: On the first quire of the first edition of the *Liber Elementorum Euclidis*. La Bibliofilía 115: 525–552.

Baxandall, Michael. 1963. A dialogue on art from the court of Leonello d'Este: Angelo Decembrio's *De politia litteraria pars LXVIII*. *Journal of the Warburg and Courtauld Institutes* 26 (3/4): 304–326.

Baxandall, Michael. 1986. *Giotto and the orators: Humanist observers of painting in Italy and the discovery of pictorial compositioni*. Oxford: Oxford University Press.

Bennett, Jim. 2006. Instruments and ingenuity. In *Robert Hooke: Tercentennial studies*, ed. Michael Cooper and Michael Hunter, 65–76. Aldershot: Ashgate.

Biagioli, Mario. 1989. The social status of Italian mathematicians. *History of Science* 27 (1): 41–95.

Bietenholz, Peter G. 1971. *Basle and France in the sixteenth century*. Geneva: Droz.

Blume, Dieter. 2014. Picturing the stars: Astrological imagery in the Latin west, 1100–1550. In *A companion to astrology in the Renaissance*, ed. Brendan Dooley, 333–398. Leiden: Brill.

Blume, Dieter, Mechthild Haffner, and Wolfgang Metzger. 2012. *Sternbilder des Mittelalters: Der gemalte Himmel zwischen Wissenschaft und Phantasie, Band I: 800–1200.* Berlin: Akademie.

Blume, Dieter, Mechthild Haffner, and Wolfgang Metzger. 2016. *Sternbilder des Mittelalters und der Renaissance: der gemalte Himmel zwischen Wissenschaft und Phantasie, Band II: 1200–1500.* Berlin: De Gruyter.

Bowen, Barbara C. 1979. Geofroy Tory's "Champ Fleury" and its major sources. *Studies in Philology* 76 (1): 13–27.

Boy, Madeleine. 1971. *Science et astrologie au XVIe siècle. Oronce Fine et son horloge planétaire, 22 Novembre-22 Décembre. Catalogue.* Paris: Bibliothèque Sainte-Geneviève.

Bruni, Flavia. 2018. Peace at the Lily. The De Franceschi section in the stockbook of Bernardino Giunti. In *Selling & collecting: Printed book sale catalogues and private libraries in Early Modern Europe*, ed. Giovanna Granata and Angela Nuovo, 265–281. Macerata: Eum.

Burnett, A. David. 1998. *The engraved title-page of Bacon's Instauratio Magna.* Durham: Thomas Harriot Seminar.

Considine, John. 2008. *Dictionaries in early modern Europe: Lexicography and the making of heritage.* Cambridge: Cambridge University Press.

Cormack, Lesley B., Steven A. Walton, and John A. Schuster, eds. 2017. *Mathematical practitioners and the transformation of natural knowledge in Early Modern Europe.* Cham: Springer.

Corsten, Severin. 1987. Universities and early printing. In *Bibliography and the study of 15th-century civilization*, ed. Lotte Hellinga and John Goldfinch, 83–123. London: British Library.

Crowther, Kathleen M., and Peter Barker. 2013. Training the intelligent eye: Understanding illustrations in early modern astronomy texts. *Isis* 104 (3): 429–470.

Delft, Marieke van. 2010. Illustrations in early printed books and manuscript illumination. The case of a Dutch book of hours printed by Wolfgang Hopyl in Paris in 1500. In *Books in transition at the time of Philip the Fair*, ed. Hanno Wijsman, 131–164. Turnhout: Brepols.

Eisenstein, Elizabeth L. 1980 [orig. 1979, in 2 vols.]. *The printing press as an agent of change: Communications and cultural transformations in Early-Modern Europe.* Cambridge: Cambridge University Press.

Eisenstein, Elizabeth L. 2012. *Divine art, infernal machine: The reception of printing in the West from first impressions to the sense of an ending.* Philadelphia: University of Pennsylvania Press.

Emison, Patricia A. 2004. *Creating the 'divine' artist: From Dante to Michelangelo.* Leiden: Brill.

Folkerts, Menso. 1970. *"Boethius" Geometrie II, ein mathematisches Lehrbuch des Mittelalters.* Wiesbaden: Steiner.

Grafton, Anthony. 2018. Philological and artisanal knowledge making in Renaissance natural history: A study in cultures of knowledge. *History of Humanities* 3 (1): 39–55.

Graheli, Shanti. 2019. *Buying and selling: The business of books in Early Modern Europe.* Leiden: Brill.

Kaoukji, Natalie, and Nicholas Jardine. 2010. "A frontispiece in any sense they please"? On the significance of the engraved title-page of John Wilkins's *A Discourse Concerning A NEW World & Another Planet*, 1640. *Word & Image* 26 (4): 429–447.

Johns, Adrian. 1998. *The nature of the book: Print and knowledge in the making.* Chicago: University of Chicago Press.

Kecskeméti, Judit, Jean Céard, Bénédicte Boudou, and Hélène Cazes, eds. 2003. *La France des humanistes: Henri II Estienne, éditeur et écrivain.* Turnhout: Brepols.

Klibansky, Raymond, Erwin Panofsky, and Fritz Saxl. 1964. *Saturn and melancholy: Studies in the history of natural philosophy, religion and art.* London: Thomas Nelson and Sons.

Limbach, Saskia. 2019. Life and production of Magdalena Morhart. A successful business woman in sixteenth-century Germany. *Gutenberg-Jahrbuch* 94: 151–172.

Long, Pamela O. 2011. *Artisan/Practitioners and the rise of the new sciences, 1400–1600.* Corvallis, OR: Oregon State University Press.

Long, Pamela O. 2015. Trading zones in early modern Europe. *Isis* 106 (4): 840–847.

Marr, Alexander, ed. 2009. *The worlds of Oronce Fine. Mathematics, instruments and print in Renaissance France.* Donington: Shaun Tyas.

Marr, Alexander, ed. 2011. *Between Raphael and Galileo: Mutio Oddi and the mathematical culture of late Renaissance Italy.* Chicago: University of Chicago Press.

Marr, Alexander, Raphaële Garrod, José Ramón. Marcaida, and Richard Oosterhoff. 2018. *Logodaedalus: Word histories of ingenuity in Early Modern Europe.* Pittsburgh: Pittsburgh University Press.

Murdoch, John E. 1984. *Album of science: Antiquity and the Middle Ages.* New York: Scribner's Sons.

Oosterhoff, Richard J. 2015. A book, a pen, and the *sphere*: Reading Sacrobosco in the Renaissance. *History of Universities* 28 (2): 1–54.

Oosterhoff, Richard J. 2016. Lovers in paratexts: Oronce Fine's *Republic of Mathematics. Nuncius* 31 (3): 549–583.

Oosterhoff, Richard J. 2018. *Making mathematical culture: University and print in the circle of Lefèvre d'Étaples.* Oxford: Oxford University Press.

Oosterhoff, Richard J. 2020. A lathe and the material sphere: Astronomical technique at the origins of the cosmographical handbook. In *De Sphaera of Johannes de Sacrobosco in the early modern period: The authors of the commentaries*, ed. Matteo Valleriani, 25–52. Cham: Springer. https://doi.org/10.1007/978-3-030-30833-9_2.

Oosterhoff, Richard J. *Forthcoming.* Visualising observation in tables: Data management in astronomy before Galileo. In *Coping with copia: Epistemological excess in Early Modern Europe*, ed. Fabian Krämer and Itay Sapir. Amsterdam: Amsterdam University Press.

Pabel, Hilmar M. 2008. *Herculean labours: Erasmus and the editing of St. Jerome's Letters in the Renaissance.* Leiden: Brill.

Pantin, Isabelle. 1993. Une École d'Athènes des astronomes? La representation de l'astronome antique sur les frontispieces de la Renaissance. In *Images de l'Antiquité dans la littérature: Le texte et son illustration*, ed. Emmanuèle Baumgartner and Laurence Harf-Lancner, 87–100. Paris: PUF.

Pantin, Isabelle. 2009. *Altior incubuit animus sub imagine mundi*: L'inspiration du cosmographe d'après un gravure d'Oronce Fine. In *Les méditations cosmographiques à la Renaissance*, 69–90. Paris: Presses de l'Université Paris-Sorbonne.

Pantin, Isabelle. 2012. The first phases of the *Theoricae Planetarium* printed tradition (1474–1535): The evolution of a genre observed through its images. *Journal for the History of Astronomy* 43: 3–26.

Pattison, Mark. 1865. Classical learning in France: The great printers Stephens. *Quarterly Review* 117: 323–364.

Pedersen, Olaf. 1978. The decline and fall of the Theorica Planetarum: Renaissance astronomy and the art of printing. In *Science and history: Studies in honor of Edward Rosen,* 157–185. Wroclaw: Polish Academy.

Pettegree, Andrew. 2010. *The book in the Renaissance.* New Haven: Yale University Press.

Proot, Goran. 2018. Prices in Robert Estienne's booksellers' catalogues (Paris 1541–1552): A statistical analysis. *JLIS.It* 9 (2): 192–221.

Proot, Goran. 2019. Shifting price levels of books produced at the Officina Plantiniana In Antwerp, 1580–1655. In *Crossing borders, crossing cultures: Popular print in Europe (1450–1900)*, ed. Massimo Rospocher, Jeroen Salman, and Hannu Salmi, 89–108. Berlin: De Gruyter.

Remmert, Volker R. 2006. Docet parva pictura, quod multae scripturae non dicunt. Frontispieces, their functions, and their audiences in seventeenth-century mathematical sciences. In *Transmitting knowledge: Words, images, and instruments in early modern Europe*, eds. Sachiko Kusukawa and Ian Maclean, 239–271. Oxford: Oxford University Press.

Remmert, Volker R. 2011. *Picturing the scientific revolution: Title engravings in early modern scientific publications.* Trans. Ben Kern. Philadelphia: Saint Josephs University Press.

Renouard, Antoine August. 1843. *Annales de l'imprimerie des Estienne; ou, Histoire de la famille des Estienne et de ses éditions.* Paris: J. Renouard et Cie.

Renouard, Philippe. 1894. *Bibliographie des éditions de Simon de Colines, 1520–1546*. Paris: E. Paul, L. Huard et Guillemin.

Rice, Eugene F., ed. 1972. *The prefatory epistles of Jacques Lefèvre d'Étaples and related texts*. New York: Columbia University Press.

Rouse, Richard H., and Mary A. Rouse. 1988. The book trade at the University of Paris, ca. 1250–ca. 1350. In *La production du livre universitaire au Moyen Âge: exemplar et pecia: actes du symposium tenu au Collegio San Bonaventura de Grottaferrata en mai 1983*, eds. Louis Bataillon, Bertrand G. Guyot, and Richard H. Rouse, 41–114. Paris: Editions du CNRS.

Smith, Margaret M. 2000. *The title page: Its early development 1460–1510*. London: Oak Knoll Press.

Stein, Henri. 1891. *L'atelier typographique de Wolfgang Hopyl à Paris*. Fontainebleu.

Stein, Henri. 1891. 1895. Nouveaux documents sur les Estienne, imprimeurs parisiens, 1517–1665. *La Société de l'histoire de Paris et de l'Île-de-France* 22: 249–295.

Söderlund, Inga Elmqvist. 2010. *Taking possession of astronomy: Frontispieces and illustrated title pages in 17th-Century books on astronomy*. Stockholm: Center for History of Science at the Royal Swedish Academy of Sciences.

Swerdlow, Noel M. 1990. Regiomontanus on the critical problems of astronomy. In *Nature, experiment, and the sciences*, ed. William A. Wallace, Trevor H. Levere, and William R. Shea, 165–195. Dordrecht: Springer.

Verdier, Philippe. 1969. L'Iconographie des arts libéraux dans l'art du moyen âge jusqu'à la fin du quinzième siècle. In *Arts Libéraux et philosophie au Moyen Âge. Actes du quatrième congrès international de philosophie médiévale*, 305–355. Paris: Vrin.

White, Paul. 2013. *Jodocus Badius Ascensius: Commentary, commerce and print in the Renaissance*. Oxford: Oxford University Press.

Winn, Mary Beth. 1997. *Anthoine Vérard, Parisian Publisher, 1485–1512*. Geneva: Droz.

Richard J. Oosterhoff is lecturer in early modern history at the University of Edinburgh. He has held fellowships at the University of Cambridge and the University of Notre Dame, as well as briefer ones at Harvard, the Huntington Library, and the Warburg Institute (University of London). He has written *Making Mathematical Culture: University and Print in the Circle of Lefèvre d'Étaples* (Oxford, 2018) and is a co-author of *Logodaedalus: Word Histories of Ingenuity in Early Modern Europe* (Pittsburgh, 2018). With Alexander Marr and José Ramón Marcaida, he is also editing *Ingenuity in the Making: Matter and Technique in Early Modern Europe* (Pittsburgh, 2021).

Chapter 3
Erhard Ratdolt's Edition of Sacrobosco's *Tractatus de sphaera*: A New Editorial Model in Venice?

Catherine Rideau-Kikuchi

Abstract The aim of this paper is to investigate the construction process of Sacrobosco's *Tractatus* as a successful venture in the early publishing market and the seminal role of some of its editions. Venice is a good case study since Venetian printers regularly printed the early editions of the *Sphaera* and fashioned the way the text was laid in print. In 1478, in the context of aggressive competition with Erhard Ratdolt, Franz Renner chose a traditional conceptual approach to the text and printed it in a new formal adaptation. Ratdolt responded by emphasizing the importance of the illustrations and by printing Sacrobosco's treatise with other texts from contemporary scholars Georg Peuerbach and Regiomontanus. His editions could be found across Europe. His choices also inserted Sacrobosco's thirteenth-century treatise into contemporary academic debates. Finally, Ratdolt's edition set a formal standard, completed with Santritter's 1488 edition, and copied in Venice and across Europe. In the following years, many Venetian printers copied the publishing solution developed by Ratdolt and fully realized it in Johann Lucilius Santritter's 1488 edition in an environment of harsh competition. However, other models were developed and coexisted in Venice, probably targeting different audiences and different reading practices.

Keywords Johannes de Sacrobosco · Early modern cosmology · Erhard Ratdolt · Early modern Venice · Book history

1 Introduction

Johann of Sacrobosco's (d. 1256) *Tractatus de sphaera* was one of the earliest best sellers. By examining the catalog of Sacrobosco's thirty-eight known incunabula editions, one can distinguish three main production centers: Paris, Leipzig, and Venice. Venice, in particular, led the early stages of Sacrobosco's printing history, with four editions existing before 1480. The aim of this paper is to investigate the

C. Rideau-Kikuchi (✉)
Laboratoire Dynamiques patrimoniales et culturelles, Université de Versailles Saint-Quentin, Versailles, France
e-mail: catherine.kikuchi@uvsq.fr

© The Author(s) 2022
M. Valleriani and A. Ottone (eds.), *Publishing Sacrobosco's* De sphaera *in Early Modern Europe*, https://doi.org/10.1007/978-3-030-86600-6_3

construction process of Sacrobosco's *Tractatus* as a successful venture in the early publishing market. The seminal role of some editions has to be examined: Which formal solutions did they offer? How were these solutions and works copied or adapted by subsequent printers and publishers? Indeed, the key to this successful reception was the publishing solutions implemented in Venice, some of which were copied over the years and proved successful. Three aspects may help us understand Sacrobosco's diffusion in print:

– first, the specific position of individual printers in the Venetian printing industry;
– second, the position of Venetian printers in the European book trade and academic market;
– finally, the way Venetian printers adapted this text in an actualized *mise en livre* (the layout) (Martin 2000; Chartier 1997), as well as the different texts printers chose to combine Sacrobosco's *Tractatus de sphaera* with.

This study will investigate these three elements to consider the early printing diffusion of the *Tractatus de sphaera* and the part played by some editions, as a materially, socially, and economically situated configuration of economic and technical actors, authors, and texts. Specifically, Erhard Ratdolt's (1442–1528) 1482 and 1485 editions are often considered as important milestones for Sacrobosco's printing reception. On that point, bibliographical sources, as well as archival sources, will be used. This documentation will enable us to examine the social and economic background of Sacrobosco's printers. I wish to examine his editorial choices in relation to his position in the Venetian printing world, and as a publishing strategy targeting the European academic public. These editorial choices must be questioned in relation to the choices of previous and successive Sacrobosco printers. The first European book producers did not like innovation for its own sake: innovations could be a great economic risk if the public was confused or did not recognize what it expected in a specific book. One should ask oneself how these various choices were made possible and considered viable in a competitive market; we can then examine their impact on the construction of a publishing model for Sacrobosco's *Tractatus de sphaera*.

This paper will examine Venetian editions of the *Tractatus de sphaera* before 1520, with a special focus on Erhard Ratdolt's editions. Other Italian editions of the earlier times will be examined, but other production centers of the same period will only be mentioned. The leading position of Venice and the fact that local models were often constructed through multiple local influences justifies that scale of analysis, even if a broader study would be useful. However, by the beginning of the sixteenth century, it is no longer justified to focus solely on Venice—this is when our study will end.

Table 1 List of known Sacrobosco's *Tractatus de sphaera* editions before 1485

Format	Year	Place	Other authors	Producers	Main type of characters	General layout
in 4°, 24 fol., 3 quires	1472	Ferrara	Anonymus, *Ratio dierum secundum ordinem planetarum septem* (1 fol.)	Belfortis, Andreas	Roman	One column
in 4°, 42 fol., 4 quires	1472	Venice		de Argentina, Florentinus	Roman	One column
in 4°, 28 fol., 4 quires	1475–1477	Venice		Pietro, Filippo di	Roman	One column
in 4°, 18 fol., 2 quires	1478	Milan		Lavagna, Filippo da	Roman	One column
in 4°, 48 fol., 6 quires	1478	Venice	Gerardus Cremonensis	Renner, Franz	Roman	One column
in 4°, 16 fol., 2 quires	1478	Venice	Often found with Gerardus Cremonensis	Adam of Rottweil	Rotunda	One column
in 4°, 20 fol., 3 quires	1480	Bologna	Gerardus Cremonensis	Fusco, Domenico	Roman	One column
in 4°, 60 fol., 8 quires	1482	Venice	Georg Peuerbach, Regiomontanus	Ratdolt, Erhard	Rotunda	One column
in 4°, 58 fol., 7 quires	1485	Venice	Georg Peuerbach, Regiomontanus	Ratdolt, Erhard	Roman	One column

1.1 Before Erhard Ratdolt: Sacrobosco in Italy

Before 1482, Venetian printers issued four editions of the *Tractatus de sphaera* and two others were published in Northern Italy (Table 1).[1]

They were all rather small formats (in-4°, between sixteen and 48 folios, between two and six quires). Their similarities illustrate the original perspective on the *Sphaera*. The very first editions printed Sacrobosco's text alone—the Ferrarese edition only printed one sheet of *Ratio dierum secundum ordinem planetarum septem*,

---

[1] The editions listed in Table 1 are: (Sacrobosco 1472a, b, 1475–1477, 1478a, b, c, 1480; Sacrobosco et al. 1482, 1485).

but no commentaries and no other treatise. This "stand-alone" type of editions was relatively rare in the course of the publishing history of the *Tractatus de sphaera*: only eighteen of such editions were produced, and all before 1515 (Sphaera Corpus Tracer). In 1478, Franz Renner (1450–1486) introduced a shift of perspective: he printed the *Tractatus* with Gerardo Cremonensis' (1114–1187) work, *Theorica planetarum* (Sacrobosco 1478b). The same year, Adam Burckhardt, also known as Adam de Rottweil (b. ca. 1470), seems to have also produced his edition of the *Sphaera mundi* (Sacrobosco 1478c) at the same time as an edition of Gerardo Cremonensis' *Theorica planetarum*, but book historians are unsure whether it was the same edition or two simultaneous editions.[2]

In 1478, Franz Renner introduced some formal innovations as shown in (Shank 2012). He maintained the general layout with large margins and a single column, which was also a common format for manuscripts containing Sacrobosco's *Sphaera* and his commentators (Thorndike 1949). However, Renner was the first to introduce Sacrobosco's text with woodcuts and decorated initials. The first incunabula editions had no illustration. Printers left some blank spaces (Pantin 2020) and large margins, so that some readers drew their own diagrams, as in the copy of (Sacrobosco 1472a) preserved in the Biblioteca Nazionale Centrale di Firenze.[3] As far as the diagrams go, one could argue that Renner referred in part to the manuscript tradition, which consisted mostly of non-illustrated volumes, but also included a significant minority of illustrated ones (Pantin 2020).[4] The illustrations of Sacrobosco's *Tractatus de sphaera* used by Renner—the elemental and celestial sphere, the terrestrial zones, and the lunar and solar eclipse—already had a long manuscript tradition, and their use dated back to late antiquity (Pantin 2020; Obrist 2004; Müller 2008). Nine half-page diagrams also illustrate the *Theorica planetarum*. Renner drew inspiration from Johannes Regiomontanus' (1436–1476) editions in Nuremberg, especially the edition of the *Disputationes contra deliramenta cremonensis* and of *Theoricae novae planetarum* by Georg Peuerbach (1423–1462) printed in 1475 (Peuerbach 1475). Both editions presented a similar layout, even if Renner chose slightly larger margins and characters. Renner's diagrams are also much simpler, but the integration of illustrations in the course of the argumentation seems to be a direct inspiration of Regiomontanus' edition. It presents a significant change in comparison to the previous editions, in which only the text was presented to the buyer, provided that he would fill in the blanks for the initials and main illustrations and draw his own diagrams.

In Renner's case, the form was new but the content was old: Renner used Regiomontanus' technical innovations to print the *Theorica planetarum*, which was

---

[2] See the Incunabula Short Title Catalog (ISTC) and Gesamtkatalog der Wiegendrucke (GW) records for the Sacrobosco and Gerardo Cremonensis editions (Sacrobosco 1478c; Cremonensis 1478).

[3] Call number Magl. A.5.46. The treatise is available online: https://archive.org/details/ita-bnc-in1-00001011-001. Accessed 04 June 2021. In the same year, another edition was also published: (Sacrobosco 1472b).

[4] Seven of the twenty manuscripts of *De sphaera* examined by Lynn Throndike in (Thorndike 1949) were illustrated.

precisely the text against which Regiomontanus wrote his *Disputationes contra deliramenta cremonensis*. Michael Shank seemed surprised by this editorial choice, mixing Regiomontanus' formal innovations with the traditional, criticized text attributed to Gerardo Cremonensis (Shank 2012). On the contrary, this publishing strategy is pragmatic and paradigmatic of the printing and book trade during these first years. Four editions of the *Tractatus de sphaera* were issued in Italy before Renner. There was definitely a market for this textbook, but it was a competitive and occupied market. Given this situation, Renner had to take a stand, to distinguish himself. Renner's in-quarto edition, with its Roman type, clear layout, and headlines, with its illustrations and elegant technical solutions, was clearly meant to be a practical textbook for students and scholars. The choice of the *Theorica planetarum* seems to have occurred at two different Venetian printers at the same time for good reasons. The association between the *Tractatus de sphaera* and the *Theorica planetarum* was useful in the context of *quadrivium* classes in universities and was customary in both the academic curriculum and in the manuscript tradition (Thorndike 1949). On the one hand, Sacrobosco's *Tractatus de sphaera* "gave a general introduction to the spherical astronomy and astronomical geography" but very little on the motion of the planets (Pedersen 1981, 114). It could also be used as a manual for the use of the armillary sphere (Valleriani 2020). On the other hand, the *Theorica planetarum* gave students a necessary insight into the motion of the planets, with the help of models and diagrams (North 1994, 235). Without being a commentary of the treatise, the *Theorica planetarum* offered a practical and mathematical systematization of the more theoretical aspects developed in the *Tractatus* and was, therefore, a useful complement for the study of the *quadrivium*.

This "old wine in new skin" strategy was not unusual for the early years of printing, at a time (especially in Venice at the end of the 1470s) when the European and especially Venetian book market began to be congested (Zorzi 1986). In the hope to survive and to make readers buy new editions of texts already on the market, printers and publishers had to insist on the editorial work that brought the old text up to date. The paratexts and the texts printed in the same volume played an important part in emphasizing the novelty and therefore the desirability of a given edition (Chap. 10). This is what Ezio Ornato called the "rhétorique de la nouveauté;" these rhetorical and advertising methods were used by printers and booksellers to create a need for new books (Ornato 1997). Therefore, Renner's strategy was to print texts that the public was familiar with and that was needed in academic curricula. But he printed them wrapped in a new layout inspired by the manuscript tradition, taking advantage of the technical possibilities offered by printing and the lowering of production costs. He positioned himself on the academic market as a printer offering useful texts and relatively secure innovations that did not disrupt the public but enabled readers to buy an illustrated text in print that would have been much more expensive in manuscript form.

Franz Renner's 1478 edition (Sacrobosco 1478b) was an important step in the reception of Sacrobosco's *Tractatus de sphaera* in print. Renner was probably a very well-established printer at the time. Active in Venice since 1471, he was specialized not only in liturgical and religious books—i.e., Bibles, sermons, etc.—but also in academic books, such as the works of Thomas Aquinas (1225–1274) and the commentaries of Aristotle (385–323 BCE). When he published Sacrobosco's *Sphaera mundi*, Renner already had experience in publishing academic texts and also had the commercial network that enabled him to sell these books across Italy and Europe. He had very close connections with Florentine booksellers. His employee Simone de Verde seemed to have been entrusted with a large number of books to sell to merchants from Lucca and Genoa (Ridolfi 1967, 60–61). Lorenz Böninger recently reconstructed Renner's commercial network between Venice, Florence, Lucca, and Genova. The bookseller, in association with Venetian patrician Leonardo Donà (1536–1612), had significant sales figures across northern Italy and Tuscany (Böninger 2020). In addition to his own commercial network, Franz Renner had a very close relationship with one prominent Venetian bookseller. His daughter, Cristina Fontana, was married to Francesco de Madii, one of the main Venetian booksellers at the time.[5] The journal of de Madii's shop has been studied by Martin Lowry (Lowry 1979), and more recently by Cristina Dondi and Neil Harris (Dondi and Harris 2013, 2014). In four years, thirteen thousand books passed through his shop; more than thirteen hundred were on sale at the same time, Venetian editions as well as books printed elsewhere in Italy or in Germany (Nuovo 2003, 40). Franz Renner had a close relationship with this major figure in the Venetian book trade at the time and had first-hand experience with the export of this kind of book. He was aware of the demand for books in Venice and in Europe and had ways to distribute them efficiently. On top of that, as with many German printers in Venice, he probably maintained a very close connection with German cities, even after many years in Italy, which allowed him to remain aware of Regiomontanus' editions and technical innovations. His position between the German and Venetian environment enabled Renner to adapt these innovations to his perception of the academic book market.

His solution seemed to have worked out. Adam of Rottweil used the same layout for his own edition of Sacrobosco and of Gerardo Cremonensis, as did Domenico Fusco, a Bolognese printer, for his Sacrobosco and Gerardo Cremonensis edition of 1480 (Sacrobosco 1480). The illustrations were copied from Renner's edition, but much more poorly executed. Franz Renner's publishing choices seemed to have been a good strategy, imitated by both Venetian and Bolognese printers in the following years. It is not an intellectually innovative model, but successful editions at that time rarely were.

---

[5] Cristina Fontana was mentioned in 1490, after Franz Renner's death, as "*Dona Crestina relicta de quondam ser Francesco de Mazi da Como et fie che fo del quondam ser Francesco Fontana*" (ASV, Giudici del Procurator, Sentenze a legge, b. 12, fol. 24, December 17, 1490). Cristina soon also became an important figure in the Venetian printing industry, having the inheritance of her father, raising her brother Benedetto Fontana to be a printer himself, marrying for the second time with the printer Paganino Paganini, and raising the illegitimate son of her husband, Alessandro, also to be a major printer in Venice. See in particular (Nuovo 1990).

2 Erhard Ratdolt's Editions in the Venetian Context

2.1 Renner Versus Ratdolt

Erhard Ratdolt, like Franz Renner, had been active in Venice for some years when he printed the *Sphaera mundi*. He arrived from Augsburg in 1476, where he was a bookbinder just before he left for Venice (Redgrave 1894; Gerulaitis 1970). He began printing with two other German associates: Bernard Maler (d. 1477) and Peter Löslein (d. ca. 1487). The debate is still open as to whether Maler was responsible for the artistic quality of the company's editions. Their first editions in Venice were Regiomontanus' *Calendarium* in Latin and in Italian. Ratdolt, Maler, and Löslein were the first to publish the German astronomer apart from his own editions in Nuremberg. It is not likely that Ratdolt was in Nuremberg when Regiomontanus printed there between 1473 and 1475, since he is last mentioned in Augsburg's tax books in 1474, and he mentions in an autobiographical document that he came to Venice on September 15, 1474, "for the last time."[6] However, Nuremberg and Augsburg had close links at the end of the fifteenth century. Therefore, it is not at all surprising that Ratdolt had good working knowledge of contemporary astronomical debates and more specifically of Regiomontanus' scientific and publishing work. Ratdolt, Maler, and Löslein's company ended in 1480, but Ratdolt continued to print texts related to the *quadrivium*. By 1482, Ratdolt already had a great deal of experience printing illustrated texts and academic books.

During his Venetian career, alone or in collaboration with others, Erhard Ratdolt dedicated almost a third of his production to mathematical, geometrical, astronomical, or alchemistical texts such as Pietro Borgo's (d. after 1494) *Aritmetica mercantile* (Borgo 1484), Gaius Julius Hyginus' (64–17 BCE) *Poetica astronomica* (Hyginus 1482), and of course Euclid's (4th–3rd cent. BCE) *Elementa geometrica* (Euclid 1482). He managed this ambitious publishing agenda by diversifying his publications and publishing some breviaries and other liturgical books. These editions were likely to sell quickly and safely, which enabled him to have more peculiar or even risky projects, like refined illustrated editions. From the beginning, Ratdolt positioned himself as a specialist in publications related to mathematics, natural science, and astronomy. In that context, Renner's 1478 edition may be seen as a provocation. Before that date, Renner did not seem to have taken an interest in that specific kind of work. However, in 1478, not only did Renner published the *Tractatus de sphaera*, he also printed Dionysius Periegetes' (b. 290) *De situ orbis* (Periegetes 1478), and Pomponius Mela's (b. 15) *Cosmographia* (Mela 1478a). Erhard Ratdolt had already published the first one in 1477 (Periegetes 1477) and the latter in 1478 (Mela 1478b).

---

[6] The hypothesis of Ratdolt's stay in Augsburg was suggested in (Redgrave 1894) but refuted by (Gerulaitis 1970) on the basis of this autobiographical document, "*Notae biographicae et geneo-logicae, quas Erhardus Ratdolt, primus Venetiarum reipublice typographus, de se suisque propria manu conscripsit, a. 1462–1524 de currentibus, Germ.,*" last edited in (Diehl 1933): "*1474 ady 15 setemer. Item ich bin dass lest mal gen Venedig kumen.*"

If we take a closer look at these editions, one can see that the relationship between the two printers is more intricate than a simple overlap of editorial strategies. If the two *De situ orbis* editions are already very similar, Renner's edition of Pomponius Mela is a direct line-by-line copy of Ratdolt's 1478 edition—as has already been noted by Redgrave (Redgrave 1894, 14). The only difference is that Ratdolt's edition has bigger margins, the woodcut initials are different, and of course, the colophons differ. Moreover, the Roman characters used in these two editions are extremely similar and can be traced thanks to the *Typenrepertorium der Wiegendrucke* (TW).[7] The characters used by Ratdolt when he first arrived in Venice were very close (but not identical) to Renner's (TW 1:109R), which he had been using since 1471. This kind of Roman type was probably inspired by Regiomontanus' work in Nuremberg (TW 1:94R), Wendelin de Spira in Venice (TW 4:85R), and Nicholas Jenson (TW 1:115R) (Redgrave 1894). When Ratdolt first arrived in Venice in 1476, and while he was in partnership with Maler, he used Roman characters very similar to those of prominent printers at the time. He changed after 1480 and used mainly *rotunda* types in the years that followed.[8] However, at the end of the 1470 s Ratdolt's types do not seem to have gone unnoticed. In 1478, for his edition of Dionysius Periegetes' *De situ orbis*, Renner changed his Roman font and used a slightly different type (TW 5:109R), extremely close to Ratdolt's (TW 1:109R), which he used in his own *De situ orbis* edition in 1477. We can only make assumptions, but it seems very likely that Renner copied Ratdolt's types to pursue the same market for scientific editions.

It is highly improbable that these similarities were due to a peaceful collaboration between the two German printers. First, had it been an agreement between both printers, Renner would not have copied Ratdolt's characters but would have borrowed or rented them. Moreover, the publication of strictly identical editions, as were the *De situ orbis* and the *Cosmographia* editions, is a commercial nonstarter. Two workshops could publish similar editions while collaborating: for example, in 1477, Johann of Cologne published the second part of Antonin of Florence's *Summa theologiae* (Florentinus 1477a) while Nicholas Jenson (d. ca. 1480) was simultaneously printing the third part (Florentinus 1477b). This led Martin Lowry to believe that an agreement existed between the two firms, before their formal merger in 1480 (Lowry 1981). In this case, the two editions were complementary and did not concern the same text, contrary to Renner's and Ratdolt's case. Their simultaneous editions would not be bought twice by the same reader and could risk flooding a still fragile and unstable book market (Zorzi 1986). The similarities between Ratdolt's and Renner's editions were likely the result of direct and aggressive competition for the market of academic books.

At the end of the 1470s, the Venetian printing industry was a highly competitive one and printers fought to exist in the shadow of the two main typographical companies, Johann of Cologne and Johann Manthen on the one hand, and Nicholas Jenson

---

[7] The characters used in both editions of *De situ orbis* are Renner's 5-109R and Ratdolt's 1:109R according to the TW.

[8] According to the TW, Ratdolt used a variety of *rotunda* characters between 1480 and 1482: 4:56G, 5:155G, 6:76G. This formal change in Ratdolt's production will be examined later on.

on the other. Franz Renner and Erhard Ratdolt saw a commercial opportunity in astronomical treatises, a kind of niche publication. This led to their direct confrontation and to what we can clearly identify as Renner's piracy of Ratdolt's publications between 1476 and 1478. Ratdolt's 1482 edition of the *Tractatus de sphaera* can be seen as a response to this attack, on an editorial but also on an intellectual level.

2.2 Ratdolt's Reinterpretation of Sacrobosco's Tractatus de sphaera

Erhard Ratdolt's 1482 edition (Sacrobosco et al. 1482) is a milestone for Sacrobosco's reception from different points of view. First, the 1482 edition was the first one that associated Sacrobosco's *Sphaera mundi* with Georg Peuerbach's *Theoricae novae planetarum* and Regiomontanus' *Disputationes contra Cremonensia deliramenta*. As we mentioned before, Regiomontanus saw Peuerbach's text as a replacement for the old *Theorica planetarum* (Horst 2019). Erhard Ratdolt clearly adopted this point of view by choosing to print Sacrobosco without any of the traditional medieval commentaries or treatises, but with the new treatise on the motion of the planets, alongside Regiomontanus' plea for the new *Theoricae* and criticism of the old *Theorica planetarum*. By doing so, Ratdolt inserted the *Tractatus de sphaera* in the intellectual debates of the time on Ptolemaic and Aristotelian models (Pedersen 1981); these debates were particularly vivid in the humanist and intellectual circles of southern Germany and in Austria (Horst 2019). Owen Gingerich stated that binding Sacrobosco's *Sphaera mundi* with Peuerbach's *Theoricae novae planetarum* had been common since the beginning of the fifteenth century (Gingerich 1999). However, the manuscripts studied by Michela Malpangotto only contain Peuerbach's treatise, not the *Sphaera*, even in Viennese copies (Malpangotto 2012). It does not appear that the habit of reading Sacrobosco with Peuerbach was very well established (Pedersen 1975). Ratdolt's choice can be explained by his knowledge of the learned debates in southern Germany, but could also be reproduced in a Venetian and Italian context. Peuerbach's work was supported by Cardinal Bessarion (1403–1472), who himself possessed a copy of the *Theoricae novae planetarum* (Horst 2019). Bessarion being close to the political and intellectual Venetian elite, the reception of Peuerbach's treatise was probably facilitated thereby. Moreover, Bessarion also acted as pontifical legate across Italy and Europe. He protected Regiomontanus in Rome in the 1460s, and the German astronomer died there in 1476. The decision to print Peuerbach's treatise and Regiomontanus' text alongside the *Tractatus de sphaera* can be understood in this intellectual environment.

Given these various elements, Ratdolt must have felt that Peuerbach's treatise was likely to have a good reception both in German-speaking regions and in the Italian peninsula. His personal knowledge of it is not surprising given the tight links he maintained with the German-speaking region and, at the same time, his close connections to the Venetian patriciate. He also received the support of some officials, such as Matthias Corvin and Michael Turon, bishop of Milkow and suffrage bishop of Esztergom in Hungary, who commissioned him a breviary for the diocese Esztergom in 1480 (*Breviarium* 1480). His connections with Augsburg authorities must have still been important: in 1486, he returned to his city, called upon by the bishop of Augsburg. He also remarried in Augsburg in 1485, while he was still active in Venice.[9] In Venice, apart from his close partners Maler and Löslein, and a collaboration with the German printer Nicolas of Francfort (1473–1524), Ratdolt does not seem to have had many professional relations in the Venetian book market, especially outside the German community. However, he was not without support: some of his editions included luxury exemplars, printed on vellum and in gold. The letter of dedication and the remaining exemplars lead us to believe that the edition was probably completed in part with the support of the doge Mocenigo (Carter et al. 1983; Baldasso 2013). While choosing to print Peuerbach's and Regiomontanus' texts, Ratdolt was probably already counting on the support of part of both the south-German and Venetian elite, some of whom were already well aware of these works and their intellectual significance.

2.3   Sacrobosco and the Italo-German Comparison

The direct influence of Regiomontanus' editions and the choice of German authors to print in complement to Sacrobosco's *Tractatus de sphaera* can also be understood in relation to the position of Ratdolt between Venice and German cities, and more generally the German market. Ratdolt's catalog already presented a certain number of German authors: Werner Rolewinck (1425–1502) *Fasciculus temporum* (Rolewinck 1480; Rolewinck 1481), Paul of Middelburg's *Prognostico* (Middelburg 1481–1482). After 1482, he also published Johann Danck's commentary of the Alphonsine tables (Alfonso 1483), Mark of Lindau's *Buch der zehn Gebote* (Lindau 1483), and two more editions of Rolevinck's *Fasciculus* after 1482 (Rolewinck 1484, 1485). These publishing decisions can be linked to the close ties he maintained in German cities and to his targeting of German markets. Moreover, for German scholars, Regiomontanus must have been a particularly good selling argument. In the humanist circles of Augsburg, Nuremberg, or Vienna, in the universities of Leipzig or Cologne, demand for such books and the appeal of names well known to local scholars such as Peuerbach or Regiomontanus was likely to be high. Later sources underlined the prestige associated with that name. Back in Augsburg, Ratdolt called Regiomontanus "the ornament of Germans" (*germanorum decor*) in a 1488 edition of the almanac.[10] He was not the only one. At that time, German authors and humanists often cited the

---

[9] Ratdolt wrote: "*1485 ady 27 setemer in Augspurg. Item ich hab mein weib fronica genumen den obgeschriben dag und hab huchczeit mitt ir gehabtt. 1485 ady 14 nobemer*" (Diehl 1933) (Vienna, Osterreichische Nationalbibliothek *15473*).

[10] "*Johannis de Monte Regio, germanorum decoris, etatis nostre astronomorum, principis Ephemerides*" (Author's emphasis) (Regiomontanus 1488, 1r).

German astronomer as an object of pride: Hartmann Schedel used a similar expression, "honor of Germans" (*germanorum decus*), to qualify Regiomontanus (Zinner 1990, 187–188). At a time when the German-speaking territories were divided into a multitude of political entities, these expressions can nonetheless be interpreted as a manifestation of a German conscience of worth in the intellectual contemporary debate, which thrived especially thanks to the emulation of Italian scholarship. The comparison with Italian writers was a real concern for German humanists, between admiration, emulation, and competition, as has been shown, among other examples, in (Bertalot 1975; Dörner 1999; Gier 2010). This phenomenon was not limited to academic controversies. The genre of works dedicated to specific cities, such as Conrad Celtis' (1459–1508) *Norimberga*, also displayed such a comparison with Italian cities and local pride (Celtis 2000; Buchholzer-Rémy 2006). In the arts, Albrecht Dürer's (1471–1528) life and writings are a good example of the ambivalent sentiment German artists could have toward Italy, between admiration and conscience of self-worth (Vaisse 1995).

The intellectual context of emulation between German and Italian scholars laid an interesting backdrop for printers in search of a public. In the early years, Venetian printers of Germanic backgrounds often addressed their origin as an object of pride. The very first printer in the city, Johann de Spira (d. 1470), clearly stated in his colophons the profit German printers brought to Venice,[11] and his brother Vindelinus even prophesied that their hometown, Spira, would be as celebrated as Mantua, Virgil's (70–19 BCE) motherland.[12] In this context, Peuerbach's and Regiomontanus' publication could also be seen as a way to assert the worth of German scholarship, printed by German printers in an Italian city. The link between these publications and a German affirmation of self-worth in Venice is explicit in the 1488 Sacrobosco edition, financed by Johann Lucilius Santritter (1460–1498). He was a scholar as well as an investor. In his 1488 edition of Sacrobosco, the colophon is in the form of a poem praising his work, presenting himself as "Joannes Lucilius Santritter from the city of Heilbronn," acknowledging Girolamo de Sanctis' work as a printer. In a second paragraph, Santritter underlined the "German genius" (*ingenio germanico*), which enabled the completion of his book—he might be talking of himself as well as

---

[11] Johan de Spira's colophon reads: "*Primus in Adriaca formis impressit aenis/Urbe libros Spira genitus de stirpe Johannes/In reliquis sit quanta, vides, spes, lector, habenda/Quom labor hic primus calami superaverit artem*" (Cicero 1469); "*Hesperiae quondam Germanus quisque libellos/Abstulit: en plures (plura) pise daturus adest./Namque vir ingenio mirandus et arte Joannes/Exscribi docuit clarius aere libros./Spira favet Venetis: quarto nam mense peregit/Hoc tercentenum his Ciceronis opus*" (Cicero 1470). This affirmation of a proud German identity is particularly developed in Venice and had an effect on the way these German printers were perceived in the Italian context; see (Amelung 1964; Kikuchi 2018c).

[12] Vindelino de Spira wrote: "*Vindelinum.../Cui tantum debes urbs spira superba nepoti/Quantum Virgilio mantua clara suo*" Niccolò Tedeschi (Nicolaus Panormitanus de Tudeschis), *Lectura super primo et secundo Decretalium*, vol. 3: (Tedeschi 1472); and again in 1473, "*Supra tua est virtus italias jam nota per urbes/Ore tuum nomen posteritatis erit*" (Caracciolo 1473). For a complete study of the German community in Venice, see (Braunstein 2016).

the authors published.[13] These points, in an edition containing two texts of German scholars and financed by a German publisher in Venice, are to be understood in this context of competition as an affirmation of the role of German craftsmanship, commerce, and scholarship.

Not only astronomy should be considered to understand Ratdolt's and Santritter's publishing choices. The emphasis on the German origin of printers and scholars in the context of early printing is not only a question of national pride but a commercial one. The German market played an important part in the development of early Venetian printing, and it certainly played a part in Ratdolt and Santritter's strategy, whether they had personal opinions on the intellectual content or not. One way or another, Erhard Ratdolt took advantage of the situation to fashion himself as a thorough scientific publisher and printer on the new and unstable market of academic books.

2.4 Ratdolt's Reinterpretation

Erhard Ratdolt was in the right position to print Peuerbach's and Regiomontanus' texts with Sacrobosco's *Tractatus de sphaera*. As a well-known Venetian printer, aware of scholarly debates in Germany and in Italy, and capitalizing on the emulation between German and Italian scholars, he made the opposite choice of his opponent, Franz Renner. Regiomontanus' text is a clear and ferocious criticism of the errors and "*deliramenta*" contained in the old *Theorica planetarum*. Ratdolt positioned himself in a learned debate as well as in an economic market. To that end, he also adapted the form of his edition, going further than Renner. The specific *mise en livre* he chose played a large part in the success of these editions.

The material analysis of Erhard Ratdolt's edition of the *Sphaera mundi* is not new and we will only rehearse the main aspects of it here. The 1482 edition is an in-quarto volume of 60 folios and eight quires, printed in black and red. The layout is dense but leaves a large place for illustrations, which are of high quality. The sketches and tables are printed within the text blocks, as part of the demonstration itself. Owen Gingerich underlined the apparition in Ratdolt's edition of two large figures, an armillary sphere and a sketch of the geocentric universe, which became typical in subsequent editions (Gingerich 1999). He also followed the circulation and transformation of some specific illustrations, as did Jürgen Hamel (Hamel 2006). Ratdolt used all of Regiomontanus' technical innovations to illustrate Sacrobosco as well as the two other texts assembled with the *Sphaera mundi*: tables, geometrical diagrams, models for the eclipse and the movement of the planet, decorated initials,

---

[13] Johann Lucilius Santritter wrote: "*Carmina impressorum huius opusculi laudem//Uranie quantum debere fatentur/Cuncta canopeo: cognitaque astra viro/Santritter helbonna lucili ex urbe Joannes/Schemata sic debet ipsa reperta tibi/Naec minus haec tibi de Sanctis Hieronume debent/Quam socio: nanque hic invenit: ipse secas.//Hoc quoque sideralis scientiae singular opusculum/mirifica illa arte nuper* ingenio germanico/*in lucem prodita impression videlicet/Prididie calenda Aprilis./Anno Salutis./M. cccc.lxxxviii/completum est/Venetiis.*" (Author's emphasis): Sacrobosco, *Sphaera mundi*: (Sacrobosco et al. 1488).

etc. Some of the sketches were directly copied from Regiomontanus' edition of the *Theoricae novae planetarum* (Shank 2012). The layout is fuller than in Renner's edition, but the accent is on the beauty, legibility, and diversity of the illustrations. This is truer for the 1485 edition (Sacrobosco et al. 1485), in which the illustrations are even more abundant and diverse (Fig. 1).

Isabelle Pantin (Pantin 2012) underlined how the introduction of diagrams and illustrations in Ratdolt's edition created a standard for the printers and the public. We will see at a later stage how printers reacted to this new standard. After Ratdolt, readers largely expected the *Tractatus de sphaera* to be illustrated, even if it meant a larger investment. It also gave more commercial value to the editions; the vitality of the study of mathematics at the time made the investment worthy of the risk. For Peuerbach's treatise, the illustrations were also needed, not only as a demand from the public but also for the pedagogical value of the sketches (Pantin 2012, 2013). In Ratdolt's strategy, Sacrobosco's edition must of course be understood in parallel with the 1482 edition of Euclid's *Elementa*. It was the first attempt to publish Euclid, and Ratdolt immediately set a standard by presenting this richly decorated and illustrated version of the text. The format is different since it is an in-folio; the layout of the illustrations is slightly different since they are not inserted in the block text. However, illustrated scientific books and their pedagogical use were something Ratdolt had been taking seriously during these years.

The first edition sold well; Ratdolt was able to reissue a second version only three years later, a very short period in terms of the delay of profitability in the fifteenth-century printing industry.[14] He took advantage of the reissue to change the form of the edition significantly. The illustrations offered a new pedagogical, visual, and more practical way of conveying these texts. The choice of texts accompanying the classical treatise of Sacrobosco integrated the thirteenth-century *Tractatus de sphaera* in the actual practice and study of contemporary astronomy.

Not only were the illustrations richer and more elaborate in the 1485 edition, but Ratdolt also used a Roman character, rather than the *rotunda* he used in 1482. Rotunda types were widely used in Europe for medical, juridical, or mathematical publications. However, Ratdolt's first edition of the *Tractatus de sphaera* was an anomaly among the early editions of Sacrobosco: all were printed in Roman characters except Adam of Rottweil's 1478 edition. If one examines Ratdolt's production in 1482, none of the main sets of types of his publications during this period are Roman. That year, Ratdolt also used a *rotunda* type in printing Alchabitius' *Libellus isagogicus* (Alchabitius 1482) and Jacobus Publicius' *Artes orandi* (Publicius 1482) but reissued them both in 1485 with Roman types (Alchabitius 1485; Publicius 1485). He chose to print texts from the *artes* curriculum in gothic types when other printers would have printed them in Roman.

Was it a deliberate choice or the consequence of the circumstances? Ratdolt seemed to cease using Roman fonts after the end of his partnership with Maler and Löslein, in 1480. Did his partners leave with the material? It seems unlikely

---

[14] For an example of delay between production and return on investment, see (Pettas 1973). For the example of Aldus Manutius, see (Kikuchi 2018a).

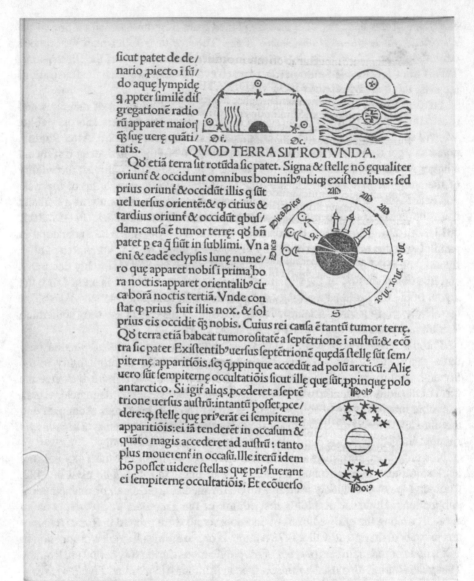

ſicut patet de de/
nario ‚piecto i ſu̅/
do aquę lympidę
q̟ ‚ppter ſimile̅ diſ
gregatione̅ radio
ru̅ apparet maior|
q̅z ſuę uerę quáti/
tatis.

QVOD TERRA SIT ROTVNDA.

Q̊ etiã terra ſit rotũda ſic patet. Signa & ſtellę no̅ ęqualiter oriu̅t̅ & occidunt omnibus bominib̕ ubiꝗ̉ exiſtentibus: ſed prius oriu̅t̅ &occidũt illis q ſũt uel uerſus oriente̅:& ꝗ citius & tardius oriu̅t̅ & occidũt qbuſ/ dam: cauſa e̅ tumor terrę: q̊ bñ patet ꝑ ea q̅ ſiũt in ſublimi. Vn a eni & eade̅ eclypſis lunę nume/ ro quę apparet nobiſ i prima‚bo ra noctis: apparet orientalib̕ cir ca‚borã noctis tertiã.Vnde con ſtat ꝗ prius ſuit illis nox.& ſol prius eis occidit q̅z nobis. Cuius rei cauſa e̅ tantũ tumor terrę. Q̊ terra etiã babeat tumoroſitate̅ a ſepte̅trione i auſtrũ:& eco̅ tra ſic patet Exiſtentib̕ uerſus ſepte̅trione quędã ſtellę ſũt ſem/ piternę apparitio̅is.ſꜫ q̅ppinque accedũt ad polũ arcticũ. Alię uero ſũt ſempiternę occultatio̅is ſicut ille q̅ ſũt‚ppinquę polo antarctico. Si igit̅ aliqs‚pcederet a ſepte̅ trione uerſus auſtrũ: intantũ poſſet‚pce/ dere:ꝗ ſtellę quę pri̕ erãt ei ſempiternę apparitio̅is: ei iã tendere̅t in occaſum & quáto magis accederet ad auſtrũ : tanto plus moue̅e̅t in occaſũ.Ille iteru̅ idem bo̅ poſſet uidere ſtellas quę pri̕ fuerant ei ſempiternę occultatio̅is. Et eco̅uerſo

---

Fig. 1 Erhard Ratdolt's editions provide a close text-image relationship, facilitating the reading of the treatise while offering small informative but also decorative figures, that were copied across Europe. Santritter replicated the same kind of layout in his 1488 edition with a new set of woodcuts. (Sacrobosco et al. 1485, 4v). München, Bayerische Staatsbibliothek—4 Inc.c.a. 430. https://nbn-resolving.org/urn:nbn:de:bvb:12-bsb00036841-7 CCBY-NC-SA4.0

that Ratdolt did not have the ability to secure a new Roman set since in the same years he had important investments for his Euclid's *Elementa*. But sets of characters were expensive, and it is possible that he chose to continue his activity with his *rotunda* characters while waiting for the right moment to acquire new Roman sets. It was obviously the case in 1485 when he issued a series of publications in Roman types, including a reprint of some of the previously *rotunda*-printed works. While this evolution may be partly due to technical, financial, and material issues in Ratdolt's workshop, it could also indicate how the categorization of works and the use of Roman or Gothic characters were still very unstable. For the same text, and from the same printer in a short period of time, formal aspects could be modified to appeal to different audiences. Rotunda characters tended to be more common for university law and medicine textbooks, while Roman characters were used for classical texts or *studia humanitatis* treatises, the kind of works that were also read outside teaching institutions. The importance of Sacrobosco's academic reception is undeniable, but it is also plausible that the use of Roman characters was also a way to target a larger audience. The ambiguity is still present in Venetian production at the end of the fifteenth century, in the production of Ratdolt's successors.

2.5 An Actualization of the Tractatus de sphaera

While Ratdolt's edition was a key part of a coherent and comprehensive publishing strategy concerning mathematics and natural sciences, it also offered an actualization of the thirteenth-century *Tractatus de sphaera*. In comparison to the previous editions in Venice and in Italy, Ratdolt managed to offer a new material object as well as intellectual content and integrated them into the intellectual debates of his time. It could be used in a traditional teaching context as a standard textbook for students (Chap. 12), offering pedagogical help and a lower cost; it could also be read at a higher level, by scholars interested in the latest astronomical debates. Ratdolt's coherent publishing strategy was able to reach a large audience.

The actualized form and content of the *Sphaera mundi* also had important consequences on the status of those involved. Erhard Ratdolt used the reputation he had already acquired before 1482 to promote this new publishing achievement. In return, he also consolidated his trademark as an academic up-to-date printer, by associating himself with a classical textbook used in universities and with works written by contemporary scholars. The status of the three authors also changed in the process. Thanks to his association with one of the most read textbooks in astronomy, Peuerbach and Regiomontanus consolidated their status as important authors in the academic field. The aura of Sacrobosco was used as a "label" in the sense Matteo Valleriani gave to the expression (Valleriani 2017, 430). The association with Sacrobosco probably allowed both authors to gain legitimacy. The next editions of the *Tractatus de sphaera* included the same compilation of works as Ratdolt's, which enforced the association in the publishing market and intensified the legitimization and labeling of both contemporary authors. Peuerbach was probably the one who gained the most in the exchange. This edition was the first step toward the development of his own publishing reputation. In 1495, Peuerbach's *Theoricae planetarum* was published in Venice for the first time on its own since Regiomontanus' 1474 edition, and no longer as an addendum to Sacrobosco's or Regiomontanus' works (Peuerbach 1495). He continued to be published on his own and with commentaries during the sixteenth century, in German printing centers as well as in Paris (Pantin 2013). Regiomontanus continued to be printed occasionally in Venice, but mainly in German centers such as Nuremberg and Basel.

But the most important consequence would be for Sacrobosco's text: instead of being relegated as a text from a time gone by, his work was completely integrated with the new discussion, and therefore into the editorial programs of printers all over Europe. While Sacrobosco maintained his position as an important author in academic curricula, he also gained a sense of novelty, thanks to the association with the new form and recent authors. The idea that Sacrobosco was relevant was an extremely important selling point on the new academic book market, as we already mentioned. The *mise en livre* Ratdolt introduced allowed the text to be reinterpreted along with the evolution of the study of mathematics in Renaissance Europe: as later editions show, it became customary to associate Sacrobosco's *Tractatus de sphaera* with treatises and commentaries of other contemporary scholars. Ratdolt gave way to that kind of construction while discarding the old *Theoricae planetarum* and giving his preference to Regiomontanus' and Peuerbach's works. For that reason, Ratdolt's editions were an important step in the reception and posterity of Sacrobosco's text, enabling it to be read not only for itself but also in relation to the texts of contemporary scholars.

In that sense, Sacrobosco was not only a label; the perception of his name changed according to the context in which it was used. The presence of multiple names, be they authors, commentators, editors, or printers, alters the perception of all of them individually and collectively. This co-presence created something more than the simple addition of individuals and texts. A compilation that associates these different entities created a new identity and a narrative through the interactions between them.[15] The prestige of some authors or printers may serve in return to enhance the prestige of the other actors associated with a given compilation. This phenomenon may explain why some authors kept being published, while others never really made it to the new printing industry, and some new names managed to emerge from the multitude of authors and commentators present in the book market. This is one explanation why Sacrobosco continued to be published in the following years and decades, rarely alone but in association with other authors, some of the contemporary scholars.

---

[15] This analysis is inspired by Harrison White's network analysis, especially in (White 2011) and I worked on a case study in a recent article (Kikuchi 2018b). Recent studies have also shed light on the intricate relationships between every name involved in a compilation and the implication of the compilation or collection format (Ouvry-Vial and Réach-Ngô 2010; Réach-Ngô 2014; Furno and Mouren 2012). See also (Grafton 2011).

3 Diffusion and Reinterpretation of Ratdolt's Editorial Model

3.1 Diffusion of Ratdolt's Editions in Europe

As we have examined how Ratdolt's editions offered a new editorial model, at a formal and intellectual level it is now necessary to turn to the actual diffusion of the exemplars produced. This kind of investigation is always a difficult one since we have to rely on the remaining exemplars, which do not always give us relevant information as far as their fifteenth-century ownership is concerned. Thanks to the progress in cataloging from the last decades, it is now possible to have an almost complete overview of the surviving incunabula and their places of conservation. Given this information, present in the ISTC, Ezio Ornato investigated the localization of the surviving copies and inferred the areas these books could reach also based on the global commercial tendencies of the time. His hypothesis is that the circulation of books after their first sales, through collectors, sales of private collections, etc., rarely traveled beyond regional boundaries and does not prevent us from reaching some partial conclusions regarding the commercial networks of printers in the fifteenth century. If one except the bias introduced by great libraries such as the Bibliothèque Nationale de France in Paris or the British Library in London, and the displacement of books in non-European countries, the data might be interpreted with caution (Ornato 2017). The overview of the exemplars remaining in Europe today offers a first overview of the possible diffusion of both editions (Table 2). The 1482 edition has 113 exemplars remaining, 104 of which are in Europe or Russia. The 1485 edition has 110 exemplars remaining, 91 of which are in Europe or Russia. We have presented the distribution of the remaining copies in their holdings, according to contemporary political entities.

Unsurprisingly, the data seems to indicate that both the Italian and the German markets were important for the commercialization of Ratdolt's books. The number of copies is also higher in France for the second edition, but five of those are in the Bibliothèque nationale de France, which might give us false indications.

This method can give us an initial overview; however, it seems too unreliable, especially for editions such as Ratdolt's, which could have interested bibliophiles from the sixteenth century until today. Venetian printed books were under the eye of collectors from all over Europe. Examples present in France today could have been brought back by eighteenth-century bibliophiles.[16] The contemporary localization of exemplars is offered here as a first attempt, but, in our opinion, it cannot be used as a robust argument.

---

[16] See for instance François-Xavier Laire, a French collector, who went to Venice in 1789–1790: He encountered Jacopo Morelli, librarian at the San Marco library, who wrote to Angelo Maria Bandini that Laire talked to him endlessly of Quattrocento books: "Qui mi trovo frequentemente col p. Laire da Lei raccomandatemi e si parla di libri del Quattrocento sine fine." The same Laire was considering doing a catalog of all Venetian Quattrocento editions (Ruffini 2012). This kind of collector was very likely to buy books during a trip.

Table 2 Localization of surviving copies of 1482 and 1485 Ratdolt editions in Europe and Russia

	Austria	Croatia	Belgium	Czech Republic	Denmark	Finland	France	Germany	Hungary
1482	4		1	1	1	1	3	20	2
1485	3	2					11	23	2

	Italy and Vatican	Netherlands	Poland	Portugal	Russia	Spain	Sweden	UK	Switzerland
1482	44	2	1	2	1	3	2	16	
1485	25	3		1	3	5	1	11	1

Another method to track the circulation of books is to examine the books themselves for possession marks and annotations that may indicate their location at different times. The systematic examination of the surviving copies of incunabula editions considerably diminishes the information available but also limits the risks of over interpreting the results. Cristina Dondi and her ERC Project *15c Booktrade* aimed to make a systematic catalog of all remaining incunabula by examining all the marks that could reveal information on their circulation until today (Dondi 2013). This project gave birth to *Material Evidence in Incunabula* (MEI), a database "which provides copy-specific information on some of the copies listed in ISTC," according to the presentation. For the 1482 and 1485 Ratdolt editions cataloged in the ISTC, only sixteen and eight copies out of 113 and 110, respectively, are present in the MEI database. This is a drastic reduction of data, but one which can nevertheless yield interesting results. The MEI visualization tool (https://15cbooktrade.ox.ac.uk/visual ization/) shows that Ratdolt's 1482 edition was probably well distributed in Italy, with five exemplars that can be traced in the peninsula before 1500. However, some copies arrived in Switzerland, Austria, and England not long after the publishing date. The information concerning the 1485 edition is scarcer, but one can still observe that one copy was in Spain around 1500.

These observations are confirmed by the examination of copies in other libraries that are not yet included in the MEI database. For example, the copies currently at the Bayerische Staatsbibliothek were bought by German institutions at the end of the fifteenth or the beginning of the sixteenth century. One copy of (Sacrobosco et al. 1482),[17] has an *ex libris* from the Benedictine convent of the Holy Virgin in Scheyern, in Bayern. A copy of (Sacrobosco et al. 1485),[18] belonged to Johann Albrecht Widmannstetter, a German humanist of the sixteenth century. Outside of the German world, the study of remaining copies seems to confirm a commercial distribution toward France. A copy of (Sacrobosco et al. 1485) preserved at the Bibliothèque Sainte Geneviève in Paris[19] shows many marginal annotations from the end of the fifteenth or beginning of the sixteenth century, including some in French. The copy of (Sacrobosco et al. 1485) preserved at the Bibliothèque Mazarine[20] used to belong to the Collège de Sorbonne, probably bought not long after the publication date.[21]

The information we have on the distribution of these volumes is always very fragmentary. However, it seems important to underline the wide spectrum of Ratdolt's distribution in Europe, in Italy, and the German world—which is expected considering the background of Erhard Ratdolt—but also in France and in Spain, toward religious institutions and universities. In the 1480s, before his departure from Venice, Erhard Ratdolt must have had a very stable commercial network in Europe, which

---

[17] Bayerische Staatsbibliothek, 4 Inc c.a. 256.

[18] Bayerische Staatsbibliothek, 4 Inc c.a. 430.

[19] Bibliothèque Sainte-Geneviève Paris, OEXV 762 (2) RES (P.2).

[20] Bibliothèque Mazarine, Inc 412–413.

[21] See Catalogue Régionaux des Incunables Informatisé (CRII), http://www.bvh.univ-tours.fr/inc unables.asp. Accessed 08 June 2021.

enabled him to launch some very ambitious editions, such as the 1482 and 1485 editions of Sacrobosco, as well as the 1482 edition of Euclid's *Elementa*. Contrary to Franz Renner, we have few archival sources about Ratdolt's commercial networks. The distribution of his books together with Ratdolt's close connections with Venetian patricians and German cities tend to indicate that he had the means to distribute his production efficiently. However, it is difficult to reach any conclusion about the various markets targeted by the first and second editions.

3.2 Adaptations of Ratdolt's Model

Erhard Ratdolt left Venice in 1486 to go back to Augsburg, where he continued publishing Regiomontanus' work and other scientific texts. His departure left space for other printers to occupy the scientific market in Venice (Table 3).[22]

The first to take advantage of the opportunity was Johann Lucilius Santritter. Santritter hailed from Heilbronn like Renner, but he never worked with him. Instead, he worked with Ratdolt on five editions between 1481 and 1485. He must have seen Ratdolt working on Sacrobosco and copied his work in an edition published in 1488 (Sacrobosco et al. 1488). Some of the initials are very similar, but were not printed from the same material: Santritter or his printer, Girolamo de Sanctis (fl. 1487–1494), probably copied the initials from Ratdolt's edition. Ratdolt's 1485 edition also directly inspired their diagrams, with some minor transformations, even if the 1488 edition offers a few new diagrams as well. The general disposition of illustration and text is very close in both editions; Santritter obviously followed the formal model introduced by Ratdolt. The *Tractatus de sphaera* is illustrated in the margins and in the space left by the blocks of text; illustrations are small woodcuts but directly linked to the text they refer to (Figs. 1 and 2). Peuerbach's *Theoricae novae* are printed in big, unified blocks of text and illustrated with half-page, sober, factual diagrams, whose purpose was to convey the physical and mathematical movements of the celestial bodies. The 1485 edition was the model retained and copied, not only in Venice, but also in other printing centers and from various points of view: Ratdolt's 1485 edition was also the model for Wolfgang Hopyl's (fl. 1489–1523) 1489 edition in Paris, as far as the text and layout were concerned, but with only two illustrations (Pantin 2013, 23). Ratdolt's illustrations also provided a model for the Leipziger editions studied by Richard Kremer in this volume (Chap. 12).

Santritter's 1488 Venetian edition of Sacrobosco is not an epiphenomenon in his publishing portfolio. The same year and in collaboration with the same printer, Girolamo de Sanctis, he also printed John Buridan's (ca. 1300–ca. 1358) *Quaestiones in libros Physicorum Aristotelis* (Buridan 1488); the next year, he printed Johannes Eschuid's *Summa astrologiae judicialis* (Eschuid 1489). Alongside other editions aimed at the academic market, Santritter took an interest in *quadrivium* publications.

---

[22] The editions listed in Table 3 are: (Sacrobosco and Borro 1494; Sacrobosco et al. 1488, 1490, 1491, 1499, 1501, 1508, 1513, 1518a, b, 1519; Sacrobosco and Ferraris 1500).

Nevertheless, Sacrobosco's *Tractatus* seems to have been his only illustrated publication, which he was probably able to do at a reasonable cost since he copied many woodcuts from Ratdolt's edition.

Table 3 List of known Venetian Sacrobosco's *Tractatus de sphaera* incunabula editions after 1485

Format	Year	Other authors	Producers	Main type of characters	General layout of the *Sphaera*
In 4°, 69 fol., 8 quires	1488	Georg Peuerbach, Regiomontanus	Girolamo de Sanctis for Johann Lucilius Santritter	Roman	Single column
In 4°, 48 fol., 6 quires	1490	Georg Peuerbach, Regiomontanus	Boneto Locatello for Ottaviano Scoto I.	Roman	Single column
In 4°, 48 fol., 6 quires	1491	Georg Peuerbach, Regiomontanus	Guglielmo Anima Mia da Trino	Roman	Single column
In 4°, 64 fol., 8 quires	1494	Gasparino Borro	Boneto Locatello or Bartholomeo de Zanis for Ottaviano Scoto I.	Roman	Single column, text interspersed throughout commentary
In 2°, 150 fol., 27 quires	1499	Georg Peuerbach, Regiomontanus, Cecco d'Ascoli, Franciscus Capuanus de Manfredonia, Jacques Lefèvres d'Etaples	Simone Bevilacqua	Roman	Single column, text surrounded by commentary
In 2°, 146 fol., 26 quires	1499	Georg Peuerbach, Regiomontanus, Cecco d'Ascoli, Franciscus Capuanus de Manfredonia, Jacques Lefèvres d'Etaples	Simone Bevilacqua	Roman	Single column, text surrounded by commentary
In 4°, 26 fol., 6 quires	1500	Georgius de Ferrariis	Giovanni Battista Sessa I. or Jacopo Pincio for Giorgio de Monteferrato	Roman	Single column, text surrounded by commentary
In 4°, 47 fol., 6 quires	1501	Georg Peuerbach, Regiomontanus	Giovanni Battista Sessa I	Roman	Single column

(continued)

Table 3 (continued)

Format	Year	Other authors	Producers	Main type of characters	General layout of the *Sphaera*
In 2°, 159 fol., 22 quires	1508	Georg Peuerbach, Regiomontanus, Bartolomeo Vespucci, Robert Grosseteste, Pierre d'Ailly, Franciscus Capuanus de Manfredonia, Jacques Lefèvres d'Etaples	Giovanni Rosso and Bernardino for Giuntino Giunta	Roman	Single column, text surrounded by commentary
In 4°, 47 fol. 6 quires.	1513	Georg Peuerbach, Regiomontanus	Melchior Sessa I	Roman	Single column
In 2°, 233 fol., 31 quires	1518	Pseudo-Ptolemy, Campano da Novara, Pierre d'Ailly, Cecco d'Ascoli, Gerardo Cremonensis, Theodosius de Bithynia, Francesco Capuano di Manfredonia, Jacques Lefèvre d'Etaples, Michael Scot, Robert Grosseteste, Regiomontanus	Lucantonio Giunta	Rotunda	Two columns
In 2°, 233 fol., 31 quires	1518	Pseudo-Ptolemy, Campano da Novara, Pierre d'Ailly, Cecco d'Ascoli, Theodosius of Bithynia, Francesco Capuano di Manfredonia, Jacques Lefèvre d'Etaples, Michael Scot, Robert Grosseteste, Regiomontanus	For the heirs of Ottaviano Scotto I	Rotunda	Two columns

(continued)

Table 3   (continued)

Format	Year	Other authors	Producers	Main type of characters	General layout of the *Sphaera*
In 4°, 47 fol., 6 quires	1519	Georg Peuerbach, Regiomontanus	Giacomo Pincio	Roman	Single column

As Isabelle Pantin showed (Pantin 2020), Santritter's first complete set of "Venetian Sacrobosco diagrams" was in turn copied by many printers in Venice and elsewhere in Europe. However, some copies diverged in the technical and formal innovations. For instance, in the 1490 edition printed by Ottaviano Scotto (fl. 1479–1499) (Sacrobosco et al. 1490) (Fig. 3) and the 1491 edition printed by Gugliemo da Trino (Sacrobosco et al. 1491), the diagrams are the same as in the 1488 edition but the layout is much more cluttered and less clear. Their editions also consisted of one-third fewer folios than Santritter's. To gain space, paper, and money, the diagrams were no longer closely linked to the text they illustrated; sometimes the reader had to search among diagrams to find the right one, instead of simply having it by the text. Pedagogy does not seem to have been at the heart of the conception of these editions, and there was no, or little, thinking on the conception of the diagrams since they were all directly taken from Santritter's edition. The fact that printers now copied illustrations (even though they were not the core of their occupations, and while trying to minimize cost at the expense of legibility and the practical use of the book) confirms that these treatises—the *Tractatus de sphaera*, the *Theoricae novae planetarum*, and the *Disputationes*—were now considered illustrated books. Readers expected diagrams and illustrations. It was no longer acceptable for printers to print the text alone, as in the early stage of Sacrobosco's printing history. The construction of this public expectation is a long process that began with Renner's edition but was emphasized by Ratdolt's and completed with Santritter's.

Moreover, the set of woodcuts used for the diagrams in Scotto's 1490 edition were the exact same as Santritter's (Figs. 2 and 3). The damage spots are identical in both editions. It seems Ottaviano Scotto borrowed or rented Santritter's blocks, through what was probably a commercial agreement.[23] The same woodcuts appear once again in the 1494 edition (Sacrobosco and Borro 1494)[24] printed by Ottaviano Scotto, which is no surprise since Ottaviano Scotto commissioned them both. We do not have any trace of another collaboration between Santritter and Scotto, but there seems to have been a long-running understanding between the two publishers. The circulation of woodcut sets was not unusual in the early years of printing; publishers and booksellers who had the means to pay for such production often reused them or rented them out to profit from their investment (Chap. 5) (Bonicoli 2015). It is therefore plausible that Santritter owned the woodblocks and rented them to Scotto

---

[23] See also Saskia Limbach's and Richard Kremer's studies (Chaps. 5 and 12).

[24] See the New York Public Library record for (Sacrobosco and Borro 1494): https://catalog.nypl.org/record=b14346745~S1. Accessed 04 June 2021.

DIFFINITIO SPHAERAE ET DE QuIBVSDAM
PRINCIPIIS GEEOMTRICIS SVPPONENDIS

Phæra igitur ab Euclide fic defcribitur. Sphæra eft tran
fitus circunferentiæ dimidii circuli quotiens fixa diame
tro quoufq̃ ad locum fuum redeat.
circunducitur.id.eft.Sphæra eft tale
rotundum & folidum quod defcri/
bitur ab arcu femicirculi circũducto.
Sphæra etiam a Theodofio fic defcribitur. Sphæ/
ra eft folidum quoddam una fuperficie contentum
in cuius medio punctus eft:a quo omnes lineæ du/
ctæ ad circunferentiam funt æquales. Et ille pũctus
dicitur centrum fphæræ. Linea uero recta tranfiens
per centrum fphæræ applicans extremitates fuas ad
circunferentiam ex utraq̃ parte dicitur axis fphæ/
ræ.Duo quidem puncta axem terminantia dicun/
tur poli mundi.

De quibufdá principiis geometricis
fupponendis quæ funt addita

In primis quidé uidentur fupponen/
da & intelligenda ab eis qui nefciunt geo
metricam difciplinam quibus dicenda le
uius capere poffint: & funt quæiam fub/
fcribimus & proximioribus nouitiorum
gratia addidimus.
Punctus in re quanta:eft quid indiui/
fibile:uel cuius non eft aliqua pars quæ ui
fu percipiatur.

A v

Fig. 2 In Ratdolt and Santritter's editions, diagrams were laid in front of the corresponding text, in a clear layout with large margins that enabled students to take notes (Sacrobosco et al. 1488, A5r). Image courtesy History of Science Collections, University of Oklahoma Libraries; copyright the Board of Regents of the University of Oklahoma

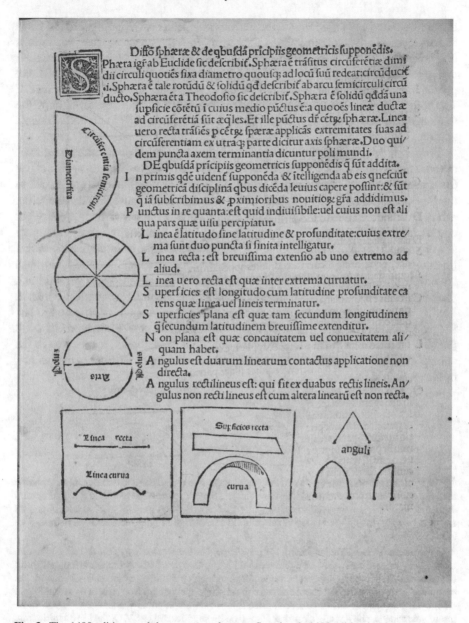

Diffo fphæræ & de qbufdã pricipiis geometricis fupponēdis.
Phæra igr ab Euclide fic defcribit.Sphæra e trafitus circūferētiæ dimi
dii circuli quotiēs fixa diametro quoufq; ad locū fuū redeat:circūducē
.i.Sphæra e tale rotūdū & folidū qd defcribit ab arcu femicirculi circū
ducto.Sphæra ēt a Theodofio fic defcribit.Sphæra ē folidū qddã una
fupficie cōtētū ĩ cuius medio pūctus ē:a quo oēs lineæ ductæ
ad circūferētiā fūt æq les.Et ille pūctus dr cētg fphæræ.Linea
uero recta trāfiēs p cētg fpæræ applicās extremitates fuas ad
circūferentiam ex utraq; parte dicitur axis fphæræ.Duo qui
dem puncta axem terminantia dicuntur poli mundi.
   DE qbufdã pricipiis geometricis fupponēdis q fūt addita.
In primis qdē uidenf fupponēda & itelligenda ab eis q nefciūt
geometricā difciplinā qbus dicēda leuius capere poffint:& fūt
q ĩa fubfcribimus & pximioribus nouitiog gra addidimus.
Punctus in re quanta:eft quid indiuifibile:uel cuius non eft ali
   qua pars quæ uifu percipiatur.
   Linea ē latitudo fine latitudine & profunditate:cuius extre
      ma fūt duo puncta fi finita intelligatur.
   Linea recta : eft breuiffima extenfio ab uno extremo ad
      aliud.
   Linea uero recta eft quæ inter extrema curuatur.
   Superficies eft longitudo cum latitudine profunditate ca
      rens quæ linea uel lineis terminatur.
   Superficies plana eft quæ tam fecundum longitudinem
      q fecundum latitudinem breuiffime extenditur.
   Non plana eft quæ concauitatem uel conuexitatem ali
      quam habet.
   Angulus eft duarum linearum contactus applicatione non
      directa.
   Angulus rectilineus eft: qui fit ex duabus rectis lineis.An
      gulus non recti lineus eft cum altera linearū eft non recta.

Fig. 3 The 1490 edition used the same woodcuts as Santritter's 1488 edition but in a much more condensed layout, which allowed the printer to gain space but made the page fuller and less legible. The same kind of layout was used in the 1491 edition (Sacrobosco et al. 1491, 5r). Courtesy of the Library of the Max Planck Institute for the History of Science

for the 1490 and 1494 editions.[25] Illustrations being expensive, printers found ways
to curb costs, but it could also lead to less acceptable solutions. The illustrations
from the 1491 edition were not printed by means of the same woodblocks but their
layouts are identical to those of Scotto's 1490 edition. This is probably a case of direct
plagiarism among Venetian printers, a habit that they were very eager to denounce.[26]
The competition in Venice was harsh and many printers did not hesitate to copy the
work of others without permission. It was probably the case of Guglielmus da Trino's
edition.

From Santritter's point of view, even if he himself copied most of Ratdolt's edition,
the situation was dangerous at a moment when he was trying to distinguish himself on
the academic market. Santritter searched for a way to protect his work and turned to
the Venetian institution of privileges, which was thriving in those years. A Venetian
privilege could not prevent copies from being made outside Venetian territory but
could protect it from fellow Venetian printers. In 1498, Santritter was the beneficiary
of a privilege from the Venetian authorities for a series of texts, including astronomical
instruments, treatises of geometry, and other mathematical and astronomical texts
"that were never printed in Venice."[27] He seems to have wanted to take over Ratdolt's
former position as leader of the market in mathematical and astronomical books in
Venice, a decade after his edition of Sacrobosco. The privilege was supposed to last
for ten years, but his last known edition is a Regiomontanus' *Ephemerid* of 1498
(Regiomontanus 1498). He seems to have abandoned his project, maybe due to the
competition and the growing influence big publishers such as the Scotto, the Giunta,
or the Sessa had in Venice.

---

[25] The hypothesis that Santritter owned the woodblocks and rented them to Scoto is more probable
than the notion that Santritter's printer, Girolamo de Sanctis, owned the blocks and rented or sold
them to Scotto. De Sanctis' production indicates that he was a typographer without a lot of capital
and that he most likely lent his technical printing knowledge to publishers such as Santritter or
others.

[26] For instance, in the petitions asking for privileges, Gaspar Dinslach wrote: "In order to avoid,
after having printed said works at great cost and labor, that some other competitor might reprint
them and sell them at a lower price, as it often happens, which would be a ruin and a damage for the
petitioner" (Non volendo, che dapoi che cum grandissima spesa et faticha l'havera facto stampar le
dicte opere che qualche altro a concorrentia le fesse restampir et poi le vendesse a vil pretio come
molto vole achade, che tornaria a ruina et damno de lui supplicante) (ASV, Collegio Notatorio, reg.
14, image 312, Avril 18, 1497).

[27] Santritter's privilege reads: "The works to be printed are the following: *Astrolabium Instrumentum* with canons, *Ephemeridem perpetuum, Scotum Super animam, Jordanum in Geometrica*,
and some other works of astronomy and geometry not yet printed by others in the city of Venice"
(Opera autem imprimenda sunt ista videlicet Astrolabium Instrumentum ipsum cum canonibus
suis, Ephemeridem perpetuum, Scotum super animam, Jordanum in Geometrica, autem nonnulla
alia opera et astronomica et geometrica ab aliis non impressa in hac civitate venetiarum). ASV,
Collegio Notatorio, reg. 14, image 368_194r, November 13, 1498.

3.3 Model Replaced: Sacrobosco's Posterity in Venice and in Europe

A parallel tendency in Venetian publishing appeared with the publication of Sacrobosco's *Tractatus de sphaera* by Simone Bevilaqua (1450–1518) in 1499 (Sacrobosco et al. 1499) (Fig. 4). In opposition to what was usually done until then, Bevilacqua chose to print the *Tractatus de sphaera* in an in-folio format, around 33 cm high, whereas the previous editions, from Ratdolt to Guilelmus da Trino, were in-quartos around twenty to 22 cm high. Regiomontanus' and Peuerbach's texts are still present, but there are many other commentaries as well, some of them surrounding the main text they refer to. Among these texts some are common commentaries of the *Sphaera* already present in the manuscript tradition (Thorndike 1949), like Cecco d'Ascoli (1257–1327), others were of fifteenth-century scholars such as Francesco Capuano di Manfredonia (d. ca. 1490) and Jacques Lefèvre d'Etaples (1455–1536). The layout is dense, with the commentaries surrounding the main text. Thus, the publishing model changes drastically, integrating Sacrobosco into the corpus of glossed texts. The integration of this text in the academic curriculum made it useful to add linear commentaries to use in courses. This kind of *mise en livre* is of significance to the status of the author: Sacrobosco is printed as an authority, whose words have to be commented upon and expanded by teachers for their students.

This also changes the general equilibrium between text and images. Here the illustrations are very small in comparison to the page and to the space dedicated to the texts, and the legibility suffers. Moreover, these are not dedicated woodcuts, made for this particular edition and format. Instead, the printers reused in part the same set of woodcuts from the in-quarto 1491 Guglielmo da Trino edition, as can be proven by various damage spots one can identify in both editions (Figs. 4 and 5). Simone Bevilacqua was probably aiming to reduce costs for this already heavy volume, while still presenting a decorated edition. However, while Santritter and Ratdolt made these designs according to their position on the page and the space allocated to them, the layout used in the 1499 edition tends to give more importance to the commentary than to the sketches as a tool for the reader in understanding the text.

The presentation of the *Sphaera* as a glossed text is not unprecedented in the manuscript tradition, even if it is not the most common format (Thorndike 1949). For instance, a manuscript of the Bodleian Library presents a gloss on Sacrobosco's *Tractatus de sphaera* surrounding the original text, though the commentary was probably added in the margins later on.[28] Similar to what Isabelle Pantin already stated concerning Parisian editions (Pantin 2013, 24), this kind of arrangement mirrored scientific manuscripts: in-folio or big in-quarto, dense typography, few and small geometric sketches, and not very legible. This format, layout, and choice of texts and authors can be interpreted as a form of integration of Sacrobosco's *Tracatus de*

---

[28] Bodleian Library. MS. Canon. Misc. 161.

Fig. 4 The edition published in 1499 shows a glossed layout, presenting Sacrobosco's text surrounded by commentaries. The space left for illustrations is scarce. The accent is laid on the commentaries rather than on the diagrams (Sacrobosco et al. 1499, 8r). München, Bayerische Staatsbibliothek—2 Inc.c.a. 3386. https://nbn-resolving.org/urn:nbn:de:bvb:12-bsb000 54721-1 CCBY-NC-SA4.0

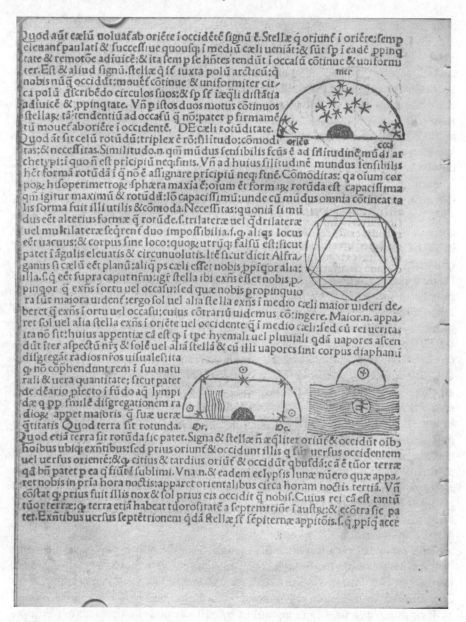

Fig. 5 This diagram, printed for the first time in Ratdolt's 1485 edition shows the effects of refraction and the false hypothesis of a flat sky (Pantin 2020). This 1491 edition copies the illustrations of Ratdolt and Santritter, but without using the exact same material (Sacrobosco et al. 1491, 7v). München, Bayerische Staatsbibliothek—4 Inc.c.a. 856. https://nbn-resolving.org/urn:nbn:de:bvb:12-bsb00083188-9 CCBY-NC-SA4.0

Sphaera into the more common publishing model for academic textbooks, in particular the *summae* and commentaries used in scholastic teaching frequently printed in Venice. Simone Bevilacqua's choice probably appealed to the actual practice of some readers, especially in university classes. However, Ratdolt and Santritter initiated a very different approach, which was more practical and encouraged a more linear and comprehensive reading of the treatise. It was also perhaps a kind of *mise en livre* adapted to more personal reading and to practical scholarly use. The importance given to the illustrations emphasized the more practical aspects of astronomical reflection and inserted Sacrobosco's more theoretical propositions in space and materiality, not only in logical discourse.

Some printers and publishers followed and adapted Bevilacqua's in-folio model: Giovanni and Bernardino Rosso (fl. 1482–1519) for Giuntino Giunti (1477–1521) in 1508 (Sacrobosco et al. 1508), who published works of Pierre d'Ailly (1351–1420) and Robert Grossetete (1175–1253) in the same edition; Lucantonio Giunta and the heirs of Ottaviano Scotto in 1518 (Sacrobosco et al. 1518b, 1518a), who both added other commentaries and treatises, this time from authors from antiquity: Pseudo-Ptolemy and Theodosius de Bithynia (160–100 BCE). Both 1518 editions chose to print Sacrobosco's *Sphaera* in a two-column format and in a gothic font (Chap. 8). Other editions of the same time held to the model Ratdolt had initiated (in-quartos, no gloss, and the association with Peuerbach's and Regiomontanus' texts): for example, the 1501 edition by Giovanni Battista Sessa (fl. 1489–1505) (Sacrobosco et al. 1501), the 1513 edition by Melchior Sessa (fl. 1505–1565) (Sacrobosco et al. 1513), and the 1519 edition by Giacomo Pincio (fl. ca. 1486–1527) (Sacrobosco 1519). Both models were printed simultaneously in Venice, probably targeting different audiences. Indeed, both publishing choices could apply to different scholarly practices: in universities, schools, and the academic milieu, or for scholars and astronomers outside teaching institutions, who probably had a very different way of using this text than did a professor in Paris. These practices of reading and the uses of a text contribute to creating epistemic communities (Jacob 2007; Meyer and Molyneux-Hodgson 2011; Valleriani et al. 2019). There are no unique uses of a book, and evolutions are often not linear. These very different choices in bookmaking remind us to pay attention to their variety and their cohabitation (Martin and Vezin 1990; Martin 2000; Grafton 2011; Pantin 2008).

Since the end of the fifteenth century, several paths had been available for the publication of Sacrobosco's treatise. Most of them were possibilities already present in the manuscript tradition but adapted to new formats and techniques. Each printing center implemented its own solution, printers being influenced by the solutions offered by their immediate neighbors as well as by editions that circulated across Europe. Hopyl's 1494 edition, for example, is also an in-folio format with commentaries, but the layout is significantly different from the 1499 Venetian edition; the commentaries do not surround the main text; however, they are printed with types of different sizes.

The Parisian context became particularly important to understanding the formal solutions offered by printers. Richard Oosterhoff investigates the pedagogical solutions Parisian printers offered while printing with the university public in mind and in relationship with major scholars of the time (Chap. 2). Jacques Lefèvre d'Etaples introduced many innovations in Paris at the beginning of the sixteenth century (e.g., long lines, a dedicated space for illustrations, tools to facilitate the reading). He did not choose the traditional gloss layout but promoted a format that emphasized the specificity of these treatises: the importance of illustration as a tool for the reader. His solution enhanced the relationship of the texts to formalized diagrams and their pedagogical use. The production of mathematical books in Paris responded to an important demand, in part from the *calculatores* current studied by Alissar Levy (Chap. 13). At that point, formal innovations around the *Tractatus de sphaera* were closely related to the contemporary studies of mathematics in Paris and the work of contemporary mathematicians, especially Oronce Finé (1494–1555). The editions he completed from 1532 on went even further in the direction initiated by Lefèvre d'Etaples as he established a new but enduring design: a continuous presentation of mathematical propositions (*mise au point d'un exposé continu*) (Pantin 2013). After Venice, Paris became one of the leading and most innovative producers of Sacrobosco's *Tractatus de sphaera*.

4   Conclusion

Sacrobosco's *Tractatus de sphaera*'s first editions are a good case study for understanding the beginning of the European book market. Its printing adaptations allow us to observe the expectations of printers regarding readership, as well as the progressive evolution of the expectations of the readers themselves. These early editions also enable us to comprehend the publishing history of one specific work in the economic and social milieu of the European book market at the time.

Printers had to anticipate their production and make choices in a very unstable economic environment. The competition between Franz Renner and Erhard Ratdolt in Venice—and in the wider market for academic books—is paradigmatic of this situation. They both had strong ties with German cities and had a robust commercial network across Italy and Europe. Renner chose a traditional approach from an intellectual point of view, printing Sacrobosco with the *Theoricae planetarum*, while implementing some formal adaptations. This enabled him to brand his edition as new and innovative in comparison to the previous incunabula editions, yet still identify with manuscript models known to his public. He did so while directly attacking Ratdolt on his specialty: mathematical and astronomical publications. Traces of this fierce competition can be found in the materiality of their editions.

Ratdolt replied by emphasizing Regiomontanus' formal innovations and by printing Sacrobosco's treatise with different texts: Peuerbach's *Theoricae novae planetarum* and Regiomontanus' criticism of the old *Theorica planetarum*. Ratdolt's 1482 and 1485 editions, despite their differences, had a major impact on multiple levels. First, they allowed Ratdolt to symbolically assert his superiority as an academic publisher in this specific market, to bring discredit on Renner's edition, to consolidate his own position in Venice, and to gain wider distribution in Europe. The publication of Regiomontanus and Peuerbach was also a good selling point in the German book market. Moreover, Ratdolt's editions allowed Regiomontanus and Peuerbach to benefit from the association with Sacrobosco, which enhanced their perceived legitimacy. In return, it also folded Sacrobosco's thirteenth-century treatise into contemporary academic debates and actualized its significance in the European book market. Finally, Ratdolt's edition set a formal standard, completed with Santritter's 1488 edition and copied in Venice and across Europe.

Following editions of Sacrobosco's texts in Venice, highlight the fiercely competitive market printers had to face. The production of Sacrobosco's text now had to be illustrated, which was more expensive than the plain editions of the early years. Printers, therefore, managed to rent one another's woodcuts—a channel we can often track, though some actors did not hesitate to copy the illustrations of previous editions. The Venetian system of privileges was small protection against a culture of imitation that existed not only within Venetian borders but also at a European level. However, other models were developed and coexisted in Venice that targeted different audiences. Venice progressively lost its leadership; eventually, it was no longer the main source of formal and intellectual innovation in the publication of the *Sphaera*. Paris soon became the most influential center of production, thanks to the collaboration of mathematicians and local presses.

The dynamics between book producers and between book producers and readers are at the heart of the transformations of books and book markets. Sacrobosco's *Tractatus de sphaera* is but one paradigmatic example. Thanks to the multiple aspects of its publication—cost, distribution, text-image relationship, formalization, target markets, partnership with investors and scholars—it highlights some of the main issues in the history of the beginning of printing.

Abbreviations

Digital Repositories

GW	Gesamtkatalog der Wiegendrucke. Stiftung Preußischer Kulturbesitz. https://www.gesamtkatalogderwiegendruck e.de. Accessed 07 June 2021
ISTC	Incunabula Short Title Catalog. Consortium of European Research Libraries. https://data.cerl.org/istc. Accessed 07 June 2021
MEI	Material Evidence in Incunabula. Consortium of European Research Libraries. https://data.cerl.org/mei. Accessed 07 June 2021
Sphaera Corpus*Tracer*	Max Planck Institute for the History of Science. https://db.sphaera.mpiwg-berlin.mpg.de/. Accessed 07 June 2021

TW   Typenrepertorium der Wiegendrucke. Staatsbibliothek zu Berlin. https://tw.staatsbibliothek-berlin.de/. Accessed 07 June 2021

Archives and Special Collections

ASV   Archivio di Stato di Venezia

References

Primary Literature

Alchabitius. 1482. *Libellus isagogicus*. Venice: Erhard Ratdolt. ISTC ia00362000.

Alchabitius. 1485. *Libellus isagogicus*. Venice: Erhard Ratdolt. ISTC ia00363000.

Alfonso, King of Castile. 1483. *Tabulae astronomicae*. Venice: Erhard Ratdolt. ISTC ia00534000.

Algorismus; De sphaera; Compotus. Etc. Oxford, Bodleian Library Ms. Canon. Misc. 161.

Borgo, Pietro. 1484. *Aritmetica mercantile*. Venice: 1484. ISTC ib01034000.

Breviarium Strigoniense. 1480. Venice: Erhard Ratdolt for Johann Cassis. ISTC ib01183050.

Buridan, John of. 1488. *Quaestiones in libros Physicorum Aristotelis*. Venice: Girolamo de Sanctis and Johann Lucilius Santritter for Peter Benzon and Pietro de Plasiis. ISTC ij00355000.

Caracciolo, Roberto. 1473. *Sermones quadragesimales*. Venice: Vindelinus of Spira. ISTC ic00172000.

Celtis, Conrad. 2000. *Norimberga*. Ed. Gerhard Fink. Nuremberg: Verlag Nürnberger Presse.

Cicero, Marcus Tullius. 1469. *Epistolae ad familiares*. Venice: Johann of Spira. ISTC ic00504000.

Cicero, Marcus Tullius. 1470. *Epistolae ad familiares*. Venice: Vindelinus of Spira. ITSC ic00506000.

Cremonensis, Gerardus. 1478. *Theorica planetarum*. Venice: Adam of Rottweil. ISTC ig00163000.

Diehl, Robert. 1933. *Erhard Ratdolt: ein Meisterdrucker des XV. und XVI. Jahrhunderts*. Vienna: H. Reichner.

Eschuid, Johann. 1489. *Summa astrologiae judicialis*. Venice: Johann Lucilius Santritter, for Francesco Bolano. ISTC ie00190000.

Euclid. 1482. *Elementa geometriae*. Venice: Erhard Ratdolt. ISTC ie00113000.

Florentinus, Antoninus. 1477a. *Summa theologica (Pars II)*. Venice: Johann of Cologne and Johann Manthen. ISTC ia00868000.

Florentinus, Antoninus. 1477b. *Summa theological (Paris III)*. Venice: Nicholas Jenson. ISTC ia00872000.

Hyginus, Gaius Julius. 1482. *Poetica astronomica*. Venice: Erhard Ratdolt. ISTC ih00560000.

Lindau, Mark of. 1483. *Buch der Zehn Gebote*. Venice: Erhard Ratdolt. ISTC im00261000.

Mela, Pomponius. 1478a. *Cosmographia, sive De situ orbis*. Venice: Franz Renner. ISTC im00450000.

Mela, Pomponius. 1478b. *Cosmographia, sive De situ orbis*. Venice: Bernard Maler, Erhard Ratdolt and Peter Löslein. ISTC im00449000.

Middelburg, Paul of. 1481–1482. *Prognosicon anni 1482*. Venice: Erhard Ratdolt. ISTC ip00185000.

Periegetes, Dionysius. 1477. *De situ orbis.* Venice: Bernhard Maler, Erhard Ratdolt and Peter Löslein. ISTC id00253000.

Periegetes, Dionysius. 1478. *De situ orbis.* Venice: Franz Renner. ISTC id00254000.

Peuerbach, Georg. 1474. *Theoricae novae planetarum.* Nuremberg: Regiomontanus. ISTC ip01134000.

Peuerbach, Georg. 1495. *Theoricae novae planetarum.* Venice: Simon Bevilaqua. ISTC ip01135000.

Publicius, Jacobus. 1482. *Artes orandi, epistolandi, memoranda.* Venice: Erhard Ratdolt. ISTC ip01096000.

Publicius, Jacobus. 1485. *Artes orandi, epistolandi, memoranda.* Venice: Erhard Ratdolt. ISTC ip01097000.

Tedeschi, Niccolò. 1472. *Super libros Decretalium, I-III.* Venice: Vindelinus of Spira. ISTC ip00058000.

Regiomontanus, Johannes. 1488. *Almanach ad annos XVIII calculatum.* Augsburg: Erhard Ratdolt. ISTC ir00108000.

Regiomontanus, Johannes. 1498. *Ephemerides, sive Almanach perpetuum.* Venice. Peter Liechtenstein for Johann Lucilius Santritter. ISTC ir00110000.

Rolewinck, Werner. 1480. *Fasciculus temporum.* Venice: Erhard Ratdolt. ISTC ir00261000.

Rolewinck, Werner. 1481. *Fasciculus temporum.* Venice: Erhard Ratdolt. ISTC ir00264000.

Rolewinck, Werner. 1484. *Fasciculus temporum.* Venice: Erhard Ratdolt. ISTC ir00270000.

Rolewinck, Werner. 1485. *Fasciculus temporum.* Venice: Erhard Ratdolt. ISTC ir00271000.

Sacrobosco, Johannes de. 1472a. *Ioannis de Sacrobosco anglici v[iri] c[larissimi] spaera mondi feliciter incipit.* Ferrara: Andreas Belfortis. https://hdl.handle.net/21.11103/sphaera.101122.

Sacrobosco, Johannes de. 1472b. *Tractatum de spera.* Venice: Florentinus de Argentina. https://hdl.handle.net/21.11103/sphaera.100685.

Sacrobosco, Johannes de. 1475–1477. *Tractatum de spaera.* Venice: Filippo di Pietro https://hdl.handle.net/21.11103/sphaera.100262.

Sacrobosco, Johannes de. 1478a. *Sphera magistri Io. de Sacroboscho.* Milan: Filippo da Lavagna. https://hdl.handle.net/21.11103/sphaera.100263.

Sacrobosco, Johannes de. 1478b. *Ioannis de sacrobosco anglici viri clarissimi spera mundi feliciter incipit.* Venice: Franz Renner. https://hdl.handle.net/21.11103/sphaera.100686.

Sacrobosco, Johannes de. 1478c. *Ioannis de sacrobosco anglici viri clarissimi spera mundi feliciter incipit.* Venice: Adam of Rottweil. https://hdl.handle.net/21.11103/sphaera.100687.

Sacrobosco, Johannes de. 1480. *Ioannis de sacrobusto anglici viri clarissimi spera mundi.* Bologna: Dominico Fusco. https://hdl.handle.net/21.11103/sphaera.100264.

Sacrobosco, Johannes de and Gasparino Borro. 1494. *Commentum super tractatum sphaerae mundi.* Venice: Boneto Locatello or Bartholomaeus de Zanis for Ottaviano Scotto. https://hdl.handle.net/21.11103/sphaera.100272.

Sacrobosco, Johannes de, Pierre d'Ailly, Robert Grosseteste, Johannes Regiomontanus, Jacques Lefèvre d'Étaples, Francesco Capuano di Manfredonia, and Bartolomeo Vespucci. 1508. *Nota eorum quæ in hoc libro continentur.* Venice: Giovanni Rosso and Bernardino Vercellense for Giuntino Giunta. https://hdl.handle.net/21.11103/sphaera.100915.

Sacrobosco, Johannes de, Cecco d'Ascoli, Jacques Lefèvre d'Étaples, and Francesco Capuano di Manfredonia. 1499. *Sphera Mundi.* Venice: Simone Bevilacqua. https://hdl.handle.net/21.11103/sphaera.100273.

Sacrobosco, Johannes de and Georgius de Ferraris. 1500. *Figura sphere: cum glosis Georgii de Monteferrato.* Venice: Giovanni Battista Sessa or Jacopo Pincio for Georgio de Monteferrato. https://hdl.handle.net/21.11103/sphaera.100275.

Sacrobosco, Johannes de, Johannes Regiomontanus, and Georg von Peuerbach. 1482. *Novicijs adolescentibus.* Venice: Erhard Ratdolt. https://hdl.handle.net/21.11103/sphaera.100692.

Sacrobosco, Johannes de, Johannes Regiomontanus, and Georg von Peuerbach. 1485. *Novicijs adolescentibus.* Venice: Erhard Ratdolt. https://hdl.handle.net/21.11103/sphaera.101123.

Sacrobosco, Johannes de, Johannes Regiomontanus, and Georg von Peuerbach. 1488. *Spaera Mundi Compendium foeliciter inchoat*. Venice: Girolamo de Sanctis for Johann Lucilius Santritter. https://hdl.handle.net/21.11103/sphaera.100822.

Sacrobosco, Johannes de, Johannes Regiomontanus, and Georg von Peuerbach. 1490. *Sphaera mundi compendium foeliciter inchoat*. Venice: Boneto Locatello for Ottaviano Scotto I. https://hdl.handle.net/21.11103/sphaera.100885.

Sacrobosco, Johannes de, Johannes Regiomontanus, and Georg von Peuerbach. 1491. *Sphaera mundi*. Venice: Guglimo da Trino. https://hdl.handle.net/21.11103/sphaera.100824.

Sacrobosco, Johannes de, Johannes Regiomontanus, and Georg von Peuerbach. 1501. *Sphaera mundi*. Venice: Giovanni Battista Sessa I. https://hdl.handle.net/21.11103/sphaera.100892.

Sacrobosco, Johannes de, Johannes Regiomontanus, and Georg von Peuerbach. 1513. *Sphaera Mundi*. Venice: Melchior Sessa I. https://hdl.handle.net/21.11103/sphaera.101023.

Sacrobosco, Johannes de, Johannes Regiomontanus, and Georg von Peuerbach. 1519. *Sphaera Mundi*. Venice: Giacomo Pinctio. https://hdl.handle.net/21.11103/sphaera.100992.

Sacrobosco, Johannes de, Ptolemy (ps.), Campano da Novara, Pierre d'Ailly, Cecco d'Ascoli, Robert Grosseteste, Theodosius of Bithynia, Michael Scot, Johannes Regiomontanus, Jacques Lefèvre d'Étaples, and Francesco Capuano di Manfredonia. 1518a. *Sphera cum commentis in hoc volumine contentis*. Venice: Heirs of Ottaviano Scotto I. https://hdl.handle.net/21.11103/sphaera.101057.

Sacrobosco, Johannes de, Ptolemy (ps.), Campano da Novara, Pierre d'Ailly, Cecco d'Ascoli, Gerad of Cremona, Robert Grosseteste, Theodosius of Bithynia, Michael Scot, Johannes Regiomontanus, Jacques Lefèvre d'Étaples, and Francesco Capuano di Manfredonia. 1518b. *Sphera mundi*. Venice: Lucantonio Giunta. https://hdl.handle.net/21.11103/sphaera.100047.

Secondary Literature

Amelung, Peter. 1964. *Das Bild des Deutschen in der Literatur der italienischen Renaissance (1400–1559)*. Munich: Max Hueber.

Baldasso, Renzo. 2013. Printing for the Doge: On the first quire of the first edition of the *Liber elementorum Euclidis*. *La Bibliofilía* 115 (3): 525–252.

Bertalot, Ludwig. 1975. *Studien zum Italienischen und Deutschen Humanismus*. 2 vol. Rome: Edizioni di Storia e Letteratura.

Bonicoli, Louis-Gabriel. 2015. La production du libraire-éditeur Antoine Vérard (1485–1512). Nature, fonctions et circulation des images dans les premiers livres illustrés. PH.D. Thesis: Université Paris Ouest–Nanterre La Défense.

Böninger, Lorenz. 2020. Da Vespasiano da Bisticci a Franz Renner e Bartolomeo Lupoto. Appunti sul commercio librario tra Venezia, la Toscana e Genova (ca. 1459–1487). In *Printing R-Evolution and Society 1450–1500. Fifty Years that Changed Europe*, ed. Cristina Dondi, 623–648. Venice: Edizioni Ca'Foscari.

Braunstein, Philippe. 2016. *Les Allemand à Venise (1380–1520)*. Rome: École française de Rome.

Buchholzer-Rémy, Laurence. 2006. *Une ville en ses réseaux. Nuremberg à la fin du Moyen Âge*. Saint-Étienne: Belin.

Carter, Victor, Lotte Hellinga, Tony Parker, and Jane Mullane. 1983. Printing with gold in the fifteenth century. *The British Library Journal* 9 (1): 1–13.

Chartier, Roger. 1997. Du livre au lire. *Réseaux. Communication—Technologie—Société* 1 (1): 271–290.

Dondi, Cristina. 2013. "15c Booktrade." An evidence-based assessment and visualization of the distribution, sale and reception of printed books in the Renaissance. *La Gazette du livre médiéval* 60: 83–101.

Dondi, Cristina, and Neil Harris. 2013. Best selling titles and Books of hours in a Venetian bookshop of the 1480's: The Zornale of Francesco de Madiis. *La Bibliofilia* 115 (1): 63–82.

Dondi, Cristina, and Neil Harris. 2014. Exporting books from Milan to Venice in the fifteenth century: Evidence from the Zornale of Francesco de' Madiis. *La Bibliofilia* 116: 121–148.

Dörner, Gerald. 1999. *Reuchlin und Italien.* Stuttgart: Thorbecke.

Furno, Martine, and Raphëlle Mouren, ed. 2012. *Auteur, traducteur collaborateur, imprimeur… qui écrit?* Paris: Classiques Garnier.

Gerulaitis, Leonardas V. 1970. A fifteenth-century artistic director of a printing firm: Bernard Maler. *Papers of the Bibliographical Society of America* 64 (3): 324–332.

Gier, Helmut. 2010. Italienrezeption im Augsburger Humanismus. In *Schwaben und Italien: zwei europäische Kulturladschaften zwischen Antike und Moderne,* 223–238. Augsburg: Wissner.

Gingerich, Owen. 1999. Sacrobosco illustrated. In *Between demonstration and imagination: Essays in the history of science and philosophy: Presented to John D. North,* eds. Lodi Nauta and Arjo J. Vanderjagt, 211–224. Leiden: Brill.

Grafton, Anthony. 2001. Le lecteur humaniste. In *Histoire de la lecture dans le monde occidental,* eds. Guglielmo Cavallo and Roger Chartier, 209–248. Paris: Seuil.

Grafton, Anthony. 2011. *Humanists with inky fingers: The culture of correction in Renaissance Europe.* The Annual Balzan Lecture 2. Florence: Olschki.

Hamel, Jürgen. 2006. Johannes de Sacroboscos Sphaera: Text und frühe Druckgeschichte eines astronomischen Bestsellers. *Gutenberg-Jahrbuch* 81: 113–136.

Horst, Thomas. 2019. The reception of Cosmography in Vienna: Georg von Peuerbach, Johannes Regiomontanus, and Sebastian Binderlius. Preprint. https://www.mpiwg-berlin.mpg.de/preprint/ reception-cosmography-vienna-georg-von-peuerbach-johannes-regiomontanus-and-sebastian.

Jacob, Christian, ed. 2007. *Lieux de savoir,* 2 vol. Paris: Albin Michel.

Kikuchi, Catherine. 2018a. Concurrence et collaboration dans le monde du livre vénitiens, 1469– début du XVIe siècle. *Annales. Histoire, Sciences Sociales* 73 (1): 185–212. https://doi.org/10. 1017/ahss.2018.114.

Kikuchi, Catherine. 2018b. L'imprimerie en réseau: la construction de l'édition comme marché économique et culturel (Venise, 1469–1500). *Temporalités. Revue de sciences sociales et humaines* 27. https://doi.org/10.4000/temporalites.4371.

Kikuchi, Catherine. 2018c. L'imprimeur allemand dans les premiers temps des presses européennes : modèle et contre-modèle. *Source(s)* 13: 9–22.

Lowry, Martin. 1979. *The World of Aldus Manutius: business and scholarship in Renaissance Venice.* Ithaca: Cornell University Press.

Lowry, Martin. 1981. The social world of Nicholas Jenson and John of Cologne. *La Bibliofilia* 83: 193–218.

Malpangotto, Michela. 2012. Les premiers manuscrits des *Theoricae novae planetarum* de Georg Peurbach: présentation, description, évolution d'un ouvrage. *Revue d'histoire des sciences* 65 (2): 339–80.

Martin, Henri-Jean. 2000. *La Naissance du livre moderne : mise en page et mise en texte du livre français, XIVe–XVIIe siècles.* Paris: Éditions du Cercle de la Librairie.

Martin, Henri-Jean, and Jean Vezin, ed. 1990. *Mise en page et mise en texte du livre manuscrit.* Evreux: Promodis.

Meyer, Morgan, and Susan Molyneux-Hodgson. 2011. "Communautés épistémiques:" une notion utile pour théoriser les collectifs en sciences? *Terrains travaux* 18 (1): 141–154.

Müller, Kathrin. 2008. *Visuelle Weltaneignung. Astronomische und kosmologische Diagramme in Handschriften des Mittelalters.* Göttingen: Vandenhoeck und Ruprecht.

North, John. 1994. *Medieval and early Renaissance Europe.* London: Fontana Press.

Nuovo, Angela. 1990. *Alessandro Paganino : 1509–1538.* Padoue: Antenore.

Nuovo, Angela. 2003. *Il Commercio librario nell'Italia del Rinascimento.* Milan: Franco Angeli.

Obrist, Barbara. 2004. *La cosmologie médiévale. I. Les fondements antiques.* Florence: Sismel.

Ornato, Ezio. 1997. Les conditions de production et de diffusion du livre médiéval (XIIIe–XVe siècles). Quelques considérations générales. In *La Face cachée du livre médiéval. L'histoire du livre vue par Ezio Ornato, ses amis, ses collègues*, 97–116. Rome: Viella.

Ornato, Ezio. 2017. L'Europe des sermons à la fin du XVe siècle: concurrence ou évitement? Lecture at the *Rencontres Renouard 2017: le livre à Paris au XVIe siècle*. Paris, March 31, 2017.

Ouvry-Vial, Brigitte, and Anne Réach-Ngô, ed. 2010. *L'Acte éditorial: publier à la Renaissance et aujourd'hui*. Paris: Classiques Garnier.

Pantin, Isabelle. 2008. Mise en page, mise en texte et construction du sens dans le livre moderne: où placer l'articulation entre l'histoire intellectuelle et celle de la disposition typographique? *Mélanges de l'école française de Rome* 120 (2): 343–361. https://doi.org/10.3406/mefr.2008. 10550.

Pantin, Isabelle. 2012. The first phases of the *Theoricae Planetarum* printed tradition (1474–1535): The evolution of a genre observed through its images. *Journal for the History of Astronomy* 43 (1): 3–26. https://doi.org/10.1177/002182861204300102.

Pantin, Isabelle. 2013. Oronce Finé mathématicien et homme du livre: la pratique éditoriale comme moteur d'évolution. In *Mise en forme des savoirs à la Renaissance. À la croisée des idées, des techniques et des publics*, 19–40. Paris: Armand Colin.

Pantin, Isabelle. 2020. Borrowers and innovators in the history of printing Sacrobosco: the case of the in-octavo tradition. In *De Sphaera of Johannes de Sacrobosco in the Early Modern Period: the authors of the commentaries*, ed. Matteo Valleriani, 265–312. Cham: Springer International Publishing. https://doi.org/10.1007/978-3-030-30833-9_9.

Pedersen, Olaf. 1975. The *corpus astronomicum* and the traditions of mediaeval latin astronomy. *Studia Copenicana* III: 58–96.

Pedersen, Olaf. 1981. The origins of the *Theorica Planetarum*. *Journal for the History of Astronomy* 12: 113–123.

Pettas, William A. 1973. The cost of printing a Florentine incunable. *La Bibliofilia* 75: 67–86.

Réach-Ngô, Anne, éd. 2014. *Créations d'atelier: l'éditeur et la fabrique de l'oeuvre à la Renaissance*. Paris: Classiques Garnier.

Redgrave, Gibert Richard. 1894. *Erhard Ratdolt and his work at Venice*. Berlin: Chiswick Press.

Ridolfi, Roberto. 1967. Francesco della Fontana, stampatore e libraio a Venezia. In *Studi bibliografici*, 53–66. Florence: Olschki.

Ruffini, Graziano. 2012. *La Chasse aux livres : bibliografia e collezionismo nel viaggio in Italia di Étienne-Charles de Loménie de Brienne e François-Xavier Laire (1789–1790)*. Florence: Firenze University Press.

Shank, Michael H. 2012. The Geometrical Diagrams in Regiomontanus's Edition of His Own *Disputationes* (c. 1475): Background, Production, and Diffusion. *Journal for the History of Astronomy* 43 (1): 27–55. https://doi.org/10.1177/002182861204300103.

Thorndike, Lynn. 1949. *The Sphere of Sacrobosco and its commentators*. Chicago: Chicago University Press.

Vaisse, Pierre. 1995. *Albrecht Dürer*. Paris: Fayard.

Valleriani, Matteo. 2017. The tracts on the Sphere. Knowledge restructured over a network. In *The structures of practical knowledge*, ed. Matteo Valleriani, 421–473. Dordrecht: Springer.

Valleriani, Matteo. 2020. Prolegomena to the Study of Early Modern Commentators on Johannes de Sacrobosco's *Tractatus de Sphaera*. In *De Sphaera of Johannes de Sacrobosco in the early modern period: The authors of the commentaries*, ed. Matteo Valleriani, 1–23. Cham: Springer International Publishing. https://doi.org/10.1007/978-3-030-30833-9_1.

Valleriani, Matteo, Florian Kräutli, Maryam Zamani, Alejandro Tejedor, Christoph Sander, Malte Vogl, Sabine Bertram, Gesa Funke, and Holger Kantz. 2019. The Emergence of Epistemic Communities in the Sphaera Corpus. *Journal of Historical Network Research* 3 (November): 50–91. https://doi.org/10.25517/jhnr.v3i1.63.

White, Harrison C. 2011. *Identité et contrôle. Une théorie de l'émergence des formations sociales*. Paris: Éditions de l'EHESS.

Zinner, Ernst. 1990. *Regiomontanus: his life and work*. Amsterdam: North-Holland.
Zorzi, Marino. 1986. Stampatori tedeschi a Venezia. In *Venezia e la Germania: arte, politica, commercio, due civiltà a confronto*, 115–133. Milan: Electa.

Catherine Rideau-Kikuchi is associate professor at Paris-Saclay—Université de Versailles Saint-Quentin since 2018. She completed her PhD at Sorbonne University on the economic and social history of early Venetian printing between 1469 and 1530. A publication followed this study (*La Venise des livres*. 2018). She is now interested in the construction of a book market in Northern Italy and the circulation of book producers in the region. Her research also aims to integrate the study of material aspects of the book production, and the social and economic realities of this commerce and industry, as can be seen through archivistical sources. She was also involved in several popularization and pedagogical projects related to history and medieval history in particular.

Chapter 4
Printers, Booksellers, and Bookbinders in Wittenberg in the Sixteenth Century: Real Estate, Vicinity, Political, and Cultural Activities

Insa Christiane Hennen

Abstract In the sixteenth century, Wittenberg developed into one of the most important centers of printing in Germany. The works of Martin Luther and Philipp Melanchthon became exceptional bestsellers. Sacrosbosco's *Sphaera* appeared between 1531 and 1600 in more than forty editions. Economically successful printers such as Georg Rhau, Hans Lufft, and Johann Krafft, as well as the booksellers and publishers Moritz Goltz, Christoph Schramm, Conrad Ruehel, Bartholomäus Vogel, and Samuel Selfisch—all of whom were involved in the Wittenberg editions of the *Sphaera*—influenced local affairs as members of the town council and confidants of the prince electors. Analyzing tax lists, contracts, and historic town maps makes it possible to identify the houses they owned and to reconstruct their locations. Mapping the real estate of printers, booksellers, and bookbinders who lived in Wittenberg in the sixteenth century shows this group's economic success quite plainly and points to professional and private relationships. Some of the houses still exist. They are three-dimensional documents of the taste and home decor of a new elite.

Keywords Wittenberg · Archeology · Early modern topography · Protestant Reformation · Book history · Art history

1 Wittenberg: An Intellectual Center in the Sixteenth Century

Numerous scholars have shown how Wittenberg became one of the most important printing places in Germany beginning around 1520 (Reske et al. 2015; Pettegree 2016; Oehmig 2015; Rothe 2013). On the one hand, the University of Wittenberg, founded by Frederic III the Wise (1463–1525) in 1502, launched this development. On the other hand, the Lutheran Reformation, which started in 1517, caused an

I. C. Hennen (✉)
LEUCOREA, Wittenberg, Germany
e-mail: hennen.hennen@t-online.de

© The Author(s) 2022
M. Valleriani and A. Ottone (eds.), *Publishing Sacrobosco's* De sphaera *in Early Modern Europe*, https://doi.org/10.1007/978-3-030-86600-6_4

increasing demand for prints. The Reformation had a significant unintentional influence over the development of the town and the university, and the growth of the printing industry as well.

Prince Elector Frederic III, son of Ernest of Wettin (1441–1486), had established the new university—the *Leucorea* (Greek "white mountain" for Wittenberg)—because the existing Saxon academy at Leipzig had become part of the Albertine duchy when Saxony was divided into two territories in 1485. Frederic, heir of the Ernestine territory between 1486 and 1525, made Wittenberg the cultural and spiritual center of his electorate. He raised a splendid new castle, called famous artists and craftsmen to Wittenberg, systematically modernized the structure and infrastructure of the town, and built new colleges tailored to the needs of the new university. In 1504 a new town constitution—which had been worked out by scholars of the university and skilled craftsmen involved in the prince elector's large building projects—was enacted by the town council and confirmed by the prince elector. This corpus of legislation regulated nearly every part of life, from measures and weights, rights to use land and meadows, and brewing beer to stockbreeding and inheritance law. It contained a series of regulations concerning the execution of construction work, like the duty to use tiles instead of straw for the roofs of the new dwellings and workshops. The *Statuta* had been an important instrument in modernizing the town in terms of both the living environment of its inhabitants and the urban society itself (Hennen 2015, 2017a, b, 2017d, 2020b, 2020d). Expecting an increase in population, it had been necessary to optimize hygiene. A new graveyard was opened outside the walls, and the use of the runnels in town was regulated.

Frederic and his court tarried in town every now and then during religious festivals. His celebrated collection of relics, which was shown regularly at the castle church until 1521, made Wittenberg a relevant place of pilgrimage.

The well-educated prince elector, who had traveled to the Netherlands, Italy, and the Holy Land, had a clear idea of a modern intellectual and spiritual center, and he had the right touch in bringing together capable people and creating an innovative climate. He managed to install a new elite. Experts from outside, many of them from Franconia, some from Italy—like the painter Jacopo de' Barbari (1450–1516)—shaped the singular buildings that were erected by order of the prince elector and influenced the restructuring of the social life in town as well.

During the early years of the reign of Frederic III, the sculptor Claus Heffner (named at Wittenberg for the first time in 1491/1492, d. 1539), who became the first master builder appointed by the council (*Ratsbaumeister*); the painter, "designer," and entrepreneur Lucas Cranach the Elder (1472–1553); and a number of talented administrators settled in the town at the river Elbe. They worked closely together to model a kind of ideal city. The comparatively small size of the town, which around the year 1500 housed less than 2,000 inhabitants, may have been one reason for choosing Wittenberg for the project.[1] The locals did not resist the reshaping of their

---

[1] Other reasons had been that Wittenberg was the capital of the so-called *Kurkreis*—the territory to which the dignity of the prince elector was bound—that also the Ascanians had buried their ancestors there, and that the capitulars of the All Saints collegiate could take over functions at the

town. Many of them profited from the new achievements, though others lost their influence and social status.

Frederic's propaganda rested upon a sophisticated concept of visual communication. It had been easy to make a great visual impact on the manageable framework of the city. Because of the small size and simple structure of the town, all transformations became immediately visible. The ground plan of the town, characterized by two long roads parallel to the river Elbe and a few crossroads, was regulated and harmonized. Several measures were taken to optimize hygienic conditions (Hennen 2015, 2020a). New roads were traced out, and the central marketplace was enlarged to become a stage for big events like tournaments, homages to the prince elector when he visited, pageants, or processions. The western side of the *Schmergasse*, a little road parallel to the eastern flank of the marketplace, was broken down. The last owner of a house to be razed received compensation in 1570 (Hennen 2015, 336–340). A new town hall was built by the town council. The appearance of the town was radically reshaped: the monumental and splendid castle in the west of the town prospect received an optical and semantic counterbalance with the large university colleges in the east (Fig. 1). Inhabitants who wanted to build new, possibly ambitious houses (in the town records they are termed *stattliche Gebäude*) enjoyed tax privileges. The prince elector evoked a climate of dynamic innovation in the town, and in all respects, Cranach, Heffner, and their colleagues granted a high aesthetic standard.

Fig. 1 Cranach workshop, View of Wittenberg, woodcut, 1558. Stiftung Luthergedenkstätten in Sachsen-Anhalt, Lutherhaus Wittenberg

---

new academy. Besides this, most of the towns in the electorate were of a similar size and did not have more than 2,000 inhabitants. See *Wittenberg-Forschungen 1–5*, especially (Lang and Neugebauer 2017).

Within two generations the town had changed completely. The number of inhabitants increased from less than 2,000 in 1500 to 4,000 or more in 1550. Fifty percent of the inhabitants were students, many international, who stayed in town only for some months or a few years. Many of them lived in private houses inside and outside the walls, numerous at the colleges. Between 1500 and 1550 the number of houses inside the wall increased from about 390 to about 470 (Hennen 2011, 139, 2017b, 428). In 1644, 504 roofed parcels existed. For each house, an average of four or five residents can be assumed at a minimum, but many could accommodate more. From the 1530s, the development of single parcels was intensified, and side wings and adjoining buildings were erected to house students and employees. Extant examples of that building boom include the stately homes of Cranach himself, the so-called *Cranachhöfe* at the Wittenberg marketplace, and Samuel Selfisch's (1529–1615) house nearby (*Markt* 3; see below). In a register compiled around 1638, many houses with one or more side wings and more than ten heated living rooms (*Stube*) are mentioned (RatsA Wittenberg, *Urbarium* 7; Hennen 2013a, 33–54). Most of them had presumably existed around 1550–1560. The most splendid buildings were erected at the marketplace and in the long main street between the castle and the *Collegium Fridericianum*.

The university needed capable printers. After 1517 the works of Martin Luther (1483–1546) became bestsellers. Under the rule of the Catholic George, Duke of Saxony (1471–1539), Protestant literature was banned in Leipzig, whereas Wittenberg, only 70 km north, ascended as a center of printing. In 1518, Philipp Melanchthon (1497–1560) joined the University of Wittenberg and became its most influential scholar propagating the classic languages, history, and the natural sciences. His activities further stimulated Wittenberg book production.

It is possible to reconstruct many details of life in the emerging town by analyzing the annual records of the town council, by interpreting contracts, and by studying town maps from 1623 (Fig. 2) and 1742 (Fig. 3), which are still preserved in the town archive. The records of the *Common Chest*, the treasure of the parish installed in 1526, offer more information of this kind. Besides the archive documents, archaeological findings like types and the surviving houses built by persons who were involved in the occurrences during the sixteenth century allow us to draw conclusions concerning aesthetic conceptions, technical processes, and social networks (Hennen 2014). Additional information and denominations used especially in contracts make it possible to differentiate between printers and booksellers respective to publishers and bookbinders. In particular, some of the booksellers invested money in printing projects and acted as publishers.

This minute study is based upon results from the research project *Das ernestinische Wittenberg*, which ran at the Leucorea foundation from 2009 to 2018. This project had the goal of determining the situation in Wittenberg around 1500 and the changes that ensued during the following five decades leading to the Schmalkaldic War (1547), either because of the Lutheran Reformation or by other reasons. The book industry had been only a secondary aspect of this research, which dealt with the whole town, including all its inhabitants and social groups. The project was oriented toward written and printed documents, as well as buildings and artifacts. A lot of

Fig. 2 Andreas Goldmann, Map of Wittenberg, drawing, 1623. Städtische Sammlungen Wittenberg

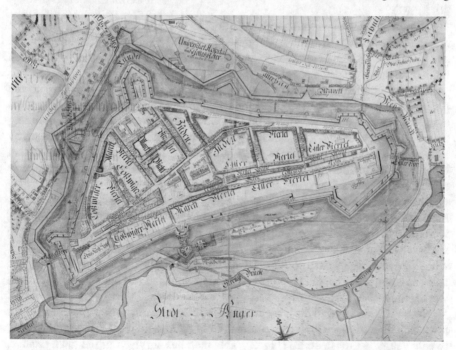

Fig. 3 Schmidt, Map of Wittenberg, drawing, 1742. Städtische Sammlungen Wittenberg

data concerning the real estate of the citizens was collected and analyzed. On the basis of these data, it is possible to identify the exact places where printers, booksellers, publishers, and bookbinders lived and presumably worked in Wittenberg. The present text, which was inspired by Saskia Limbach (Chap. 5), focuses on the printers of Sacrobosco's *Sphaera*. Revealing examples of vicinity ("neighborhood") are quoted. Some surviving houses, which were used for living, working, and trading, are described more in detail.

2 Printers, Booksellers, and Bookbinders in Wittenberg: An Overview on Their Real Estate

By means of secondary sources, Christoph Reske and Josef Benzing have listed forty-one printers who worked in Wittenberg in the period from 1500 to 1600 (Reske and Benzing 2015, 1075–1103). Vicky Rothe identified ninety-five printers, twenty-three booksellers or publishers, and eighty-five bookbinders who were active in Wittenberg during the sixteenth century (Rothe 2013, 81). A few more key figures of the book industry have been identified in the meantime by incidental finds in archive documents. Some printers and bookbinders could be located in the suburbs (Lang 2015, 122).

According to the town council's tax records, forty-one printers identified by Rothe and the author of this work, owned houses in town, as did thirty-nine bookbinders and sixteen publishers or booksellers. As owners of houses, the members of the book industry also had to pay an annual property tax (*Schoß*).

The tax records specify the names of the owners of houses respective to the parcels, and the individual amount anyone had to pay annually. In 1490, the annual amounts that had to be paid for each parcel were fixed. From that point on, building a new, larger house did not imply a new, higher tax. An obstacle to build was thereby removed. At the beginning of the sixteenth century, it was moreover decided that builders of new dwellings or/and workshops were relieved from paying the property tax for several years.

The exact positions of the houses—the addresses—are not mentioned in the tax lists. Designation systems consisting of the name of the road and a street number were established later at Wittenberg, but not until the nineteenth century. But the sequence of names and amounts in the lists mirrors the route the tax collectors took when moving from house to house, with the result that the sequences of names and/or amounts of successive years are similar. Separate tax books (*Schoßbücher*) were launched in 1556.

Not until 1644 were all the parcels numbered. These numbers are documented in a town map that had been drawn by Andreas Goldmann, a field surveyor from Torgau, twenty years before (in 1623) (Fig. 2). This map has survived in the town archive, as well as a second one from the year 1742 (Fig. 3), which contains a mark at the beginning and the end of the tour of the tax collectors (Fig. 4). Different colors mark the four quarters in which the town had been divided: the *Coswiger Viertel* in the west

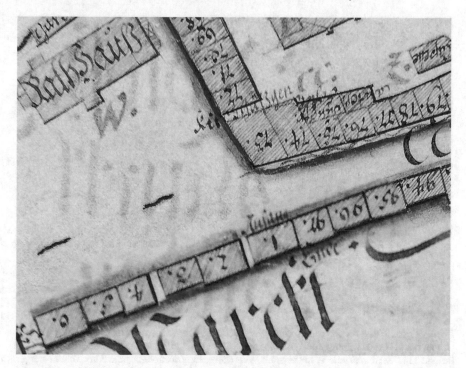

Fig. 4 Detail from the map from 1742 (Fig. 3) with the beginning and the end of the route of the tax collectors. Städtische Sammlungen Wittenberg

(Coswig is a small town situated 20 km west of Wittenberg), the central *Marktviertel* (market quarter), the *Elsterviertel* in the east (Elster, another small town, is located in the eastern direction), and the northern *Jüdenviertel* (Jewish quarter).[2] If a property was sold or the owner died, the name of the new owner was recorded next to the amount of tax due on the property (which remained the same). Considering these correlations and copying the lists from 1644 backward to 1490 offers the possibility of reconstructing the positions of single parcels in town, locating the names of the owners, and mapping the results of this work (Hennen 2011, 2014).[3] Besides, sale contracts often describe the estates, providing sufficient details to locate them in the context of the town.

Comparing the situation of 1520–1550 (Map 1) to that of 1550–1575 (Map 2), it is apparent that the different trades of the book industry could at first be found in all quarters aside from the area near the castle and the collegiate church. Since the fourteenth century, the capitulars had lived there, and since around 1490, prominent craftsmen who had been involved in the construction of the new castle had settled

---

[2] Until now the life of Jews in Wittenberg, especially in the Middle Ages, is scarcely investigated. For further reference, see (Hennen 2020c; Titze 2020).

[3] This work is not done. The project of a *Häuserbuch*, a register of all the owners of parcels in town between 1450 and 1650, is currently being prepared at the Leucorea foundation.

Fig. 5 Types, found on the parcel *Marktviertel* 32, the former location of Georg Rhau's house. Landesamt für Denkmalpflege und Archäologie Sachsen-Anhalt. Schrifttypen und Blindstücke aus Befund 1141 der Fundstelle 2, Lutherstadt Wittenberg, Franziskanerkloster; Foto: Daniel Berger. Landesamt für Denkmalpflege und Archäologie Sachsen-Anhalt, Daniel Berger

there too. After the Reformation, scholars of the university had the tendency to relocate to this area.[4]

Around 1550–1575, most of the printers were sitting in the northern district, the *Jüdenviertel*, while most of the bookbinders were near the university in the *Elster-viertel*. This separation is even more obvious at the end of the sixteenth century (Map 3). It is also visible that the houses of the successful publishers and booksellers were situated near the marketplace and the parish church (Schramm, Cranach, Vogel, Selfisch) or near the university colleges (Goltz, Ruehel).

It seems that the various craftspeople, especially the printers and bookbinders, sought to live close together, to cluster with colleagues. Perhaps the printers cooperated now and then to execute larger orders. Unfortunately, no distinct proof for such a collaboration can be found in archival documents.

Presumably, the bookbinders primarily searched for customers in the academic setting: scholars and students of the university and the university itself. It should be noted, however, that most of the parcels in the two western quarters of the town were already occupied by around 1520, while in the *Jüdenviertel* and the *Elsterviertel* areas still were available for construction. The *Neustraße*, which connects the *Mittelstraße*

---

[4] The capitulars, like other priests who remained in the "old faith," were allowed to use their houses for the rest of their lives. Many of them nevertheless left the protestant town and territory in time, see (Hennen 2013a, Vol. 2.1, 40).

Fig. 6 Vault in the house of Bartholomäus Vogel, *Mittelstraße* 5. Photo: I. C. Hennen

and the *Jüdenstraße*, for example, was only partly defined in the 1520s; some parcels on its eastern side were empty until that time. Here the theologian and pastor Johannes Bugenhagen (1485–1558), the mathematician Erasmus Reinhold (1511–1553), and the printer Veit Kreuzer (d. 1578) took residence. Kreuzer held a second parcel in the *Jüdenstraße* (see below). The (later) wealthy publisher and bookseller Christoph Schramm the Elder (d. 1549) first owned a house between the church yard (*Kirchhof*) and the *Collegienstraße* (*Marktviertel* 78), and later we see that he paid taxes for a house situated on the eastern side of the marketplace (*Marktviertel* 70).[5] Only one year later, in 1537, he moved to one of the most representative parcels at the south side of the marketplace (see below), where he became a neighbor of Lucas Cranach (*Marktviertel* 2), Christian Beyer (1482–1535) (*Marktviertel* 1)—who was the chancellor of the prince elector—and other notables.

Parts of the area surrounding the convent of the Gray Friars were privatized following the liquidation of the convent after 1525. Here, in the northern part of the town, for example, Georg Rhau (1488–1548) settled in 1541 (*Marktviertel* 32), as did Johannes Krafft (d. 1578) in the mid-1550 s (*Jüdenviertel* 25).

---

[5] The parcel (*Marktviertel* 70) originally belonged to the *Schmergasse*, whose western side was pulled down between 1521 and 1570, when the marketplace was enlarged. In 1535, Schramm knew that the house would become a house at the marketplace soon.

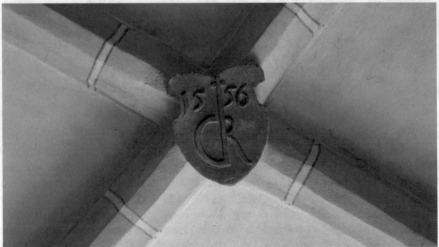

Fig. 7 Vault with decorations and keystone in the house of Conrad Ruehel, *Collegienstraße,* 62a. Photo: I. C. Hennen

2.1 Publishers, Printers, Booksellers, and Bookbinders

Contracts, which were executed on different occasions and which are preserved in the town archive, also allow us to deduce the professions of the signatories as well as professional and private relationships. Because of corresponding information in a contract concluded in 1533, we know that Christoph Schramm the Elder had been the seller of prints by Joseph Klug (ca. 1490–1552) (RatsA WB, 113 (Bc 101),

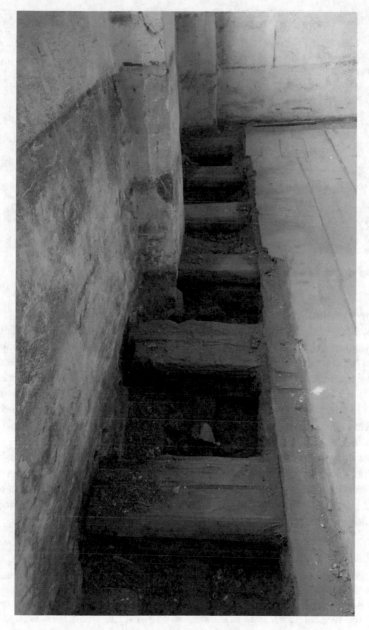

Fig. 8 Ceiling construction in the side wing of *Markt* 3. Photo: I. C. Hennen

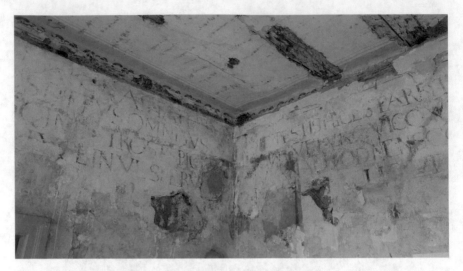

Fig. 9 Inscriptions in the former *studiolo* of Samuel Selfisch. Photo: I. C. Hennen

fol. 133r–v).[6] But Schramm also acted as publisher when he was financing Klug's projects. Klug and Schramm came to a compromise concerning outstanding debts and payments, which is documented as follows, "…to clear what he [Klug] sold, vouched for and what money he added…."[7] The contract was signed in the house of Conrad Rastger, in the presence of Klug's and Schramm's colleagues Moritz Goltz (ca. 1495–1548), Hans Lufft (1495–1584), and Bartholomäus Vogel (1504–1569).[8]

In 1538 and 1540, Joseph Klug and Moritz Goltz came to another mutual agreement; Klug there called himself *Buchdrucker* (printer), Goltz is called *Buchhändler* (bookseller) (RatsA WB, 113 (Bc 101), fol. 197r–198v).[9]

In the contract concerning the sale of Christoph Schramm's house to Samuel Selfisch from 1564, Schramm junior (d. 1579) as well as Selfisch are referred to as *Buchführer* (bookseller/publisher), Gabriel Schnellboltz (fl. 1562–1571) in another contract is named as printer and *Illuminator* (RatsA WB, 120 (Bc 109), fol. 141v (Schramm), fol. 142r (Selfisch), fol. 184v (Schnellboltz)).[10]

---

[6] The *Gerichtsbücher* contain two systems for numbering the sheets. The newer one (i.e., in most cases the higher number) is cited.

[7] "…*zcu vorrechnen was ehr vorkaufft, vorburgett, vnd was an gelde daran gegeben hat*"(RatsA WB, 113 (Bc 101), fol. 133r). The text mentions an edition of "the small" *Episteln Pauli* in Latin and German.

[8] Conrad Rastger was responsible for the wine cellar of the town council (*Weinschenk*).

[9] With titles of books like the *Großes Gesangbuch* und *Kleines Gesangbüchlein*, Melanchthon's *Grammatic* and *Syntaxis*, each of them printed in 3,000 copies, and the *Dialogus Urbani Regii* in 2,000 copies (RatsA WB, 113 (Bc101), fol. 205r–206r).

[10] A second entry in (RatsA WB, 111 (Bc 99), fol. 107r–108r) concerns the final payment on May 23rd, 1573.

3 Wittenberg Printers of the *Sphaera* and Their Real Estate

Josef Klug, who printed the first Wittenberg edition of the *Sphaera* in 1531, owned house number 162 in the *Jüdenviertel* from 1526 to 1552. It was situated at the southeastern corner of the church square (*Kirchplatz*, today *Mittelstraße* 1). The building was replaced in the 1920s. Where Klug lived and worked before, we do not know—maybe in one of the houses, Lucas Cranach possessed, presumably in *Marktviertel* 2 (*Markt* 5).[11] Around 1523, the painter and successful businessman owned several properties: *Marktviertel* 2 (*Markt* 5), *Coswiger Viertel* 1 (*Schloßstraße* 1), *Jüdenviertel* 146 (*Neustraße* 8), and the house at the southeastern corner of the *Neustraße* (*Elsterviertel* 12).[12] Cranach had a print shop with Christian Döring (ca. 1490–1533); evidence suggests that it was located at *Markt* 5 and that Klug followed Melchior Lotter the Younger (ca. 1490–1542) there as an employee of the two businessmen. We also know that Cranach and Döring later owned the house *Jüdenviertel* 161, built by Hans Schotte around 1532 and situated next to Joseph Klug's print shop at the churchyard (*Jüdenviertel* 162). In 1535, Cranach and the widow of Döring sold it to Peter Schorer (RatsA WB, 147 (Bc 107), fol. 67r–v, *Sontag nach Conversionis Pauli 1535*). The contract indicates that Schotte had pledged the house to Cranach and Döring. In the tax lists, Cranach and Döring's widow does not appear as owners of this parcel (*Jüdenviertel* 161), since Schotte paid the tax. Only the contract gives the information that Schotte had committed the house to them so that they could act as sellers. Peter Schorer paid 170 Gulden for the house with ground and appurtenances.[13]

After Klug's death, a person named Kilian Krumbfuß became the owner of the house *Jüdenviertel* 162 for a short time, then it was the property of the bookseller Johann Schröter. A hundred years later, in 1638, the estate was described as a large complex consisting of a main building and two side wings, offering ten rooms with stoves on three floors (RatsA WB *Urbarium* 7 (Bb 4)). The presses and types of his father were taken over by Thomas Klug (fl. 1551–1563) in 1553. He compensated his brothers and sisters (RatsA WB, 147 (Bc 107), fol. 236r–238r). The complete equipment, including presses (*preßn*), Greek and Latin types (*buchstaben*), and stencils (*Matricen*), had been evaluated by Hans Lufft, Veit Kreutz(ig)er, and Hans Krafft,

---

[11] We know that Melchior Lotter worked and lived in *Marktviertel* 2, not in *Coswiger Viertel* 1, as Reske and elder authors presume (Hennen 2015, 318; Reske et al. 2015, 1082). Even though Klug followed Lotter on *Marktviertel* 2, it is revealing that Cranach invested in the two houses in *Neustraße* at the same time; he later sold *Jüdenviertel* 146 to Johannes Bugenhagen, and *Elsterviertel* 12 to the bookbinder Barthel Lieberau. See (Hennen 2015, 321–322).

[12] Some contracts document participation on other houses and financial transactions with a lot of his coevals (Hennen 2015, 313–350).

[13] A similar relationship appears in a contract concluded between some other notabilities from the book industry. It strikes a house in *Bürgermeistergasse* (*Jüdenviertel* 43), situated between Mag. (Sebastian) Dietrich and Christoph Wilken. It is sold by Zacharias Moller, the legal guardian of the heirs of Ludwig Soliman. The buyer is Johann Krafft who was acting for Lamprecht Pfeifer from Kemberg. Krafft committed the house directly to Samuel Selfisch (RatsA WB, 110 (Bc 98), fol. 345v–346v (1568)). See attached list of houses.

and priced at 284 Gulden (RatsA WB, 147 (Bc 107), fol. 236r–v). Thomas Klug seems to have continued his father's workshop but had not had the money to buy the house, too. Johann Schröter owned *Marktviertel* 83, a house nearby as well. Perhaps he first left the workshop to Thomas Klug, who died in 1563.[14]

Between 1534 and at least 1541, Peter Seitz senior (fl. 1534–1548) can be located at *Marktviertel* 42 (today *Juristenstraße* 16a).[15] He acquired this property for 400 Gulden from Wolle Kersten.[16] Later his heirs possessed the parcel, and M. Simon Grünberger was the owner from 1576 until around 1600. Grünberger printed the *Elementa doctrinae de circulis coelestibus, et primo motu* here in 1587. Peter Seitz II (fl. 1557–1593) also owned *Marktviertel* 49 in the *Scharrengasse* nearby. Neither house is mentioned in the register from 1638, therefore it is unknown how many rooms with heating they contained.

Veit Kreutzer first owned the house *Jüdenviertel* 131 (*Neustraße* 13), and, from 1561, the house *Jüdenviertel* 151 (today *Jüdenstraße* 26) as well. In 1568, Kreutzer retired and sold his printing press. Up to his death in 1578 he lived at *Jüdenviertel* 151. The contract concerning the sale of his print shop is preserved and specifies types, two printing presses, and other materials (RatsA WB, 120 (Bc 109), fol. 393r–v). From this document, we know that the printers Clemen Schleich (fl. 1569–1588) and Anton Schöne (fl. 1569–1585) bought Kreuzer's equipment—for which they paid 385 Gulden—in 1573, the year they launched *Novae quaestiones sphaerae*. Kreuzer received the final installment in 1575.[17] The building *Jüdenviertel* 131 is not mentioned explicitly in the contract. But Schleich is to be found next door in *Jüdenviertel* 133 from 1573 on, while Schöne appears as the owner of the house *Jüdenviertel* 131 in the period from 1577 to 1586. Meanwhile, the parcel with the workshop building (presumably situated on *Jüdenviertel* 131) was possessed by Jacob Lehman, Henricus and Merten Henrich, and Burchardt Winner.[18] The print shop may have been rented to the two printers first. The house *Jüdenviertel* 20 (*Klosterstraße*), which Schleich owned between 1560 and 1570, seems to have been sold in order to move nearer to his print shop.

---

[14] Concerning Thomas Klug see (Reske et al. 2015, 1088). Possibly Thomas Klug bought a short time before he died the house *Jüdenviertel* 78 (*Mauerstraße* 9). He does not appear in the tax lists, only in (RatsA WB, *Schoßbuch* 1556–1565).

[15] The town records for the years 1542 to 1546 are lost.

[16] *Juristenstraße* this time was still named *Brüdergasse*, the contract is in (RatsA WB 147, Bc 107, fol. 69v–70v). Seitz bought the house with the brewery equipment.

[17] Many contracts are concerning *Erbkäufe*, which means that the buyer paid in several installments over a couple of years.

[18] Concerning the possession of the parcel, see (RatsA WB, Schoßbuch 1556–1565, fol. 55v, 238r (JV 131), fol. 240v (JV 151), Schoßbuch 1566–1570, fol. 55r Veit Kreutzer (JV 151), Schoßbuch 1571–1576, n. fol., *Jüdenviertel* 131 Merten Henrich/Bernhard Winner, Jüdenviertel 133 Clemen Schleich, *Jüdenviertel* 151 Veit Creutzer. Schoßbuch 1577–1582, fol 73v–74r Anthonius Schöne *Jüdenviertel* 131, fol. 74v–75r Clemen Schleich *Jüdenviertel* 133, fol. 78v–79r Veit Creutzer/Gürge Zuckerteigk).

After Schleich's death, *Jüdenviertel* 133 became the property of Georg Müller (fl. 1590–1624) and Lorenz Säuberlich (fl. 1597–1613), who later moved to *Elsterviertel* 4. Müller and Säuberlich were also printers.

Johann Krafft the Elder and his heirs could be found at *Jüdenviertel* 25 (*Bürgermeisterstraße* 5) between 1554 and 1582. This parcel had originally been part of the ground belonging to the Franciscan Convent that was locked between 1525 and 1537. Krafft bought it for 360 Gulden from Severin Weiß, who had become pastor of Dobin, a village near Wittenberg, sometime before (RatsA WB, 147 (Bc 107), fol. 234r–235r). In 1638, the house and a single adjoining building possessed six heated rooms (RatsA WB, *Urbarium* 7 (Bb 4)).

Clemen Schleich, as has been said, owned *Jüdenviertel* 20—a small house with two heated rooms—until 1570, while Hans Lufft was already residing nearby at *Jüdenviertel* 15 (*Bürgermeisterstraße*) from 1528 to 1582.[19] Later this estate was owned by the publisher Wolff Stauffenbuel (1580–1619). In 1556, Lufft also owned a house in *Marstallgasse* (*Coswiger Viertel* 64).

Matthäus Welack (d. 1593), who printed five editions of the *Sphaera* between 1576 and 1591 (Blebel 1576, 1576–1577, 1588; Dietrich 1583, 1591), had been—in exactly this period—the owner of *Elsterviertel* 109 (today *Collegienstraße* 36), a parcel close to the university and the houses of Conrad Ruehel (1528–ca. 1579). Welack's house in 1638 is described as a big timber frame building with nine heated rooms on two floors.

Other printers who are of interest in our research can be located too. Georg Rhau can first be placed in *Coswiger Viertel* 61 (today *Marstallgasse*, 1528–1541), then on *Marktviertel* 32 (1541–1548), a parcel at the end of *Juristenstraße* that had belonged to the convent of the Gray friars as well. Rhau built a new house there and profited from tax exemption for the first time in 1541 (RatsA WB, Kämmereirechnung 1541, fol. 16v (20 gr *Sommerschoß*)). After Rhau's death, his wife possessed the parcel and the house until 1571.

Without a doubt, Rhau used the new building for his print shop. In 2011, during a large archaeological excavation, a high number of types were found in this area, which was later unambiguously identified as his workshop (Meller 2014, 5–27) (Fig. 5).

Similar findings were made at the parcel where Johann Krafft and his heirs' workshop had been situated (*Jüdenviertel* 25/*Bürgermeisterstraße* 5), as well as from the site of the workshop of Peter Seitz (and heirs), and later that of Simon Grünberger (*Marktviertel* 42, *Juristenstraße* 16a) (Meller 2014). These findings prove that Rhau, Seitz I, and Krafft used their houses as both working and living spaces. Neither Peter Seitz I nor Johann Krafft held other houses. Rhau sold his first house in the *Coswiger Viertel* when he got the opportunity to build the new one on the land of the former convent.

---

[19] Lufft's house is not mentioned in the register from 1638 (RatsA WB, *Urbarium* 7 (Bb 4)), but the cellar is still extant and gives an impression of quite a large stone house.

Johann Rhau-Grunenberg (d. 1529) possessed *Elsterviertel* 134 (1520/21) and *Elsterviertel* EV 96 (1520–1529), two houses situated in the *Collegienstraße*, each with only three rooms with stoves (Schirmer 2015, 170).[20]

Melchior Lotter the Younger, after his separation from Cranach and before he returned to Leipzig, could be found at the northern corner of the *Pfaffengasse* and the *Marstallgasse* at *Marktviertel* 27 for a short period.[21]

4   Examples of Vicinities: Hints to Social and Professional Networks

The zone at the beginning of the *Mittelstraße* and the southern part of the church square was one of the hubs of printing and the book trade that evolved in the town in the sixteenth century. Joseph Klug's house, *Jüdenviertel* 162, which had been his estate between 1526 and 1552, has already been mentioned, as well as the fact that Cranach and the widow of Christian Döring had granted credit to Hans Schotte, the neighbor of Klug and builder of a new house in *Jüdenviertel* 161, in the 1530s.[22] From 1532 to 1550, the printer Hans Weiß possessed *Jüdenviertel* 168 (RatA WB, 147 (Bc 107), fol. 384r), currently *Mittelstraße* 60. Hans (Johann) Schröter later owned *Jüdenviertel* 162 and *Marktviertel* 83 nearby.

In 1541, the bookseller and publisher Bartholomäus Vogel bought *Elsterviertel* 1. The family owned it up to 1589. Later, Christoph Cranach (d. 1596) became the owner of this estate before he sold it to his colleague, the bookseller and publisher Zacharias Schürer (fl. 1607–1624). While Schürer was active, Lorenz Säuberlich printed next door (*Elsterviertel* 4).

Christoph Schramm senior owned *Marktviertel* 78 (today *Collegienstraße* 5/*Kirchplatz* 16) in 1528–1529. Between 1530 and 1534 he is not listed as a taxpayer in the records of the town council.[23] In 1535–1536 he paid for *Marktviertel* 70, and from 1537 for *Marktviertel* 4 (see Appendix).

In 1533 Schramm I, Vogel, and Goltz collectively purchased technical equipment, books, and Döring's privilege to print and sell the Luther Bible shortly before Döring's death (RatsA WB, 147 (Bc 107), fol. 63r–64r; Schirmer 2015). His widow received 800 Gulden from the sale. Acquiring the privilege may have been the ultimate goal of this business. The heirs of the goldsmith and publisher owned a house

---

[20] Later, another Johann Rhaw (maybe a relative of the printer) owned *Marktviertel* 18 (Juristenstraße 10; 1537–1547), a large estate built of stone with adjoining buildings. He bought it in 1538 from the baker Matthes Mose. The price had been 900 Gulden. The final installment was paid in 1544, see (Ratsarchiv Wittenberg, 147 (Bc 107), Gerichtsbuch 1520–1555, fol. 82r–v). The contract is dated "Dienstag nach Hilarii" in 1538. Brewery equipment had been included in the price.

[21] The location of the house in the Juristenstraße made by Kühne and cited by (Reske et al. 2015, 1080) is wrong.

[22] See the contract with Peter Schorer in (RatsA WB, 147 (Bc 107), fol. 67r–v).

[23] The records of the years 1531 and 1533 are lost.

in *Schloßstraße* (*Coswiger Viertel* 4) near the marketplace at this time. The bookshop of Cranach and Döring was probably situated in *Marktviertel* 2 (*Markt* 5) until 1527–1528.[24]

Vogel had possessed *Marktviertel* 82 (*Collegienstraße* 9/*Kirchplatz* 12 a) before he bought the splendid estate in the *Mittelstraße* (*Elsterviertel* 1) in 1541. From the beginning, he and Christoph Schramm the Elder had lived (and worked?) as next-door neighbors at a short distance to Klug. Even later, the distance between their houses had not increased to a noteworthy extent. It had always been possible to communicate immediately. The small size of the town encouraged a quick exchange of ideas and goods.

Schramm the Elder died in 1549. He had started to rebuild the house *Markt* 3 (*Marktviertel* 4). His son and successor, Christoph Schramm junior, enlarged it further and furnished it with wall paintings and other rich decorations, perhaps with the help of his neighbor, Lucas Cranach the Younger (1515–1586). In Vogel's house, vaults from the beginning of the sixteenth century still exist (Fig. 6).

In the second half of the sixteenth century, the bookbinders Paul Droscher (*Elsterviertel* 2) and Heinrich Blume (*Jüdenviertel* JV 165) settled in the area near Vogel's house, as did the booksellers Andres Hofmann (*Jüdenviertel* 167) and Wenzel Lob (fl. 1592–1600) (*Jüdenviertel* 170) (Map 2). Until 1576, Hans Schröter had his bookshop at *Jüdenviertel* 162. Later, the furrier Philipp Olschläger owned this parcel. He probably provided the leather for the bindings.

A similar cluster could be recognized near the university and the houses of Martin Luther—the former convent of the Augustinian hermits at the eastern end of the *Collegienstraße* (today *Collegienstraße 54*)—and Philipp Melanchthon (*Elsterviertel* 92/*Collegienstraße* 60). Here Moritz Goltz on *Elsterviertel* 102 kept his bookshop from 1528 to 1550. The printer Nickel Schirlentz possessed *Elsterviertel* 104 from 1535 to at least 1541.[25] When Goltz settled on *Elsterviertel* 102 in 1528, Johann Rhau-Grunenberg was still living and presumably working nearby on *Elsterviertel* 96 (1520–1529).

---

[24] See below what is said concerning Lotter.

[25] The annual records of the town council (*Kämmereirechnungen*) for the years 1541 to 1546 are lost. Schirlentz probably died in 1549, see (Oehmig 2015, 158) contrary to (Reske et al. 2015, 1080–81). In the beginning of his activity in Wittenberg (1521) Schirlentz printed in the house of Andreas Karlstadt (Reske et al. 2015, 1080). Karlstadt can be located at *Coswiger Viertel* 61 between 1512 and 1516. Later this house became the property of a person named Licentiat Otho. In 1513, Karlstadt paid for *Marktviertel* 27 as well. The house in the *Marstallgasse* or the one in the *Pfaffengasse* was probably still named "the house of Karlstadt." On the other hand, Karlstadt paid *Winterschoß* (5 gr) for *Coswiger Viertel* 25 (or a house in this area; here the parceling in the early sixteenth century is not clear) in 1521, see (RatsA WB, *Kämmereirechnung* 1521, fol. 32v). So, Schirlentz may have printed there. The parcel is only mentioned in the *Winterschoß* list. In 1520 and 1522 (before and after Karlstadt) Ludewig Henrich paid, see (*Kämmereirechnung* 1520, fol. 32v, *Kämmereirechnung* 1522, fol. 34v). In 1526, Schirlentz was the owner of a house at the marketplace, which may have been broken down soon thereafter in order to enlarge the place; there are no other payments to be found, see purchase of 10 gr in (RatsA WB, *Kämmereirechnung* 1526, fol. 45 r). In 1535, Schirlentz owned EV 20 and EV 104; see attached list of houses.

Three decades later, in 1553, Conrad Ruehel, son-in-law of Moritz Goltz, was the owner of the bookshop (*Elsterviertel* 102) as well as of the house *Elsterviertel* 101, which he modernized and may have rebuilt completely, in order to relocate the bookshop. The vault in *Elsterviertel* 101 is decorated with elaborated paintings (Fig. 7 top). The keystone shows the initials of Ruehel and the date 1556 (Fig. 7 bottom). Between 1561 and 1582, the printer Johann Schwertel owned the house *Elsterviertel* 105, while Ruehel's direct neighbor, the smith (*Kleinschmied*) Merten Metko, added his new building (*Elsterviertel* 103) in 1569 at the western gable of *Elsterviertel* 101 (RatsA WB, *Schoßbuch* 1566–1570, fol. 74r, *Merten Metka* 1569 and 1570 (*frey*)). Builders of new houses were tax exempt for a couple of years.

It is possible that Ruehel and Metko, and maybe Schwertel too, worked together, and that Ruehel was interested enough in strengthening their cooperation that he agreed to allow Metko to use his wall; Metko did not build a second gable wall for his house but fitted it to Ruehel's wall. Metko may have provided types or casting molds to produce them, or he may have fabricated fittings for covers in return. Unfortunately, no contract or other document can be found that allows us to draw conclusions concerning the precise relationship between Ruehel, the bookbinders in his neighborhood, and the smith Metko. At any rate, it is certain that Metko needed Ruehel's permission to use the gable of the house where the bookshop was located. Even if the bookseller Ruehel and his neighbors did not work together at all, their clients would have profited from the wide variety of goods produced by the book industry in that small area.

In the period from 1563 to 1576 the bookbinder Georg Bernutz settled on *Elsterviertel* 95. Hans Reinisch followed Bernutz around 1580. Thomas Krüger, bookbinder owned *Elsterviertel* 96 from 1566 to 1600. Before him, his colleague Thomas Saup had held that house while Hans Dietz, also a bookbinder, owned *Elsterviertel* 108, later *Elsterviertel* 90. Around 1569, Hans Dietz moved to *Marktviertel* 75 (see Appendix).

Conrad Ruehel had been the instructor of Samuel Selfisch (Schirmer 2015). He, the son of a bookseller at Erfurt, came to Wittenberg in 1545. First, the young man was educated by Bartholomäus Vogel, then he became an apprentice to Ruehel (Rüger 1978, 8). So, from the very beginning, he was in close contact with the heart of the Wittenberg book trading network, the so-called Bible consortium. In 1557, Selfisch married Maria Ruehel (d. 1580), the sister of the bookseller. At this time, around 1556/1557, Conrad Ruehel owned *Elsterviertel* 101 and 102 on the opposite side of the road; in 1571, he also owned *Elsterviertel* 100. Presumably in one of these houses, in *Elsterviertel* 101 or 102, Selfisch and his wife settled before they moved to the marketplace scarcely ten years later.

5   Samuel Selfisch

Samuel Selfisch was the most influential Wittenberg bookseller and publisher of the last four decades of the sixteenth century and the beginning of the seventeenth century (Leonhard 1902; Schirmer 2015, 181–189). In 1564 he took over the estate *Markt 3/Marktviertel* 4 from the heirs of Christoph Schramm the Elder.[26] In the contract, it is noted that a number of books from the property of Schramm (senior?), which "were stored in barrels," were sold to Selfisch.[27] Selfisch paid for the house and the books 4,050 Gulden. Shortly thereafter he added a side wing 20 meters in length with a large vaulted cellar and three floors onto the western side of the parcel. The beams of the ceilings in this narrow building (the distance between the eastern and the western external wall is less than five meters) are very strong and placed close to each other (Fig. 8). The building was obviously constructed to store heavy loads like stacks of imprints. This detail gives a strong architectural indication that this was a warehouse; archival material offers further evidence that Josef Klug used his house near the parish church as a print shop, as do the archeological findings from the grounds where Rhau, Seitz, and Krafft had settled and worked.

Selfisch enhanced the main building by adding a separate staircase, similar to the stair towers that are characteristic of the courtly architecture of the time, and a belvedere. These parts of his estate were visible from outside the town wall, from the ships operating on the river Elbe and the trading route coming from Leipzig. The courtyard south of the house at the marketplace became a splendid room itself: the facades of the buildings were richly fashioned. Inside the house, a room on the second floor was decorated with an extended cycle of inscriptions, a combination of antique texts with passages from the Apocrypha and the Old and New Testaments (Jäger 2011); most of this is preserved (Fig. 9).

In 1902, Hans Leonhard provided a detailed description of Selfisch's personal life and his trading connections (Leonhard 1902, 18). Several showrooms are named as parts of the estate at the Wittenberg marketplace. Selfisch sold books at Wittenberg, did business at the fairs of Leipzig and Frankfurt (Chap. 6), invested in editions, traded with paper, and granted credit to colleagues and other professionals. Even a catalog of the works he published has survived, as well as some fair catalogs—and we do not know about all of his projects and products.

It is certain that Selfisch was connected to some of the printers of the *Sphaera*. He worked together with Hans Lufft, Johann Krafft, Georg Rhau, and Lorentz Säuberlich (Leonhard 1902, 18). Leonhard presumes that Säuberlich worked for Selfisch and did not possess a print shop of his own. Säuberlich however had been first the owner of *Jüdenviertel* 133, then of *Elsterviertel* 4, as mentioned above (Leonhard 1902, 18). The latter is described in 1638 as a large complex with two side wings.

In 1596 Selfisch bought the print shop of Matthes Welack (Leonhard 1902, 17). The widow sold the four presses, the types, the matrices, and lots of other implements

---

[26] The sales agreement is available at (RatsA WB, 120 (Bc 109), fol. 141v–143v, date: 13.02.1564).

[27] "…und damit auch die Bücher inhals (!) eines Inventarii in etzlichen Feßern eingepast…" (RatsA WB, 120 (Bc 109), fol. 142r). The inventory could not be retrieved until now. When Christoph Schramm the Elder died in 1549, the heirs taxed his property on 10,000 Gulden, Elisabeth, the widow, inherited half of this, and the other part went to the four sons. Initially Christoph Schramm II continued the business of his father.

for the remarkable price of 1,500 Gulden. Some of the materials had originally stemmed from the printing shop of Johann Schwertel. Welack had spent 500 Gulden for two presses and twenty-one center types when he bought them from the heirs of Johann Schwertel in 1578.

6 Printers, Bookbinders, and Booksellers as Members of the Town Council

Two of the pioneers of the book industry in Wittenberg, Lucas Cranach the Elder and the goldsmith Christian Döring, entered the town council in 1519.[28] Döring, who died in 1533, officiated until 1531, Cranach until 1544 or 1547.[29]

In 1541, Georg Rhau became a member of the town council. One year later, so did Hans Lufft; in 1543 so did Moritz Goltz; as did Christoph Schramm the Elder in 1545. Lufft officiated until 1582, the others only for a few years.[30] There is some evidence that Rhau, Goltz, and Lufft—together with the other members of the town council—commissioned the new altar screen for the town church in 1547, the famous *Reformationsaltar* finished by Cranach the Elder (?), Cranach the Younger, and their workshop in 1548 (Hennen 2015, 351–361).[31] Conrad Ruehel, member of the council from 1553 to 1575, Barthel Vogel (1554–1569), the bookbinder Hans Cantzler (1564–1579), and Hans Krafft (1567–1576) became successors of Rhau, Goltz, and Schramm the Elder. Christoph Schramm junior officiated from 1561 to 1579, Samuel Selfisch in 1569 for the first time, and in 1613 for the last (RatsA WB, *Kämmereirechnung* 1613, fol. 1r).

Later also, the booksellers and publishers Wolf Stauffenbuel and M. Johann Ruehel (1585–1591) were members of the town council (Kettner 1734, 120–121).

In the scope of this paper, it is impossible to show all the more or less far-reaching decisions the councilors made during their terms. I describe elsewhere what was undertaken in order to modernize the town in the first half of the sixteenth century when Cranach the Elder deeply participated in building the new town hall, enlarging the marketplace, tracing out the *Scharrengasse*, and developing the ground of the former convent of the Grey Friars to an attractive housing and commercial area— where Georg Rhau and Johann Krafft the Elder among others settled and practiced

---

[28] If Johann Ritter had been a bookseller (*bibliopola*), which seems likely but is not certain, he would have been the first from the book branch in the town council. Ritter was a member of the council from 1509 to 1535. He was the owner of *Marktviertel 4*/*Markt 3* before Schramm and Selfisch and was connected to Hans Lufft. His widow owned the house *Jüdenviertel 15* before Hans Lufft, see appendix.

[29] Cranach the Elder left Wittenberg after the Schmalkaldic War, definitely in 1550, and followed Duke John Frederic to Weimar, see (Hennen 2015, 318, 417–418). Lucas Cranach the Younger entered the council in 1549 or 1552.

[30] Goltz was a member of the council until 1546, Rhau until 1547, Schramm until 1548.

[31] A list of the members of the town council between 1504 and 1550 is also available: (Hennen 2015, 351–361).

their business (Hennen 2020a). Obviously, the successful representatives of the book and media industry were constantly involved in the local affairs and administration of the parish as well. Georg Rhau and Hans Cantzler administrated the *Common Chest*, the treasure of the church, as well as Hans Schröter, Hans Krafft, and the bookbinder Paul Thilo. In 1565, Conrad Ruehel was also a member of this body. In 1568 and 1570 the book trader (*bibliopola*) Heinrich Hesse, the printer Hans Krafft, and the bookbinders Hans Cantzler and Frobenius Hempel together managed the *Common Chest*, along with only two other persons from other professions, Hans Mengewein and Christoph Grumme. What they worked on we do not know.

During these years a large and politically relevant building project of prince Elector August (1526–1586, r. 1553–1586) was realized: the so-called *Ordinandenstube*. This unique room was erected by raising a second story above the sacristy on the north side of the town church. It was used as the location of the final exam that young Lutheran priests had to pass before they were ordained in the parish church and afterward sent to their first rectorate.

Prince Elector August appointed Conrad Ruehel and Samuel Selfisch as his attorneys. Together they provided 1,000 Gulden for this building measure (PfarrA WB, A I, 674 *Kirchenbaw*). Five years earlier, in 1564, the prince elector had renewed the book trading privilege of Selfisch, Ruehel, and Vogel. This fact seems to be the background of that generous engagement.

The result of this effort had been an unrivaled room concept without any precedent. The *Ordinandenstube* and most parts of its interior still exist. It was furnished with benches, an oval table, and armoires that contained all the relevant confessional documents. Most of these books had been printed at Wittenberg, where Selfisch and Ruehel held the privilege to print and sell the Lutheran Bible and the texts authored by the Reformer.[32] Besides these documents, the annual records of the parish were stored there, in order to show the practical side of the thorough and godly administration of the Reformed parish. One wall of this room was decorated with a large inscription, quoting verses from the book *Jesus Sirach*. The inscription had been recovered during the last refurbishment between 2014 and 2016. The *Ordinandenstube* was designed as a room of the word of God and the writings of his community, which should be regarded as sound documents proving the godliness of the responsible persons.

Building the *Ordinandenstube* consecrated the church where Martin Luther had preached as a central place of Lutheranism, a confession based on the word. Outside, on the southern facade of the church, this function was expressed by additional

---

[32] In 1575 Wolff Stauffenbuel Buchfuhrern was paid for each of four copies of the Nachtmahl Lutheri, Schmalkaldische Artickell, Sprüche der Altväter (?), Agenda Herzogk Heinrich, Kirchenordnung Wittenberg, and one copy of an Agenda Herzog Heinrichs, see (PfarrA WB, A I 129, Gemeiner Kasten 1575, n. fol., Pos. Ausgabe vor Underhaltung der Pfar Kirchen). Stauffenbuel received all in all 5 Gulden 5 Groschen 6 Pfennige. Hans Cantzler made the bindings. In 1560 xxviii gr were paid fur ein buch Corpus doctrine, which was stored in the sacristy, see (Gemeiner Kasten 1560, n. fol., Pos. Ausgabe vor unterhaltung der pfarkirchen). In 1570, Samuel Selfisch received 1 Gulden 10 Groschen 6 Pfennige…fur eine gespaltene Biblia, 1 fl 15 gr fur eine deutzsche Biblia inn viertel gebund mit puckeln beschlagen, see (Gemeiner Kasten 1570, n. fol. Pos. Ausgabe zum Kirchenbaw und unterhaltung der pfarkirchen; Hennen 2020c, nn. 63, 64).

inscriptions (Hennen 2020c). To the room above the sacristy, whence the young
Lutheran priests were sent out into "the world" for more than four centuries, was
attached the same significance as the so-called *Lutherstube* at the former dwelling
of the Reformer, and his tomb in the castle church.

In his own house, Samuel Selfisch had used sophisticated inscriptions to convey
his personal beliefs; in the case of the *Ordinandenstube* he and his colleague and
brother-in-law Conrad Ruehel—who acted to express the confession and the image
of the Prince Elector—indeed functioned as a mouthpiece of August.[33] In the second
half of the sixteenth century, inscriptions written on facades or walls of rooms became
a common vehicle of religious messages. It was a popular form of expression and a
common decorative element in the cityscape as well as in private dwellings (Hennen
2013b).

7 Conclusion

The leaders of the book industry that emerged in Wittenberg in the first half of
the sixteenth century rapidly became part of the new social and commercial elite
in town. They profited from the university, founded in 1502—where some of them
matriculated—and particularly from the Lutheran Reformation. The basis of the
extraordinary period of prosperity Wittenberg underwent during the sixteenth century
was the innovative and communicative climate Prince Elector Frederic the Wise
had created during his reign. He transformed an average town into an intellectual
and cultural center. Together with artists, like Lucas Cranach the Elder and Claus
Heffner, and with scholars from the university, he developed a sophisticated concept
of visual and written communication, which his successors, especially Prince Elector
John Frederic (1503–1553, r. 1532–1547) and Prince Elector August enhanced. The
specialty goods of the Wittenberg print industry fitted this concept perfectly and
allowed the town's publishers to enjoy a large output and high profits. Successful
businessmen like Cranach, Schramm, Goltz, Ruehel, Vogel, and Selfisch (but also
printers like Klug, Rhau, Lufft, Krafft, and Seitz senior and junior) invested much of
their earnings in real estate properties. The sequences of properties' ownership and
the contracts concerning those sales and credits are testaments to the dense network
of book industry professionals in Wittenberg.

---

[33] Selfisch spent a lot of money on social welfare. For example, he donated the ground for the new
graveyard (Leonhard 1902).

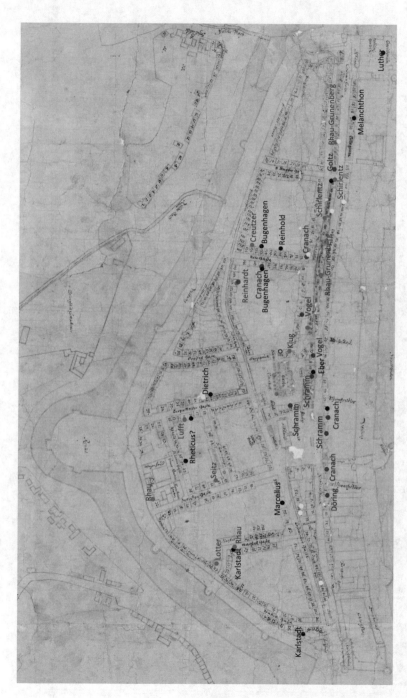

Map 1 Houses of printers, publishers, bookbinders and booksellers, 1520–1550. Red: publishers/booksellers; blue: printers; yellow: bookbinders, green: providers (for example smiths); black: scholars. Author's marks. Map: Städtische Sammlungen Wittenberg

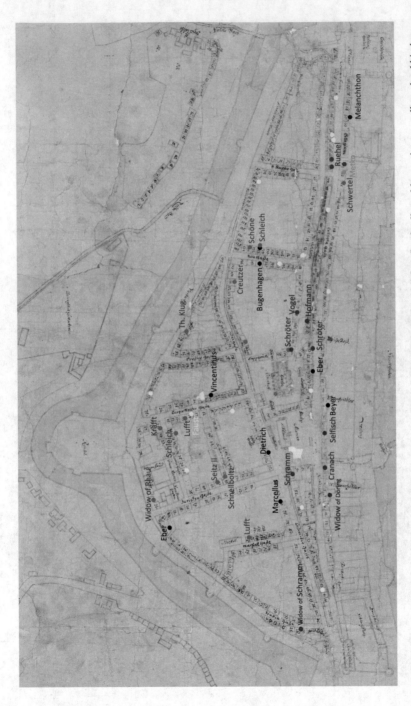

Map 2 Houses of printers, publishers, bookbinders and booksellers 1550–1575. Red: publishers/booksellers; blue: printers; yellow: bookbinders, green: providers (for example smiths); black: scholars. Author's marks. Map: Städtische Sammlungen Wittenberg

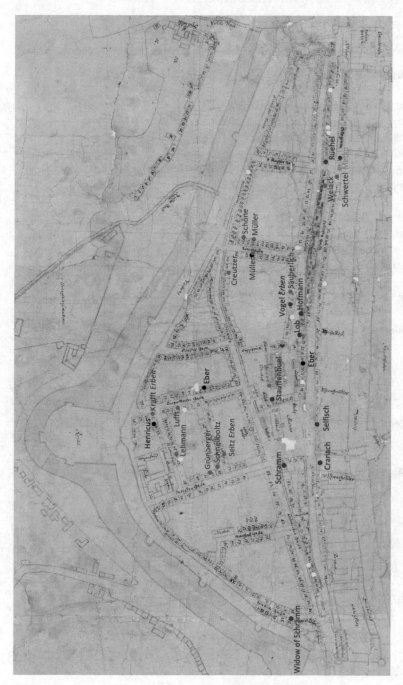

Map 3 Houses of printers, publishers, bookbinders and booksellers 1575–1600. Red: publishers/booksellers; blue: printers; yellow: bookbinders; green: providers (for example smiths); black: scholars. Author's marks. Map: Städtische Sammlungen Wittenberg

Appendix: List of Houses Mentioned in the Text

The parcels are denoted by the historic quarter number and the modern address. The annual payment (*Schoß*) is specified; at least one entry in the town records is cited.

Abbreviations

CV   *Coswiger Viertel* (Coswig quarter)
MV   *Marktviertel* (market quarter)
JV   *Jüdenviertel* (Jewish quarter)
EV   *Elsterviertel* (Elster quarter)
KR   [year] Ratsarchiv Wittenberg, Kämmereirechnung [year]
fol.   Sheet
n.   fol. sheets not numbered
ß   *Schock,* 1ß = 60 *Groschen*
fl   *Gulden,* 1 fl = 21 *Groschen*
gr   *Groschen,* 1 gr = 12 *Pfennige*
d   *denar, Pfennig*

Coswiger Viertel

CV 1/Schloßstraße 1

2 ß

KR 1518, fol. 7r

(Hennen 2015, 318–320, 2017c, 446, 2020a, 469)

1518–1606   Lucas Cranach the Elder and heirs, see MV 2, JV 146, JV 151, JV 161, EV 12

CV 9/Schloßstraße 4

50 gr,

KR 1518, fol. 7r, KR 1547, fol. 11r

(Hennen 2017c, 446, 2020a, 469)

1512–ca. 1562   Christian Döring, goldsmith and printer(?)/publisher, and heirs

CV 25 (?)/Pfaffengasse (1?)

5 gr

KR 1520, 32v, KR 1521, fol. 32v, KR 1522, fol. 34v, KR 1523, fol. 33v

(Hennen 2017c, 448)

1517–1520   Ludewig Henrich/Hernich

1521   Andreas Bodenstein called Karlstadt, theologian, see CV 61, MV 27
1522–1523   Ludewig Henrich/Weiman

CV 26/Pfaffengasse (2)

14 gr

KR 1564, fol. 6r, KR 1566, fol. 16r

(Hennen 2017c, 448)

1564   *das Rathhaus* (i.e., property of the town council)
1566–ca. 1594 *die Alte Schrammin* (i.e., the widow of Christoph Schramm I) and
                heirs, see MV 4, MV 10, MV 70, MV 78

CV 43/Coswiger Straße 14/Schloßstr. 19

22 gr

KR 1549, fol. 11r, Schoßbuch 1580–1589, fol. 45r

1547–ca. 1582   Gregor Dietz, bookbinder, see MV 37, EV 108

CV 44/Coswiger Straße/Schloßstr. 21

30 gr

Schoßbuch 1580–1589, fol. 45r

(Hennen 2017c, 450)

1580–ca. 1589   Hans Dietz, bookbinder, see MV 75, EV 90, EV 108

CV 61/Marstallgasse 8

34 gr

KR 1512, fol. 7r, KR 1517, fol. 8r, KR 1522, fol. 12r, KR 1523, fol. 6r

KR 1526, fol. 67r, KR 1528, fol. 7r, KR 1547, fol. 16v

(Hennen 2013a, 59, 2017c, 451, 2020a, 472; Reske et al. 2015, 1080; Oehmig 2015, 127)

1512–1516   M. Andreas Karlstadt, see CV 25, MV 27
1517–1522   Magister/Lic. Otho
1523–1526   Johann Agricola (*Magister Eysleben*), theologian
1528–1541[34]   Georg Rhau, printer, see MV 32
1547   Donat Fischer

---

[34] The records of 1542 to 1546 are lost.

CV 64/Marstallgasse 14

4 gr

KR 1550, fol. 13r, KR 1553, fol. 15v, KR 1554, fol. 13r, KR 1564, fol. 10r

KR 1566, fol. 20r

1550–1553	Wolf Krumbholz
1554–1564	Hans Lufft, printer, see JV 15
1566–1577	Matthes Krafft

CV 79/Coswiger Straße 7

18 gr

KR 1523, fol. 6v, KR 1528, fol. 7v, KR 1536, fol. 5 (b)v

(Hennen 2017c, 451, 2020a, 473)

1523–1524	Andreas Bernutz, bookbinder, see CV 80
1528	Wenzel Salbach, tin caster
1536	Georg Schrotter/Schröter, bookbinder, see MV 70

CV 80/Coswiger Straße 6

56 gr

KR 1526, fol. 7v

1526–ca. 1550   Andreas Bernutz, bookbinder, see CV 79

CV 86/Coswiger Straße 31

1 ß 30 gr

KR 1547, fol. 18r, KR 1549, fol. 16r, KR 1550, fol. 14v

(Hennen 2020a, 478)

1547–1550   Mgr. Marcellus

Marktviertel

MV 1/Markt 6

1 ß 30 gr

KR 1512, fol. 13v, KR 1559, fol. 17v

(Hennen 2017c, 451)

1512–1559   D. Christian Beyer, chancellor, and heirs

MV 2/Markt 5

54 gr

Cranach's bookshop?

KR 1512, fol. 6r, KR 1517, fol. 7r, KR 1518, fol. 7r, KR 1521, fol. 9r, KR 1522, fol. 11r, KR 1547, fol. 26r

(Schirmer 2015, 173–175; Hennen 2015, 342, 2017c, 452, 2020a, 473)

1512–1517	Lucas Cranach I., painter, bookseller/publisher, enterpriser, see CV 1, JV 146, JV 151, JV 161, EV 12
1518–1521	Valten Mellerstadt
1522–1546	Cranach d. Ä.; Melchior Lotter works here, later Josef Klug too, see JV 162
1547–1550	Caspar Pfreundt, son-in-law of Cranach I.

MV 4/Markt 3

1 ß 12 gr;

KR 1504, fol. 177r, KR 1512, fol. 6r, KR 1536, fol. 5r, KR 1537, fol. 6r, KR 1547, fol. 26r, KR 1550, fol. 15v, KR 1563, fol. 12v, KR 1564, fol. 12v, KR 1572, fol. 34v

(Hennen 2017c, 452, 2020a, 473; Schirmer 2015, 181)

1492–1512	Hans Fehrmeister and widow, see JV 15, EV 76
1515–1536	Johann Ritter, *Ratsbaumeister* (master builder of the council), Buchführer (*bibliopola*)? See JV 15
1537–1564	Christoph Schramm the Elder, bookseller/publisher, heirs, see MV 70, MV 78, CV 26, MV 10
1564–1564	Samuel Selfisch and heirs, bookseller/publisher, heirs also EV 102

MV 10/Coswiger Straße 1

15 gr

KR 1570, fol. 36v

1570–1600	Christoph Schramm II, publisher/bookseller, widow, see MV 4, MV 70, MV 78, CV 26

MV 18/Juristenstraße 10

45 gr

KR 1537, fol. 8r, KR 1564, fol. 14r

(Hennen 2017c, 452)

1537–1547	Johann Rhau, brother of Georg Rhau
1550–1564	Johann Rhauin

MV 22/Pfaffengasse (near the corner Pfaffengasse/Juristenstraße, today part of the road)

8 gr

KR 1553, fol. 22r, KR 1556, fol. 15v (now two houses on the parcel, 8 gr and 4 gr)

(Hennen 2020a, 474)

1553–1560   M. Paulus Eber, mathematician, theologian, see MV 79, JV 44

MV 27/Pfaffengasse 24

21 gr (11 gr Walpurgis plus 11 gr Michaelis)

KR 1513, fol. 33r, KR 1520, fol. 9v, KR 1526, fol. 8r, 39r

(Hennen 2015, 318, 324, 343, 2017c, 453)

1513   D. Karlstadt, see CV 25, CV 61
1515–1524   D. Torgau (i.e., Matthäus Beskau)
1526–1528   Melchior Lotter II., printer

MV 32/Juristenstraße 14

40 gr

KR 1538, fol. 9r, KR 1541, fol. 8r, 16v, KR 1547, fol. 67v

Because of his new building, Rhau was tax exempt in 1541 for the first time. The records from 1542 to 1546 are lost.

1538–1548   Georg Rhau, printer, see CV 61
1547–1571   Georg Rhauin, widow and heirs, owners of the print shop

MV 36/Klosterstraße (today part of the Arsenalplatz, near the former Franciscan church)

12 gr

KR 1541, fol. 8r, KR 1549, fol. 21r, KR 1576, fol. 41r, KR1583, fol. 46v, KR 1584, fol. 38v

(Hennen 2017c, 454, 2020a, 475)

1537–1541   Magister Feldkirch (= Joachim Rheticus?)
1549–1550   Severin Weiß, see JV 25
1554–1583   Magister *Feldkirchen haus/Feldkirchen erbe*
1584–1589   Zacharias Lehmann, printer

MV 37/Klosterstraße (lost, today part of the square)

12 gr

KR 1554, fol. 19v, Schoßbuch 1571–1576, n. fol., *Marktviertel* (37)

(Hennen 2017c, 454, 2020a, 475)

1554–1576   Gregor Dietz, bookbinder, heirs, see CV 43, EV 108

MV 42/Juristenstraße 16a (northern part)

37 gr/30 gr

KR 1534, fol. 68v, KR 1541, fol. 8v, KR 1547, fol. 21r, KR 1549, fol. 22r, KR 1550, fol. 20r, KR 1575, fol. 33v, KR 1576, fol. 42v, KR 1583, fol. 47v

1533–1541   Peter Seitz I., printer
1547–1549   Peter Seitzin, widow, owner of the print shop
1550–1560   Peter Seitz I. *Erben* (heirs), owners of the print shop
1561–1575   Peter Seitz II., printer
1576–1600   M. Simon Grünberger, printer, see JV 51

MV 43/Juristenstraße 16a (southern part)

15 gr

KR 1563, fol. 28r, Schoßbuch 1577–1582, fol. 30v, 31r

(Hennen 2017c, 455, 2020a, 475)

1563–1582   Gabriel Schnellboltz, printer, see EV 54

MV 49/Scharrengasse 5

4 gr

KR 1575, 33v (Seitz), fol. 34v (Jericke), KR 1589, fol. 30v, KR 1590, fol. 32v

1575–1589   Georg Jericke
1590–ca. 1612   Peter Seitz I. *Erben* (heirs), owners of the print shop, see MV 42

MV 57/Markt 20

1ß 5 gr

KR 1561, fol. 20r, KR 1562, fol. 37r

(Hennen 2017c, 455, 2020a, 476)

1561–1574   M./D. Sebastian Dietrich, mathematician, see JV 44

MV 67/Markt 14/Kirchplatz 6

15 gr

KR 1581, fol. 49r

(Hennen 2020a, 477)

1577–1582   Wolf Staufenbuel, bookseller/publisher, see JV 15

MV 70/Markt 12/Kirchplatz 3

30 gr

KR 1529, fol. 15r, 44r, KR 1530, fol. 16v, 47r, KR 1532, fol 16v, 56r, KR 1534, fol. 15v,

KR 1535, fol. 14v, KR 1536, fol. 13v, KR 1537, fol. 15v

(Hennen 2017c, 456, 2020a, 477)

1529	Gregor Bogh
1529–1534	David Schotte, see JV 161
1535–1536	Christoph Schramm I., publisher/bookseller, see MV 4, MV 10, MV 78, CV 26
1537–ca. 1547	Georg Schröter, bookbinder, see CV 79

MV 75/Collegienstraße 3

13 gr

KR 1569, fol. 62r, Schoßbuch 1580–1590, fol. 67v

(Hennen 2020a, 478)

1569–1589   Hans Dietz, bookbinder, see CV 44, EV 90, EV 108

MV 78 (80?)/Collegienstraße 5/Kirchplatz 16

18 gr;

KR 1520, fol. 14v, KR 1524, fol. 15r, KR 1526, fol. 15r, KR 1528, fol. 15r, 44r, KR 1529, fol. 15r, fol. 44r, KR 1530, fol. 16r, KR 1536, fol. 13r, KR 1547, fol. 24r, KR 1550, fol. 24v, KR 1572, fol. 43v

(Hennen 2017c, 457, 2020a, 478)

1520–1526	Christoph Cleynschmidt, smith
1528–1529	Christoph Schramm I, bookseller, see MV 4, MV 10, MV 70, CV 26
1529–1589	Simon Schulz, rope maker

MV 79/Collegienstraße 6/Kirchplatz 15

12 gr

KR 1541, fol. 15r, KR 1556, fol. 22v, Schoßbuch 1580–1589, fol. 68r.

(Hennen 2017c, 457, 2020a, 478)

1541–1589   M. Paulus Eberus, see MV 22, JV 44, and heirs

MV 82/Collegienstraße 9/Kirchplatz 12a

10 gr

KR 1532, fol. 16r, KR 1536, fol. 13r, KR 1539, fol. 15r, KR 1540, fol. 14r, KR 1541, fol. 15r

(Hennen 2017c, 458)

1532–1540   Bartholomäus Vogel, publisher, see EV 1
1541–1556   Georg Blochinger, verger of the Castle Church

MV 83/Collegienstraße 10

13 gr

KR 1566, fol. 32r, KR 1572, fol. 44r

(Hennen 2020a, 478)

1566–1572   Hans (Johann) Schröter, bookseller, see JV 162

Jüdenviertel

JV 3/Bürgermeisterstraße 3 (northern part)

6 gr

KR 1550, fol. 27v

1550–1570   Paul Thilo/Thiele, bookbinder, see JV 93

JV 14/Bürgermeisterstraße (lost, today shopping mall)

21 gr

KR 1528, fol., 9r, KR 1541, fol. 9r, KR 1564, 25v, KR 1566, fol. 35v, KR 1575, fol. 41r

(Hennen 2017c, 459, 2020a, 479)

1528–1543   Matthias Aurogallus, hebraist
1547–1564   Thomas Lehmann and widow
1566–1575   Hans Faust, bookbinder

JV 15/Bürgermeisterstraße (lost, today shopping mall)

50 gr

KR 1512, fol. 8v, KR 1518, fol. 9v, KR 1526, fol. 9r, KR 1528, fol. 9r, KR 1547, fol. 28r, KR 1581, fol. 55r, KR 1583, fol. 56r, KR 1584, fol. 49r, KR 1585, fol.40r, KR 1586, fol. 40r; *Schoßbuch* 1583–1588, fol. 49v, 50r

(Schirmer 2015, 181; Hennen 2017b, 459, 2020a, 479–480)

1512–1515	Hans Ritter = Johann Ritter? *Ratsbaumeister*, see MV 4
1518–1524	Hans Fehrmeisterin, siehe MV 4, EV 76
1526	Thomas Hesse
1528–1583	Hans Lufft, printer, see CV 64
1584–1585	Hans Lufft's heirs
1586	ca. 1610 Wolf Staufenbuhl, publisher/bookseller (for a short time Michael Schaller), see MV 67

JV 20/Klosterstraße (today shopping mall)

4 gr

KR 1559, fol. 31v, KR 1564, fol. 26v, KR 1566, fol. 36v, KR 1570, fol. 49v, KR 1571, fol. 43v

(Hennen 2017c, 460, 2020a, 480)

1559–1564	Conrad Fuchs, pastor at Loeben (?)
1565–1570	Clemen Schleich, printer, see JV 133
1571	*Custos zu Prata*

JV 24/Klosterstraße (part of the former convent)

15 gr

KR 1583, fol. 57r, KR 1587, fol. 41r, KR 1590, fol. 41r

(Hennen 2017c, 460)

1583–1588	M. Martinus Henricus, hebraist
1587–1589	Zacharias Krafft, printer
1595–1600	Zacharias Lehman

JV 25/Bürgermeisterstraße 5

15 gr

KR 1553, fol. 36r, KR 1554, fol. 29v, KR 1559, fol. 31v, KR 1566, fol. 36v, KR 1583, fol. 57r, KR 1587, fol. 41r, KR 1590, fol. 41r

(Hennen 2017c, 460)

1547–1553	Severin Weiß, see MV 36
1554–1582	Johann Krafft d. Ä., printer
1583–1610	Hans Kräfftin/M. Johannes Krafft d. J., printer

JV 33/Mauerstraße 15

4 gr

KR 1568, fol. 38v, KR 1571, fol. 44v

1568–1571   Wenzel Dorfler, bookbinder, see EV 86

JV 42/Bürgermeisterstraße 16 (southern part)

8 gr

KR 1556, fol. 29v, KR 1563, fol. 28v, KR 1568, fol 39v, *Schoßbuch* 1563–1565, n. fol. *Jüdenviertel*

(Hennen 2017c, 460)

1556–ca. 1600   Christoph Wilckau

JV 43/Bürgermeisterstraße 17

30 gr

KR 1556, fol. 29v, KR 1563, fol. 28v, KR 1568, fol 39v

Schoßbuch 1556–1560, n. fol. *Jüdenviertel*, *Schoßbuch* 1563–1565, n. fol. *Jüdenviertel*

1556–1564   Merten Thilo, Amtsschreiber, (Abraham Moritz, Lamprecht Pfeiffer)
1563–1565   Peter Schliebener/Lamprecht Pfeiffer (*Schoßbuch*)
1568   Peter Schliebener

JV 44/Bürgermeisterstraße 18

13 gr

KR 1556, fol. 30r, KR 1562, fol. 47r, KR 1563, fol. 29r, KR 1568, fol. 40r

Schoßbuch 1563–1565, n. fol., *Jüdenviertel*

(Hennen 2020a, 480)

1550–ca. 1560   M. Sebastianus Dietrich, mathematician, see MV 57
1562–ca. 1565   M. Petrus Vincentinus, dialectician and ethician
1568–ca. 1577   D. Paulus Eberus, physician, later theologian, and heirs, see MV 22, MV 79

JV 51/Bürgermeisterstraße 21 (south)

8 gr

KR 1582, fol. 48r

Schoßbuch 1580–1589, fol. 80v, *Schoßbuch* 1589–1594, fol. 49v, 50r, *Schoßbuch* 1595–1600, fol. 56v, 57r

| 1582–ca. 1590 | Hans Schütze, goldsmith, widow |
| 1591–1600 | Valentin Hempel, *M. Grunin* (i.e., widow of Simon Grünberger, see MV 42) |

JV 78/Mauerstraße 9

12 gr

KR 1560, fol. 33r, KR 1563, fol. 32r, KR 1566, fol. 42r

Schoßbuch 1563–1565 (1563–1571), n. fol., *Jüdenviertel*

(Hennen 2020a, 482)

1560–1563	Peter Bartolomäus
1563	Thomas Klug (only in the *Schoßbuch*), son of Josef Klug, see JV 162
1566–1567	Alexander Dufft

JV 93/Jüdenstraße 10 (or 11)

32 gr

KR 1541, fol. 10v

| 1537–1547 | Paul Thilo/Thiele, bookbinder, see JV 3 |

JV 114/Fleischerstraße 14 (eastern part)

5 gr

KR 1566, fol. 45v, KR 1576, fol. 61r, KR 1577–1579 lost, 1586, fol. 50r

(Hennen 2020a, 483)

| 1566–ca. 1586 | Thomas Saup, bookbinder, see EV 96 |

JV 131/Neustraße 13

12 gr

Schoßbuch 1556–1560, fol. 55v, *Schoßbuch* 1565–1570, fol. 52v

KR 1549, fol. 44r (*in KR 1547 some names are missing because people had left the town during the war; KR 1548 is lost*), KR 1560, fol. 40r, KR 1564, fol. 37r, KR 1566, fol. 47v, KR 1576, fol. 63r, KR 1580, fol. 40r, KR 1583, fol. 68r, KR 1584, fol. 60r

(Hennen 2017c, 461)

1549–1564	Veit Kreuzer, printer, see JV 151
1565–1583	Henricus and Merten Henrich/Jacob Cunradt, Burchardt Winner
1574–ca. 1586	Thone Schöne, printer

JV 133/Neustraße 14

8 gr

KR 1523, fol. 9r, KR 1571, fol. 53r, KR 1572, fol. 60r, KR 1588, fol. 51r, KR 1590, fol. 52v

Schoßbuch 1595–1600, fol. 74v, 75r

(Hennen 2017c, 461)

1523–1571	Johannes Bugenhagen, theologian, and widow
1572–1589	Clemen Schleich, printer, see JV 20
1590–ca. 1595	M. Georg Müller, printer, see JV 146
1595	Lorenz Säuberlich, printer, see EV 4

JV 137/Neustraße 18

8 gr

KR 1541, fol. 11r, KR 1550, fol. 40r

(Hennen 2017c, 461)

1541–1550   M. Erasmus Reinhold, mathematician

JV 143/Neustraße 6

7 gr

KR 1551, fol. 43r, KR 1564, fol. 38r, KR 1566, fol. 48v

1551–1564	Hans Cantzler, bookbinder, see EV 139
1564	Peter Küchenschreiber

JV 146/Neustraße 8 (today part of a hotel)

16 gr

KR 1534, fol. 11r, KR 1540, fol. 10r, KR 1541, fol. 11r, KR 1550, fol. 41r, KR 1570, fol. 62r, KR 1572, fol. 61r

Schoßbuch 1595–1600, fol. 77v, 78r, *Schoßbuch* 16191524, fol. 83v, 84r, *Schoßbuch* 1625–1630, fol. 83r,v

(Hennen 2015, 321–322, 348, 2020a, 484)

1521–1540	Cranach, painter, printer/publisher, see CV 1, MV 2, JV 151, JV 161, EV 12
1541–ca. 1599	Johannes Bugenhagen, theologian, and heirs, see JV 133
1599–1624	M. Georg Müller, printer, and heirs, see JV 133
1619–44	Hiob Wilhelm Fincellius, printer

JV 151/Jüdenstraße 26

15gr/30 gr

KR 1528, fol. 11v (*Simprecht*), KR 1529, fol. 11v, KR 1570, fol. 62v, KR 1580, fol. 42r

Schoßbuch 1556–1565, fol. 240v, *Schoßbuch* 1565–1570, fol. 53r,

(Lang 2015; Hennen 2015, 323, 349, 2020a, 484)

1528	Symphorian Reinhardt, printer
1529	Lucas Cranach, painter, see CV 1, MV 2, JV 146, JV 161, EV 12
1561–1580	Veit Creutzer (d.1578), printer, see JV 131

JV 154/Jüdenstraße 29

48 gr

Schoßbuch 1595–1600, fol. 78v, 79r, *Schoßbuch* 1613–1618, fol. 86v, 87r

1580–ca. 1616	Andreas Ruehel
1617–1624	Hans Frömichen

JV 161/Kirchplatz 10

6 gr

KR 1523, fol. 38r, KR 1526, fol. 42r, KR 1528, fol. 41r, KR 1532, fol 12v, KR 1534, fol. 11v, KR 1535, fol. 10v

(Hennen 2017c, 462, 2020a, 485)

1523–1526	David Schotte
1528–1530	Andres Juchen
1532–1534	Hans Schotte (owner: Cranach, Döring, see CV 1, MV 2, JV 146, JV 151, EV 12; CV 9)
1535	Peter Schorer

JV 162/Kirchplatz 11–12/Mittelstraße 1

50 gr/44 gr

KR 1524, fol. 12r, KR 1526, fol. 11v, KR 1551, fol. 45r, KR 1552, fol. 42r, KR 1553, fol. 51r, KR 1559, fol. 45v, KR1560, fol. 43v, KR 1561, fol. 41v

Schoßbuch 1556–1565, fol. 61v, *Schoßbuch* 1571–1579, fol. 91v

1496–1524	Gores Schneider
1526–1552	Josef Klug, printer
1553–ca.1560	Kilian Krumbfuß
1561–ca.1576	Johannes/Hans Schröter, bookseller, see MV 83
1577	Philipp Olschläger, furrier (*Kürschner*)

JV 165/Mittelstraße 4

30 gr

KR 1585, fol. 56v, KR 1586, fol. 56v, KR 1587, fol. 56v

(Hennen 2020a, 485)

1585   Christoph Rahn, see EV 98
1586–ca. 1610   Heinrich Blume, bookbinder, see EV 98

JV 167/Mittelstraße 59

8 gr

Schoßbuch 1571–1576 (see *Elsterviertel* following EV 3 and Nr. 144 Moritz Köselitz), *Schoßbuch* 1595–1600, fol. 84v, 85r

(Hennen 2017c, 462)

1571–ca. 1600   Andres Hofmann, publisher/bookseller

JV 168/Mittelstraße 60

8 gr

KR 1532, fol. 13r, KR 1536, fol. 10r, KR 1547, fol. 39v, KR 1550, fol. 44v

(Hennen 2020a, 485)

1532–1550   Hans Weyse/Weiß (II?), printer

JV 170/Mittelstraße 61/Collegienstraße 11

20 gr

Schoßbuch 1580–1589, fol. 99r

1580–ca. 1600   Wenzel Lob, bookseller/publisher

Elsterviertel

EV 1/Mittelstraße 5

1ß 20 gr

KR 1504, fol. 194v, KR 1512, fol. 11r, KR 1518, fol. 11v, KR 1536, fol. 10r, KR 1541, fol. 12r,

KR 1547, fol. 39v, KR 1550, fol. 44v, KR 1572, fol. 64r, KR 1585, fol. 56v

Schoßbuch 1595–1600, fol. 83v, 84v, *Schoßbuch* 1607–1612, fol. 82v, 83r

RatsA WB, 147 (Bc 107), *Gerichtsbuch* 1520–1555, fol. 117r–118r

(Hennen 2015, 328, 349, 2017c, 462, 2020a, 485; Schirmer 2015, 181)

1492–1512	Andreas Eberhard and widow, tailor, WBF 5
1518–1537	Valten Eberhard u. Witwe, tailor, WBF 5
1541–1589	Barthel Vogel, bookseller/publisher, and heirs, see MV 82
1595–1600	Christoph Cranach
1607–1524	Zacharias Schürer, publisher/bookseller

EV 2/Mittelstraße 57

12 gr

Schoßbuch 1583–1588, fol. *Schoßbuch* 1589–1594, fol. 99v, 100r, *Schoßbuch* 1595–1600, fol. 117v, 118r, *Schoßbuch* 1607–1512, fol. 115v, 116r

1583–ca. 1587	Paul Droscher, bookbinder
1588	Jakob Kurtzschin
1589–ca. 1620	Hans Klug, painter

EV 4/Mittelstraße 7

1 ß

KR 1547, fol. 40v, KR 1550, fol. 44v, *Schoßbuch* 1595–1600, fol. 83v, 84r, *Schoßbuch* 1601–1606, fol. 82v, 83r, *Schoßbuch* 1613–1618, fol. 91r, 93r (!)

(Hennen 2017c, 462, 2020a, 485)

1541–1650	Barthel Bluhme, *Ratsbaumeister* (master builder of the council)
ca. 1580–1603	Wilhelm Adam
1604–ca. 1616	Lorenz Säuberlich, printer, see JV 133

EV 12/Mittelstraße 12

40 gr

KR 1521, fol. 13r, KR 1541, fol. 12r, KR 1547, fol. 41r, KR 1555, fol. 46r

(Hennen 2015, 322–323, 349, Hennen 2020a, 486)

1521–1542?	Cranach the Elder, see CV 1, MV 2, JV 146, JV 151, JV 161
1547–1554	Barthel Lieberau, bookbinder
1555–1582	Nickel Moller, bookbinder

EV 20/Mittelstraße 43a

5 gr

KR 1535, fol. 11v, KR 1540, fol. 11v

(Hennen 2020a, 486)

1535–1540　　Nickel Schirlentz, printer, see EV 104, (CV 61)

EV/former convent of the Augustinian hermits

Free

No entries in the records of the council (*Schoßregister*)

KR 1554, fol. 54v; *Schoßbuch* 1556–1560, fol. 79r *D. Martinussin Erben* is related to a separate house built in the vicinity of the former convent. Around 1570 it becomes part of the *Collegium Augusteum*.

1508–1546　　(with interruptions) Martin Luther

EV 86/Collegienstraße 49/Mittelstraße 34b

4 gr

KR 1571, fol. 68r, KR 1572, fol. 73r

(Hennen 2017c, 465, 2020a, 488)

1570–1571　　*der Probst zu Schlieben*
1571–ca. 1576　　Wenzel Dörffer, bookbinder

EV 90/Collegienstraße 47/Mittelstraße 35

6 gr

KR 1554, fol. 56r, *Schoßbuch* 1556–1560, fol. 81v

1554–1570　　Hans Dietz, bookbinder, see EV 108, CV 44, MV 75

EV 92/Collegienstraße 60

15 gr/later 30 gr

KR 1520, fol. 37r, KR 1560, fol. 54v, KR 1561, fol. 52v

(Hennen 2017c, 466, 2020a, 488)

1520–1560　　Philipp Melanchthon, philologist, theologian

EV 95/Collegienstraße 44

6 gr

KR 1571, fol. 69r, KR 1576, fol. 77v

1571–1576　　Georg Bernutz, bookbinder

EV 96/Collegienstraße 42/43 (in front of Collegium Fridericianum)

3 gr/5 gr

KR 1520, fol 13r (*Johann Buchdrucker*), KR 1520, fol. 37r, KR 1528, fol. 13v, KR 1529, fol. 13v, KR 1530, fol. 14v, KR 1531 lost, KR 1532, fol. 14v, KR 1550, fol. 55v, KR 1570, fol. 58v, KR 1571, fol. 70r, *Schoßbuch* 1566–1570, fol. 72v

(Schirmer 2015, 170)

1520–1529	Johann Rhau-Grunenberg, printer, see EV 134
1530	Er Simon Fungk
1532–1541	Paul Francke
1550–1570	Thomas Saup, bookbinder, see JV 114
1571–ca. 1600	Thomas Krüger, bookbinder

EV 98/Collegienstraße 41 (western part)

6 gr/9gr

KR 1585, fol. 67v, KR 1586, fol. 67v, KR 1587, fol. 67v

1585	Heinrich Blume, bookbinder, see JV 165
1586	Christoph Rahn
1587	Christoph Rahn's heirs

EV 100/Collegienstraße 39

4 gr/free (*frey*)

KR 1566, fol. 63v, KR 1567, fol. 71v, *Schoßbuch* 1571–1576, n. fol. (JV 100)

(Hennen 2017c, 466)

1566–1576	Conrad Ruehel, publisher/bookseller, see EV 101, EV 102

EV 101/Collegienstraße 62a

1 ß

KR 1541, fol. 14r, KR 1549, fol. 63r, KR 1550, fol. 56v, KR 1553, fol. 66r, KR 1554, fol. 56v, KR 1566, fol. 63v, KR 1567, fol. 71v, *Schoßbuch* 1563–1571, n. fol. (JV 101), *Schoßbuch* 1571–1576, n. fol. (JV 101, *frey*); *Schoßbuch* 1583–1588, fol. 107v, 108r, *Schoßbuch* 1589–1594, fol. 90v, 91r, *Schoßbuch* 1595–1600, fol. 106v, 107r

(Hennen 2017c, 466)

1537–1549	Andres Lehmann
1550–1553	Urban Molman
1554–1588	Conrad Ruehel, publisher/bookseller, see EV 100, EV 102
1589–1600	M. Johann Ruehel, publisher

EV 102/Collegienstraße 38/Mittelstraße 40b

13 gr

KR 1518, fol. 12v, KR 1526, fol. 14r, fol. 44r, KR 1528, fol. 14r,

KR 1541, 14r, 42v, KR 1551, fol. 58v, KR 1553, fol. 65v, KR 1556, fol. 54r, KR 1566, fol. 63v, KR 1567, fol. 71v, KR 1569, fol. 74v, KR 1570, fol. 76v

Schoßbuch 1563–1571, *Schoßbuch* 1571–1576, n. fol. (JV 102: Conrad Rhuel)

(Schirmer 2015; 181, Hennen 2017c, 466, 2020a, 488)

1518–1526	Clemen Tischer, stonemason
1528–1550	Moritz Goltz, publisher
1553–1582	Conrad Ruehel, publisher, see EV 100, EV 101

EV 103/Collegienstraße 63

30 gr

KR 1556, fol. 54v, *Schoßbuch* 1563–1571, n. fol. (JV 103)

RatsA WB, 147 (Bc 107), fol. 179r–180r,

(Hennen 2017c, 466, 2020a, 488)

1549–1600	Merten Metko, *Kleinschmied* (smith)

EV 104/Collegienstraße 37

16 gr

KR 1517, fol. 12v, KR 1534, fol. 14r, KR 1535, fol. 13r, KR 1541, fol. 14r, KR 1549, fol. 62v

(Oehmig 2015, 134, 142–144; Hennen 2017c, 467)

1517–1532	Mag. Bernhard (Gessner?)
1535–1549	Nickel Schirlentz, printer, see EV 20, CV 61
1550–1582	Sigmund Frank

EV 105/Collegienstraße 64

KR 1566, fol. 64r

(Hennen 2017c, 467)

1566–1582	Johann Schwertel, printer

EV 108/Collegienstraße 66

11 gr

KR 1549, fol. 64r, KR 1550, fol. 57v, KR 1551, fol. 60r

1549	Gores/Georg Dietz, bookbinder, see CV 43, MV 37
1550	Hans Dietz, bookbinder, see EV 90, CV 44, MV 75

1551   Matthes Schlor

EV 109/Collegienstraße 36

KR 1576, fol. 80r, *Schoßbuch* 1571–1579, fol. 114r

1576–1610   Matthäus Welack and widow, printer

EV 116/Collegienstraße 32

7 gr

KR 1549, fol. 63v, KR 1574, fol. 71r

(Hennen 2017c, 467)

1537–1574   Cunz/Conrad Neidel, bookbinder, widow

EV 134/Collegienstraße 21

5 gr

KR 1520, fol. 13v

(Schirmer 2015, 170)

1520–1521   Johann Rhau-Grunenberg (Johann von Groningen), see EV 96

EV 139/Collegienstraße 79

30 gr

KR 1560, KR 1566, fol. 68r, KR 1581, fol. 87v

(Hennen 2020a, 490)

1566–1581   Hans Cantzler, bookbinder, see JV 143

Abbreviations

Digital Repositories

Sphaera Corpus*Tracer*   Max Planck Institute for the History of Science. https://db.
                          sphaera.mpiwg-berlin.mpg.de/. Accessed 07 June 2021.

Archives and Special Collections

PfarrA WB.   Pfarrarchiv Wittenberg

– A I 129,   *Gemeiner Kasten* (records of the parish/Common Chest) 1560, 1570,
              1575

– A I, 674   *Kirchenbaw* (building of the church), 1569–1571.

RatsA WB.   Ratsarchiv Wittenberg

–   *Kämmereirechnungen* (records of the town council) 1500–1600, 1613 [lost: 1531, 1533, 1542–1546, 1548, 1577–1579]
–   110 (Bc 98), *Gerichts- und Handelsbuch* (book of contracts), 1556–1571
–   111 (Bc 99), *Gerichts- und Handelsbuch* 1572–1590
–   113 (Bc 101), *Gerichts- und Handelsbuch* 1523–1551
–   147 (Bc 107), *Gerichts- und Handelsbuch* 1523–1554
–   120 (Bc 109), *Gerichts- und Handelsbuch* 1555–1575
–   *Schoßbuch* (tax register) 1556–1565
–   *Schoßbuch* 1563–1571)
–   *Schoßbuch* 1565–1570
–   *Schoßbuch* 1566–1570
–   *Schoßbuch* 1571–1576
–   *Schoßbuch* 1571–1579
–   *Schoßbuch* 1577–1582
–   *Schoßbuch* 1580–1589
–   *Schoßbuch* 1583–1588
–   *Schoßbuch* 1589–1594
–   *Schoßbuch* 1595–1600
–   *Schoßbuch* 1601–1606
–   *Schoßbuch* 1607–1612
–   *Schoßbuch* 1613–1618
–   *Schoßbuch* 1619–1624
–   *Schoßbuch* 1625–1630
–   *Urbarium* 7 (Bb 4), fol. 41r–652r, *Einquartierungsverzeichnis* (short description of the single houses in order to use rooms for quartering),1638

References

Primary Literature

Blebel, Thomas. 1576. *De sphaera et primis astronomiae rudimentis libellus ad usum Scholarum trivialium, ut vocant, maximè accomodatus, & in gratiam studiosae iuventutis Curianae, ex artificum libris accurata methodo & brevitate conscriptus a M. Thoma Blebelio Budiss: Ludiliterarij Curiensis collega.* Wittenberg: Matthaeus Welack. https://hdl.handle.net/21.11103/sphaera. 101095.

Blebel, Thomas. 1576–1577. *De sphaera et primis astronomiae rudimentis libellus ad usum scholarum triuialium, ut uocant, maximè accommodatus, & in gratiam studiosae iuuentutis Curianae, ex artificum libris accurata methodo & breuitate conscriptus a M. Thoma Blebelio rudissimi Budiss: Ludiliterarij Curiensis collega.* Wittenberg: Matthaeus Welack. https://hdl.handle.net/ 21.11103/sphaera,101306.

Blebel, Thomas. 1588. *De Sphaera et primis astronomiae rudimentis libellus, ad usum Scholarum maximè accomodatus: accurata methodo & brevitate conscriptus, ac denuò editus A M. Thoma Blebelio Budissino.* Wittenberg: Matthaeus Welack. https://hdl.handle.net/21.11103/sphaera.101055.

Dietrich, Sebastian. 1583. *Novae quaestiones sphaericae, hoc est, circulis coelestibus et primo mobili, in gratiam studiosae iuventutis scriptae, A M. Sebastiano Theodorico Vuinshemio, Mathematum Professore.* Wittenberg: Matthaeus Welack. https://hdl.handle.net/21.11103/sphaera.101100.

Dietrich, Sebastian. 1591. *Novae quaestiones sphaericae, hoc est, de circulis coelestibus & primo mobili, in gratiam studiosae iuventutis scriptae, a M. Sebastiano Theodorico Vuinshemio. Mathematum Professore.* Wittenberg: Matthaeus Welack. https://hdl.handle.net/21.11103/sphaera.100201.

Secondary Literature

Berger-Schmidt, Diana, Insa-Christiane Hennen, Thomas Schmidt, and Isabelle Frase. 2011. Das Wohn- und Geschäftshaus Markt 3: Zeugnis der Glanzzeit der Universität und Stadt Wittenberg. In *Das ernestinische Wittenberg: Universität und Stadt (1486–1547)*, eds. Heiner Lück, Enno Bünz, Leonhard Helten, Dorothée Sack, und Hans-Georg Stephan, 191–197. Petersberg: Imhof.

Hennen, Insa Christiane. 2011. Universität und Stadt: Einwohner, Verdichtungsprozesse, Wohnhäuser. In *Das ernestinische Wittenberg: Universität und Stadt (1486–1547)*, eds. Heiner Lück, Enno Bünz, Leonhard Helten, Dorothée Sack, und Hans-Georg Stephan, *Wittenberg-Forschungen*, Vol. 1, 135–145. Petersberg: Imhof.

Hennen, Insa Christiane. 2013a. Reformation und Stadtentwicklung. Einwohner und Nachbarschaften. In *Das ernestinische Wittenberg: Stadt und Bewohner*, eds. Heiner Lück, Enno Bünz, Leonhard Helten, Armin Kohnle, Dorothée Sack, und Hans-Georg Stephan, *Wittenberg-Forschungen*, Vol. 2, 33–76, Petersberg: Imhof.

Hennen, Insa Christiane. 2013b. Der Wittenberger Schulbau von 1564 bis 1567 im städtebaulich-historischen Kontext. In *Das ernestinische Wittenberg: Stadt und Bewohner*, eds. Heiner Lück, Enno Bünz, Leonhard Helten, Armin Kohnle, Dorothée Sack, und Hans-Georg Stephan, *Wittenberg-Forschungen*, Vol. 2, 175–186, Petersberg: Imhof.

Hennen, Insa Christiane. 2014. Quellen des Wittenberger Häuserbuchs: Schoßregister, Stadtkarten, archäologische Relikte, Bauten und Ausstattungen. In *Mitteldeutschland im Zeitalter der Reformation*, ed. Harald Meller, Forschungsberichte des Landesmuseums für Vorgeschichte Halle 4, 101–110. Halle: Landesamt für Denkmalpflege und Archäologie Sachsen-Anhalt—Landesmuseum für Vorgeschichte Halle.

Hennen, Insa Christiane. 2015. 'Cranach 3D:' Die Häuser der Familie Cranach in Wittenberg und das Bild der Stadt. In *Das ernestinische Wittenberg: Spuren Cranachs in Schloss und Stadt*, eds. Heiner Lück, Enno Bünz, Leonhard Helten, Armin Kohnle, Dorothée Sack, und Hans-Georg Stephan, *Wittenberg-Forschungen*, Vol. 3, 313–362. Petersberg: Imhof.

Hennen, Insa Christiane. 2017a. Der Umbau des Closters zum Augusteum. Repräsentation und Gedenken unter Johann Friedrich I. und August von Sachsen. In *Das ernestinische Wittenberg: Die Leucorea und ihre Räume*, eds. Heiner Lück, Enno Bünz, Leonhard Helten, Armin Kohnle, Dorothée Sack, und Hans-Georg Stephan, *Wittenberg-Forschungen*, Vol. 4, 171–202. Petersberg: Imhof.

Hennen, Insa Christiane. 2017b. Gelehrtenwohnungen und Studentenbuden in Wittenberg. Ein Schlaglicht auf das Jahr 1520. In *Das ernestinische Wittenberg: Die Leucorea und ihre Räume*, eds. Heiner Lück, Enno Bünz, Leonhard Helten, Armin Kohnle, Dorothée Sack, und Hans-Georg Stephan, *Wittenberg-Forschungen*, Vol. 4, 427–468. Petersberg: Imhof.

Hennen, Insa Christiane. 2017c. Häuserliste ‚Universität'. In *Das ernestinische Wittenberg: Die Leucorea und ihre Räume*, eds. Heiner Lück, Enno Bünz, Leonhard Helten, Armin Kohnle, Dorothée Sack, und Hans-Georg Stephan, *Wittenberg-Forschungen*, Vol. 4, 445–68. Petersberg: Imhof.

Hennen, Insa Christiane. 2017d. Wittenbergs Stadtbild in der Reformationszeit. In *Initia Reformationis. Wittenberg und die frühe Reformation*, ed. Irene Dingel, Leucorea-Studien zur Geschichte der Reformation und der Lutherischen Orthodoxie [LStRLO] 33, 121–148. Leipzig: Evangelische Verlagsanstalt.

Hennen, Insa Christiane. 2020a. Bauen im „Ernestinischen Wittenberg:" Auftraggeber, Handwerker, Organisatoren. In *Das ernestinische Wittenberg: Residenz und Stadt*, eds. Leonhard Helten, Enno Bünz, Armin Kohnle, Heiner Lück, and Ernst-Joachim Waschke, *Wittenberg-Forschungen*, Vol. 5, 119–141. Petersberg: Imhof.

Hennen, Insa Christiane. 2020b. Ruhmreiche Gottesstadt, Sitz und Burg der wahren Lehre, heiligster Ort—Wittenberg als „Bühne" der Reformation. *Orte und Räume reformatorischer Kunstdiskurse in Europa*, ed. Bruno Klein, 34–47. Leipzig: Sächsische Akademie der Wissenschaften, Abh. d. SAW, Philolog. Klasse, Bd. 84, Heft 4.

Hennen, Insa Christiane. 2020c. Juden in Wittenberg und lutherische Judenfeindlichkeit. Zur Wirkungsgeschichte des „schweinischen Steingemähldes." In *Die „Wittenberger Sau." Entstehung, Bedeutung und Wirkungsgeschichte des mittelalterlichen Reliefs der sogenannten „Judensau" an der Stadtkirche Wittenberg*, eds. Jörg Bielig, Johannes Block, Harald Meller, and Ernst-Joachim Waschke, 65–91, Kleine Hefte zur Denkmalpflege 15. Halle/Saale: Landesamt für Denkmalpflege und Archäologie Sachsen-Anhalt.

Hennen, Insa Christiane. 2020d. Residenz—Universitätsstadt—Modell. Das Stadtbild Wittenbergs im 16. Jahrhundert. In *Die Stadt im Schatten des Hofes? Bürgerlich-kommunale Repräsentation in Residenzstädten des Spätmittelalters und der Frühen Neuzeit*, eds. Matthias Müller and Sascha Winter, 55–74. Ostfildern: Jan Thorbecke Verlag.

Jäger, Franz. 2011. Die Stube des gelehrten Bürgers. Zu den Wandinschriften im Haus Markt 3. In *Das ernestinische Wittenberg: Universität und Stadt (1486–1547)*, eds. Heiner Lück, Enno Bünz, Leonhard Helten, Dorothée Sack, und Hans-Georg Stephan, *Wittenberg-Forschungen*, Vol. 1, 191–197. Petersberg: Imhof.

Kettner, Paul Gottlieb. 1734. *Historische Nachricht von dem Raths-Collegio der Chur-Stadt Wittenberg*. Wolfenbüttel: Meißner.

Lang, Thomas. 2015. Simprecht Reinhart: Formschneider, Maler, Drucker, Bettmeister. In *Das ernestinische Wittenberg: Spuren Cranachs in Schloss und Stadt*, eds. Heiner Lück, Enno Bünz, Leonhard Helten, Armin Kohnle, Dorothée Sack, und Hans-Georg Stephan, *Wittenberg-Forschungen*, Vol. 3, 93–138. Petersberg: Imhof.

Lang, Thomas and Anke Neugebauer. 2017. Die Leucorea, Wittenberg und das Reich: eine Universitätsgründung und ihr kulturelles, personelles und politisches Umfeld. In *Das ernestinische Wittenberg: Die Leucorea und ihre Räume*, eds. Heiner Lück, Enno Bünz, Leonhard Helten, Armin Kohnle, Dorothée Sack, und Hans-Georg Stephan, *Wittenberg-Forschungen*, Vol. 4, 11–52. Petersberg: Imhof.

Leonhard, Hans. 1902, *Samuel Selfisch. Ein deutscher Buchhändler am Ausgange des XVI. Jahrhunderts*. Dissertation. Leipzig: Jäh & Schunke.

Meller, Harald, ed. 2014. *Heavy metal. Bewegliche Lettern für bewegende Töne*. Halle a. d. Saale: Landesamt für Denkmalpflege und Archäologie Sachsen-Anhalt—Landesmuseum für Vorgeschichte Halle.

Oehmig, Stefan, ed. 2015. *Buchdruck und Buchkunst im Wittenberg der Reformationszeit*. Schriften der Stiftung Luthergedenkstätten in Sachsen-Anhalt, 21. Leipzig: Evangelische Verlagsanstalt.

Pettegree, Andrew. 2016. *Die Marke Luther*. Trans. Ulrike Bischoff. Berlin: Insel.

Reske, Christoph, and Josef Benzing. 2015. *Die Buchdrucker des 16. und 17. Jahrhunderts im deutschen Sprachgebiet. Auf der Grundlage des gleichnamigen Werkes von Josef Benzing*, 2. ed. Wiesbaden: Harrassowitz.

Rothe, Vicky. 2013. Wittenberger Buchgewerbe und -handel im 16. Jahrhundert. In *Das ernestinische Wittenberg: Stadt und Bewohner*, eds. Heiner Lück, Enno Bünz, Leonhard Helten, Armin Kohnle, Dorothée Sack, und Hans-Georg Stephan, *Wittenberg-Forschungen*, vol. 2.1, 2.2, 77–90. Petersberg: Imhof.

Rüger, Conrad Alfred. 1978. *Samuel Selfisch d. Ä. Typographus et Bibliopola Vitebergensis 1529–1615*, Masch. (Copy of the Deutsche Bücherei Leipzig).

Schirmer, Uwe. 2015. Buchdruck und Buchhandel im Wittenberg des 16. Jahrhunderts. Die Unternehmer Christian Döring, Hans Luft und Samuel Selfisch. *Buchdruck und Buchkultur im Wittenberg der Reformationszeit*, ed. Stefan Oehmig, 169–189, Schriften der Stiftung Luthergedenkstätten in Sachsen-Anhalt, 21. Leipzig: Evangelische Verlagsanstalt.

Titze, Mario. 2020. Die Sau an der Kirche. Kunsthistorische Fragen an ein viel diskutiertes mittelalterliches Bildwerk. In *Die „Wittenberger Sau." Entstehung, Bedeutung und Wirkungsgeschichte des mittelalterlichen Reliefs der sogenannten „Judensau" an der Stadtkirche Wittenberg*, eds. Jörg Bielig, Johannes Block, Harald Meller, and Ernst-Joachim Waschke, 65–91, Kleine Hefte zur Denkmalpflege 15. Halle/Saale: Landesamt für Denkmalpflege und Archäologie Sachsen-Anhalt.

Insa Christiane Hennen is an art historian, postdoctoral research fellow of Stiftung LEUCOREA Wittenberg and an independent scholar. She researches on the history of Reformation and its consequences, especially in respect of the arts, the urban society and development of Wittenberg as a central place of the Lutheran Reformation. Moreover, her research develops in the field of preservation of ancient monuments and buildings. She is member of the Commission of the art history of Mitteldeutschland of the Saxon Academy of Sciences and Humanities Leipzig, of ICOMOS, and of the association of Förderverein Hofgestüt Bleesern e.V.

Chapter 5
Scholars, Printers, and the Sphere: New Evidence for the Challenging Production of Academic Books in Wittenberg, 1531–1550

Saskia Limbach

Abstract This chapter introduces those printers and publishers who were involved in the process and considers the economics of the local print industry, which was, at the time, the fastest-growing in the entire Holy Roman Empire. By analyzing the university's interactions with book producers, especially with respect to Melanchthon's letters, which reveal his close ties to the book industry, I argue that even in this dominant center of printing, the relationship between academics and printers/publishers could be rather fraught; authors and editors even referred to the producers of their books as "beasts," "harpies," and "men of iron." Drawing on hitherto unexplored sources, I also shed light on the prices of academic books, their print runs, and the reuse of illustrations in different editions. Finally, I establish how students in sixteenth-century Wittenberg could obtain academic books for their studies and how expensive the *Sphaera* was in comparison to other books and commodities.

Keywords Academic book · Wittenberg · *Tractatus de sphaera* · Philipp Melanchthon · Book market

1 Introduction[1]

In 1562, the mathematician Georg Joachim Rheticus (1514–1574) reflected bitterly on his work on the *Sphaera*. He had edited the book in 1538, which, after its publication in Wittenberg, had become very influential. Yet, the interaction with publisher Moritz Goltz (ca. 1495–1548) on this occasion had tainted Rheticus's memories: for his tedious work, Rheticus had only received a small reward. The money had not

---

[1]I wish to thank Matteo Valleriani and Andrea Ottone for their very helpful comments and the other contributors to this volume for a thought-provoking discussion at the workshop in Berlin. I would also like to express my gratitude to the Reformationsgeschichtliche Forschungsbibliothek in Wittenberg for awarding me a four-month fellowship, which allowed me to shed more light on the production of academic publications in sixteenth-century Wittenberg.

S. Limbach (✉)
Georg-August-Universität Göttingen, Göttingen, Germany
e-mail: saskia.limbach@theologie.uni-goettingen.de

M. Valleriani and A. Ottone (eds.), *Publishing Sacrobosco's* De sphaera *in Early Modern Europe*, https://doi.org/10.1007/978-3-030-86600-6_5

even covered the cost of the beer he had consumed on the job, Rheticus claimed in 1562. Resentfully, he determined that publishers were "beasts" who were used to "getting everything for nothing" (Rosen 1974, 247).

Although the text of the *Sphaera* has been well studied, little is known about the conditions of its actual production. In the first half of the sixteenth century, Wittenberg was by far the predominant center for publishing Johannes de Sacrobosco's (1195–1256) influential book (Valleriani 2017, 443; Valleriani et al. 2019). Between 1531 and 1550 a new edition was published nearly every second year. The many books were produced by different printers, as Wittenberg did not yet employ a designated university printer who would produce for the institution in return for an annual salary. Therefore, many printers shared the market, and when the business of books started booming, external publishers became increasingly involved. Both printers and publishers cared first and foremost about the salability of their products, a fact that often upset authors and editors alike. Martin Luther's (1483–1546) colorful insults to "money-grabbing" and "incompetent" printers have already attracted attention from historians (Grafton 1980, 278). As we will see in the following, other academics in Wittenberg shared these negative sentiments.

This paper draws on hitherto unexplored sources, such as university announcements, and considers the economic aspects of producing the *Sphaera* in Wittenberg. First, it will provide a short overview of the print industry in Wittenberg, focusing in particular on publishers and payments to contextualize Rheticus's utterance about his poor reward. It will then concentrate on the publication of academic books and will analyze agreements and problems between the university and the print industry. The differences between academics and printers/publishers are even more tangible in Philipp Melanchthon's (1597–1560) extensive correspondence. In his letters, the illustrious professor makes abundant remarks about the local printing houses, the production process of particular books, and the book trade more generally. Therefore, a close examination of the letters follows, which shows vividly that Melanchthon often acted on behalf of other authors, which made him an important intermediary between the scholarly world and the print industry. These interesting details about Melanchthon's vital connections to the print industry have not yet received much attention from historians and will thus be discussed here more extensively.

Melanchthon's comments about book production in Wittenberg shed more light on the printers of the *Sphaera* and their sometimes difficult economic situation. Hence, the next section will introduce those who were involved in the print production of Sacrobosco's key text. A closer look at their finances, their production, as well as their relation to the university will help us understand why so many different printers were involved in the production of the *Sphaera*. It will also pinpoint the relationship the producers had to the university, especially Melanchthon, who initiated the first edition of the *Sphaera* in Wittenberg. Finally, this paper will address the economics of the academic book trade, giving rare information about the price of a copy of the *Sphaera*. This allows us to assess how affordable the text was in relation to other books and commodities at the time. The last section will also include an estimation of the print run and will explore how students could obtain a copy of the *Sphaera* if they could not afford the latest printed edition. Taken together, this new evidence will offer us a unique window into the rough business of producing academic books in Wittenberg.[2]

2 Printers, Publishers, and Payments

The rapid rise of the print industry in sixteenth-century Wittenberg is intriguing. In 1500, Wittenberg was a relatively small town with about 2,000 inhabitants and no print shop (Lück 2011, 15). Some fifty years later, the number of inhabitants had doubled, and by the end of the century Wittenberg was one of the most productive print centers in the entire Holy Roman Empire (Chap. 4).[3] Over the course of the century, nearly 10,000 titles were printed in this small town in the northeastern part of the empire—more than, for instance, in Augsburg (6,000).[4] In both centers, the first half of the 1520s was a very productive period (Kaufmann 2019, 226); yet, the difference becomes more obvious when one considers the volume of production by the total number of sheets required for each edition.[5] In the first half of the sixteenth century, there were some peak years in which the printers in Wittenberg collectively produced over 2,000 sheets; their colleagues in Augsburg never reached such a number (Thomas 2018, 519, 2021, 166–200). The year Martin Luther died (1546) is particularly striking: Wittenberg's annual output of printed sheets rose to over 2,200, whereas Augsburg's was only 500.

These many books were produced by a number of different printers: taken together, thirty-eight print shops existed in Wittenberg in the sixteenth century (Reske 2015, 35). Again, this number was higher than in many other places in the empire, including major cities like Vienna, which had some 50,000 inhabitants. There, only thirty print shops operated during the entire century. Beginnings in Wittenberg were moderate

---

[2] Throughout the paper the following denominations will occur: 1 Gulden = 21 Groschen.

[3] USTC searching Holy Roman Empire and 1500–1600—Wittenberg: 9,929 editions; Cologne: 8,891 editions; Nuremberg: 8,705 editions; Leipzig: 8,599 editions. Note: VD16 lists lower figures for the same period, as it only has limited coverage of books in libraries outside modern-day Germany and does not include single-sheet items. During her time in the USTC project, the author added hundreds of broadsheets to the USTC database, such as for Cologne (350 editions), see (Limbach 2021).

[4] USTC lists 9,929 editions for Wittenberg 1500–1600 (VD16 lists 9,632). For Augsburg, USTC lists 6,007 editions (VD16 lists 5,564).

[5] Depending on the format of the book, one printed sheet was folded into two leaves (folio), four leaves (quarto), or eight leaves (octavo). With the exact bibliographical details on the number of leaves and the format we can therefore reconstruct how many printed sheets were used to make up one copy of one edition. Although this approach neglects print runs, which could vary significantly (but for which we unfortunately still lack evidence), it provides us with a much better estimation of the production output than just counting the editions (which gives equal consideration to a large Bible and a small pamphlet).

and the shops were small, with one or two presses. In August 1521, for instance, Luther spoke of a total of six presses in his hometown, two belonging to each of the three printers (Kaufmann 2019, 277). In 1534, however, the successful printer Hans Lufft (1495–1584) alone had six presses in his print shop (Schirmer 2015, 178), and in May 1587 there were seven different print shops operating at the same time (Friedensburg 1926, 547).

The outstanding productivity in sixteenth-century Wittenberg is the result of the Reformation movement and one of its most important figures, Martin Luther. With the publication of his Ninety-Five Theses and the ensuing battle with the Catholic Church, Luther became one of the most productive writers in the early modern period. His texts were eagerly awaited by an ever-growing audience, and printers did their best to keep up with this demand. Within only a few years, Luther had become the most published author in the empire, with printers in many different cities reproducing his texts (Pettegree 2015, 105). Printers from other towns flocked to Wittenberg to work closely with the reformer and his colleagues, and, as a result, the book trade as a whole prospered. Soon the number of members of the trade grew: whereas the number of masters and journeymen working as bookbinders remained more or less the same—twenty-four in 1560 and twenty-six in 1590—the number of men working as printers (both as masters and journeymen) rose from twenty-seven (1560) to thirty-nine (1590).[6] The number of type-casters even quadrupled from two (1560) to eight (1590) (Schirmer 2015, 187).

This buzzing activity also attracted the attention of publishers—men who selected texts, financed their print production, and often sold the finished books. For sixteenth-century Germany, the differentiation between printers and publishers (and booksellers for that matter) is not always clear cut.[7] In many cases, printers acted themselves as publishers, printing a text on their own initiative or interacting directly with the author or an intermediary, such as Melanchthon (see Sect. 4). In other cases, certain books were commissioned by external publishers who acquired the texts, calculated how many copies of the books could be sold, and finally paid the printer for the printed copies. Unfortunately, these interactions are hard to trace for Wittenberg, as the books often lack concrete evidence in the colophons. By contrast, the successful publisher Antoine Vérard (fl. 1485–1512) in Paris frequently used the colophons of his books for "self-fashioning" and for providing buyers with the addresses of his shops (Mullins 2014, 81–82). Without such information, the activities of local publishers only become visible in letters, contracts, and other documentary evidence.

These sources show us just how powerful and rich publishers could become in sixteenth-century Wittenberg. Over the course of the century, at least twenty-three men earned their living as publishers or booksellers in Wittenberg, and some of

---

[6] By way of contrast—the number of bookbinders was never as high in Leipzig: in 1506, there were only three; in 1529 there were fifteen; in 1554 there were only ten; and in 1558 there were twelve (Schirmer 2015, 172).

[7] All Wittenberg publishers discussed in this chapter were also working as booksellers and are referred to as *bibliopolis* in primary sources.

them made a fortune (Rothe 2013, 89). A tax list reveals, for instance, that the three richest men in Wittenberg in 1542 were Martin Luther, with a total capital of some 7,000 Gulden, closely followed by the two publishers Bartholomäus Vogel (1504–1569), with 6,000 Gulden, and Moritz Goltz, with 5,000 Gulden. The latter even lent Luther money (Clemen 1942–1943, 166). Both publishers were also part of the local government for years—Goltz from 1543 to 1546, and Vogel even from 1554 to 1569 (Chap. 4).

This success came at a price. Early in the century, in the mid-1520s, the printer Georg Rhau (1488–1548) remarked in a letter that Wittenberg publishers had become very powerful and imposed harsh conditions on the local printers (Claus 2002, 79). Publishers were scrooges, Rhau complained, and left printers hanging for weeks until—all of the sudden—they demanded that certain work should be published immediately. If he did not need the money so urgently, Rhau stated, he would not work for any publisher in Wittenberg. The fact that this statement was not made by a smaller printer, but one of the most famous, further underlines how dependent printers were on publishers in the small university town.

Similarly, publishers could be hard on the editors who prepared a text before it went to the print shop. As we noted above, Georg Joachim Rheticus stated that he received very little compensation for his work on the *Sphaera* and the *Computus* (Burmeister 1968, 162–164). In fact, the publisher Moritz Goltz offered a compensation so small that Rheticus claimed he could not even cover the expense of the beer he drank during his editorial work. Rheticus concluded that the "beasts" (i.e. the publishers) were used to getting "everything for nothing." Unfortunately, we do not know how much Rheticus earned exactly, but other examples from the first half of the sixteenth century show that such work was indeed not very well paid. In 1528, a corrector in Wittenberg received two Groschen per printed sheet; if more effort was required, such as preparing the manuscript for print or producing translations or compilations, the pay rose to 15 Groschen per sheet (Clemen 1941, 177–178). By contrast, a Wittenberg messenger earned between five Groschen for delivering a letter to Weimar (fifty kilometers distance) or 6 Groschen for transporting Lucas Cranach's (1471–1553) artwork to the same destination; and a chancery scribe spent 15 Groschen on room and board for three nights in Wittenberg while he supervised the print production of mandates concerning coins (Lang 2015, 123).[8]

The payments of two Groschen and 15 Groschen respectively were, however, paid by the printer Joseph Klug (d. 1552). If publishers were involved in the publication the prices could differ, and reimbursements were lower and later than expected. In 1544, the publishers Goltz, Vogel, and Christoph Schramm (d. 1549) asked Georg Rörer (1492–1557) and Veit Dietrich (1506–1549) to work on Luther's lecture on Genesis (Volz 1963, 115).[9] Rörer was supposed to fulfill the main task, receiving 21

---

[8] Thanks to Thomas Lang who also informed me that the weekly rent for a chamber suited for four people cost 5 Groschen and a pair of simple shoes cost between 2 and 5 Groschen.

[9] The work was delayed, mostly because of the Smalkaldic War, and eventually appeared in Nuremberg (Luther and Dietrich 1550).

Groschen, and Dietrich was to correct everything Rörer was not able to, receiving 14 Groschen per sheet.

In theory, this was more than the printer Klug had offered his editors—but the reality looked rather different. The publishers did not pay in a timely manner, and when they finally did pay, it was less than expected, as they shamelessly stated they had no money.[10] This caused the two editors Rörer and Dietrich to fight over the little money they had received. In 1545, Dietrich's mentee Hieronymus Besold (1520–1562) had to step in and mediate between the two editors.[11] In his letter, Besold gently pointed out to Dietrich that he, Dietrich, had made a mistake in his calculations; he had, in the end, agreed to the rate of 12 Groschen per sheet (Besold was careful to emphasize that Rörer had shown him three (!) of Dietrich's letters to double-check this lower rate). Besold continued that it was nearly impossible to get more than this rate from the publishers, and added as a consolation that Rörer was also "treated not very gently by those harpies."[12] This was an obvious dig at the publishers. Although they had made an agreement that detailed the exact rates for the editorial work, the publishers underpaid their editors. Once again, it seems, the "harpies" had gotten away with "everything for nothing."

The utterance that 12 Groschen per sheet was the maximum Goltz would pay an editor allows us to speculate on the payment Rheticus may have received for his work on the *Sphaera*. The 1538 edition was printed in octavo, consisting of 108 leaves. This meant at least thirteen and a half sheets for the production, probably fourteen sheets, as we need to consider one or two extra leaves for the additional material of the volvelles (see Sect. 6). If Rheticus was indeed paid the rate mentioned above, he would have received 168 Groschen or 8 Gulden respectively. This was indeed not much, especially when compared to the high prices for beer in 1538. Later the same year, the guests at a peasant wedding in Hohendorf consumed no less than 28 Gulden worth of beer on this single occasion (Friedensburg 1926, 198). Admittedly, this was an extreme example, cited by Wittenberg academics to illustrate the outrageous beer prices that year, which they blamed on the alleged profit-seeking peasants. The comparison shows, nevertheless, that Rheticus was indeed not able to buy much beer for his work on the *Sphaera*, especially since he was particularly fond of the beverage (Burmeister 2015, 19).

The publisher Moritz Goltz was one of the key players in Wittenberg. Together with two of his colleagues, Goltz had secured a privilege for the Bible from the Elector of Saxony in 1533, allowing them to be the only publishers of Bibles in the entire territory (Claus 2002, 89). This provided him with great wealth, and four years later he was able to even buy the bookshop of one of his colleagues for the impressive sum of 4,000 Gulden (Volz 1964, 637–638). Goltz had a strong network in Wittenberg: his daughter married a publisher, and her sister-in-law later married

---

[10] In this instance, Goltz claimed he simply lacked cash. In another instance, Schramm stated he had no money at all (Clemen 1942–1943, 116).

[11] See the letter written by Hieronymus Besold to Veit Dietrich on September 13, 1545, edited in (Albrecht and Flemming 1913, 170–173).

[12] The original Latin text reads: "non leniter exercetur ab harpijs illis."

Samuel Selfisch (1529–1615), another publisher who even traded with paper, owned a bookbinder's shop as well as a print shop, and acted as a creditor (Rothe 2013, 85; Schirmer 2015, 189). We know that Goltz also worked as a bookseller and that he had his shop in Elsterviertel 102 (Chap. 4). As we have seen above, he acted as a creditor, too, at least for Luther. Similarly, we can assume that he also traded with paper (Schirmer 2015, 181).

Although Goltz was one of the biggest publishers in Wittenberg, his name appears only very rarely in the colophon of the books he financed. Both in 1540 and 1541, Goltz financed editions of various parts of the Bible in Low German, printed by Hans Lufft in Wittenberg (Luther 1540, 1541a, b, c; Luther et al. 1541a, b). A few years later, he also commissioned two Low German Bibles in Magdeburg (Luther et al. 1545a, b). Taken together, only eight publications specifically mention his involvement in the colophon. Naturally, he published much more than this. Apart from the Bible privilege he held with his colleagues, he was also involved in the production of grammar books and song books, as we can see from an agreement he made with the printer Joseph Klug (Rothe 2013, 86). In 1539, Goltz wanted Klug to print 2,000 copies of Urbanus Rhegius's (1489–1541) *Dialogus* (Rhegius 1539), 3,000 copies of Melanchthon's grammar, 3,000 copies of a large song book (in octavo), 3,000 copies of a small song book, 3,000 copies of Melanchthon's *Syntaxis*, and 3,000 copies of a certain "sixteenth chapter."[13]

Goltz was certainly targeting the academic market and hired—as we have already seen—Rheticus to edit Sacrobosco's *Sphaera*. In this context, it is particularly telling that Goltz's name also comes up in the university announcements, the *scripta publice proposita*. This was a collection of what were originally handwritten announcements or printed single sheets at the University of Wittenberg. They were quite diverse, including ordinances, obituary notices, lecture advertisements, feast day announcements, invitations to disputations, congratulations, poems, and announcements for charity collections (Domtera-Schleichardt 2014, 565). From 1545 onward, the professor Johannes Marcellus (1510–1551) collected these announcements and had them produced as books (*Scripta* 1545–1546, 1548, 1549a, 1549b, 1551). After Marcellus's death, Michael Maius (ca. 1530–ca. 1572) continued the splendid collection, issuing new volumes as well as editing the old ones (*Scripta* 1553, 1556, 1559, 1560, 1561, 1562, 1564, 1568, 1570, 1572).[14]

In the edition of 1553, which covers the announcements of the years 1540–1553, Erasmus Reinhold (1511–1553) refers specifically to Goltz (*Scripta* 1553, leaves Er–E2r). In 1542, Reinhold started an announcement with the statement that many students wanted to hear him lecture on arithmetic but the lack of necessary books prevented him from doing so (Burmeister 2015, 40). Therefore, Reinhold had suggested to have Gemma Frisius's (1508–1555) work published, a textbook which would prove very popular, with twenty-two editions printed in Wittenberg alone (Burmeister 2015, 229; Frisius 1542). Reinhold continued his announcement

---

[13] The contract between Goltz and Klug was later annulled, see below.

[14] Editing the old collection comprised the inclusion of new texts, rearranging the collection and assigning different authors to the texts, see (Domtera-Schleichardt 2014, 568).

by saying that the copies of the book would be finished "the next day," and everyone who was interested in the subject could obtain a copy from "Moritz" (*apud Mauricium*). This can only refer to Moritz Goltz since the book was printed by Joseph Klug and no other publisher in Wittenberg was named Moritz at that time. Moreover, the fact that the publisher is mentioned only by his first name shows that he must have been very familiar to the students. They probably frequently visited his shop to buy their books. Given that Goltz was also financing the production of the *Sphaera*, it seems likely that Goltz not only sold Frisius's book but was also involved in its production. Presumably, Reinhold had contacted Goltz in the first place about printing the book, and then the publisher had it prepared and produced.

The actual labor of printing university texts, such as this arithmetic book, could be undertaken by various printers, provided they had the skill and the necessary equipment. As there was no designated university printer in sixteenth-century Wittenberg who printed solely for the institution and received an annual salary in return, many printers shared the market, as was the case in Basel at the time (Limbach 2017, 397). On the one hand, this was an advantage, as the competition between printers prevented the inflation of prices. In Louvain, for instance, where there was no shared market, many academics resented the high cost of printing dissertations (Walsby 2017, 357). On the other hand, it meant that printers in Wittenberg were free to choose the texts they produced and could reject manuscripts if they deemed the publication to be potentially less profitable than others. As we shall see in the next section, in 1587 the members of the medical faculty in Wittenberg experienced such rejection first hand.

3 Printing for the University of Wittenberg: Texts Between Intellectual and Economic Ambitions

Before the groundbreaking reforms of the Wittenberg professors, it was mostly private individuals who ran print shops in the university town. These privately operated print shops (*Privatpressen*) were financed by scholars who printed texts more out of ideological than economic interests (Reske 2017, 38). The first printer in Wittenberg, Nikolaus Marschalk (ca. 1470–1525), serves as a good example. He was primarily a Greek teacher who also studied law. When he was asked by Elector Frederick the Wise (1463–1525) to set up a print shop—in the same year the university was founded (1502)—Marschalk readily obliged (Reske 2015, 1076). Marschalk employed a printer who had produced a total of thirteen publications in quarto format by the end of the following year (Reske 2017, 43). Among these publications were, however, some texts criticizing the scholastic teachings at the university; Marschalk was, in turn, criticized for publishing them. It is no surprise that once Marschalk received his doctorate in jurisprudence he did not think long before leaving Wittenberg and making a living somewhere else. The print shop remained in the hands of his student, Hermann Trebelius (b. ca. 1475) (Reske 2015, 1077).

For the better part of the 1500s and 1510s, the production of printed documents was limited in Wittenberg. Trebelius, as well as Wolfgang Stöckel (ca. 1473–ca. 1541)—another printer who had come to Wittenberg—could not keep their presses running for long. There was also temporarily a printing press in the Wittenberg castle (Lang 2015, 125–127), but it was used infrequently and primarily for official documents, such as mandates and instructions. It was Johannes Rhau-Grunenberg (d. 1529), who had come to Wittenberg in 1508, who kept a print shop open for nearly two decades. He was the first to produce Luther's books.

After a while, however, there were again problems between scholars and printers. When Rhau-Grunenberg became ill in 1513, a number of students complained to the Duke of Saxony that they were not able to get enough books for their studies (Friedensburg, 1917, 71–72), and when the reformatory movement gained momentum, the small shop of Rhau-Grunenberg soon proved inadequate to cater for the needs of the prolific author. Therefore, Luther called another printer to the university town—Melchior Lotter the Younger (ca. 1490–ca. 1545). In the following years, Luther occasionally entrusted Rhau-Grunenberg with the production of texts, but not without a certain degree of resentment. When the printer produced Luther's *Von der Beicht* (1521), the author was furious with the result. In his eyes, the text was barely legible—the paper quality was low and the typesetting dirty (Luther 1521; Reske 2015, 1078). On top of that, Luther grieved, the printer was exceptionally slow and simply useless.

The newly arrived Melchior Lotter was, therefore, more than welcome. One of his first major achievements was the print production of the New Testament, finished in September 1522. To produce it, Lotter had to acquire another press as the two presses in his print shop did not suffice (Schirmer 2015, 173). The book was produced in 3,000 copies, of which three were sent to the Elector of Saxony for the price of 15 Groschen each (Schirmer 2015, 173). The book contained twenty-one large woodcuts, each covering an entire page, and the binding of the loose sheets cost another 7 Groschen (Schirmer 2015, 173–174). The remaining copies were sold so quickly that a second edition was produced with much haste. As it required significant investment, the goldsmith Christian Döring (d. 1533) as well as the famous artist Lucas Cranach entered the market as publishers. They started printing on their own the following year, leaving Lotter's position increasingly vulnerable. It certainly did not help that at some point Lotter struggled with a local bookbinder and, when the situation escalated, punctured the bookbinder's nose with a stitching awl (Schirmer 2015, 174). Eventually, Lotter left Wittenberg. Döring and Cranach continued their work with the printers Joseph Klug and Hans Lufft, who both came to Wittenberg between 1522 and 1523. A few years later, Cranach left the publishing business. Döring later made serious miscalculations that resulted in large losses. Luther tried to solve Döring's problems by persuading the Elector of Saxony to provide the publisher with a privilege for Luther's Bible, but it did not help (Schirmer 2015, 179). Shortly after he received it, Döring had to sell the privilege to three men who would become very successful publishers: Christoph Schramm, Bartholomäus Vogel, and Moritz Goltz.

The agreements between publishers, printers, and the university at that time are difficult to reconstruct. It was not until 1616 that the scholars fixed prices for academic publications with the printer Johann Gormann (d. 1628) (Friedensburg 1927, 31–32). According to this agreement, lecture announcements and notifications of festivities and graduations should cost no more than nine Groschen each, and the printer had to supply seventy copies (in the desired format) to the university. This way the copies could be given out at the institution. Disputations, on the other hand, should cost no more than 12 Groschen per sheet, and again Gormann had to supply seventy copies to the university. If a disputation covered more than one sheet, however, the respondent had to pay for the extra costs. Depending on the print run, the print shop, and the year, this could be quite expensive. From one of Melanchthon's letters, we know that in 1552 the printer Veit Kreutzer (d. 1578) usually charged 24 Groschen per printed sheet (presumably for a print run of several hundred copies).[15] By 1616, when the university fixed the prices for disputations, printers probably charged less than that, but it must nevertheless have meant a significant investment if the respondent wanted to print a disputation that exceeded one printed sheet.

Until 1616, the terms of university publications depended on individual agreements, probably similar to the agreements that were made in sixteenth-century Leipzig.[16] For Leipzig, three interesting examples illuminate the background of the local production and sale of academic books. In 1503, the arts faculty made an advance payment of 30 Gulden to the printer Martin Landsberg (d. 1523) (Chap. 12) for the production of John Peckham's (1227–1292) *Perspectiva communis*—a book that had not been printed in Leipzig before (it appeared a year later) (Peckham 1504). However, the book did not sell well and the printer was unable to pay back his advances. In the end, the faculty had to buy the remaining copies for a set unit price. Similarly, in the 1520s, a printer was left with a significant number of copies of a Silius Italicus (26–101) edition that he had produced nearly twenty years before (Italicus 1504). So, in 1522, the printer asked the university to focus the next compulsory rhetoric lecture on Silius Italicus so he could sell more copies (this is a rare example of a printer trying to influence university teaching). Lastly, in 1519, the arts faculty negotiated with a printer about a new Aristotle (384–322 BCE) edition. Both the *Physica* and the *Metaphysica* were to be produced in 300 copies (Aristoteles 1519, 1519–1520). Yet, the production and sale of the books proved to be far from straightforward. Not only did the corrector steal the manuscript (!) but when the sales were slower than expected, the faculty had to agree to buy twenty copies—despite the fact that it had already provided the printer with the significant fee of 80 Gulden. Surely such agreements existed in Wittenberg as well. A close examination of the documents regarding the finances of the University of Wittenberg would yield more evidence (the documents are now in the collection of the university archive in Halle).

---

[15] Philipp Melanchthon to Noah Buchholzer ([Wittenberg], July 9, 1552), MBW 6491, mentioning that the printer usually takes one Taler (= 24 Groschen) per sheet.

[16] The following Leipzig examples are taken from (Eisermann 2009, 166–167). I'd like to thank Falk Eisermann for his valuable input on my paper.

In Wittenberg, there was certainly tension between printers and academics, especially in the later sixteenth century. In May 1586, the new Elector of Saxony had some
of his councilors inspect the University of Wittenberg (Friedensburg 1917, 322). In
the report, members of the various faculties disapproved of the local printers: many of
their books were useless and should not have been printed in the first place; texts were
full of mistakes; printers did not allow deans to double-check their own announcements before they were printed; and the city council, which had jurisdiction over the
printers, did nothing to solve the problems (Friedensburg 1926, 514–527). A member
of the medical faculty even reported that none of the local printers would take on
medical publications (Friedensburg 1926, 525). In the end, the elector issued a new
ordinance for the university, regulating that printers could only produce a book if
the rector, the four deans, and the responsible faculty had previously seen the text
and signed off on it (Friedensburg 1926, 567–568). On top of that, correctors should
work more diligently, and—most importantly—printers were from then on not only
subject to city but also university jurisdiction.

There were certainly problems in the first half of the sixteenth century as well,
but most of them were probably discussed directly from face-to-face. Thus, we have
very few sources that give us insights into the interaction between scholars and
printers. In this context, Melanchthon's extensive correspondence sheds more light
on the collaboration between authors and those who produced their texts. The many
letters Melanchthon wrote when he was traveling, and the letters he received from
authors outside Wittenberg paint an interesting picture of just how problematic it
could be to publish a book (even in such a buzzing print center like Wittenberg), why
printers refused to print some publications, and why the production of books could
be delayed.

4 Melanchthon's Close Ties to the Book Industry

As soon as Melanchthon arrived in Wittenberg (1518), he was much more than
just an author who gave his manuscripts to printers. He took a great interest in
the production of texts written by many different authors. In the mid-1520s, for
instance, Melanchthon felt the need to have Lambertus's (ca. 1025–ca. 1081) chronicle printed.[17] He instructed a printer to visit the library where the manuscript was
kept, make a copy of it, and give it to a professor to revise for its print production. The
book finally appeared in August 1525 (Lambertus 1525).[18] Melanchthon's involvement with the print industry had started early in his life. Already as a student, he had
worked as a corrector and editor in the town of his alma mater, Tübingen (Widmann
1971, 33–34). There, he had looked up to the printer Thomas Anshelm (d. 1523),

---

[17] Philipp Melanchthon to Caspar Churrer in Tübingen (Wittenberg, [after March 1523–1524]),
MBW 304.

[18] Lambertus's book was eventually produced by a different printer.

who was one of the first printers in the German-speaking area to produce Greek texts (Steiff 1881, 89).

Once in Wittenberg, Melanchthon continued to work closely with local printers, as well as printers in Augsburg, Basel, Hagenau, Leipzig, and Erfurt (Kaufmann 2019, 88). This made the academic a valuable informant for outsiders. Melanchthon's extensive knowledge of the local book trade was, for instance, called upon in a criminal investigation. In 1543, the mayor of Nordhausen, Michael Meyenburg (1491–1555), was trying to track down the printer of a lampoon and asked Melanchthon for help.[19] The professor first examined the watermark, which could potentially lead him to the producer of the paper and thus narrow down the group of suspected printers who bought paper from him. After careful investigation, however, Melanchthon determined that none of the printers in Wittenberg used paper with that watermark; the publication, therefore, could not have been produced in the university town. Melanchthon even interrogated one of the Wittenberg printers he suspected (twice!), but the printer had not admitted to the deed. Melanchthon suspected that the lampoon had been produced in Erfurt or by the printer Henning Rüdem (d. 1553), who used to work in Wolfenbüttel and had moved to Hildesheim in the meantime (Reske 2015, 405–406).

This case is interesting for two reasons: first, it shows vividly just how well-informed Melanchthon was about the print industry in Wittenberg and beyond. Secondly, it seems telling that Meyenburg, as a mayor, did not rely on the city council of Wittenberg, although the local printers were at that time solely under its jurisdiction.[20] Even if Meyenburg had previously contacted the council about the issue, he was keen for Melanchthon's assessment. Presumably, Meyenburg did not fully trust the city council with the matter, as it had just accepted the publisher Moritz Goltz as its member the same year. Goltz could have been involved in the production of the lampoon and may have hindered Meyenburg's investigation. Melanchthon was therefore a much more neutral source of information.

Many from outside Wittenberg also saw Melanchthon as the ideal intermediary between the scholarly world and the print industry. On multiple occasions, he received manuscripts from authors in other cities and helped them to find a suitable printer for the texts. Such requests to other, more prolific authors were not unusual. When the scholar Joannes Moibanus (1527–1562) suffered from a fatal illness, he sent his manuscript to the renowned Conrad Gessner (1516–1565), asking him to finish and publish it (Blair 2017, 11–12). Similarly, the Italian professor Girolamo Mercuriale (1530–1606) frequently asked professor Theodor Zwinger (1533–1588) in Basel for his help in getting Italian authors (including himself) published in Basel (Siraisi 2008,

---

[19] Philipp Melanchthon to Michael Meyenburg in Nordhausen ([Wittenberg, December 1543]), MBW 3417.

[20] It was not until the late 1580s that Wittenberg printers were also subject to university jurisdiction (see below).

80–84).[21] Basel was one of the most important publishing centers in the German-speaking area, and Mercuriale hoped to reach a much larger audience for his books with their publication in the city on the Rhine (Leu 2014, 61).

Equally, many authors hoped to publish in the famous Wittenberg and asked if Melanchthon could be of assistance.[22] In 1556, for instance, David Chytraeus (1530–1600) sent a manuscript to Melanchthon asking the latter to pass it on to a local printer who possessed Greek type, as the printer in Chytraeus's home town, Rostock, did not own such material.[23] Once he had reached an agreement with a printer, Melanchthon would also send updates about progress, as for instance in 1560, when he received the funeral sermon commemorating Duke Philip of Pomerania (1515–1560) written by Jakob Runge (1527–1595).[24] In his letter to Runge, Melanchthon noted that he had given the manuscript to the printer Georg Rhau, who had already produced twelve copies, and that the remaining copies would be out in time for the Leipzig fair.

If necessary, Melanchthon also explained why there were difficulties in the print production or why the publication took longer than expected. In December 1539, for instance, Jakob Schenck (ca. 1508–1546) in Weimar inquired why his collection of sermons had not yet appeared though Melanchthon had given it to the printer nearly four months earlier.[25] Melanchthon was quick to assure him that the delay was not his fault.[26] It just took longer because the printer had given the manuscript to Martin Luther for corrections and he simply had not had time for them yet. After all, Melanchthon reminded the author, Luther was burdened with important matters at the moment. Another collection of sermons, this time written by Antonius Corvinus (1501–1553) in 1535, was delayed as well. In his letter to the author, Melanchthon argued that the printer, Georg Rhau, simply did not have a free press available to

---

[21] Many thanks to Ann Blair for this reference and for her reading suggestions on scholars acting as "publishing agents" for other scholars.

[22] E.g., Philipp Melanchthon to Joachim Camerarius in Nuremberg ([Wittenberg], January 13, [1532]), MBW 1210; Philipp Melanchthon to [Paul vom Rode in Stettin] ([Jena, before January 13, 1536]), MBW 1686a; Philipp Melanchthon to Georg Spalatin [in Altenburg] ([Wittenberg], September 2, 1542), MBW 3031; Philipp Melanchthon to Johannes Lang in Erfurt ([Wittenberg], October 24, [1542]), MBW 3075; Philipp Melanchthon to Johannes Sutelius in Schweinfurt ([Wittenberg], February 5, [1543]), MBW 3159; Philipp Melanchthon to Joachim [in Leipzig] ([Wittenberg], February 9, [1544]), MBW 3450; Philipp Melanchthon to Matthew Collinus in Prag ([Wittenberg], January 1, 1545), MBW 3780; Philipp Melanchthon to Paul Eber [in Wittenberg] (Zerbst, November 16, [1546]), MBW 4449; Philipp Melanchthon to Noah Buchholzer ([Wittenberg], July 9, 1552), MBW 6491; Philipp Melanchthon to Hieronymus Weller [in Freiberg] ([Wittenberg], January 30, [1554]), MBW 7074; Justus Jonas an Philipp Melanchthon [in Wittenberg] ([Eisfeld], December 22, 1554), MBW 7365.

[23] David Chytraeus to Philipp Melanchthon [in Wittenberg] (Rostock, March 20, 1556), MBW 7755.

[24] Philipp Melanchthon to Jakob Runge in [Greifswald] ([Wittenberg], April 14, 1560), MBW 9296; (Runge 1560).

[25] Jakob Schenck to Philipp Melanchthon [in Wittenberg] (Weimar, December 14, 1539), MBW 2329.

[26] Philipp Melanchthon to Jakob Schenck [in Weimar] ([Wittenberg], December 23, [1539]), MBW 2330.

produce it.[27] Similarly, in 1555, Melanchthon informed Konrad Heresbach (1496–1576) that it was the printer's fault that Heresbach's book had not yet appeared in print.[28]

During his continuous collaboration with printers, Melanchthon often encountered problems caused by the printers' economic ambitions. At the beginning of his career in Wittenberg, the professor confided in a letter that he actually did not desire to produce many publications in print, but that there were printers who "snooped around in his cupboards" searching for something they could turn into profit.[29] For a printer it was, of course, very lucrative to produce anything with Melanchthon's name on it—a fact often lamented by the author himself.[30] At times he even felt that he could not satisfy the printers' thirst for more manuscripts (Claus 2002, 79). Melanchthon also criticized the producers for neither wanting to wait for the final version of the text nor leaving enough space for his prefaces.[31] The latter were usually added at the very end of the production process, which provided authors with extra time to compose them. This practice, however, also meant that there was limited available space; indeed, Melanchthon once complained that he had to shorten his text considerably to fit on the free pages.[32] Given Melanchthon's grievances, it is perhaps no surprise that the first editions of some of his works actually appeared outside Wittenberg (see Sect. 5) (Kaufmann 2019, 88).

At times, Melanchthon tried in vain to persuade a printer to produce works he had himself received as manuscripts. In 1540, Melanchthon wrote to an author that he would now forward the latter's manuscript to a printer in Frankfurt because the printers in Wittenberg did not accept any other language than German at that time.[33] A few years later, Melanchthon sent a manuscript back to a Leipzig professor with the request to publish it there as soon as possible.[34] Melanchthon wanted to use the copies to teach in his courses, but there were currently no good printers in Wittenberg who were up for the job. When Wittenberg was ravaged by the plague, the city

---

[27] Philipp Melanchthon to Antonius Corvinus in Witzenhausen ([Wittenberg], January/February 1535), MBW 1534.

[28] Philipp Melanchthon to Konrad Heresbach [on Lorward] ([Wittenberg, July 21, [1555]), MBW 7536.

[29] Philipp Melanchthon to Fabian Gyrceus [in Basel?] ([Wittenberg, 1524?]), MBW 363.

[30] E.g., Philipp Melanchthon to Johannes Musler in Leipzig ([Wittenberg, 2nd half of January 1525]), MBW 375; Philipp Melanchthon to unknown person ([at the latest in 1543]), MBW 9368.

[31] Philipp Melanchthon to David Chytraeus [in Rostock] ([Wittenberg], January 20, [1556]), MBW 7693; Philipp Melanchthon to Justus Jonas the Younger [in Zerbst] ([Wittenberg], May 23, 1538), MBW 2043.

[32] Philipp Melanchthon to Justus Jonas the Younger ([Wittenberg, ca. March 1538]), MBW 2015, with an addendum from the editors saying that since Melanchthon still had time to shorten his preface it had to be the last part of the production process.

[33] Philipp Melanchthon to Christoph Hoffmann in Jena ([Wittenberg], January 4, [1540]), MBW 2343.

[34] Philipp Melanchthon to Joachim Camerarius [in Leipzig] ([Wittenberg], February 9, [1544]), MBW 3450.

physician of Coburg sent a treatise about healing to Melanchthon for publication.[35] In theory, this must have been a particularly good time to publish such a treatise. However, Melanchthon responded that he could not pass it on to a printer since he was currently not in Wittenberg, and—more importantly in this context—because the length of the treatise, as well as the small handwriting, were unappealing to printers.

When it came to Melanchthon's own manuscripts and older publications, however, the professor must have wished that printers were as cautious as in the abovementioned cases. As Melanchthon's reputation grew, printers often sought to reprint some of his older works, to the dismay of the author. Only rarely did printers ask for permission before they embarked on such ventures.[36] In 1548, for instance, a printer intended to produce a physics lecture Melanchthon had given twelve years earlier.[37] This request did not find the author's sympathy. In fact, Melanchthon was so upset about it that he wrote to his contact in Strasbourg instructing him to prevent any publication of this kind in that city and beyond. As the old version of the lecture manuscript contained outdated astrological examples, the publication could not only be harmful to the printer; it could ultimately also hurt the author himself.

On other occasions printers did not ask Melanchthon for permission and just produced unauthorized editions of his works, such as his criticism of the Augsburg Interim.[38] Repeatedly, the professor complained about the printers' insatiable thirst for profit—he even wrote about it in the prefaces of his books.[39] In some cases, Melanchthon was able to produce an improved version of the text shortly after the undesired edition appeared.[40] This was for instance the case in 1530 when Melanchthon wrote in another preface that he produced the edition because two months earlier an "enterprising" printer had produced a "spoiled" version of the text.[41] Printers even boasted of the fact that they would soon publish Melanchthon's latest work, although the author was not involved in the production. Yet there was little the professor could do besides warning his readers about the "liars who just used [his] name to make profit."[42]

---

[35] Philipp Melanchthon to Christoph Stathmion [in Coburg] ([Torgau], October 8, [1552]), MBW 6592.

[36] E.g., Philipp Melanchthon to Konrad Embecanus at [Johannes] Gymnicus's in Cologne ([Bonn], June 11, [1543]), MBW 3259.

[37] Philipp Melanchthon to Martin Bucer in Strasbourg ([Wittenberg], October 1, [1548]), MBW 5310.

[38] E.g., Philipp Melanchthon to [Elector Moritz of Saxony] ([Wittenberg, September 3, 1548]), MBW 5280.

[39] E.g., in the preface to his *Solomonis sententiae, versae ad Hebraicam veritatem* (1525), MBW 394.

[40] E.g., Philipp Melanchthon to Wilhelm Reiffenstein [in Stolberg], ([Wittenberg], June 2, [1529]), MBW 789; in the preface to his *Catechesis puerilis* (1543), MBW 3418.

[41] In the preface to the *Confessio Augustana* (1531), MBW 1103.

[42] Philipp Melanchthon to Wilhelm Hausmann [in Augsburg] ([Wittenberg], November 10, [1554]), MBW 7329.

5   Printing the *Sphaera*

Following the groundbreaking success of Luther's publications, printers often came
to Wittenberg to start a prosperous business. Among them was, as we have already
seen above, Joseph Klug, the first to produce the *Sphaera* in the university town. He
came to Wittenberg in the early 1520s and started to work for Lucas Cranach (Reske
2015, 1082). Shortly thereafter he started to publish under his own name and printed
books for both Luther and Melanchthon. Over the years, Klug gained the trust of
Melanchthon, who not only gave him permission to print his works but also confided
in him with sensitive matters. When, in 1525, Melanchthon expected letters from
Frankfurt, he asked Klug to bring them with him when he returned from the book
fair.[43] This was a strong sign of confidence: out of all the Wittenberg booksellers
and printers who regularly attended the fair, Melanchthon entrusted Klug with this
sensitive matter.

The relationship between Melanchthon and Klug continued to flourish. In the
later 1520s, Klug produced various books for the scholar and even helped him when
there was trouble with another printer. Since his time in Tübingen, Melanchthon
had continued to work with Thomas Anshelm, even after the printer moved to
Hagenau (Rhein 1997, 71). This fruitful relationship between Melanchthon and
the printing house in Hagenau lasted even after Anshelm had left the shop to his
successor Johannes Setzer (ca. 1478–1532). On numerous occasions, Melanchthon
wrote a preface to works that appeared in Hagenau (Scheible 2010a, 309). Over the
years, however, Melanchthon grew more and more dissatisfied with the works from
Hagenau (Scheible 2010a, 309–310). At that time, it was Klug in Wittenberg who was
able to help. Two publications, in particular, had angered the professor: his commentary on the Colossians and his *Dialectics* (Melanchthon 1527, 1528). In one of his
letters, Melanchthon quite explicitly wrote that he was no longer willing to put up
with the bad quality of the texts Johannes Setzer produced.[44] Instead, Melanchthon
stated that he would ask Joseph Klug to print an improved version of the text. This
Wittenberg edition was much more to the liking of the author, and, in later letters,
Melanchthon recommended this edition to his readers.[45]

It was also Joseph Klug whom Melanchthon entrusted with the production of
the first Wittenberg edition of the *Sphaera*, when he reissued various classic textbooks for the university (Omodeo 2014, 67; Sacrobosco and Melanchthon 1531).
The book had been on the list of required books for the study of mathematics from
1514 when it had become an independent subject of study (Friedensburg 1917, 106).
Despite this fact, however, no copies had been produced in Wittenberg in the 1510s

---

[43] Philipp Melanchthon to Georg Spalatin [in Altenburg] ([Wittenberg, September 28, 1525]), MBW
424.

[44] Philipp Melanchthon to Johannes [Koch in Wittenberg] ([Weimar], November 2, [1528]), MBW
720. The original text reads: "Secerius mea patientia inepte abutitur, at viderit, ut diu tolerare hanc
tantam negligentiam possim".

[45] Philipp Melanchthon to Wilhelm Reiffenstein [in Stolberg] ([Wittenberg], June 2, [1529]), MBW
789.

and 1520s, as the necessary copies could easily be imported from other production sites, such as Ingolstadt and Leipzig (Chap. 12). At the beginning of the 1520s, Melanchthon perceived the need to improve several key texts for the study of mathematics, including Sacrobosco's fundamental study (Reich 1998, 110). In a letter that would become the preface of the *Sphaera* edition, Melanchthon described his interest in mathematics and astronomy, which provided him with important clues for the study of the past (Pantin 2020, 279).[46] The letter must have been written before August 17, 1531.[47] It could then have been included in the publication just in time for the new academic year.

This first Wittenberg edition was a game changer. A well-written preface from one of Europe's most famous professors made the book rise in value. This important sign of approval was quickly taken up by other printers, even outside Germany. Just one year after the first Wittenberg edition had appeared, the preface—as well as the design of the book—was replicated by a printer in Venice (Sacrobosco and Melanchthon 1532; Valleriani et al. 2019). This shows how much value Melanchthon's preface had added to the text—after all, the printer could have copied one of the Sacrobosco editions that had already appeared in Venice (Pantin 2020, 283). Instead, he chose the new Wittenberg version.

In the following years, Klug produced a new edition nearly every other year (Sacrobosco and Melanchthon 1534, 1536, 1538, 1540), with important revisions to the text and new features from 1538 onward. This included, most notably, the addition of another treatise by Sacrobosco named *De anni ratione* (often referred to as the *Computus*). It has long been assumed that Klug produced both of these treatises in 1538. However, the analysis of the typographical material used for the *Computus* reveals that it was actually Hans Lufft who printed that edition (Claus 2014, 727). The two treatises have separate prefaces and individual signatures, but only the *Sphaera* has a colophon (mentioning Klug). Since the preface of the *Computus* was written in August, it may indicate that the decision to include the text was made relatively late so that another printer needed to be involved if the whole publication was to be available by the start of the new semester. The 1538 edition of the *Sphaera* also contained additional illustrations from various Sacrobosco editions that had appeared in Venice, Leipzig, and Ingolstadt, as well as from other books (Pantin 2020, 294). With the inclusion of the *Computus*, the total number of illustrations grew from 45 to 79 woodcuts (see Sect. 6).

Each of Klug's editions of the *Sphaera* included Melanchthon's preface. It is, however, unclear how much the professor was actually involved in the production of the book. Instead, Georg Joachim Rheticus did most of the textual editing, at least in 1538 (Pantin 2020, 291).[48] As we saw above, in 1562 Rheticus confirmed his role in editing both of Sacrobosco's works for the press in the same letter in which he

---

[46] For an English translation of Melanchthon's letter on mathematics and astronomy, see (Sacrobosco and Melanchthon 1999, 105–112).

[47] See the additional notes for Melanchthon's letter in MBW 1176.

[48] It had been assumed that Rheticus prepared the 1550 edition. However, Matteo Valleriani has recently shown that he actually worked on the 1538 edition (Valleriani et al. 2022).

complained about the poor reward from Moritz Goltz. Additionally, in the preface of the *Computus*, Melanchthon argued that it was Rheticus's idea to have the two treatises put together (Rosen 1974, 245).

Rheticus, who would later be instrumental in the publication of Copernicus's (1473–1543) masterpiece, came to Wittenberg in 1532 (Gingerich 2005, 135). There he worked closely not only with Melanchthon but also with Johannes Volmar (ca. 1480–1536), who taught mathematics and astronomy at the university (Schöbi 2014, 41). After Volmar died, Rheticus was asked to succeed his teacher and the former student obliged—if only hesitantly. In his inaugural lecture, Rheticus described his inner conflict: an academic torn between teaching publicly and doing research in the privacy of his study.[49] Some of Rheticus's works were printed by Klug, such as this inaugural lecture (1536) and, as we have seen, the new edition of Sacrobosco's text in 1538.

After the publication of the 1538 edition, it is, however, unlikely that Rheticus was still involved in the editing process of Sacrobosco's text. First, the sloppy mistake of including the wrong date in the preface is an indicator that the 1540 edition was not overseen by the rigorous proofreader Rheticus.[50] Second, Rheticus spent most of his time during that year at Copernicus's side, preparing *De Revolutionibus* for its publication and seeing his *Narratio prima* through the press (Rheticus 1540). It is therefore implausible that Rheticus returned briefly to Wittenberg to fulfill his teaching duties in 1540 (Burmeister 2006, 10–11).

The assumption that Rheticus returned to Wittenberg in 1540 was based on three lecture announcements. They can all be found in the first volume of the *Scripta publice proposita*, which was edited and reprinted in 1560.[51] Two of the lectures specifically refer to Sacrobosco. All of them are undated, but the position of one of them (a lecture on Ptolemy) indicates that it was probably given in February 1540. The notification is wedged between two university announcements that actually reveal dates: January 27 and February 1. It has been assumed that the announcement from February 1 is an invitation to a graduation ceremony of *Magistri* in mid-April since the text also mentions the dean of arts, Christian Neumair (d. 1543) (Burmeister 2006, 10). During Neumair's time as dean, only two *Magistri* graduation ceremonies took place, and since the text refers to "next Thursday," it has been concluded that the announcement could only mean the graduation on April 15, since February 5 was a Friday. However, this assumption needs to be corrected. First, if an announcement were to be produced

---

[49] The authorship of this inaugural lecture remains unclear: it has been assumed that it was written entirely by Melanchthon. But the recovery of a lecture script from Rheticus shows that it was most likely Rheticus who wrote the mathematical content of *In Arithmeticen* (1536), whereas Melanchthon concentrated on the references to Greek philosophy, see (Deschauer 2003, V; Reich 2017, 562). In bibliographies, such as VD16 and USTC, *In Arithmeticen* (1536) is still attributed solely to Melanchthon.

[50] Isabelle Pantin has pointed out that this mistake occurred in the print shop, as it was often the practice to match the date of the preface to the date of its print production (Pantin 2020, 281).

[51] (*Scripta* 1560, leaves C4v–C5v, C7r–C8r, and E4v–E5r). I'd like to express my gratitude to Christiane Domtera-Schleichardt for her help with the *Scripta publice proposita*. For an English translation of all three lecture announcements, see (Kraai 2000, 197–202).

more than two months in advance, it would more likely contain the precise date and not the expression "next Thursday" (Limbach 2017, 387). Second, 1540 was in fact a leap year, as we can see from a letter written to Melanchthon on February 29.[52] As we also know that March 1, 1540, was a Monday, it turns out that February 5 was indeed a Thursday—the day on which fifteen *Magistri* graduated (Burmeister 2015, 415).[53] With Rheticus's lecture announcements, however, it is important to keep in mind that such notifications are by no means an indication that he actually gave the lectures or even that the lectures took place at all. This lecture might have been taken over by a colleague or could have been canceled. The announcements are therefore no proof that Rheticus was in Wittenberg at the time.

Let us return to the printer Joseph Klug. The 1540 edition of the *Sphaera* was the last that Klug produced. Although the printer worked in Wittenberg until the early 1550s, the later editions of the *Sphaera* (i.e. 1543, 1545, 1549, and 1550) were produced by other printers. The reason for this is most likely the fact that Klug experienced increasing financial difficulties from the 1530s onward (Claus 2002, 99). At the time of his death, Klug's financial situation was so dire that even his house was dilapidated. Melanchthon comments on this fact in one of his letters, after a colleague has asked if he should buy Klug's house. Melanchthon remarks that the upper part of Klug's house was completely ramshackle and that his colleague should rather choose the house Hans Lufft was offering.[54] This shows that Klug was never able to recover from his financial difficulties—the rough printing business had once again taken its heavy toll.

The reason for Klug's downfall was his involvement with the two powerful publishers Christoph Schramm and Moritz Goltz. In 1533, Klug made a contract with Schramm and later also worked for Goltz (Stiegler 1989, VII). Over the years, the printer repeatedly owed Goltz money, which at some point amounted to a large sum of 243 Gulden (Rothe 2013, 86). As Goltz was involved in the production of the *Sphaera*, some of the losses may have been attributed to it (just as the Leipzig printers suffered losses with their academic publications). To clear his debts, Klug agreed to print a number of editions for Goltz in 1539 (see the list above). But the conditions of the contract were very hard: for the production, Klug would receive some money to cover his expenses, yet this was only a loan that had to be repaid (Rothe 2013, 86). Eventually, Klug's wife had to step in and argue before the city council that Klug could not meet the requirements, especially concerning the high numbers of copies for every book (2,000–3,000). The contract between Klug and Goltz was annulled in 1540 and the printer had to agree to a new contract that required him and his wife to pay 20 Gulden at the time of each trade fair in Leipzig. The couple even used

---

[52] Wenzeslaus Linck, Dominicus Schleupner, Andreas Osiander, Veit Dietrich, and Thomas Vena-torius to Martin Luther, Justus Jonas, Johannes Bugenhagen, and Melanchthon in Wittenberg (Nuremberg, February 29, [1540]), MBW 2383.

[53] See the information on dating the letter of Philipp Melanchthon to Justus Syringus in Weilburg [or Waldeck] ([Schmalkalden], March 1, [1540]), MBW 2384.

[54] Philipp Melanchthon to [Paul Eber in Torgau] ([Wittenberg], December 26, [1552]), MBW 6684.

their house at the Kirchplatz as a guarantee (Chap. 4). This hardly improved Klug's situation.

The long financial insecurity left traces in Klug's production, which seems to have started after the summer of 1536. On August 14, Rheticus gave his very first lecture at the university, teaching arithmetic. Thanks to a surviving lecture script, diligently composed by a student, we know that Rheticus recommended to his listeners the "small books which Joseph Klug had recently printed."[55] This most likely included Peuerbach's *Elementa Arithmetices* (Peuerbach 1534); a copy of this book was annotated by the same hand who composed the lecture script (Deschauer 2003, 5). The "small books" probably also included Peuerbach and Vogelin's *Elementa Geometriae* (Peuerbach and Vögelin 1536).

Then the trouble began and the quality of Klug's books declined. A few months later in the same year, Melanchthon was once again approached by an author to find a suitable printer for his text. In his answer, Melanchthon deemed Klug inadvisable—even though the printer had already agreed to produce the text.[56] Melanchthon advised the author to go against this agreement, stating that Klug's publications were simply full of mistakes and suggesting that he, Melanchthon, could pass the manuscript along to another printer. Some years later, in 1541, Klug's financial problems were also discussed in Melanchthon's correspondence. By that time, the printer already had problems with Goltz and was desperately trying to meet the requirements of his new deal with the publisher. In April, Melanchthon was informed that the printer was currently jobless and eagerly looked forward to new manuscripts so that he could produce them.[57]

Despite Melanchthon's strong statement about the quality of Klug's books in 1536, the printer continued to work for the university and for Melanchthon. In 1543, Melanchthon was asked to find a printer for a text, to which he responded that he would negotiate with various producers in Wittenberg concerning the conditions of the production.[58] It was Klug who eventually received the order, indicating that he must have made the best offer and probably also had a free press.[59] The professor also entrusted Klug with the production of the university statutes and the first four volumes of the *Scripta publice proposita* mentioned above (*Scripta* 1545–1546, 1548, 1549a, b).[60] In the first volume, we find evidence that Klug also printed a textbook for one

---

[55] Like Goltz in the example above, Joseph Klug was invoked by his first name alone, "Iosippus," indicating a certain familiarity of the students with the printer. For the transcription of, and a commentary on, the manuscript, see (Deschauer 2003).

[56] Philipp Melanchthon to Johannes Stigel [in Wittenberg] ([Wittenberg, ca. mid-November 1536?]), MBW 1809.

[57] Paul Eber to Philipp Melanchthon in Regensburg ([Wittenberg], April 15, 1541), MBW 2669.

[58] Philipp Melanchthon to Johannes Sutelius in Schweinfurt ([Wittenberg], February 5, [1543]), MBW 3159.

[59] The book comprises 112 leaves and was printed in quarto format: (Sutell [1543]).

[60] Philipp Melanchthon to Paul Eber [in Wittenberg] ([Halle], December 23, [1545]), MBW 4102.

of Bugenhagen's lectures in 1545; in the announcement, the students are prompted to buy the book for the lecture in Klug's shop.[61]

Although Klug continued to produce academic books, he missed out on some opportunities. Ever since Klug had helped Melanchthon during his quarrel with the printer Johannes Setzer (who had produced a faulty edition of the *Dialectics*), Melanchthon had repeatedly entrusted Klug with the production of new editions of the work (Melanchthon 1529, 1531, 1533, 1534a, b, 1536). Then the Schmalkaldic War in 1546–1547 affected Wittenberg and even put a temporary end to the teaching activities at the university (Lück 2011, 14). During that time, Klug lost the opportunity to produce another *Dialectics* edition. In February 1547, Melanchthon sent parts of the new edition to his colleague Paul Eber (1511–1569) with the order to give it to Klug.[62] Although Eber's response has not survived, he must have reported that Klug was unable to deal with Melanchthon's request, possibly because the printer had left the university town during the war. In any case, Eber advised Melanchthon to entrust Hans Lufft with the production of the *Dialectics* instead and Melanchthon agreed.[63] The work finally appeared a few months later in Lufft's shop.[64]

It is also telling that Klug lost his grip on the market for printed music. From the late 1520s onwards, the printer had produced Luther's important hymnal in several editions—(1529, 1533, 1535, 1543–1544)—with the explicit approval of the author in the form of a woodcut (Volz 1957, 153; Luther 1533, 1535, 1543–1544).[65] Another printer who produced song books in Wittenberg did not have this sign of Luther's trust. Soon, however, the market for printed music became competitive, and eventually, Klug's competitor Rhau became the most successful producer in the field (Heidrich 2015, 192). On top of that, Klug did not produce smaller music publications. Many song pamphlets were produced in Wittenberg, especially in the course of the Schmalkaldic War, but Klug did not print any of them (Nehlsen 2015, 211). Again, it seems that the financial problems Klug experienced at that time had an undeniable effect on this part of the business.

As we will see (see Sect. 6), when Peter Seitz (d. 1548) took over the production of the *Sphaera*, he used the same woodblocks for the illustrations that Klug had used before him. By the time the new edition of Sacrobosco's textbook appeared (Sacrobosco and Melanchthon 1543), Seitz had been producing in Wittenberg for nearly ten years (Reske 2015, 1087). The printer had strong family ties to Klug's competitor Georg Rhau (Claus 2002, 100). Despite this, however, Seitz never became one of the top printers and remained a rather unimportant figure in the Wittenberg print industry. He produced only some fifty editions in the course of fourteen years, a relatively low

---

[61] The book in question is Augustinus *De spiritu et littera,* SPP 1, 132b–133a (*Corpus Reformatorum* 1835, 5, 810f. (no. 3236)), MBW 3973.

[62] Philipp Melanchthon to Paul Eber [in Wittenberg] (Zerbst, March 26, [1547]), MBW 4668.

[63] Philipp Melanchthon to Paul Eber [in Wittenberg] (Zerbst, April 22], 1547), MBW 4720.

[64] The book comprises 264 leaves and was printed in octavo format: (Melanchthon 1547).

[65] No copy of the 1529 edition has survived, see (Volz 1957, 153).

number for such a long period.[66] Seitz is rarely mentioned in Melanchthon's letters and most likely did not interact with him as frequently as other printers did.

Two years later, in 1545, the next edition of the *Sphaera* appeared (Sacrobosco and Melanchthon 1545). This time it was printed by a more prosperous producer, Veit Kreutzer. Again, he used the same woodblocks as the previous two printers and produced two more editions, one in 1545 and another in 1549. Unlike Seitz, Kreutzer had just opened his shop in Wittenberg a few years before his first Sacrobosco edition came out. And unlike Seitz, Kreutzer stayed in the business much longer. He produced for over two decades and printed some 340 editions, with a particular focus on the works of Melanchthon (Reske 2015, 1086–1087).

Interestingly, Kreutzer had a strong connection to Klug, the first printer of the *Sphaera*. It seems that he had been working in Klug's shop since 1538 (Claus 2014, 726). In that year, Melanchthon had one of his many books printed by Klug, and it was Kreutzer who delivered the proofs. Presumably, he was acting on his master's behalf. During his time in Klug's print shop, Kreutzer could have familiarized himself with the production of the *Sphaera*. After all, Klug produced one edition of Sacrobosco's text in 1538 and another in 1540.

When he opened his own shop, Kreutzer collaborated with Melanchthon from the very beginning. In April 1542, the professor wrote in a letter that he had worked very intensively with Kreutzer on the production of his prominent *Loci theologici*.[67] It was one of the works that Melanchthon reissued multiple times. By that time the work comprised no less than 740 pages (Melanchthon 1542). Just like the *Sphaera*, the previous editions of the *Loci theologici*, both printed in 1535, had first been produced in Klug's workshop (Melanchthon 1535a, b). Then, after Melanchthon had written a new preface for the theology students in Wittenberg, the new edition appeared in Seitz's workshop in 1541 (Melanchthon 1541). Finally, Melanchthon edited the work, translated it, and wrote another new preface, which he then gave to Kreutzer to produce (Melanchthon 1542). This similar production history seems to suggest that Melanchthon may have been involved in finding a suitable printer for the next *Sphaera* edition as well. When it became clear that first Klug and then Seitz were not able to produce the next edition of Sacrobosco's text anymore, Melanchthon probably suggested to Goltz that he should employ Kreutzer for the job.

In 1550, the woodblocks for Sacrobosco's text once again changed hands (Sacrobosco and Melanchthon 1550). The new edition appeared in the print shop of Johannes Krafft (ca. 1510–1578), who, like Klug, was one of the most important printers in Wittenberg. The printer operated a print shop for over 30 years and produced no less than 840 known editions (Reske 2015, 1087). The production of the *Sphaera* remained firmly in his hands for over two decades until the mid-1570s. By that time he had produced no fewer than six editions of Sacrobosco's fundamental work.

Krafft was an ambitious printer with very good connections. Through his wife, he was related to Hans Lufft, and he had himself worked for Georg Rhau at some point

---

[66] USTC searching Seitz, Wittenberg and 1534–1548.

[67] Philipp Melanchthon to Veit Amerbach [in Wittenberg] ([Wittenberg, April 1542]), MBW 2949.

(Reske 2015, 1087). This involvement with two of the most important printers in Wittenberg allowed Krafft to acquire a good understanding of the market. When his master Rhau died, Krafft set up his own business and produced from 1549 onwards. The first years were trying, especially the early 1550s. In August 1551, Melanchthon informed one of his correspondents about the outbreak of the plague in Magdeburg.[68] It did not take long before it was in Leipzig, too.[69] The problem became so imminent that in July 1552 the University of Wittenberg sent away its students.[70] Some print shops even closed down their businesses, but Krafft was able to continue printing in Wittenberg despite the precarious situation.[71]

Krafft also took a firm stand on the frequent last-minute corrections authors wanted to incorporate into their manuscripts. In his letters, we find multiple instances when Melanchthon wanted to look over the text one more time before it was given to the printer. The production of the university statutes serves as a good example. The manuscript was corrected by at least two of Melanchthon's colleagues. Then Melanchthon asked another colleague to correct the proofs, but before giving them to the printer, Melanchthon wanted to have a final look at them.[72] These rigorous correction practices must have tested the patience of many printers, who needed to stick to a tight publication schedule if they wanted to remain in business. Therefore, when Melanchthon once again wanted to correct a manuscript, Krafft simply proceeded with its publication.[73] Needless to say, Melanchthon was not amused, but he nevertheless provided Krafft with many more manuscripts after this instance (Claus 2014, 2801–2803).

6   Prices, Print Runs, and the Wittenberg Set of Woodblocks

One of the copies of the 1545 edition reveals just how expensive the *Sphaera* was, compared to other university books at the time. This particular copy is part of the collection of the Archenhold Observatory in Berlin (Fig. 1).[74] On the title page, a contemporary owner has noted the price of 18 Groschen.

An overview listing seven university books from a few years earlier helps to contextualize this price. The list was drawn up by Simon Wilde (ca. 1520–ca.

---

[68] Philipp Melanchthon to Michael Meienburg in Nordhausen ([Wittenberg], August 5, 1551), MBW 6158.

[69] Philipp Melanchthon to Michael Meienburg in Nordhausen ([Wittenberg], September 3, 1551), MBW 6191.

[70] Philipp Melanchthon to Hieronymus Baumgartner in Nuremberg ([Wittenberg], July 6, [1552]), MBW 6486.

[71] Philipp Melanchthon to Heinrich Buscoducensis ([Wittenberg], June 18, 1552), MBW 6472; Philipp Melanchthon to Noah Buchholzer ([Wittenberg], July 3, [1552]), MBW 6483.

[72] Philipp Melanchthon to Paul Eber [in Wittenberg] ([Halle], December 23, [1545]), MBW 4102.

[73] Philipp Melanchthon to David Chytraeus [in Rostock] ([Wittenberg], January 20, [1556]), MBW 7693.

[74] I'd like to thank the director, Dr. habil. Felix Lühning, for his help.

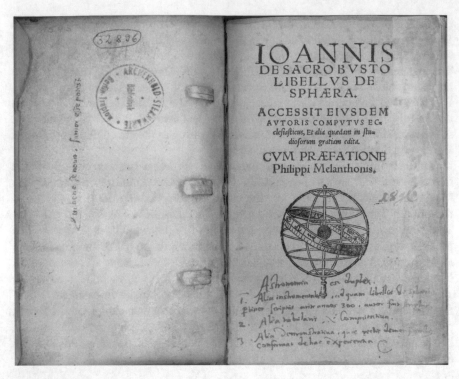

Fig. 1 The title page of the *Sphaera* copy from (Sacrobosco and Melanchthon 1545). A contemporary owner has noted the price of 18 Groschen. Courtesy of the Archenhold-Sternwarte, Berlin

1560) who included it in a letter to his uncle and benefactor in 1540, asking him for more financial support.[75] The books varied quite significantly in price: four books were about two Groschen each—Melanchthon's *Commentarius de anima* and three *Dialectics* (written respectively by Melanchthon, Jodocus Willich (1501–1552), and Johannes Caesarius (ca. 1468–1550)). Another two books were priced at 4 and 6 Groschen respectively—the *Physica* of Johannes Velcurio (1490–1534) and Melanchthon's *Loci communes*.[76] The most expensive book on the list was a Bible for 15 Groschen, which was thus more than twice the cost of the other books. Wilde explained that he needed the Bible for Melanchthon's lecture on the prophet Daniel. Although the 15 Groschen was a high price in contrast to other university books, it

---

[75] The Latin letter was edited in (Buchwald 1894–1902, 86–88) and partly translated into German in (Burmeister 2015, 41–42). Although Wilde already possessed a copy of Melanchthon's *Dialectica*, he insisted on buying the latest version, since many parts had been added or changed. The changes were so significant, stated Wilde, that one could even doubt the two versions were written by the same author.

[76] Interestingly, at that time not all of these six books had been printed in Wittenberg: Velcurio's *Physica* had only been printed in Tübingen (1539, 1540) and Willich's *Dialectica* had only appeared in Strasbourg (1540): (Velcurio 1539, 1540; Willich 1540).

was still low in comparison to the prices we know from other Wittenberg Bibles. In fact, the price might indicate that Wilde either bought only one part of the Bible, such as the Old Testament, or just the Prophetic Books. He might also have bought a used copy or even a manuscript copy. After all, in 1522 a printed copy of the New Testament alone had cost 15 Groschen (Schirmer 2015, 173), and a copy of the full Bible in 1534 had even cost 2 Gulden and eight Groschen—the price of 17 geese (Krieg 1953, 22). Although the price must have decreased to some extent by the 1540s, a whole Bible would probably still have cost more than 15 Groschen. In light of all these prices, however, it becomes clear that the *Sphaera*, at 18 Groschen, was one of the more expensive university books that students could buy in Wittenberg in the 1540s.[77]

Student life in Wittenberg was generally quite expensive. This is especially true for the mid-sixteenth century. In late 1538, Melanchthon mentions that the price for food had risen significantly and that the costs for student accommodation had doubled from 3 to 6 Gulden per year (Friedensburg 1926, 220). In 1544, the city council of Wittenberg even asked the elector to intervene, as the prices for food were driven up by "greedy" peasants who wanted to make more profit (Friedensburg 1926, 240). A letter written by the student Philipp Bech (1521–1560) in 1542 illustrates how that affected student life (Burmeister 2015, 89).[78] Bech had previously studied in Basel, and after spending only one month in Wittenberg he wished he could return to his former alma mater. According to his report, Wittenberg had many disadvantages: the water was undrinkable, the food was inedible, and the local beer caused scabies and fever. Bech had to pay 6 Gulden for accommodation (just like Melanchthon had noted) and added that it had been extremely hard to find housing, given the many students in town. Bech also noted that the cheapest food cost 18 Gulden per year.[79]

At first glance, these high costs of living, as well as the considerable price for a copy of the *Sphaera* (more than half a month's supply of low-cost food), do not indicate that the printers/publishers could expect to sell a large number of copies. But how large were the print runs approximately? This is rather difficult to determine. On the one hand, academic books seem to have had print runs of a few hundred copies. As we have seen above (see Sect. 3), the two Aristotle editions printed in Leipzig in 1519 were manufactured in runs of 300 copies each; but they did not sell well, and the university eventually agreed to buy twenty copies. As both the universities of Leipzig and Wittenberg grew significantly, the print runs increased as well, albeit

---

[77] As mentioned above, Wilde planned to buy books that had appeared outside Wittenberg. For the remaining books it is not clear which exact editions he referred to. It is therefore sadly not possible to assess the prices per printed sheet for these books and compare them to the *Sphaera*.

[78] The Latin letter was edited in (Kolde 1883, 380–382) and partly translated into German in (Burmeister 2015, 88–89).

[79] Equal to 378 Groschen per year and 31.5 Groschen per month.

slowly.[80] When Melanchthon wanted to discuss a certain publication in his lecture, he asked for 400 to 500 copies to be sent from Leipzig to Wittenberg.[81]

On the other hand, Wittenberg also catered to the international market. Best sellers among the university textbooks were produced in several thousand copies and surely found their way to readers in many places outside Wittenberg. As we have seen from the agreement between Goltz and Klug from 1539 (see Sect. 2), the publisher urged the printer to produce Melanchthon's grammar and his *Syntaxis* in 3,000 copies each. A few years later, Melanchthon claimed that 3,000 copies of his *Dialectics* had been sold.[82] He wrote the letter on October 18, 1547, referring most likely to the edition that had just appeared on September 1 (which Lufft had printed instead of Klug).[83] If Melanchthon was not exaggerating, it would mean that 3,000 copies were sold within the space of just six and a half weeks.

Does this mean the *Sphaera* was also produced in several thousand copies per edition? Probably not. Melanchthon's books sold extraordinarily well, given that he was one of the professors who attracted the many students to come to Wittenberg in the first place. On top of that, his *Dialectics* cost only one ninth of what the *Sphaera* cost. We also need to remember that students could have acquired their *Sphaera* copy in other ways: they could have bought a manuscript copy or even copied the text themselves. One of Melanchthon's letters reveals that he sometimes relied on copyists himself.[84]

Students could also have obtained a used copy. Although sources are scarce for the sixteenth century, there was probably a lively secondhand book market in and around Wittenberg.[85] In 1537, for instance, Melanchthon asked the preacher of Naumburg for a favor: the preacher was about to help a widow sell her husband's private library, and Melanchthon requested that one of his former students, Heinrich Scheidewein (ca. 1510–1580) be allowed to buy the legal books in the collection.[86] At that time,

---

[80] Between 1532 and 1542, 3,000 students matriculated at Wittenberg; in Leipzig about 2,500 students matriculated at a slightly later time period (1542–1551). The actual number of students was even higher than that, since many tried to avoid the matriculation fee and therefore do not show up in the records—a problem the university addressed with an announcement in 1540, urging all students to matriculate (Burmeister 2015: 24–25). In a letter from 1542, the student Philipp Bech estimated that 2,300 students were presently in Wittenberg (Kolde 1883, 381). Kolde's work is partly translated into German in (Burmeister 2015, 88–89).

[81] Philipp Melanchthon to Joachim Camerarius [in Leipzig] ([Wittenberg], January 25, [1545]), MBW 3806.

[82] Philipp Melanchthon to Johannes Koch [in Nordhausen] ([Wittenberg], October 18, [1547]), MBW 4927.

[83] According to the date of the preface, MBW 4875: (Melanchthon 1547).

[84] Philipp Melanchthon to the city council of Soest (Bonn, Juni 20, 1543), MBW 3266. In this instance it is interesting to see Melanchthon saying that since he had no one to copy the document, he just had it printed.

[85] At the university of Helmstedt private libraries were auctioned after a professor died. Often lectures were cancelled for this event, see (Nelles 2001, 172). The *Scripta publice proposita* may also contain notices for such auctions in sixteenth-century Wittenberg.

[86] Philipp Melanchthon and Justus Jonas to Nikolaus Medler in Naumburg (Stolberg, August 25, 1537), MBW 1934.

Scheidewein was in desperate need of books—he had lost his own book collection, housed in his mother's home, in a fire. Acquiring books this way was much cheaper than buying the latest printed editions.

It seems much more likely, therefore, that the *Sphaera* editions ran a few hundred copies, maybe even a thousand, but presumably not more. Sacrobosco's text was certainly not as popular as Melanchthon's *Dialectics*, but it was one of the two books that formed the basis of the compulsory lecture for bachelor students studying mathematics from 1514 onwards (Kathe 2002, 39). The other book on the syllabus was the *Computus*, which was printed together with the *Sphaera* after Rheticus suggested the expansion of the book in 1538. Thus, the bachelor students had everything they needed for the lecture contained in one book.

The *Sphaera* owed its high price in part to the many woodcut illustrations. For books printed in the Officina Plantiniana in Antwerp between 1580–1655, for instance, the inclusion of woodcuts raised the price by 50% on average (Proot 2019, 104). The first Wittenberg edition included no fewer than forty-five illustrations. Out of those illustrations, only two were produced from the same woodblock: the armillary sphere appeared both on the title page and at the beginning of Chap. 2. When the Wittenberg *Sphaera* was revised in 1538, it was illustrated even more lavishly, nearly doubling the amount of illustrations. If we consider the seven woodcuts shown in the *Computus* as well, the whole publication contained at least seventy-nine illustrations, including, most notably, four volvelles.[87] The total amount of images can only be guessed since the parts for the four-wheel charts were printed on a separate sheet and users had to cut them out to assemble them, as we can see from another edition, printed in Antwerp in 1547 (Hamel 2014, 47). However, all digitized copies of the 1538 edition lack such a separate sheet, which makes it more difficult to count the total number of woodcuts. Most of the illustrations that accompany the text of the *Sphaera* were either inspired by or copied from the edition printed by Peter Apian (1495–1552) three years before the first Wittenberg edition came out; for the 1538 edition, the producers also borrowed from a book outside the *Sphaera* tradition: the *Protomathesis* by Oronce Finé (1494–1555), which had been printed in Paris in 1532 (Pantin 2020, 294, 297–298).

In the sixteenth century, woodcut illustrations were created from reliefs, requiring several steps: First, an artist created a template, then the template was copied onto a piece of wood. Finally, a woodcutter (*Formschneider*) cut away everything except the lines of the illustrations (Fig. 2). This way when the woodblock was inked and pressed on the paper, only the lines would be printed. The block fit conveniently into the same form as the type, so that illustrations could be printed alongside text on the same page in the same impression (unlike copperplate engravings, which usually required a special press and—if accompanied by text—double impression).

In general, printers had several options when working with woodcuts—they could either borrow, buy, or copy the woodcuts necessary for the production (Pollard 1902, 73). This of course depended on the illustrations. In 1563, a professor of medicine

---

[87] Note: the first woodcut in the *Computus* is the same woodcut from Chapter one of the *Sphaera* (elementary and celestial spheres).

Fig. 2 A rare glimpse into the work behind a woodcut. For Leonhart Fuchs's herbal, an artist created a template that was later copied on a piece of wood. A woodcutter cut away everything except for the lines of the illustration. From (Fuchs 1543, BB7v). Basel, Universitätsbibliothek—Lo I 6. https://doi.org/10.3931/e-rara-1698 Public domain

in Tübingen, Leonhart Fuchs (1501–1566), wanted to publish a new book following the success of his herbal in 1543 (Fichtner 1968, 80). The new work was meant to include over 1,500 illustrations and cost no less than 3,000 Gulden. However, the printer deemed the project to be too risky and abandoned it, selling some of the already created woodblocks to another printer. In other cases, woodcuts could be borrowed. If it was, for instance, a particularly popular image, such as the Virgin Mary, the woodblocks could travel to print shops within the same region and even to print shops in other countries (Bellingradt 2019, 27).

Although the *Sphaera* contained far fewer woodcuts than Leonhart Fuchs's herbal, it was still lavishly illustrated when compared to other university books. Since the *Sphaera* was meant to be a textbook used for bachelor students, the academics in Wittenberg may have been involved in the design of the necessary woodblocks. Although we lack the sources, it seems very likely that a specialist designed the templates for the mathematical figures, as it happened in Paris. There, the mathematician Finé was involved in the creation of the illustrations (Pantin 2009). It may have even been the case that the university contributed financially to the creation of the woodblocks, especially when the *Sphaera* was produced for the first time in 1531. When, in 1538, the publisher Moritz Goltz became involved in the production, he may have bought the woodcuts from the university, covered the costs for the new ones, and passed on the whole set of woodblocks to the printers whenever a new edition was printed.

The woodcuts for the *Sphaera* editions printed in Wittenberg in the first half of the sixteenth century derive from the same woodblocks. Their repeated use eventually took its toll. Most woodblocks were intact for either the 1531 or the 1538 edition that Klug produced. Over time, however, we can see deterioration. These usage marks are clearly visible in the editions produced in 1543 (Seitz), 1545 (Kreutzer), and 1550 (Krafft). In the following, I will show one example with particularly obvious damaged spots, but the damages are also visible in the illustration of circles for measuring astronomical risings (Chap. 3); the lunar eclipse (Chap. 4); and—in the editions from 1538 onwards—the zodiac circle (Chap. 1) (Fig. 3).

The illustration of the solar eclipse depicts the earth on a relatively thin horizontal line in the middle of a circle. To produce this thin line, the woodblock was carved in such a way that all the wood was removed around it. What was left was a relatively fragile strand that could break easily if the printers did not handle it with extra care. This is exactly what has happened here. Over the many years that the woodblock was used, the line did indeed break on the right side close to the middle part. In the first edition from 1531 we can see that the line is already a little bit wobbly. Presumably, the line was already produced slightly flawed. By the time the woodblock was used for the 1543 edition, the line had chipped, revealing a little hole in the printed area. This exact hole can be seen in the editions of 1545 and 1550.

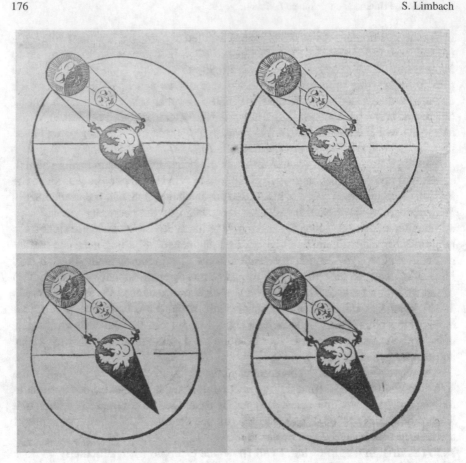

Fig. 3 Top left: Woodcut of the solar eclipse from (Sacrobosco and Melanchthon 1531, F1v). The thin line in the middle is still intact. Austrian National Library—72.N.50, http://data.onb.ac. at/rep/10AF970A. Top right: Woodcut of the solar eclipse from (Sacrobosco and Melanchthon 1538, G4v). The thin line in the middle is broken on the right. Bavarian State Library Munich, Astr.u. 154. https://nbn-resolving.org/urn:nbn:de:bvb:12-bsb10998879-9. Bottom left: Woodcut of the solar eclipse from (Sacrobosco and Melanchthon 1545, G1v). The broken line in the middle is still visible which proves that the image derives from the same woodblock as the previous editions. Austrian National Library—46.L.42, http://data.onb.ac.at/rep/108B40A8. Bottom right: Woodcut of the solar eclipse from (Sacrobosco and Melanchthon 1550, G1v). The damaged spot on the woodblock is still visible in this later edition. Austrian National Library—*48.W.42(3), http://data. onb.ac.at/rep/103BEA3D

7 Conclusion—Of Beasts, Harpies, and Men Made of Iron

In January 1525, Melanchthon complained about the questionable character of local printers. They were a separate human race made of iron, Melanchthon determined, and they neither cared about the advancement of society nor about the growth of knowledge.[88] Printers just cared about profit, the disgruntled professor concluded.[89] In the following decades, many of Melanchthon's utterances took the same line. To be sure, the professor also made plenty of positive remarks about printers and their books. The many negative comments, however, make clear that the interaction between Wittenberg scholars and printers was not as straightforward as one might think, at least in the first half of the sixteenth century. As the book business started to boom, printers produced publications without the author's consent, made many mistakes in the text, did not leave enough space for prefaces, boasted new publications without the author's knowledge (let alone his approval), and "snooped" around in Melanchthon's cupboard, searching for manuscripts that promised profit.

Not surprisingly, Melanchthon and his colleagues in Wittenberg had an ambivalent relationship with the producers of books. Local printers could be held in high esteem at one point only to lose the academics' respect at the next. Joseph Klug, who printed the first edition of the *Sphaera* in Wittenberg, had to experience this first hand. When he started to print, Melanchthon entrusted Klug with a number of publications, but when the printer faced financial difficulties and the quality of his books declined, Melanchthon shifted his attention to other printers. In the following years, Klug lost most of his print jobs, including new editions of the *Sphaera*. Instead, the production of Sacrobosco's treatise moved to more prosperous printing houses. Although the publisher Goltz was involved in the production of the *Sphaera*—at least in 1538—and therefore most likely determined which printer should produce the latest edition, it seems he followed Melanchthon's suggestions. Coinciding with the *Sphaera* editions, Melanchthon's *Loci theologici* were also first produced by Klug, then by Seitzer, and finally by Kreutzer. From 1550, the *Sphaera* was printed by Krafft—a very successful printer who had opened his shop shortly beforehand and who would continue to produce new editions of the book for more than two decades.

The *Sphaera* was a costly book. With a price of 18 Groschen, it was nearly nine times as expensive as Melanchthon's *Dialectics*. The inclusion of the many woodcuts—especially from 1538 onwards, when it was expanded to include over seventy-five illustrations—made the book a luxurious textbook. Yet, students who were already struggling with the high cost of living in Wittenberg did not necessarily need to buy the latest printed edition. As we have seen, the university town most likely already had a lively secondhand book market in the first half of the sixteenth century, and it was, of course, still possible to copy the book by hand. Given these possibilities as well as its high price, it seems unlikely that the *Sphaera* was, like Melanchthon's *Dialectics*, produced in 3,000 copies. Certainly, the *Sphaera* was aimed at a large market, and the imitation of its design in other print centers shows that copies did indeed find their way into the hands of an international audience. Still, it is more likely that the print runs for the *Sphaera* were closer to those for academic books in Leipzig, where printers who produced 300 copies of a textbook still ended up with unsold stock.

---

[88] "Ferreum genus hominum est nec publicis commodis nec dignitate rei literariae movetur." Philipp Melanchthon to Johannes Musler in Leipzig ([Wittenberg, 2nd half of January 1525]), MBW 375.

[89] For the Latin original of the text, as well as a German translation, see (Kaufmann 2019, 84).

For the production of these *Sphaera* copies, the publisher Moritz Goltz played an important role. Despite the lack of appropriate information in the colophons, we know that Goltz targeted the academic market since he prompted Klug to print several academic books. Goltz's involvement was an advantage for academics, as they could request that he publish a new book necessary for a lecture, just like Rheinhold probably did when he initiated the production of Frisius's arithmetic work. In Rheinhold's lecture announcement, the professor invited the students to visit "Moritz's" bookshop, a place they seem to have visited rather frequently.

Yet the rise of powerful publishers also had negative effects on the book industry. Publishers ran tight ships, and printers, as well as editors, scolded them for being scrooges. Academics who worked for them, like Rheticus, were underpaid (the mathematician could indeed barely pay the cost of his beer from his reward). They eloquently dubbed publishers "harpies" and "beasts," and learned to endure both of them and the men made of iron (albeit with gritted teeth) to succeed in their quest to advance both society and knowledge.

Abbreviations

Digital Repositories

Sphaera Corpus*Tracer*	Max Planck Institute for the History of Science. https://db. sphaera.mpiwg-berlin.mpg.de/. Accessed 7 June 2021.
USTC	Universal Short Title Catalogue. The University of St Andrews. https://www.ustc.ac.uk. Accessed 7 June 2021.
VD16	Bayerische Staatsbibliothek. https://www.bsb-muenchen. de/sammlungen/historische-drucke/recherche/vd-16/. Accessed 7 June 2021.

References

Primary Literature

Aristoteles. 1519. *Aristotelis philosophorum principis peri akroaseōs tēs Physikēs hoc est de auscultatione naturali libri octo.* Leipzig: Wolfgang Stöckel. USTC 611165 and VD16 A 3561.
Aristoteles. 1519–1520. *Ta meta ta physika tu aristotelus.* Leipzig: Wolfgang Stöckel. USTC 695806 and VD16 A 3491.
Bretschneider, Karl Gottlieb, ed. 1835. *Corpus Reformatorum.* Halle: Schwetschke.
Buchwald, Georg. 1894–1902. Simon Wilde aus Zwickau. Ein Wittenberger Studentenleben zur Zeit der Reformation. *Mitteilungen der Deutschen Gesellschaft zur Erforschung Vaterländischer Sprache und Alterthümer in Leipzig* 9: 61–111.

Frisius, Gemma. 1542. *Arithmeticae practicae methodus facilis, per Gemmam Frisium Medicum ac Mathematicum*. Wittenberg: Georg Rhau. USTC 613033 and VD16 G 1112.

Fuchs, Leonhart. 1543. *New Kreüterbuoch*. Basel: Michael Isengrin. USTC 602521 and VD16 F 3243.

Italicus, Silius. 1504. *Silius Jtalicus poeta insignis de secundo bello punico*. Leipzig: Martin Landsberg. USTC 693960 and VD16 S 6478.

Lambertus. 1525. *Quisquis es gloriae Germanicae et maiorum studiosus, hoc utare ceu magistro libello*. Tübingen: [Ulrich Morhart]. USTC 689530 and VD16 ZV 23175.

Luther, Martin. 1521. *Von der Beicht ob die der Bapst macht habe zu gepieten*. Wittenberg: [Johann Rhau-Grunenberg]. USTC 703380 and VD16 L 7184.

Luther, Martin. 1533. *Geistliche lieder auffs new gebessert*. [Wittenberg: Joseph Klug]. USTC 658881 and VD16 ZV 6453.

Luther, Martin. 1535. *Geistliche Lieder zu Wittenberg*. Wittenberg: Joseph Klug. USTC 658961 and VD16 G 842.

Luther, Martin. 1540. *Biblia: dat ys: de gantze Hillige Schrifft. Düdesch: Vpt nye thogerichtet vnde mit vlite corrigert*. Wittenberg: Hans Lufft for Moritz Goltz. USTC 616692 and VD16 B 2842.

Luther, Martin. 1541a. *Biblia: dat ys: de gantze Hillige Schrifft. Düdesch: Vpt nye thogerichtet vnde mit vlite corrigert*. Wittenberg: Hans Lufft for Moritz Goltz. USTC 616690 and VD16 ZV 1549.

Luther, Martin. 1541b. *De propheten alle*. Wittenberg: Hans Lufft for Moritz Goltz. USTC 631132 and VD16 ZV 1550.

Luther, Martin. 1541c. *De propheten alle*. Wittenberg: Hans Lufft for Moritz Goltz. USTC 631133 and VD16 ZV 1551.

Luther, Martin. 1543–1544. Geistliche Lieder zu Wittenberg. Wittenberg: Joseph Klug. USTC 658946 and VD16 G 850.

Luther, Martin, Veit Dietrich and Johannes Bugenhagen. 1545a. *BIBLIA Dat is de Gantze Hillige Schrift Oldes vnd Nyen Testamentes*. Magdeburg: Hans Walther for Moritz Goltz. USTC 616662 and VD16 B 2844.

Luther, Martin, Veit Dietrich and Johannes Bugenhagen. 1545b. *Biblia: Dat ys: De gantze Hillige Schrifft.... Vth der lesten Correctur mercklick vorbetert vnde mit grotē vlyte corrigert*. Magdeburg: Hans Walther for Moritz Goltz. USTC 616691 and VD16 B 2843.

Luther, Martin, Veit Dietrich and Johannes Bugenhagen. 1541a. *Dat nye Testament*. Wittenberg: Hans Lufft for Moritz Goltz. USTC 628328 and VD16 ZV 3785.

Luther, Martin, Veit Dietrich and Johannes Bugenhagen. 1541b. *Dat nye Testament*. Wittenberg: Hans Lufft for Moritz Goltz. USTC 628327 and VD16 ZV 24538.

Luther, Martin and Veit Dietrich. 1550. *In Genesin enarrationum reverendi patris domini doctoris Martini Lutheri bona fide et diligenter collectarum. Tomus secundus*. Nuremberg: Johann vom Berg and Ulrich Neuber. USTC 665772 and VD16 B 2994.

MBW. Philipp Melanchthon: Briefwechsel. Kritische und kommentierte Gesamtausgabe, Regesten, Bd. 1 f., ed. Heinz Scheible. 1977f. Stuttgart—Bad Cannstatt, https://www.hadw-bw.de/forsch ung/forschungsstelle/melanchthon-briefwechsel-mbw. Accessed 7 June 2021.

Melanchthon, Philipp. 1527. *Scholia in epistolam Pauli ad Colossenses*. Hagenau: Johannes Setzer. USTC 692352 and VD16 M 4187.

Melanchthon, Philipp. 1528. *Dialectices*. Hagenau: Johannes Setzer. USTC 635786 and VD16 M 2996.

Melanchthon, Philipp. [1529]. *De dialectica libri quatuor*. Wittenberg: Joseph Klug. USTC 629450 and VD16 M 2997.

Melanchthon, Philipp. 1531. *De dialectica libri quatuor*. Wittenberg: Joseph Klug. USTC 629451 and VD16 M 3001.

Melanchthon, Philipp. 1533. *De dialectica libri quatuor recogniti*. Wittenberg: Joseph Klug. USTC 629457 and VD16 M 3003.

Melanchthon, Philipp. [1534a]. *De dialectica libri quatuor recogniti. Anno. XXXIIII*. Wittenberg: Joseph Klug. USTC 629453 and VD16 M 3004.

Melanchthon, Philipp. [1534b]. *De dialectica libri quatuor recogniti. Anno. XXXIIII*. Wittenberg: Joseph Klug. USTC 629454 and VD16 M 3005.

Melanchthon, Philipp. 1535a. *Loci communes theologici recens collecti et recogniti a Philippo Melanthone*. Wittenberg: Joseph Klug. USTC 673154 and VD16 M 3614.

Melanchthon, Philipp. 1535b. *Loci communes theologici recens collecti et recogniti a Philippo Melanthone*. Wittenberg: Joseph Klug. USTC 673155 and VD16 ZV 25988.

Melanchthon, Philipp. 1536. *De dialectica libri quatuor recogniti*. Wittenberg: Joseph Klug. USTC 629452 and VD16 M 3008.

Melanchthon, Philipp. 1541. *Loci communes theologici recens collecti et recogniti a Philippo Melanthone*. Wittenberg: Peter Seitz I. USTC 673138 and VD16 M 3624.

Melanchthon, Philipp. 1542. *Die Heubtartikel Christlicher Lere zusamen gezogen*. Wittenberg: Veit Kreutzer. USTC 636803 and VD16 M 3673.

Melanchthon, Philipp. 1547. *Erotemata dialectices continentia fere integram artemita scripta, ut juventuti utiliter proponi possit*. Wittenberg: Hans Lufft. USTC 653659 and VD16 M 3242.

Peckham, Johannes. 1504. *Perspectiva communis*. Leipzig: Martin Landsberg. USTC 683511 and VD16 J 677.

Peuerbach, Georg von. 1534. *Elementa Arithmetices*. Wittenberg: Joseph Klug. USTC 649618 and VD16 P 2048.

Peuerbach, Georg von, and Johannes Vögelin. 1536. *Elementa Geometriae*. Wittenberg: Joseph Klug. USTC 649617 and VD16 E 4165.

Rhegius, Urbanus. 1539. *Dialogus von der schönen predigt*. Wittenberg: Joseph Klug. USTC 636000 and VD16 R 1767. Rheticus, Georg, Joachim. 1536. *In Arithmeticen*. [Wittenberg: Joseph Klug 1536]. USTC 662581 and VD16 M 3433.

Rhegius, Urbanus. 1540. *De libris revolutionum Copernici narratio prima*. Danzig: Rhode. USTC 608832 and VD16 J 268.

Rheticus, Georg Joachim. 1540. *De Libris Revolutionum Copernici Narratio Prima*. Danzig: Rhode. USTC 608832 and VD16 J 268.

Runge, Jakob. 1560. *Eine Leichpredigt vber dem Begrebnis des Durchleuchtigen Hochgebornen Fuersten vnd Herrn Herrn Philippen Hertzogen zu Stetin*. Wittenberg: [heirs of Georg Rhau]. USTC 648055 and VD16 ZV 16436.

Sacrobosco, Johannes de, and Philipp Melanchthon. 1531. *Liber de sphaera. Addita est praefatio in eundem librum Philippi Melanchthonis ad Simonem Gryneum*. Wittenberg: Joseph Klug. USTC 672736 and VD16 J 720. https://hdl.handle.net/21.11103/sphaera.100138.

Sacrobosco, Johannes de, and Philipp Melanchthon. 1532. *Liber de sphaera*. Venice: G.A. Nicolini da Sabio for Melchior Sessa. https://hdl.handle.net/21.11103/sphaera.100118.

Sacrobosco, Johannes de, and Philipp Melanchthon. 1534. *Liber de sphera*. Wittenberg: Joseph Klug. USTC 672737 and VD16 J 721. https://hdl.handle.net/21.11103/sphaera.100085.

Sacrobosco, Johannes de, and Philipp Melanchthon. 1536. *Liber de sphera*. Wittenberg: Joseph Klug. USTC 672733 and VD16 J 722. https://hdl.handle.net/21.11103/sphaera.101008.

Sacrobosco, Johannes de, and Philipp Melanchthon. 1538. *Libellus de sphaera*. Wittenberg: Joseph Klug. USTC 667768 and VD16 J 723. https://hdl.handle.net/21.11103/sphaera.101106.

Sacrobosco, Johannes de, and Philipp Melanchthon. 1540. *Libellus de sphaera*. Wittenberg: Joseph Klug. USTC 667769 and VD16 J 724. https://hdl.handle.net/21.11103/sphaera.101025.

Sacrobosco, Johannes de, and Philipp Melanchthon. 1543. *Libellus de sphaera*. Wittenberg: Peter Seitz. USTC 667490 and VD16 J 725. https://hdl.handle.net/21.11103/sphaera.101029.

Sacrobosco, Johannes de, and Philipp Melanchthon. 1545. *Libellus de sphaera*. Wittenberg: Veit Kreutzer. USTC 667491 and VD16 J 726. https://hdl.handle.net/21.11103/sphaera.100818.

Sacrobosco, Johannes de, and Philipp Melanchthon. 1549. *Libellus de sphaera*. Wittenberg: Veit Kreutzer. USTC 667492 and VD16 J 727. https://hdl.handle.net/21.11103/sphaera.100136.

Sacrobosco, Johannes de, and Philipp Melanchthon. 1550. *Libellus de sphaera*. Wittenberg: Johann Krafft. USTC 667489 and VD16 J 728, https://hdl.handle.net/21.11103/sphaera.100157.

Sacrobosco, Johannes de, and Philipp Melanchthon. 1999 [1531]. Preface to *On the Sphere (1531)*. In *Orations on Philosophy and Education*, ed. Sachiko Kusukawa and trans. C.F. Salazar, 105–112. Cambridge: Cambridge University Press.

Scripta proposita publice, liber I. 1545–1546. Wittenberg: Joseph Klug. USTC 692611 and VD16 W 3752.

Scripta proposita publice, liber II. 1548. Wittenberg: Joseph Klug. USTC 692869 and VD16 W 3753.

Scripta proposita publice, digesta in duos libros. 1549a. Wittenberg: Joseph Klug. USTC 692610 and VD16 W 3754.

Scripta proposita publice, liber III. 1549b. Wittenberg: Joseph Klug. USTC 696155 and VD16 W 3756.

Scripta proposita publice, liber IV. 1551. Wittenberg: Peter Seitz. USTC 689360 and VD16 W 3757.

Scripta proposita publice ab anno 1540 usque ad annum 1553. 1553. Wittenberg: heirs of Peter Seitz. USTC 692608 and VD16 W 3758.

Scripta proposita publice Tomus II. 1556. Wittenberg: heirs of Georg Rhau for Konrad Rühel. USTC 692620, 692621 and VD16 W 3759/W 3760.

Scripta proposita publice Tomus III. 1559, Wittenberg: heirs of Georg Rhau for Konrad Rühel. USTC 692624 and VD16 W 3761.

Scripta proposita publice ab anno 1540 usque ad annum 1553 (edited and reprinted). 1560. Wittenberg: heirs of Georg Rhau for Konrad Rühel. USTC 692631 and VD16 W 3762.

Scripta proposita publice Tomus IV. 1561. Wittenberg: heirs of Georg Rhau. USTC 692628 and VD16 ZV 15568.

Scripta proposita publice Tomus II (edited and reprinted). 1562. Wittenberg: heirs of Georg Rhau for Konrad Rühel. USTC 692629 and VD16 W 3763.

Scripta proposita publice Tomus V. 1564. Wittenberg: Hans Lufft. USTC 692625 and VD16 W 3764.

Scripta proposita publice Tomus VI. 1568. Wittenberg: Hans Lufft. USTC 692630 and VD16 ZV 15569.

Scripta proposita publice Tomus III (edited and reprinted). 1570. Wittenberg: Lorenz Schwenck for Konrad Rühel. USTC 692632 and VD16 W 3765.

Scripta proposita publice Tomus VII. 1572. Wittenberg: Clemens Schleich and Anton Schöne. USTC 692626 and VD16 ZV 15570.

Sutell, Johannes. 1543. *Historia von Lazaro aus dem XI. Cap. Des Evangelii S. Johannis*. Wittenberg: Joseph Klug. USTC 663320 and VD16 S 10311.

Velcurio, Johannes. 1539. *Commentarii in universam Physicam Aristotelis libri quatuor, diligenter recogniti*. Tübingen: Ulrich Morhart. USTC 667683 and VD16 B 2024.

Velcurio, Johannes. 1540. *Commentarii in universam physicam Aristotelis libri quatuor, diligenter recogniti*. Tübingen: Ulrich Morhart. USTC 667684 and VD16 B 2025.

Willich, Jodocus. 1540. *Erotematum dialectices libri tres*. Strasbourg: Kraft Müller. USTC 667936 and VD16 W 3241.

Secondary Literature

Albrecht, Otto and Paul Flemming. 1913. Das sogenannte Manuscriptum Thomasianum. V. Dritter Teil. Nr. 94–126. *Archiv für Reformationsgeschichte* 13 (2): 161–199.

Bellingradt, Daniel. 2019. The dynamic of communication and media recycling in early modern Europe: Popular prints as echoes and feedback loops. In *Crossing borders, crossing cultures*, eds. Massimo Rospocher, Jeroen Salman, and Hannu Salmi, 9–32. Oldenbourg: De Gruyter.

Blair, Ann. 2017. The 2016 Josephine Waters Bennett Lecture: Humanism and printing in the work of Conrad Gessner. *Renaissance Quarterly* 70: 1–43.

Burmeister, Karl Heinz. 1968. *Georg Joachim Rhetikus, 1514–1574. Eine Bio-Bibliographie III. Briefwechsel.* Wiesbaden: Pressler.

Burmeister, Karl Heinz. 2006. Magister Joachimus Aeliopolitanus. In *Beiträge zur Astronomiegeschichte 8,* ed. Wolfgang R. Dick and Jürgen Hamel, 7–18. Leipzig: AVA Akademische Verlagsanstalt.

Burmeister, Karl Heinz. 2015. *Magister Rheticus und seine Schulgesellen. Das Ringen um Kenntnis und Durchsetzung des heliozentrischen Weltsystems des Kopernikus um 1540/50.* Konstanz: UVK Verlagsgesellschaft.

Claus, Helmut. 2002. «…als ob die Engel Botenläufer gewesen seien.» Wittenberg als Druckerstadt. In *Wittenberg als Bildungszentrum 1502–2002. Lernen und Leben auf Luthers Grund und Boden,* ed. Peter Freybe, 75–103. Wittenberg: Elbe-Druckerei.

Claus, Helmut. 2014. *Melanchthon Bibliographie 1510–1560.* Gütersloh: Gütersloher Verlagshaus.

Clemen, Otto. 1941. Beiträge zur Geschichte des Wittenberger Buchdrucks in der Reformationszeit. *Gutenberg-Jahrbuch* 16: 174–185.

Clemen, Otto. 1942–1943. Weitere Beiträge zur Geschichte des Wittenberger Buchdrucks und des Buchgewerbes in der Reformationszeit. *Gutenberg-Jahrbuch* 17: 114–125.

Deschauer, Stefan, ed. 2003. *Die Arithmetik-Vorlesung des Georg Joachim Rheticus, Wittenberg 1536. Eine kommentierte Edition der Handschrift X-278 (8) der Estnischen Akademischen Bibliothek.* Augsburg: Erwin Rauner Verlag.

Domtera-Schleichardt, Christiane. 2014. Paul Ebers Beiträge in den gedruckten Wittenberger Scripta publice proposita. In *Paul Eber (1511–1569). Humanist und Theologe der zweiten Generation der Wittenberger Reformation,* eds. Daniel Gehrt and Volker Leppin, 565–586. Leipzig: Evangelische Verlagsanstalt.

Domtera-Schleichardt, Christiane. 2021. *Die Wittenberger "Scripta publice proposita" (1540–1569). Universitätsbekanntmachungen im Umfeld des späten Melanchthon.* Leipzig: Evangelische Verlagsanstalt.

Ehrig-Eggert, Carl. 2019. 'Arabische' Astronomie in der Reformationszeit: Die Sphaera mundi von Sacrobosco—ihre Quellen und ihr Vorwort von Philipp Melanchthon. In *Barther Bibliotheksgespräch,* ed. Christian Heitzmann and Falk Eisermann, 91–111. Rostock: Hinstorff.

Eisermann, Falk. 2009. Die schwarze Gunst. Buchdruck und Humanismus in Leipzig um 1500. In *Der Humanismus an der Universität Leipzig,* ed. Enno Bünz, 149–179. Wiesbaden: Harrassowitz.

Fichtner, Gerhard. 1968. Neues zu Leben und Werk von Leonhart Fuchs aus seinen Briefen an Joachim Camerarius I. und II. in der Trew-Sammlung. *Gesnerus: Swiss Journal of the History of Medicine and Sciences* 25: 65–82.

Friedensburg, Walter. 1917. *Geschichte der Universität Wittenberg.* Halle: Niemeyer.

Friedensburg, Walter. 1926. *Urkundenbuch der Universität Wittenberg. Teil 1 (1502–1611).* Magdeburg: Selbstverlag der Historischen Komission.

Friedensburg, Walter. 1927. *Urkundenbuch der Universität Wittenberg. Teil 2 (1611–1813).* Magdeburg: Selbstverlag der Historischen Komission.

Gingerich, Owen. 2015. *The book nobody read. Chasing the revolutions of Nicolaus Copernicus.* London: Arrow.

Grafton, Anthony. 1980. The importance of being printed. *Journal of Interdisciplinary History* 11 (2): 265–286.

Hamel, Jürgen. 2014. *Studien zur Sphaera des Johannes de Sacrobosco* . Leipzig: AVA.

Heidrich, Jürgen. 2015. Georg Rhau als Wittenberger Musikaliendrucker und -verleger. In *Buchdruck und Buchkultur im Wittenberg der Reformationszeit,* ed. Stefan Oehmig, 191–203. Leipzig: Evangelische Verlagsanstalt.

Kathe, Heinz. 2002. *Die Wittenberger Philosophische Fakultät 1502–1817.* Cologne: Böhlau.

Kaufmann, Thomas. 2019. *Die Mitte der Reformation. Eine Studie zu Buchdruck und Publizistik im deutschen Sprachgebiet, zu ihren Akteuren und deren Strategien, Inszenierungs- und Ausdrucksformen.* Tübingen: Mohr Siebeck.

Kolde, Theodor. 1883. *Analecta Lutherana. Briefe und Aktenstücke zur Geschichte Luthers.* Gotha: Perthes.

Kraai, Jesse. 2000. *Rheticus' heliocentric providence: A study concerning the astrology, astronomy of the sixteenth century.* Unpublished PhD thesis submitted to the University of Heidelberg, available online https://www.ub.uni-heidelberg.de/archiv/3254. Accessed 7 June 2021.

Krieg, Walter. 1953. *Materialien zu einer Entwicklungsgeschichte der Bücher-Preise und des Autoren-Honorars vom 15. bis zum 20. Jahrhundert.* Vienna: Stubenrauch.

Lang, Thomas. 2015. Simprecht Reinhart: Formschneider, Maler, Drucker, Bettmeister—Spuren eines Lebens im Schatten von Lucas Cranach d. Ä.. In *Das ernestinische Wittenberg: Spuren Cranachs in Schloss und Stadt*, eds. Heiner Lück, Enno Bünz, Leonhard Helten, Armin Kohnle, Dorothée Sack, and Hans-Georg Stephan, 93–138. Petersburg: Imhof Verlag.

Leu, Urs Bernhard. 2014. Die Bedeutung Basels als Druckort im 16. Jahrhundert. In *Basel als Zentrum geistigen Austauschs in der frühen Reformationszeit*, eds. Christine Christ-von Wedel, Sven Grosse, and Berndt Hamm, 53–78. Tübingen: Mohr Siebeck.

Limbach, Saskia. 2017. Advertising medical studies in sixteenth-century Basel: Function and use of academic disputations. In *Broadsheets. Single-sheet publishing in the first age of print*, ed. Andrew Pettegree, 376–398. Leiden: Brill.

Limbach, Saskia. 2021. *Government use of print. Official publications in the Holy Roman Empire.* Frankfurt am Main: Klostermann.

Lück, Heiner. 2011. Das ernestinische Wittenberg: Universität und Stadt (1486–1547). Ein Forschungsvorhaben der Martin-Luther-Universität Halle-Wittenberg und der Stiftung Leucorea. In *Das ernestinische Wittenberg: Universität und Stadt (1486–1547)*, ed. Heiner Lück, Enno Bünz, Leonhard Helten, Dorothée Sack, and Hans-Georg Stephan, 9–19. Petersberg: Michael Imhof Verlag.

Mullins, Sophie. 2014. *Latin Books Published in Paris, 1501–1540.* Unpublished PhD thesis submitted to the University of St. Andrews, available online https://research-repository.st-andrews.ac.uk/handle/10023/6333. Accessed 7 June 2021.

Nehlsen, Eberhard. 2015. In Wittenberg gedruckte Liedflugschriften des 16. Jahrhunderts. In *Buchdruck und Buchkultur im Wittenberg der Reformationszeit*, ed. Stefan Oehmig, 205–229. Leipzig: Evangelische Verlagsanstalt.

Nelles, Paul. 2001. Historia Litteraria at Helmstedt. Books, Professors and Students in the Early Enlightenment University. In *Die Praktiken der Gelehrsamkeit in der Frühen Neuzeit*, eds. Helmut Zedelmaier and Martin Mulsow, 147–176. Tübingen: Niemeyer.

Omodeo, Pietro Daniel. 2014. *Copernicus in the cultural debates of the Renaissance. Reception, legacy, transformation.* Leiden: Brill.

Pantin, Isabelle. 2009. Altior incubuit animus sub imagine mundi. L'inspiration du cosmographe d'après un gravure d'Oronce Finé. In *Les Méditations cosmographiques à la Renaissance*, eds. Frank Lestringant and Centre V.L. Saulnier, 69–90. Paris: Presses de l'Université Paris-Sorbonne.

Pantin, Isabelle. 2020. Borrowers and innovators in the history of printing Sacrobosco: the case of the In-Octavo tradition. In *De sphaera of Johannes Sacrobosco in the early modern period*, ed. Matteo Valleriani, 265–312. Dordrecht: Springer International. https://doi.org/10.1007/978-3-030-30833-9_9.

Pettegree, Andrew. 2015. *Brand Luther: 1517. Printing and the making of the Reformation.* New York: Penguin.

Pollard, Alfred Wiliam. 1902. The transference of woodcuts in the fifteenth and sixteenth centuries. In *Old picture books with other essays on bookish subjects*, ed. Alfred Wiliam Pollard, 73–98. London: Methuen.

Proot, Goran. 2019. Shifting price levels of books produced at the Officina Plantiniana in Antwerp, 1580–1655. In *Crossing borders, crossing cultures*, ed. Massimo Rospocher, Jeroen Salman, and Hannu Salmi, 89–107. Oldenbourg: De Gruyter.

Reich, Karin. 1998. Melanchthon und die Mathematik seiner Zeit. In *Melanchthon und die Naturwissenschaften seiner Zeit*, ed. Günther Frank and Stefan Rhein, 105–121. Sigmaringen: Thorbecke.

Reich, Ulrich. 2017. Mathematik. In *Philipp Melanchthon. Der Reformator zwischen Glauben und Wissen*, ed. Günther Frank, 457–468. Berlin: De Gruyter.

Reske, Christoph. 2015. *Die Buchdrucker des 16. und 17. Jahrhunderts im deutschen Sprachgebiet*. Wiesbaden: Harrassowitz.

Reske, Christoph. 2017. Die Anfänge des Buchdrucks im vorreformatorischen Wittenberg. In *Buchdruck und Buchkultur im Wittenberg der Reformationszeit*, ed. Stefan Oehmig, 35–69. Leipzig: Evangelische Verlagsanstalt.

Rhein, Stefan. 1997. Buchdruck und Humanismus—Melanchthon als Korrektor in der Druckerei des Thomas Anshelm. In *Philipp Melanchthon in Südwestdeutschland: Bildungsstationen eines Reformators*, eds. Stefan Rhein, Armin Schlechter, and Udo Wennemuth, 63–74. Karlsruhe: Badische Landesbibliothek.

Rosen, Edward. 1974. Rheticus as editor of the Sacrobosco. In *For Dirk Struik: Scientific, historical and political essays in honour of Dirk J. Struik*, eds. Robert S. Cohen, J.J. Stachel, and Marx W. Wartofsky, 245–248. Berlin: Springer.

Rothe, Vicky. 2013. Wittenberger Buchgewerbe und -handel im 16. Jahrhundert. In *Das ernestinische Wittenberg. Stadt und Bewohner*, eds. Heiner Lück, Enno Bünz, Leonhard Helten, Armin Kohnle, Dorothée Sack, and Hans-Georg Stephan, 77–90. Petersberg: Michael Imhof Verlag.

Scheible, Heinz. 1997. *Melanchthon. Eine Biographie*. Munich: Beck.

Scheible, Heinz. 2010a. Melanchthons Verhältnis zu Johannes Setzer. In *Aufsätze zu Melanchthon*, ed. Heinz Scheible, 309–316. Tübingen: Mohr Siebeck.

Scheible, Heinz. 2010b. Melanchthons Beziehung zu Stadt und Bistrum Breslau. In *Aufsätze zu Melanchthon*, ed. Heinz Scheible, 342–372. Tübingen: Mohr Siebeck.

Schirmer, Uwe. 2015. Buchdruck und Buchhandel im Wittenberg des 16. Jahrhunderts. Die Unternehmer Christian Döring, Hans Lufft und Samuel Selfisch. In *Buchdruck und Buchkultur im Wittenberg der Reformationszeit*, ed. Stefan Oehmig, 169–190. Leipzig: Evangelische Verlagsanstalt.

Schöbi, Philipp. 2014. Rheticus—Wegbereiter der Neuzeit. In *Georg Joachim Rheticus 1514–1574. Wegbereiter der Neuzeit. Eine Würdigung*, eds. Philipp Schöbi and Helmut Sonderegger, 35–94. Hohenems: BUCHER Verlag.

Siraisi, Nancy. 2008. Mercuriale's Letters to Zwinger and Humanist Medicine. In *Girolamo Mercuriale. Medicina e cultura nell'Europa del Cinquecento*, ed. Alessandro Arcangeli and Vivian Nutton, 77–95. Florence: Leo S. Olschiki.

Steiff, Karl. 1881. *Der erste Buchdruck in Tübingen*. Tübingen: Laupp.

Stiegler, Elke. 1989. Nachwort. In *Kurtz-gefaßte Historie des Lebens und Factorum Hanns Luffts, berühmten Buchdruckers zu Wittenberg*, ed. Gustav Georg Zeltner, I-XVI. Leipzig: Zentralantiquariat der DDR.

Thomas, Drew. 2018. *The industry of evangelism. Printing for the reformation in Martin Luther's Wittenberg*. Unpublished PhD thesis submitted to the University of St Andrews.

Thomas, Drew. 2021. *The industry of evangelism. Printing for the reformation in Martin Luther's Wittenberg*. Leiden: Brill.

Volz, Hans. 1957. Die Arbeitsteilung der Wittenberger Buchdrucker zu Luthers Lebzeiten. *Gutenberg-Jahrbuch* 32: 146–154.

Volz, Hans. 1963. Zur Geschichte des Wittenberger Buchdrucks 1544–1547. *Gutenberg-Jahrbuch* 38: 113–119.

Volz, Hans. 1964. Goltz, Moritz. In *Neue Deutsche Biographie*, 637–638. https://www.deutsche-biographie.de/pnd136049478.html#ndbcontent. Accessed 7 June 2021.

Valleriani, Matteo. 2017. The tracts of the *Sphere*: Knowledge reconstructed over a network. In *The structures of practical knowledge*, ed. Matteo Valleriani, 421–473. Dordrecht: Springer International.

Valleriani, Matteo, Florian Kräutli, Maryam Zamani, Alejandro Tejedor, Christoph Sander, Malte Vogl, Sabine Bertram, Gesa Funke, and Holger Kantz. 2019. The emergence of epistemic communities in the Sphaera corpus: Mechanisms of knowledge evolution. *Journal of Historical Network Research* 3: 50–91. https://doi.org/10.25517/jhnr.v3i1.63.

Valleriani, Matteo, Olya Nicolaeva, and Beate Federau. 2022. The hidden *Praeceptor*: How Georg Rheticus taught geocentric cosmology to Europe. *Perspectives on Science* 30(3).

Walsby, Malcolm. 2017. Cheap print and the academic market. The printing of dissertations in sixteenth-century Louvain. In *Broadsheets. Single-sheet publishing in the first age of print*, ed. Andrew Pettegree, 355–375. Leiden: Brill.
Widmann, Hans. 1971. *Tübingen als Verlagsstadt*. Tübingen: Mohr Siebeck.

Saskia Limbach received her Ph.D from the University of St. Andrews where she also worked as a long-term editor of the Universal Short Title Catalogue (USTC). She is currently a postdoctoral fellow at the University of Göttingen. Previously she worked in the ERC-funded 'Early Modern Book Trade' project at the Università degli Studi di Milano and was also a researcher in a digital humanities project at the University of Mainz.

Saskia's research focusses on the economic and legal conditions that shaped the book trade and she has published on media and conflict resolution, as well as ephemeral material, with a particular interest in ordinances and academic disputations. Her monograph *Government Use of Print. Official Publications in the Holy Roman Empire (1500–1600)* appeared in the series of the MPI for European Legal History in 2021. Currently Saskia investigates successful women-printers in the German print industry.

Chapter 6
Sacrobosco at the Book Fairs, 1576–1624: The Pedagogical Marketplace

Ian Maclean

Abstract Between 1576 and 1624, there were at least twenty declarations of Latin editions of the *De sphaera* and one of Francesco Pifferi's Italian-language version in the Frankfurt Book Fair Catalogues. These declarations are not straightforward. Some are editions, and some are reissues; some are associated with the names not of the publisher but of the bookstore through where they were on offer. In some cases, the year of declaration and the place of publication are misleading, and, in the case of the Christophorus Clavius commentaries, the claim made about the number of the edition (whether third, fourth, fifth, or "seventh") is false. The aim of this paper is to elucidate how the fairs operated, to identify which edition is in question, to place these in the context of the Catholic, Lutheran, and Reformed pedagogy of the period, and to show how producers, advertisers, and commissioning agents of the Sacrobosco *Sphaera* editions interacted, with special reference to the editions of Rome and Venice (and their publishers the Basa family and Giovanni Battista Ciotti), and those of Lyon and St. Gervais (and their publishers de Gabiano and Samuel Crespin). The final case considered is the declaration in 1624 of the edition of the Jesuit Bernardus Morisanus produced by the Reformed printer-publisher Peter Mareschal, as an element in a *cursus philosophicus*.

Keywords Frankfurt Book Fair · Book trade · Material bibliography · Book history · Johannes de Sacrobosco · Christophorus Clavius · Giovanni Battista Ciotti

1 Introduction

In this paper I set out to use information from Frankfurt Book Fair Catalogues and the resources of material bibliography to investigate the market for Johannes de Sacrobosco's (d. 1256) *De sphaera* and Christophorus Clavius' (1538–1612) commentary over the period 1576–1625, paying particular attention to the constraints imposed by the Fair, the constraints imposed by publishing practices, and the constraints

I. Maclean (✉)
All Souls College, University of Oxford, Oxford, England
e-mail: ian.maclean@all-souls.ox.ac.uk

© The Author(s) 2022
M. Valleriani and A. Ottone (eds.), *Publishing Sacrobosco's* De sphaera *in Early Modern Europe*, https://doi.org/10.1007/978-3-030-86600-6_6

arising from the nature of the market for textbooks. I shall begin with the list of all entries relating to the *De sphaera* or commentaries on the *De sphaera* that I was able to find in the various redactions of the Fair Catalogues and in an early omnibus compendium of the Fair catalogues: the *Collectio in unum corpus omnium librorum… in nundinis Francofurtensibus ab anno 1564 usque ad nundinas autumnales anni 1592… (Collectio* 1592).[1] In this list, "S" refers to the Spring, "A" to the Autumn Catalogue; all are from the "Libri philosophici" section, except three: The declaration of S1592 is in the "Libri historici" section; the Italian-language entry of A1604 is in the "Libri peregrino idiomate scripti" section (i.e., those not written in Latin or German); the declaration of Cholinus' printing in A1607 is in the section "books forthcoming at future fairs" (*Libri proximis nundinis prodituri*).[2]

A1576 Christophori Clauii Bambergensis, ex societate Iesu, in Sphaeram Ioannes de Sacro Bosco Commentarius 4. Romae.
(Willer 1972–2001): probably a reissue of (Sacrobosco and Clavius 1570).

S1578 Fr. Iunctini Florentini, sacrae Theologiae Doctoris, Com[m]entaria in Sphaeram Ioannis de Sacro Bosco accuratissima. 8. Lugduni apud Philippum Tinghium.
(Willer 1972–2001): (Sacrobosco and Giuntini 1578).

S1581 Sphaera Ioannis de Sacrobosco emendata. Eliae Vineti Santonis scholia in eadem Sphaeram, ab ipso authore restituta. Quibus accessere Scholia Heronis et aliorum. 8. Coloniae. [apud Maternum Cholinum 1581].
(Willer 1972–2001): (Sacrobosco et al. 1581).

S1582 Christophori Clauii Bambergensis ex Societate Iesu in Sphaeram Ioannis de Sacro Bosco commentari[a]us 4. Romae.
(Willer 1972–2001): (Sacrobosco et al. (1581).

[1582] Ioan. De Sacro Busto Sphaera emendata. Antwerpiae apud Bellerum 1582. V.8 (*Collectio* 1592, 500): (Sacrobosco et al. 1582).

---

[1] On the different redactions of the Fair Catalogue, see (Schwetschke 1850–1877, VII–XXXIV). The Georg Willer catalogues have been published by Bernhard Fabian as *Die Messkataloge Georg* (Willers 1972–2001), and the Catalogues printed by Ioannes Saur (S1601–S1607) and Sigismund Latomus (A1608–A1624) are to be found at www.olmsonline.de/en/kollektionen/messkataloge. Accessed 8 June 2021. The *Collectio* was revised by Joannes Clessius with the title *Unius seculi; eiusque virorum literatorum monumentis…ab anno dom. 1500 ad 1602 nundinarum autumnalium inclusive elenchus* (Clessius 1602). The only additional edition is *De Sphaera* "apud Gosvinum Cholinum 1600" (*Collectio* 1592, 470; Sacrobosco et al. 1601). On the date of this entry, see the comment about issues with sequential dates, below. The Lutz catalogue is *Catalogus novus nundinarum vernalium Francoforti ad Moenum anno M.D.LXXXXII celebratarum…apud Thobiam* (Lutz 1592).

[2] In S1612 (*Catalogus…vernalibus* 1612), the *Opera mathematica* of (Clavius 1611–1613) was declared, in which the sixth recension of his Commentary on the *De sphaera* is included; I have not referred to it here, as the Catalogue does not specify the contents of the five volumes. On the circumstances surrounding this publication, see (Clavius 1992, VI, letters no. 305, 308, 310).

[1591] [Ioan. De Sacro Busto Compendium in Sphaeram.] Accessit compendium in Sphaeram Pierij Valerianii Bellunens. Coloniae ap. Mater. Cholinum 1591 V.8.
(*Collectio* 1592, 500): (Sacrobosco et al. 1591).

S1592 [Libri historici] Christophori Clauij Bambergensis societatis Iesu in Sphaeram Ioannis de Sacrobosco commentarius tertio recognitus & locupletatus Ven. 8.
(Lutz 1592): (Sacrobosco and Clavius 1591).

A1592 Christoph. Clauij in Sphaeram Ioan. de Sacro Bosco Com[m]entarius. Editio 4. ab authore recognita. Lugd. sumptibus fratrum de Gabiano in 4. futuris nundinis ve[r]n[alibus] exponetur.
(Willer 1972–2001): (Sacrobosco and Clavius 1593) or (Sacrobosco and Clavius 1594).

S1601 Christophori Clauii commentarius in Sphaeram Ioan. de Sacrobusto, iam recognitus apud [Joannem Baptistam Ciotti] in 4.
(*Catalogus...vernalibus* 1601): (Sacrobosco and Clavius 1601a).

S1601 Sphaera Ioannis de sacro Bosco emendata, cum notis aliquot doctorum virorum. Col. In 8. Cholin.
(*Catalogus...vernalibus* 1601): (Sacrobosco et al. 1601).

A1601 Libellus de Sphaera Iohannis de Sacrobusto. Accessit eiusdem autoris computus Ecclesiasticus, & alia quaedam. Wittebergae impensis Zachariae Schureri in 8.
(*Catalogus...autumnalibus* 1601): (Sacrobosco et al. 1601).

S1602 Christophori Clauii Iesuitae commentarius in sphaeram Sacrobusti. Editio quinta. Lugduni apud Crispinum.
(*Catalogus...vernalibus* 1602): (Sacrobosco and Clavius 1602a). A reissue of (Sacrobosco and Clavius 1602b).

S1603 Christoph. Clauij in Sphaeram Ioan de sacro busto Comment. Editio septima locupletior. ap. soc. Venet. in 4. 1603.
(*Catalogus...vernalibus* 1603): (Sacrobosco and Clavius 1603). A reissue of (Sacrobosco and Clavius 1601a, b).

A1604 Sfera di Gio. Sacrobosco tradotta e dicharata da Don Francesco Pifferi Sansauino. Con nuouo aggiunte di molte cose notabili, e dilettouoli [Societ. Venet.] in 4. 1604.
(*Catalogus...autumnalibus* 1604): (Sacrobosco and Pifferi 1604).

A1607 Christophori Clauii Bambergensis S.I. in Sphaeram Ioanis de Sacro Bosco Commentarius. Lugduni apud Iacob Chouet. 4 & 8. Candon 4.
(*Catalogus...autumnalibus* 1607): Not readily identifiable. See Sect. 5.

A1607 Sphaera Clauii nova Romae. apud Arnold. Quentel.
(*Catalogus...autumnalibus* 1607): (Sacrobosco and Clavius 1606) or (Sacrobosco and Clavius 1607d).

A1607 [Libri prodituri] Sphaera Joannis de Sacro Bosco emendata: cum scholiis Vineti et Commentariis Clauii. [Coloniae] ap. [Cholin]. 8.
(*Catalogus...autumnalibus* 1607): No surviving copy; probably the 1610 edition: (Sacrobosco et al. 1610).

S1608 Christophori Clauii in Sphaeram Ioannis de sacro Bosco commentarius. Ab ipso auctore locupletatus. Accessit Geometrica atque uberrima de Crepusculis tractatio. S Gervasio ap. Sam. Crispinum. in 4.
(*Catalogus...vernalibus* 1608): (Sacrobosco and Clavius 1608).

S1610 Sphaera Ioannis a Sacro Bosco emendata, aucta & illustrata. Coloniae apud Petrum Chol. In 8.
(*Catalogus...vernalibus* 1610): (Sacrobosco et al. 1610).

A1624 [Bernardi Morisani Derensis Ibernici] Commentarius in Sphaeram Ioannis de S. Bosco ibid. in 8.
(*Catalogus...autumnalibus* 1624): (Sacrobosco and Morisanus 1625)

Before my substantive discussion of the Latin editions of this list, I should first comment on the one Italian-language entry from A1604: Francesco Pifferi's (1548–1612) *Sfera di Gio. Sacrobosco tradotta e dicharata...con nuove aggiunte di molte cose notabili, e dilettovoli* (Sacrobosco and Pifferi 1604). This quarto edition was published by Salvestro Marchetti of Siena (fl. 1594–1620) in 1604 and declared at the Fair by the *Societas Veneta* (a consortium of Francesco de' Franceschi (ca. 1530–1599), Giovanni Battista Ciotti (1564–1635) and Roberto Meietti (1572–1634), whom we shall meet again). In his dedication to Cosimo II de' Medici (1590–1621), Pifferi declares that he was appointed by Cosimo's mother, Christine de Lorraine (1565–1637), to teach Cosimo mathematics and that he had undertaken the translation and commentary to this end. In 1605, he was replaced as tutor by Galileo Galilei (1564–1642), who in 1615 published the famous pro-Copernican *Lettera* to Cosimo's mother. Pifferi was a Camaldolese monk, a professor of mathematics at Siena, a member of the Accademia degli Intronati, and an associate of the Accademia dei Lincei. His library (inspected and inventoried in 1603 by the Inquisition) contained a number of editions of the *Sphaera* from Antwerp, Rome, Cologne, Paris, and Venice, a wide range of books by Christophorus Clavius, and an impressive collection of other works on sphaeristics. Pifferi's commentary relies heavily on those of Francesco Giuntini (1523–1590) and Clavius; its approach to cosmology is functionalist, and it engages negatively with Copernicus, but Pifferi does record the argument of Andreas Osiander (1498–1552) that the work was designed to "save the appearances" (*salvar l'apparenze*): i.e., that it was no more than a hypothesis.[3]

---

[3] For Pifferi's functionalism, see (Sacrobosco and Pifferi 1604, 119); for his references to Nicolaus Copernicus, see (Sacrobosco and Pifferi 1604, 86, 392).

This is an interesting case of the vulgarization of Sacrobosco in court circles and vernacular academies, but that does not explain its advertisement at the Frankfurt Book Fair, or who the targeted purchasers there might be.[4]

Interpreting the list of Fair declarations is not straightforward, given the legal and commercial constraints under which the Fairs operated, the ambiguity of the terminology of declarations, and the complex practices of the publishers who declared their wares in its catalogues. I shall begin by giving an account of these difficulties, offering examples from the list given above, after which I shall discuss three groups of entries at greater length.

2   *Novi, emendatiores, auctiores* ("New, Improved, Enlarged"): The *modus operandi* of Fairs and Its Effect on Publishing Practices[5]

In the half-century preceding the outbreak of the Thirty Years War, the twice-yearly Fair was the preeminent European meeting place for scholars and book merchants, who went there to exchange news from the world of letters and advertise recent publications. From 1564, catalogues of the exhibited books were printed there (the most famous being the ones produced by the Augsburg bookseller Georg Willer the Elder (1514–1593)). The catalogues only became officially sanctioned publications after 1597, and even then, not all of the catalogues that were derived from its authorized version were identical. They also contained omissions and commissions: books whose titles had been submitted in advance but were not actually present at the fair, and books not declared that were available for purchase.

During this period, regulations were introduced both at the level of the City of Frankfurt and the Holy Roman Empire covering the financial, socio-economic, and politico-religious issues relating to the book market. Declarations of titles were subject to the oversight of the Frankfurt City Council and the Imperial Book Commission, which acted also for the Roman Catholic Church. Religious censorship was only applied directly to the *De sphaera* in respect of one Catholic commentator, and in the requirement in Catholic contexts that the names of Protestant commentators (such as Philip Melanchthon (1497–1560)) should be obliterated from copies (Sander 2018). It impinged on the publication of some non-theological works in a quite different way. It was very difficult to export books in all disciplines with imprints announcing that they had been produced in Protestant or Reformed cities (Frankfurt, Geneva) to countries that accepted the authority of the Roman Index. This led publishers using printers from those cities either to adopt fictitious bibliographical addresses of unimpeachable Catholicity (mainly Cologne and Lyon), or to disguise the provenance of their books by an oblique reference to it (in the case of Geneva, "Aurelia

---

[4] The transcription of the contents of Pifferi's library is in (Maranini 2000, 127–196). On philosophy in the vernacular, see (Lines and Rufini 2015).

[5] For the details of what follows, see (Maclean 2021, 6–68).

Allobrogum," "Colonia Allobrogum," and "Saint Gervais" were all used in this way).
An example of the use of Cologne as an address to disguise a printing undertaken in
Frankfurt by an author relevant to this paper is Giovanni Battista Ciotti's edition of the
De rebus naturalibus by Jacopo Zabarella (1533–1589) (Zabarella 1590a; Rhodes
2013, 41–43); Samuel Crespin (1560–1648) of Geneva used Saint Gervais as the
address for his editions of Clavius' commentary on the *De sphaera*, and published
other works as though from Lyon, e.g., Antoine Favre's (1557–1624) *Coniecturae*
of 1607 (Favre 1607).

Certain commercial constraints applied to all declarations of titles which echoed
the conditions found in licenses or privilege agreements. These are clearly set out
in the title of the Book Catalogues: *Catalogus Universalis, Pro Nundinis Francofurtensibus…Hoc Est, Designatio Omnium Librorum, Qui Istis Nundinis…, Vel novi
vel emendatiores, aut auctiores prodierunt* (Fig. 1). Ways were found to circumvent
the requirement that all exhibited books should be in some sense "either new, or
improved, or enlarged." New is here apparently unambiguous, but in publishers'
and booksellers' catalogues, it can describe any or all of the following: books new
to Frankfurt Book Fair Catalogue; books newly available in Frankfurt (listed in
a publisher's backlist affixed to their stall at the Fair); books new to the given
publisher's or bookseller's bookshop; a first printing; or any new edition. Editions
were numbered, sometimes dishonestly, to indicate that changes had occurred (see
the example of Clavius' commentary on the *De sphaera*: "editio quinta," declared by
Samuel Crespin in 1602 is in fact the fourth edition, and "editio septima" published
by Ciotti in 1603, is in fact the third edition). In both these cases, the dishonest
numbering was used to attempt to sell editions that had been superseded.

Various claims were made about improved editions. Some were based on modifications to the text ("purged of an infinite number of errors:" *ab infinitis mendis
purgatus*), here seen in the editions that claim to be "emendata." A given publication
might be augmented by additional material in the form of other texts (this applies
to the texts from Wittenberg, and Goswin Cholinus' (fl. 1588–1612) text of 1601,
for example). The *De sphaera* might have the Pedro Nuñes (1502–1578) text, or the
various *scholiae* by Élie Vinet (1509–1587), Pierio Valeriano (1477–1558), Albertus
Hero (1549–1589), Francesco Giuntini, Christophorus Clavius, or Jacques Martin (fl.
1607), printed at the same time. A book might be produced in the same press run, yet
be supplied with title pages bearing different dates to suggest revisions and updating
(e.g., the editions of Clavius' Commentary by Jean de Gabiano (1567–1618) in 1593
and 1594, and that of Giovanni Paolo Gelli (fl. 1606–1629) in 1606 and 1607, cited
below). These practices led to the emergence of four categories of declarations: (1)
genuinely new editions (e.g., the Pierre Mareschal (1572–1622) edition of Bernardus
Morisanus' (1600–1650) commentary, declared in A1624); (2) reissues disguised as
new editions (e.g., Ciotti's entry for S1603); (3) reprints disguised as new editions
(e.g., the Cholinus entry for S1591); (4) unauthorized reprintings by unscrupulous
printer-publishers, which, if these infringed the privilege system of book protection in
given jurisdictions, could justifiably be called pirated editions and be pursued in law
(this is possibly the case of Ciotti's declaration in S1592, but by 1601, the relationship with the Basa firm had been regularized). The municipal and imperial authorities

CATALOGVS VNI-
VERSALIS PRO NVN-
DINIS FRANCOFVRTENSI-
bus autumnalibus, de anno 1607.

HOC EST:

DESIGNATIO OMNIVM
LIBRORVM, QVI ISTIS NVNDINIS
autumnalibus, vel noui, vel emendatiores, aut
auctiores prodierunt.

Das iſt:

Verzeichniß aller Bücher/ſo zu Franckfurt in
der Herbſtmeß / Anno 1607. entweder gantz neuw oder
ſonſten verbeſſert/ oder auffs new widerumb auffgelegt/ in der
Buchgaſſen verkaufft worden.

FRANCOFVRTI,
Permiſſu Superiorum excudebat Ioannes Saur:

Fig. 1 Titlepage of the Autumn Book Catalogue of the Frankfurt Fair for 1607, setting out in Latin and German the conditions to be met by books entered there: namely, that they have to be either altogether new, or revised and corrected, or enlarged. From: (*Catalogus...autumnalibus* 1607). Courtesy of the Herzog August Bibliothek Wolfenbüttel. http://diglib.hab.de/drucke/254-4-quod-3s/start.htm?image=00001

were naturally keen to prevent unauthorized republication as far as possible, in order to protect the interests of the international cohort of publishers on whom the Book Fair's prosperity depended.

The usual form of declaration in the Fair catalogue included the format of a given book. This was necessary, as a good part of the trade in books was wholesale and involved merchants not only in cash transactions but also the swapping of books with those of other producers ("Tauschhandel") according to the number of sheets of the various formats. The declarations in the fair catalogues mainly used the locative preposition "apud" after the title of the book. This could introduce any of three instances of names: (1) the printer or the printer-publisher (as in the case of Cholinus, de Gabiano, and Ciotti above); (2) the book merchant at whose temporary stall in Frankfurt's Buchgasse the book in question could be found; (3) the name of an agent acting on commission. The entry for A1607—*Christophori Clauii Bambergensis S.I. in Sphaeram Ioannis de sacro Bosco Commentarius Lugduni apud Iacob. Chouet 4 & 8 & Candon 4*—is a good example of how confusing the entries can be; it will be examined in detail below. The presence of a publisher's name in a given declaration is no guarantee that that person was physically present at the Fair: he could have been represented by a factor or a colleague. The *Societas Veneta*, for example, whom we shall meet again, represented a wide variety of Italian book producers.

Many foreign book merchants kept bookstores throughout the year in the city which were full of their unsold copies of old editions. Because they could store their holdings permanently in Frankfurt, this turned Frankfurt into an immense repository of books, all of which were available for purchase officially during the period of the fair, and through local agents at other times. The non-declared books were often advertised in "nomenclaturae," or stock lists in the form of broadsides attached to a given stall, but even these did not list all the books available. The declarations in the catalogues under-represented the whole field of the books on offer, and almost certainly under-represented the number of editions of the *De sphaera* that were available.

One near-contemporary bibliography that was compiled in Frankfurt is the *Bibliotheca classica* of the bibliographer George Draudius (1573–1635). This first appeared in 1611 and was published in a much enlarged second edition in 1625 (Draudius 1611, 1625). Its publication was announced in the "libri prodituri" section of A1607, in which it is declared to be the list not only of books declared at the Fairs but also all those present elsewhere in Frankfurt ("Verzeichnis aller Bücher, so wol deren, welche je hin und wider in Buchläden gefunden warden, so in alten Bibliotheken fast gefunden warden, nach dem Alphabet ordentlich fürgestellet"). Booksellers were asked to send in details of their holdings either to the bookseller-publisher Peter Kopf (fl. 1593–1635) or to Draudius. The entry in the *Bibliotheca classica* of 1625 under the rubric "sphaerica" reveals that the following Sacrobosco editions could at some point have been found in Frankfurt (I have retained Draudius' abbreviated descriptions, and removed references to other works on sphaeristics):

Ioannis de SACROBUSTO. Paris. 1507 in fol. cum Com. Iac. FABRI. Witeb. Venet & Antuerp.82.8. cum Alberti HERONIS Eliae VINETI, & Francisci IUNCTINI Scholijs. Colon. Apud Cholinum 91.8. In hanc: Bartholomaeus VESPUCIVS. Venet 1508. Cum aliis

opusc. De Sphaera. Christoph CLAVIUS. Romae 75.4. Et ibid. apud Dominic. Basam. 81.4.
…Venet. Apud Ciottum. 90.4. CICHUS Venet. 1499. In fol. cum alijs quibusdam. Erasmus
Oswaldus SCHRECKENFUCHSIUS. Basiliae. 1569. fol. Franc. IUNCTINUS. Lugduni
1568.8. Hartmannus BEIER. Francof. 8. Iacobus FABER Venet 1405 cum aliorum Comm.
In eand. Io GLOSCOVIENSIS. Cracouiae. 1514. Petr. CIRUELLENS. Paris. 1498.fol.cum
Petri de Aliaco quaestionibus. …Bernar. MORISANVS.4. Francof. apud Mareschall. 1624.8.

Some of these dates may have been misrecorded (Faber [Stapulensis] is probably
1495, Iunctinus 1578, Ciotti 1591, Glogoviensis 1513), but allowing for this, there
are seven editions mentioned that do not appear in the Fair Catalogues, of which
six predate 1564, the year of the catalogue's first appearance. This suggests how
many editions could have been found in Frankfurt by an assiduous enquirer (as was
Draudius).

The Fair catalogues were organized by subject area, beginning with the three
senior university faculties of theology, law, and medicine; next came "libri historici,"
and thereafter "libri philosophici," in which category all the books relating to the arts
course fell, including mathematical and astronomical works such as Sacrobosco's *De
sphaera*. The Catalogues were compiled from slips sent independently by exhibitors
to the publisher of the book fair catalogue. This resulted in the presence of competing
editions of the same work, advertised at the same time by their producers (as in S1601
and S1607, above). They frequently ended with an appendix of late declarations or
contained a list of forthcoming books.

The relationships of the book merchants with each other were sometimes made
explicit in prefatory material. The Heidelberg publisher Jean Mareschal (1510–1590),
whose son we shall meet later in this essay, made this clear (possibly disingenuously)
in his edition of Jacopo Zabarella's logical works which reproduced without permission the authorized edition of the Venetian publisher Roberto Meietti, in which he
addresses the author in the following terms:

> There is no need to fear that the publisher Meietti would complain about the appearance
> of this edition, which has been brought out not for financial gain but for the public good;
> his character and probity are known to me; he is more likely to see himself as having been
> helped by [me] in the task of disseminating your excellent doctrine, for he will be able to
> sell his copies in Italy and neighboring places. Nor will there be any harm to him through
> the fact that copies of another edition are on sale in German lands, which very few copies
> of his own edition reach.[6]

The Fair catalogues contained much unauthorized publication of this kind; speculative and unscrupulous printers could use the printed editions of others as copy,

---

[6] The original text reads: "Non esse verendum, ne Meietus typographus de hac editione, quae non
lucri, sed publici boni causa instituta est, conquereretur; novi eius mores et probitatem, potius putabit
se a Mareschallo adiutum in tua praeclara doctrina disseminanda: ipse enim poterit in Italia et vicinis
locis exemplaria distrahere, nec moleste fieret, quod alia in Germania vendantur, quo paucissima ab
ipso edita perveniunt" (Zabarella 1590b, Vol. 1, 3r). In fact, Meietti targeted the same market zone
as Mareschal: See the declaration by him of his edition of Zabarella's *Liber de naturalis scientiae
constitutione*, in A1586, which appeared alongside Mareschal's declaration of Zabarella's *Opera
omnia*; Mareschal announced his competing edition of the *Liber de naturalis scientiae constitutione*
in S1587. Ciotti later sought to make inroads in Mareschal's market zone by publishing Zabarella
under the fictitious imprint of Cologne.

saving the authorial costs, the need for closely supervised composition by a qualified proof-reader or "corrector," and the expense of creating illustrations. Savings in the costs of production by the use of an existing edition as copy could be maximized by the choice of smaller formats and cheaper paper.

In broad terms, there are three possible relationships between publishers attending the Fair at this time: collaboration, peaceful coexistence, and competition. The advertisements listed above offer examples of all three relationships: (1) Collaboration could occur between publishers of different religious persuasions who allowed commercial considerations to override confessional allegiances. An advanced form of collaboration involved the sharing of an edition, which might also have entailed the sharing of typeface, etched plates or woodcuts (e.g., Ciotti and Basa 1601) (2) coexistence, as implied disingenuously in the Mareschal quotation given above, can be found in cases where different market zones were targeted (e.g., Cholinus and Ciotti 1601) (Clavius 1992, VI, letter no. 305), (3) competition in the same market zones, which applies to the rest.

One class of books—"scholastica" or school books that were mainly produced in smaller formats (8vo, 12mo, 16mo)—was quite frequently advertised in the section on philosophical books, together with the staple diet of new scholarly monographs or editions, which often were produced in larger formats. The editions of the *De sphaera* fell principally into the former category, but even the more expensive quarto editions were probably targeted at, and certainly acquired by, educational establishments.[7] The extraordinary proliferation of universities, *gymnasia*, and colleges in Europe in the confessional age (roughly 1560–1650) produced a market opportunity that was vigorously competed for, and led to the frequent advertisement of textbooks (Maclean 2009a). It was less common for editions of school textbooks implicitly or explicitly produced for local consumption to be advertised at the Fairs, as can be seen in the case for the Elzeviers' edition of the *De Sphaera* from 1626 to 1656 ("decreto illustr. et potent. DD. Ordinum Hollandiae et West-Frisiae, in usum scholarum ejusdem provinciae" (Sacrobosco and Burgersdijk 1626)). The Elzeviers were entrepreneurial publishers who were very frequent contributors to the Book Fair Catalogues and would not have passed up lightly an opportunity for international sales, but in this case clearly did not aspire to one. Other publishers of the *De sphaera* who seem also to target only local sales are the Scoto and Sessa publishing houses of Venice with eleven editions between them between 1569 and 1620, the Lyonnais Gazeau and Pillehotte who printed octavo editions between 1606 and 1617, and Spirainx of Dijon in 1619. There are also cases of sporadic advertisement at the Fairs: The Antwerp book merchants printed the *De sphaera* with additional material between 1542 and possibly 1593 but made no declarations at the Fair after 1582. One particular group of publishers—those working in Paris, whose names appear frequently in the Book Fair Catalogues for the relevant years—did not include among the items they declared the *De sphaera*, of which they produced at least ten editions between 1564 and 1619.

---

[7] Giovanni Giacomo Staserio of the Jesuit College in Naples wrote to Clavius in 1606 to request twenty copies of Clavius' Commentary published that year: See (Clavius 1992, I, letter no. 257). Quoted in (Brevaglieri 2008, fn. 147).

The probable explanation for this is that they did not need an international dimension to their trade in order to achieve sufficient sales to make a profit. The items they did declare seem to have constituted an act of cultural politics, and to have been chosen to enhance their international reputation for up-to-date high-level scholarship.

3 *Scholastica*: Lutheran Pedagogy, the Reformed Academies, and the Jesuits

The radical revisions of school and university curricula in the second half of the sixteenth century opened up a market opportunity for publishers serving the Lutheran (Protestant), Reformed (principally Calvinist), and Catholic communities. These revisions are associated with the names of Melanchthon, Petrus Ramus (1515–1572), and the Society of Jesus, and in each case they were supported by an efficient publishing operation. Melanchthon, known as the "teacher of Germany" (*praeceptor Germaniae*), was the close associate of Martin Luther (1483–1546), and deeply involved with the academic curricula of the University of Wittenberg. He collaborated in a succession of editions of the *Libellus de sphaera* which appeared from 1531 to 1629 (Chap. 5). Only one of these editions was declared at the book fairs, in the Spring of 1601. This was produced by Zacharias Schürer (1570–1626), who became an independent publisher in 1600 when he took over the business of the recently deceased Andreas Hoffmann (d. 1600) (Benzing 1977, col. 1263). He paid for the printing of the *Libellus de sphaera* by the firm of Krafft (who had produced it up to 1568, and presumably still possessed the requisite images) in 1601.[8] In the same year, he reissued the *Opera omnia* of Melanchthon. The *Libellus de sphaera* was reissued by his heirs in 1629, betokening poor sales in the intervening years, perhaps due to market saturation or changes in school curricula, or the decline of Melanchthon's reputation and influence after the Formula of Concord of 1577–1580 (Dingel 1996; Maclean 2009b).

The fortunes of Petrus Ramus as an educational writer depended at various stages in his career on the Parisian André Wechel (d. 1581) and the Basel house of Pietro Perna (1519–1582). He was forced to flee from Paris after the St. Bartholomew's eve massacre of August 1572 in which Ramus was assassinated. After Wechel had re-established himself in Frankfurt, a publishing war broke out between the two publishers of Ramus which resulted in an astonishing number of editions of Ramus' various textbooks, including those on mathematics, and the espousal of his teaching methods by a significant group of Reformed academies, colleges and *gymnasia* in the Rhineland and Northern Germany after 1580 (Ong 1958). Friedrich Beurhaus (1536–1609), the head of the Dortmund Archigymnasium, and Bernhard Copius (1525–1589), Rector of Lemgo, were among those who promoted his works.

---

[8] It is interesting to note that the online copy of this edition was previously owned by the Jesuit College of Fulda (Sacrobosco and Melanchthon 1601, sig. p8r).

Beurhaus produced Ramus' *Dialectica* with a commentary many times greater than the parent text; this was published in 1583 by the Catholic Maternus Cholinus (1525–1588), and again in 1596 by Goswin Cholinus of Cologne (Ramus and Beurhaus 1583, 1596), both of whom, as well as being zealous supporters of Counter-Reformation publications and producers of teaching manuals for the Jesuits including the *De sphaera*, supplied materials belonging to other pedagogical traditions. Ramus' curricula were particularly popular in the Hanseatic cities, whose mercantile communities appreciated his practical approach to the teaching of mathematics. Ramus' work on sphaeristics was unpublished at his death; it was developed by Wilhelm Adolf Scribonius (1550–1600) and used in the schools that followed a Ramist or "Semi-Ramist" curriculum. Other Reformed scholars who set out to provide a complete *cursus philosophicus* include Rudolphus Goclenius the Elder (1547–1628) and Bartholomäus Keckermann (1571–1609), whom we shall meet again in the context of the 1625 presentation of the *De sphaera*. Both were eclectic and irenic (Keckermann owed much to Jacopo Zabarella) and were served by significant and committed publishers who were active at the Fairs. The apogee of the encyclopedic ambitions of the Reformed community was the monumental *Cursus philosophici encyclopaedia* (1609–1620) of Johann Heinrich Alsted (1588–1638) (Alsted 1620; Hotson 2007, 25–29, 74–79, 169–273).

In the Catholic camp (Chap. 11), the swift progress of Jesuit pedagogical establishments all over Europe, and later their missionary activities, led to measures to regulate their syllabuses; the first draft *Ratio studiorum* appeared in 1586, to be followed by another draft in 1591, and the definitive version of 1598–1599. Christophorus Clavius of the *Collegio Romano*, the commentator on the *De sphaera*, wrote memoranda to his Society in 1581, 1582, and again in 1593 about their teaching program, stressing the need for mathematics and astronomy, that were in his view underrated by the Society's espousal of a strict Aristotelian conception of "scientia." In the final version of the *Ratio studiorum*, mathematics and astronomy were given a place, even if not as prominent as that for which Clavius lobbied. They also enjoyed a respectable status in the Aristotelian commentaries of the Coimbra fathers, who included the *De sphaera* in their own *cursus*.[9] Clavius' contribution to sphaeristics was such that Alsted recommended his *De Sphaera* commentary in the reading prescribed in his *Cursus philosophici encyclopaedia* (Hotson 2007, 200–202).

The other strand of German Catholic editions of the *De sphaera* was in the hands of Maternus Cholinus of Cologne, who began in the book trade in the 1540s. He was a close associate of the Jesuits, who acted sometimes as correctors in his print shop. He held valuable printing privileges (including a General Privilege for all his publications from the Imperial Chancery) and was a frequent attendee of the Frankfurt Fair (Schrörs 1808). The majority of his publications were in the service of the Roman Catholic Church, but he also produced philosophical and mathematical textbooks, of which the *De sphaera* was one. After his death in 1588, his son Goswin took over the very successful business; he was succeeded in turn by his son Peter Cholinus (d.

---

[9] On Clavius' promotion of mathematics, see (Hellyer 2005, 120–123, 276 fn. 122). See also (Lattis 1994; Casalini 2012; Marinheiro 2012).

ca. 1645) in 1610 (Reske 2015, 481–482, 494, 503). The Cholinus family printed the
De sphaera in octavo, which is consistent with the aspiration to achieve sales of the
work as a textbook, possibly for use in Jesuit schools: The editions of 1591, 1594,
1601, and 1610 bear the printer's mark of the Society of Jesus on the title page. It is
not clear whether this implies financial support from the Jesuits or their patrons, or
whether it is an *imprimatur* relating solely to the content of the book, linking it to
the teaching program of the Society.

The edition of 1601 has added material in the form of notes taken from Clavius'
commentary, probably in response to the *Ratio studiorum* of 1599.[10] The first
Cholinus edition attested by a surviving copy is 1566; this was not declared at the
Frankfurt Fair, but the editions of 1581, 1591, 1601, and 1610 were.[11] There is also
an edition of 1594, which is a reprint (not a reissue) of the edition of 1591. The
rhythm of these declarations is dictated by the General Privilege, which specified
the protection of their editions for periods of ten years. The publishing house would
have a strong interest in reprinting at the end of a decade with some modification to
meet the legal requirement of "emendatior, novus or auctior," as happened in 1601
with the addition of excerpts from Clavius' commentary. From the evidence of the
reprint in 1594, it would seem that they were supplying a steady market in *scholastica*, and benefiting from the possession of the woodcuts, which saved the cost of
providing new illustrations. From these declarations, it is possible to infer that the
Cholinus house had secured control over a given, probably relatively local, sector
of the market, and sought to extend their sales into other parts of the German lands
through declarations at the Fair.

4   The Clavius Factor: Basa and Ciotti

There are two difficulties which have to be addressed in respect of the various editions
of Clavius' commentary on the *De sphaera*. The first concerns the format: This
text is clearly targeted at Jesuit colleges and any other institution that might wish
to avail itself of it, but it is in an expensive format (quarto) and is of substantial
length. That is not consistent with the usual, more economical way in which schoolbooks were produced. Guillaume Cavellat (1500–1576) (Chap. 9), Girolamus Scotus
(1505–1572), Melchiorre Sessa (1505–1565), Cholinus, Jean Bellère (1526–1595)
and Pierre Bellère (1530–1600), and Johann Krafft the Elder (1510–1578) all chose
the format of octavo, and presumably profited from the choice, as their editions were
repeated at regular intervals. Is Clavius' text a quarto because that format alone can
accommodate the text and its illustrations for pedagogical purposes? Or did the first
edition in Rome determine the format of later editions because the illustrations could

---

[10] Various editions of the *De Sphaera* recorded by (Houzeau and Lancaster 1882–1889, 395–397,
509) are marked as doubtful in the GHTC list: 1562, 1565, 1590 (two of these could be the result
of the publisher's practice of simultaneous printing on sequential dates). See, http://www.ghtc.usp.
br/server/Sacrobosco/Sacrobosco-ed.htm. Accessed 8 June 2021.

[11] Lalande's bibliography of astronomic works lists a Cholinus edition of 1600 (Lalande 1803, 135).
See also (Marinheiro 2012; Sigismondi 2012; Price 2014).

only fit into a page of quarto or greater? Or because there was a custom in Rome to produce *scholastica* in quarto?

The second difficulty concerns reissues of the text. A reissue (where only the title page or the first gathering is changed) can bear witness to a number of commercial decisions. If the reissue is from the same press, then the motivation is most likely to be a desire to seem as up to date as possible, or to qualify for declaration at a Book Fair under the category "libri novi." If the reissue has a different bibliographical address, the second motivation could of course still apply, but it is more likely that the change of address was dictated by the targeted market zone, whether Catholic or broadly Protestant. In the case of the Clavius commentary, there are examples of the former possibility (Rome 1570, 1575; Gabiano 1593, 1594; Gelli 1606, 1607), as well as the second (Gabiano 1602; Crespin 1602). A reissue with a changed publication date can also indicate a failure in sales, as in the case of the Wittenberg 1629 edition mentioned above.

In this list, I have supplied in italics the Rome, Venice, and Lyon editions of Clavius' Commentary that were not declared at the Frankfurt Fair, in order to give coherence to the discussion that follows.

Christophori Clavii Bambergensis, ex Societate Iesu, in Sphaeram Ioannis de Sacro Bosco, commentarius, Romae: apud Victorium Helianum, 1570, 4to.
(Sacrobosco and Clavius 1570)

Christophori Clavii Bambergensis, ex Societate Iesu, in Sphaeram Ioannis de Sacro Bosco, commentarius, Romae, 1575, 4to.
(Lalande 1803, 101). No extant copy in public domain

A1576 Christophori Clauii Bambergensis, ex societate Iesu, in Sphaeram Ioannes de Sacro Bosco Commentarius 4. Romae.
No extant copy in public domain.

Christophori Clavii Bambergensis, ex Societate Iesu, in Sphaeram Ioannis de Sacro Bosco, commentarius, Romae:ex officina Dominici Basae, 1581, 4to.
(Sacrobosco and Clavius 1581)

S1582 Christophori Clauii Bambergensis ex Societate Iesu in Sphaeram Ioannis de Sacro Bosco commentar[a]ius 4. Romae.
(Sacrobosco and Clavius 1581)

Christophori Clavii Bambergensis, ex Societate Iesu, in Sphaeram Ioannis de Sacro Bosco, commentarius, Romae: ex officina Dominici Basae, 1585, 4to.
(Sacrobosco and Clavius 1585)

Christophori Clavii Bambergensis, ex Societate Iesu, in Sphaeram Ioannis de Sacro Bosco, commentarius, Venice: apud Ioan. Baptistam Ciotum 1591, 4to.
(Sacrobosco and Clavius 1591)

S1592 Christophori Clauij Bambergensis societatis Iesu in Sphaeram Ioannis de Sacrobosco commentarius tertio recognitus & locupletatus Ven. 8.
(Sacrobosco and Clavius 1591).

Christophori Clavii in Sphaeram Ioan de Sacro Bosco Com[m]entarius. Nunc quarto ab ipso Authore recognitus, et plerisque in locis locupletatus, Lyon: sumptibus Joannis de Gabiano, 1593 and 1594.
(Sacrobosco and Clavius 1593) or (Sacrobosco and Clavius 1594).

Christophori Clavii Bambergensis, ex Societate Iesu, in Sphaeram Ioannis de Sacro Bosco, commentarius, Venetiis: apud Bernardum Basam sub signo solis, 1596, 4to.
(Sacrobosco and Clavius 1596).

Christophori Clavii Bambergensis, ex Societate Iesu, in Sphaeram Ioannis de Sacro Bosco, commentarius, Venetiis: sub signo solis, 1601, 4to.
(Sacrobosco and Clavius 1601b).

Christophori Clavii Bambergensis, ex Societate Iesu, in Sphaeram Ioannis de Sacro Bosco, commentarius, Venetiis: apud Io. Baptistam Ciottum sub signo aurorae, 1601, 4to.
(Sacrobosco and Clavius 1601a).

S1601 Christophori Clauii commentarius in Sphaeram Ioan. de Sacrobusto, iam recognitus apud [Joannem Baptistam Ciotti] in 4.
(Sacrobosco and Clavius 1601a).

Christophori Clavii Bambergensis, ex Societate Iesu, in Sphaeram Ioannis de Sacro Bosco, commentarius, Venetiis: sub signo aurorae, 1603, 4to.
(Sacrobosco and Clavius 1603).

S1603 Christoph. Clauij in Sphaeram Ioan de sacro busto Comment. Editio septima locupletior. ap soc Venet. in 4. 1603.
(Sacrobosco and Clavius 1603).

Clavius' commentary on the *De sphaera* first appeared in 1570 from the presses of Vittorio Eliano (1528–1581) who began printing for the Jesuits in 1570.[12] According to Lalande and de Backer-Sommervogel, the bibliographers of the Society of Jesus, this edition was reprinted (or reissued) in Rome in 1575; although no copy of this reprint or reissue appears to have survived, confirmation of its existence is to be found in the Frankfurt Book Fair Catalogue for Autumn 1576 (Lalande 1803, 101). Thereafter, Clavius' commentary was revised by the author five times before his death in 1612: 1581 ("iterum," printed by Domenico Basa (1500–1596)), 1585 ("tertio," printed by Basa), 1593–1594 ("quarto," printed by de Gabiano), 1606 ("quinto," printed by Gelli in Rome), and 1611–1612 ("postremo," published in Mainz by Anton Hierat (fl. 1597–1627) of Cologne as part of Clavius' *Opera mathematica*

---

[12] Clavius dedicated his Commentary to Wilhelm von Bayern. His requests for permission to dedicate the book for its first and second editions, respectively, are in (Clavius 1992, II, letters no. 1, 15).

in folio). It is not clear why the fourth revision of 1593–1594 was entrusted to the brothers de Gabiano in Lyon and endowed with a ten-year Privilege of the Jesuit Order, to which was associated the Society of Jesus' General French Privilege of 1583. According to Adriaan van Roomen (1561–1619), Clavius' correspondent, there was only one Roman bookseller (Gaspar van den Wouwer (1564–1616), originally from Antwerp) present at the Frankfurt Book Fairs in 1593 (Clavius 1992, IV, letter no. 96 (11 November 1593)). Perhaps Clavius, who no doubt hoped to disseminate his commentary beyond Italy, sought a publisher who could ensure an international distribution. It is noteworthy also that the "third" editions of Basa and Ciotti in 1596 and 1601 take no cognizance of the availability of a revised fourth edition, which may suggest that the local Italian market was impervious to certain northern publications.

After the production of the first edition of Clavius' commentary on the *De sphaera* in 1570 by Eliano and its possible reissue in 1575, the subsequent second and third editions were produced by Domenico Basa in 1581 and 1585, in collaboration with Francesco Zanetti (1530–1591) (Chap. 11). Basa was one of the major publishers and printers in Rome and a close associate and friend of Paolo Manuzio (1512–1574), the scholar and grandson of the famous Aldo. Basa was very well connected with members of the religious hierarchy and was twice involved in the development of the Vatican Presses. His relations with other Roman book merchants were not however always harmonious. His management of the Vatican Presses passed on his death in 1596 to his nephew Bernardo, who represented the interests of his uncle in Venice. His bookshop "all'insegna del sole" (*sub signo Solis*) was established in 1582 and his presses became active in 1584 (Cioni 1970). His reprinting of the third edition of Clavius' commentary in 1596 and 1601 is inextricably linked to the reprintings and reissues undertaken by Giovanni Battista Ciotti, to whom I now turn.

Ciotti was a native of Siena, as was one of the earliest Italian attendees of the Frankfurt book fair, Francesco de' Franceschi who almost certainly took him under his wing in 1587. Their publishing careers were subsequently very closely allied. Both were prolific publishers of Italian imaginative literature and post-Tridentine theological and religious books; both had strong connections to the Society of Jesus; both were based in Venice, but collaborated widely with printers in other Italian cities and in Frankfurt; both had editions printed with the fictitious bibliographical address "Cologne;" both frequently imported books from the Frankfurt Book Fair; both were members of the exporting consortium (which included also their Venetian colleague Roberto Meietti) known in Frankfurt as the *Societas Veneta*, founded in 1592, which remained very active until 1613 (Maclean 2021).

Ciotti was an opportunist and a somewhat unscrupulous publisher with an eye to potentially profitable *scholastica*: For example, he reprinted Zabarella's *De rebus naturalibus*, as we have seen, in the wake of the Meietti and Mareschal editions, and declared it in the spring catalogue of 1590 (Rhodes 2013). It is therefore not surprising that he should aspire to do the same to the Clavius commentary. Ciotti's first involvement was with the third recension, which he produced from the Basa edition of 1585 in 1591. In it, he wrote a dedicatory letter to a young Veronese literary figure dated 1591, in which he claims that Clavius' commentary had been printed in Rome "three or four times;" this could either show that he only had a vague sense of the sequence of printings, or that he was aware that the lost 1575 edition was a reissue.[13] He stressed the fact that he had had new illustrations made, and that the book had undergone "diligent correction."[14] From the colophon, it appears that the book was printed for him by Francesco de' Franceschi (fl. 1558–1599) (Rhodes 2013, 110). He declared it at the Book Fair of Spring 1592. It was followed by another Venice edition of Ciotti's text in 1596 by Bernardo Basa.[15] This is a reprint of the Basa 1585 publication of the *third* edition (even though by now the *fourth* edition had appeared in Lyon in 1593–1594); the errors recorded on the errata page in the 1585 edition are incorporated into the text. What Clavius thought of this edition of a superseded recension of his text is not recorded. It contains Ciotti's dedication, given a new date of 1596, and uses the same illustrations as the 1591 edition. The presence of the re-dated dedication makes it very likely that this is a commercial collaboration with Ciotti, printed by Daniele Zanetti (fl. 1576–1606), Basa's printer in Venice. Bernardo is presumed to have died in or before 1599; his widow Isabetta was active in the book trade in 1600–1601, inscribing the few books that she produced (or reissued) with the bibliographical address "ad instantia d'Isabetta di Bernardo Basa." The shared edition of Clavius (still not recognizing the de Gabiano fourth edition) appeared in 1601 with the addresses of both her (*sub signo solis*) and Ciotti ("apud Io. Baptistam Ciottum sub signo aurorae").[16] From the style of the printing, and the presence of a "Regestum" at the end of the volume (a compositorial practice more common in Rome than Venice), I infer this to have been printed for Isabetta Basa; Ciotti either collaborated ab initio, or bought up the stock after the demise of the Venetian enterprise of the Basa family. It appears not to have sold well, as Ciotti reissued it in 1603 (its colophon reads "Venetiis 1601") and added the false claim that it was the seventh edition of the Clavius commentary. He declared it through the *Societas Veneta* at the Spring Fair of 1603.

5   Lyon and St. Gervais

This is the relevant sequence of declarations:

---

[13] Iohannes Iacobus Tonialis, on whom see (Rhodes 2013, 88).

[14] "…ut Roma testatur quae iam ter vel quater [commentarios Clavii] impressit mihi faciendum existimavi, ut hic quoque Venetiis eosdem novis figuris et diligenti correctione meis typis imprimendis curarem." It is very difficult to determine whether or not the same woodcuts are used in different editions, as those producing the cuts were so adept at making copies. Slight differences could arise from inking. Sometimes it was possible also to effect small corrections to existing woodcuts, which might otherwise seem to be evidence of an entirely new creation.

[15] Bernardo Basa declares one work at the Frankfurt Book Fair in 1593, and one in 1597. His uncle is not named in the declaration of S1582: Many Rome publications in the Catalogues published between 1570 and 1601 do not specify the printer-publisher.

[16] (Rhodes 2013, 193) was not aware that these are two states of the same issue.

S1578 Fr. Iunctini Florentini, sacrae Theologiae Doctoris, Com[m]entaria in Sphaeram Ioannis de Sacro Bosco accuratissima. 8. Lugduni apud Philippum Tinghium.
(Sacrobosco and Giuntini 1578)

A1592 Christoph. Clauii in Sphaeram Ioan de Sacro Bosco Com[m]entarius. Editio 4. ab authore recognita. Lugd. sumptibus fratrum de Gabiano in 4. futuris nundinis ve[r]n[alibus] exponatur.
(Sacrobosco and Clavius 1593) or (Sacrobosco and Clavius 1594)

Christophori Clavii in Sphaeram Ioan de Sacro Bosco Com[m]entarius. Nunc quarto ab ipso Authore recognitus, et plerisque in locis locupletatus, Lyon: sumptibus Joannis de Gabiano, 1602.
(Sacrobosco and Clavius 1602b)

Christophori Clavii in Sphaeram Ioan de Sacro Bosco Com[m]entarius. Nunc quarto ab ipso Authore recognitus, et plerisque in locis locupletatus, St. Gervais: Crispinus, 1602.
(Sacrobosco and Clavius 1602a)

S1602 Christophori Clauii Iesuitae commentarius in sphaeram Sacrobusti. Editio quinta. Lugduni apud Crispinum.
(Sacrobosco and Clavius 1602a)

Christophori Clavii in Sphaeram Ioan de Sacro Bosco Com[m]entarius. Nunc quarto ab ipso Authore recognitus, et plerisque in locis locupletatus, St. Gervais: Crispinus, 1607.
(Sacrobosco and Clavius 1607a, b)

Christophori Clavii in Sphaeram Ioan de Sacro Bosco commentarius, Rome: Sumptibus Io. Pauli Gellii, 1606.
(Sacrobosco and Clavius 1606) or (Sacrobosco and Clavius 1607d)

A1607 Sphaera Clauii nova Romae. Apud Arnold. Quentel.
(Sacrobosco and Clavius 1606) or (Sacrobosco and Clavius 1607d)

Christophori Clavii in Sphaeram Ioan de Sacro Bosco Com[m]entarius. Nunc quinto ab ipso Authore recognitus, et plerisque in locis locupletatus, Lyon: sumptibus Joannis de Gabiano, 1607.
(Sacrobosco and Clavius 1607c)

A1607 Christophori Clauii Bambergensis S.I. in Sphaeram Ioannis de sacro Busto Commentarius. Lugduni apud Iacob. Chouet 4 & 8 & Candon 4.
Not determinable

A version of the Giuntini commentary first appeared together with others in Lyon in 1564, from the Giunti presses (Sacrobosco et al. 1564).[17] Filippo Tinghi (d. 1580), a relative of the Giunti (Chap. 8), published it in two volumes in 1577–1578 with a ten-year French Royal Privilege dated 24 December 1576 and accompanied this publication with a separate printing of the plaintext corrected by Giuntini which was not announced at the Book Fair. The Fair declaration was one of only two that Tinghi made in his time as publisher, both in 1578, two years before his death. Tinghi was a very shrewd and experienced publisher with strong connections in France, Italy, and Spain. He was the promoter of Giuntini, who, like Tinghi himself, was a member of the powerful Italian community of Lyon. No doubt he declared the theologian-astronomer's work in Frankfurt to draw international attention to his work. His heir, Symphorien Beraud (fl. 1571–1586), included the commentary in the folio edition of Giuntini's *Speculum astrologiae* of 1581, which was reissued in 1583 (Baudrier 1964–1965, V, 60–61, 65–66).

As recorded above, the expanded fourth revision of Clavius' commentary was entrusted to the brothers Jean and David de Gabiano (ca. 1559–ca. 1598) in Lyon who were in partnership, and endowed with a ten-year Privilege of the Jesuit Order, to which was associated the Society of Jesus' General French Privilege of 1583.[18] It was printed in Lyon by Guichard Jullieron (d. 1627) and reissued in 1594. There are various noteworthy features of this edition and its producer, Jean de Gabiano. After their family's self-imposed exile to Calvinist Geneva in 1568, both brothers had returned to Lyon in 1581, to take over their uncle's book business. There is no indication that the brothers chose to convert back to Catholicism as the price to pay for their repatriation. In those years, the town council of Lyon seems to have been very intolerant toward members of the Reformed Faith (Baudrier 1964–1965, V, 298), but it could not afford to turn away the scions of a family of very powerful *marchands-libraires* such as the de Gabianos, who had been established in the city since the beginning of the century and had very close commercial ties with the printing industry in Geneva (Baudrier 1964–1965, VII, 207). A notable symptom of these ties was the practice, engaged in by even impeccably Catholic Lyonnais book merchants such as Filippo Tinghi, of publishing books printed in Geneva bearing title pages with Lyon addresses, all under letters patent from the king.[19] The *Consulat* (the council of twelve of Lyon) forbade this practice; they attributed the decline of book production in Lyon directly to it. On July 14, 1588, they summoned a group of influential *libraires* including David de Gabiano to answer the charge that they had severely damaged the printing industry in Lyon by using Genevan printers and by citing on their products Lyon as their bibliographical address, to ensure that

---

[17] The *De sphaera* printed in Antwerp by Heirs of Arnold Birckmannin 1566 (and Antwerp editions thereafter) also carry this commentary.

[18] For a reference by Clavius to the Lyon edition as "copiosior," see (Clavius 1992, V, letter no. 145).

[19] For the joint submission with Sébastien Nivelle of Paris, see (Baudrier 1964–1965, VI, 440–441); for the letter patent from the King addressed to Tinghi, Beraud and Michel of July 5, 1580, see (Baudrier 1964–1965, VI, 459).

their works could be sold in Catholic countries. This had led to the emigration of compositors and other print workers to Geneva. The accusation read as follows:

> That to the great detriment of the City and [its] journeymen-printers, [the named *marchand-libraires*] have destroyed the printing industry in Lyon and have transferred their printing activities to Geneva, and, what is even worse, that they declare on the title pages of the books they have printed in Geneva that they have been printed in Lyon, so that they can be put on sale in Italy, Spain and other Catholic countries, this constituting fraud and the supposititious use of a name…and that as a result printing, which once had a high status and reputation in this City of Lyon, will be altogether lost, and in order to earn their living, the said journeymen-printers will be forced for as long as this state of affairs lasts to leave Lyon and to go to Geneva, where in the course of time they become heretics.[20]

The *marchands-libraires* produced the counter-accusation that the printers had imposed ruinously high tariffs on them, which obliged them to engage in an unscrupulous commercial practice (Baudrier 1964–1965, V, 41).

Two of the three editions of Clavius' commentary that the de Gabianos declared (1593–1594, 1607) have a colophon in which the Lyon printers that they engaged to produce them are named (1593–1594: Guichard Juillieron; 1607: Jacques du Creux dit Molliard (1607–1652)). It is reasonable to suppose that this proof of Lyonnais printing was a response to the arraignment of 1588. The fourth edition produced in their name in 1602 does not bear the name of a printer (Sacrobosco and Clavius 1602b). It is identical (except for one line of the title page) to the book declared in Spring 1602 with the bibliographical address "Sancti Gervasii, apud Samuelem Crispinum" (Sacrobosco and Clavius 1602a). Samuel Crespin was the son of a prominent Genevan printer-publisher, whose earliest imprints date from the mid-1590s. He declared books with Lyon addresses and produced works in tandem with Jean de Gabiano; there is evidence that they were on good commercial terms.[21] On its title page, this edition claims to have been produced "cum privilegio." At the end of the volume, as is also the case of the 1593–1594 edition, a Jesuit privilege is reproduced, together with a short paragraph of errata which contains an expression of piety in the Catholic mode ("vale, et nostro labore ad D[ei] O[ptimi] M[aximi] maiorem gloriam fruere"). There is no indication why a Lyonnais publisher (albeit one who published other Jesuit authors)[22] rather than a Roman one was entrusted

---

[20] The original text reads: "Qu'au grand détriment de la Ville et desdits imprimeurs, ils ont détruit l'imprimerie lyonnaise et l'ont transportée à Genève et que, pis est, font mettre à la première feuille des œuvres imprimées à Genève, qu'elles l'ont été à Lyon, afin qu'elles puissent avoir cours en Italie, en Espagne et autres pays catholiques, ce qui est une fausseté et supposition de nom…que par la l'impression qui souloit avoir un grand cours et réputation en cette ville de Lyon, sera du tout perdue et seront les compagnons imprimeurs, encore que cela dure, contraints, pour gaigner leur vie, d'abandonner Lyon pour aller à Genève, où par une succession de temps, ils se rendent heretiques" (Baudrier 1964–1965, V, 298, VII, 240–242).

[21] (*Corpus omnium veterum poetarum….* 1603): (Arbour 1979–1980, no. 3709). Crespin was known as a Lyonnais book merchant to the Plantin-Moretus firm: See MPM Archief, 759, f. 69. (Baudrier 1964–1965, VII, 218) quotes a letter of proxy written by the de Gabiano brothers in favor of Crespin on September 7, 1595.

[22] (Baudrier 1964–1965, VII, 218 (September 13, 1599)) shows an agreement concluded with the printer Claude Michel of Tournon to publish the Jesuit Joannes Osorius' *Conciones de sanctis*

with these two recensions by Clavius. No Genevan publisher (Saint Gervais being a transparent cognomen for Geneva) could ever have possessed or cited a Jesuit license. The copious illustrative material of the 1602 editions is that of the 1593–1594 edition. Jean de Gabiano had sold the plates to Claude Michel of Tournon (d. ca. 1630) in June 1599, who may have printed for both Crespin and de Gabiano. The text of the two editions is identical. It would seem likely that both versions of this edition were produced either in Tournon or in Geneva. The title page correctly records that it is the fourth edition, but the declaration in the Spring Catalogue of 1602 calls it the fifth.

Crespin reprinted the 1602 edition in 1607, but its saleability was compromised by the genuine fifth edition that had appeared a year before in Rome at the expense of the Roman bookseller Gelli.[23] The Roman edition was reprinted in 1607 in Lyon by Jacques du Creux for Jean de Gabiano, using the Gelli copy of 1606. It was reprinted in turn by Crespin in 1608 and declared at the Spring Fair; it uses the same images as 1593–1594 and 1602, but the text is certainly recomposed. The Roman bookseller Gelli reissued the 1606 edition in 1607, and declared it in the Autumn of 1607, presumably to assert its right to be seen as the authorized text against de Gabiano's reprint. It was sold through the stall of Arnold Quentel (d. 1621) of Cologne, and not through the *Societas Veneta* (Fig. 2). Perhaps that association's link with the somewhat unscrupulous Ciotti made it unattractive to some Italian book merchants; it is also possible that Gelli chose Quentel because of his excellent relations with Cologne book merchants.[24]

We now come to the enigmatic A1607 entry:

Christophori Clauii Bambergensis S.I. in Sphaeram Ioannis de sacro Bosco Commentarius Lugduni apud Iacob. Chouet 4 & 8 & Candon 4 (Fig. 3)

"Apud" here can only refer to the stall at which the books were available for sale. Jacques Chouet (1626–1683) was not a Lyonnais, but a Genevan publisher and a colleague of Crespin; he and his heirs had a stall in Frankfurt.[25] I suggest, therefore, that the quarto edition here announced is that produced by Crespin of the fourth edition. The edition "apud Ca[r]don" was put on sale at the stall of

---

(Osorio 1596). (Baudrier 1964–1965, VII, 220 (June 22, 1599)) records Claude Michel buying the plates used in the production of Clavius' commentary. It is not clear whether he was able to produce his own edition, or whether he was the printer who acted in 1602 for both de Gabiano and Samuel Crespin.

[23] Giovanni Paolo Gelli (b. ca. 1569) owned the Roman bookshop "all'insegna della nave" ("ad signum navis"). He was active from ca. 1606 to ca.1624. In the Plantin-Moretus records (MPM Archief 759, f. 85), Gelli is described as being the Frankfurt agent for a Roman bookseller Gasparo Paleotti in 1605–1606 (Zannini 1980, 109; Ascarelli and Menato 1989, 106).

[24] In (Clavius 1992, VI, letter no. 305) Anton Hierat refers to Gelli as "amico meo singulari." Gelli was a bookseller; the 1606 *In sphaerum* appears to be his first incursion into publishing.

[25] (Bonnant 1999, 14) says that Crespin had a commercial outlet in Frankfurt in 1610 on the evidence of the imprint on Hieronymus Gonzalez, *Dilucidum ac perutile glossema, seu commentatio ad regulam octavam cancellariae, de reservatione mensium et alternativa Episcoporum*: "prostat Francofurti, in officina Samuelis Crispini" (Gonzalez 1610).

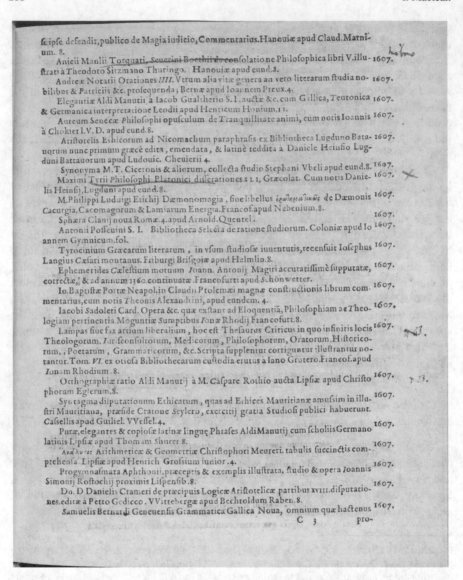

Fig. 2 A page from the *Libri philosophici* section of the Autumn Catalogue of 1607, showing the entry for the edition of Clavius' commentary on the *Sphaera* produced in Rome by Giovanni Paolo Gelli, and advertised as being on sale through the stall of Arnold Quentel of Cologne. From: (*Catalogus…autumnalibus* 1607). Courtesy of the Herzog August Bibliothek Wolfenbüttel. http://diglib.hab.de/drucke/254-4-quod-3s/start.htm?image=00021

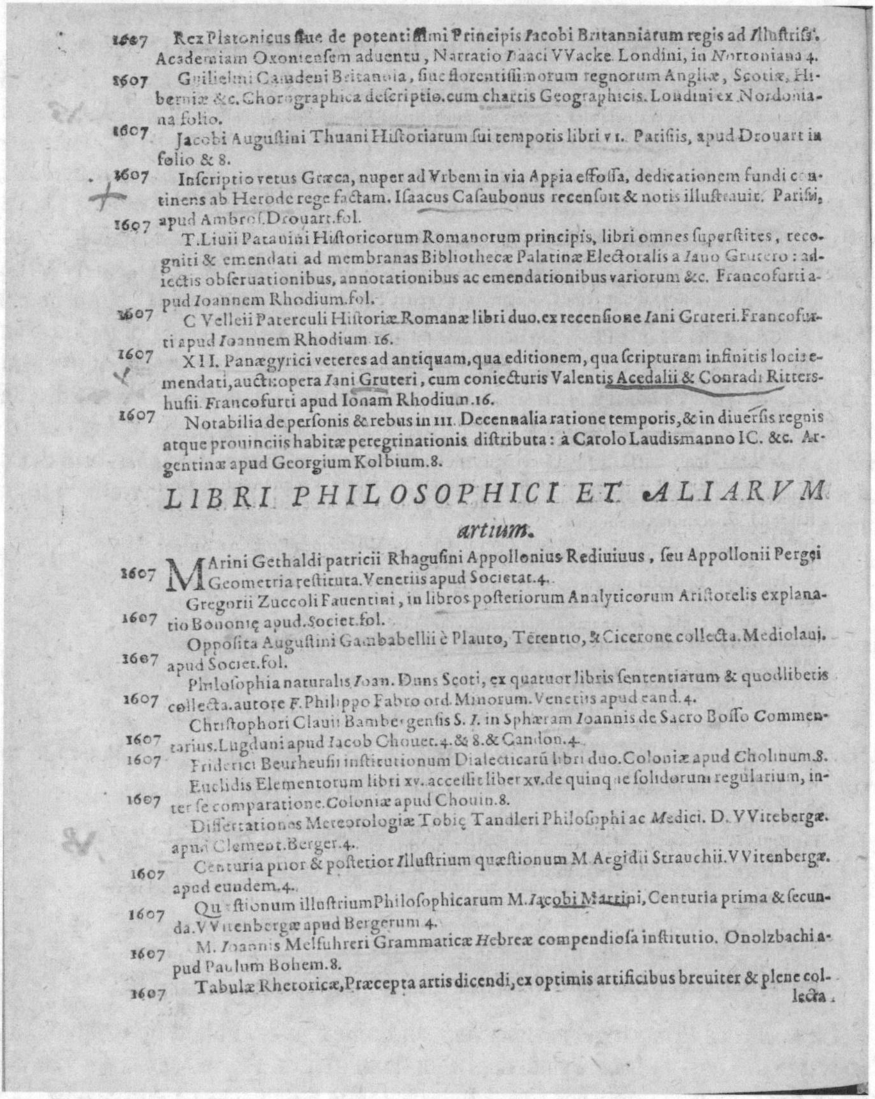

1687 Rex Platonicus ſiue de potentiſſimi Principis Iacobi Britanniarum regis ad Illuſtriſſ. Academiam Oxonienſem aduentu, Narratio Iſaaci VVacke. Londini, in Nortoniana 4.

1607 Guilielmi Camdeni Britanoia, ſiue florentiſſimorum regnorum Angliæ, Scotiæ, Hiberniæ &c. Chorographica deſcriptio.cum chartis Geographicis. Londini ex Nordontana folio.

1607 Iacobi Auguſtini Thuani Hiſtoriarum ſui temporis libri vi. Pariſiis, apud Drouart in folio & 8.

1607 Inſcriptio vetus Græca, nuper ad Vrbem in via Appia effoſſa, dedicationem fundi continens ab Herode rege factam. Iſaacus Caſaubonus recenſuit & notis illuſtrauit. Pariſii, apud Ambroſ.Drouart.fol.

1607 T.Liuii Patauini Hiſtoricorum Romanorum principis, libri omnes ſuperſtites, recogniti & emendati ad membranas Bibliothecæ Palatinæ Electoralis a Iano Grutero : adlectis obſeruationibus, annotationibus ac emendationibus variorum &c. Francofurti apud Ioannem Rhodium.fol.

1607 C Velleii Paterculi Hiſtoriæ Romanæ libri duo.ex recenſione Iani Gruteri.Francofurti apud Ioannem Rhodium 16.

1607 XII. Panægyrici veteres ad antiquam,qua editionem, qua ſcripturam infinitis locis emendati,auctiopera Iani Gruteri, cum coniecturis Valentis Acedalii & Conradi Rittershuſii. Francofurti apud Ionam Rhodium.16.

1607 Notabilia de perſonis & rebus in iii. Decennalia ratione temporis,& in diuerſis regnis atque prouinciis habitæ peregrinationis diſtributa : à Carolo Laudismanno IC. &c. Argentinæ apud Georgium Kolbium.8.

LIBRI PHILOSOPHICI ET ALIARVM
artium.

1607 Marini Gethaldi patricii Rhaguſini Appollonius Rediuiuus , ſeu Appollonii Pergæi Geometria reſtituta.Venetiis apud Societat.4.

1607 Gregorii Zuccoli Fauentini , in libros poſteriorum Analyticorum Ariſtotelis explanatio Bononiæ apud.Societ.fol.

1607 Oppoſita Auguſtini Gambabellii è Plauto, Terentio, & Cicerone collecta.Mediolani. apud Societ.fol.

1607 Philoſophia naturalis Ioan. Duns Scoti, ex quatuor libris ſententiarum & quodlibetis collecta.autore F.Philippo Fabro ord.Minorum.Venetiis apud eand.4.

1607 Chriſtophori Claui Bambergenſis S. I. in Sphæram Ioannis de Sacro Boſſo Commentarius.Lugduni apud Iacob Chouet.4.& 8.& Canſon.4.

1607 Friderici Beurheuſii inſtitutionum Dialecticarū libri duo.Coloniæ apud Cholinum.8.

1607 Euclidis Elementorum libri xv. acceſſit liber xv.de quinq; ſolidorum regularium, inter ſe comparatione.Coloniæ apud Chouin.8.

1607 Diſſertationes Meteorologiæ Tobiæ Tandleri Philoſophi ac Medici. D. VVitebergæ. apud Clement.Berger.4.

1607 Centuria prior & poſterior Illuſtrium quæſtionum M Aegidii Strauchii.VVitenbergæ. apud eundem.4.

1607 Quæſtionum illuſtriumPhiloſophicarum M.Iacobi Martini,Centuria prima & ſecunda.VVitenbergæ apud Bergerum 4.

1607 M. Ioannis Melſuhreri Grammaticæ Hebreæ compendioſa inſtitutio. Onolzbachi apud Paulum Bohem.8.

1607 Tabulæ Rhetoricæ,Præcepta artis dicendi,ex optimis artificibus breuiter & plene collecta.

Fig. 3 A page from the *Libri philosophici* section of the Autumn Catalogue of 1607, showing the enigmatic portmanteau entry for two different quarto editions and one octavo edition of Clavius' commentary on the *Sphaera*, associated with Lyon as a place of publication and the book merchants Jacques Chouët (of Geneva) and Horace Cardon (of Lyon), who kept stalls at the Fair. From: (*Catalogus…autumnalibus* 1607). Courtesy of the Herzog August Bibliothek Wolfenbüttel. http://diglib.hab.de/drucke/254-4-quod-3s/start.htm?image=00018

Horace Cardon (1566–1641), a prominent Lyonnais *libraire* and a colleague of the de Gabiano brothers. It is, I suggest, the newly produced fifth edition by de Gabiano, from the Roman edition of Gelli of 1606. It carries the General Jesuit Privilege for the Kingdom of France dated 10 May 1583, and a specific six-year Privilege dated January 11, 1607, which protected the publication of three works by Clavius: the commentary on the *De sphaera*, the *Geometrica*, and the *De crepusculis tractatio*.

The octavo "apud Jacob. Chouet" is an enigma: It may refer to one of the Crespin and de Gabiano editions, as the main text, although quarto in size, is in gatherings of eight rather than four, but this is a weak surmise. It might also (even less plausibly) be the 1606 octavo edition of the *De sphaera* (not containing the Clavius commentary) produced in Lyon by Hugues Gazeau (fl. 1584–1611). There is one other entry in the Autumn Catalogue of 1607, under the rubric of books forthcoming at the next Fair (Fig. 4) The Cholinus firm, that had declared this version of their edition in the Spring of 1601, announced a reprint, which did not in fact appear until 1610. Such premature announcements were not uncommon and were used to inform rivals of a publisher's intention to bring out a new edition, in order to deter others from unauthorized reprinting.

6 The Owl of Minerva and the *cursus philosophicus*: Mareschal and Morisanus

Under the rubric "Libri Philosophici" of the Frankfurt Book Fair Catalogue of Autumn 1624, the following two books were declared:

> Bernardi Morisani Derensis Ibernici In Aristotelis Logica, Physica, Ethica, Francofurti apud Petrum Mareschall in 4.
> (Morisanus 1625)

> Ejusdem Commentarius in Sphaeram Ioannis de S. Bosco ibid. in 8.
> (Sacrobosco and Morisanus 1625).

These entries bring together two very different figures, united by exile and an interest in the Arts Course. I shall deal with them in turn. Pierre Mareschal was the son of the Lyonnais publisher Jean Mareschal. His father was an early convert to Calvinism and became himself a refugee, having fled from Lyon after its return to Catholic control in 1563. Jean Mareschal went first to Basel, and then to Heidelberg where he became one of the accredited university booksellers, whose role it was to publish disputations and official university documents (Benzing 1977, col. 1210). On most of the books he published, the title page did not specify a place but strongly implied Mareschal's Lyonnais identity by the use of the epithet "Lugdunensis" after his name. His publications include Calvinist theological and political polemic and works by prominent scholars on the university arts course and its three senior faculties. Some of these were unauthorized reprints, as we have seen in the case of Zabarella.

Operæ horarum subcisiuarum , siue meditationes historicæ , centuria tertia Philippi Camerarii J.F JC apud Petrum Kopffium.4.

Icones seu imagines virorum literis illustrium, vna cum eorum elogiis diuersorum au-ctorum.Opus iam tertio recognitum,auctumque auctore Nic.Reusnero J C. prostabit a-pud Joan.Carolum.Argent.

S. Brunonis Institutoris ordinis Carthusiensium opera. Coloniæ apud Anton. Hierat.

Consiliorum Valentini Forsteri J.C.liber primus.Helmbstadii apud Boemium.fol.

Luce Marentii Madrigolia 6.vocum.coniunctim excusa Kauffman.4.

Danielis Lackneri flores Jessæorum quatuor vocibus.ibid.

Leo Haßlers Kirchengesäng mit vier Stimmen Contrapuncts weise ibid.

Opera sancti Hieronymi omnia.Parisiis.fol.

Der Keyserliche Papst/entgegen gesetzt/deß Conradt Andreæ Jesuiters Keyserischen Luther / durch M. Joan.Giffthail.Onolzbach.Bohem.

Theses de Triunitate,Persona Christi,& Spiritu S.M.Joannis Giffthailii : apud Paul. Bohem.

Symphorematis supplicationum pro processibus Cameræ Imperialis Tom.vi. Franco-furti apud Schonwetter.

De inope debitore creditori addicendo : deque præferentiis creditorum Casparis Ba-ietiæ JC.Hispani.apud eund.

Sermones Dominicales,Festiuales & Quadragesimales Guilielmi Pepin Theolog. Pa-risiens.Colon.apud Gymnic.4.

Conciones Dominicales Fr.Mauricii Hilareti SS.Theolog.D.Colon.apud.eund.4.

Christiani Gerson eines getaufften Juden vnd widergebornen Christen zwey Bücher von der Juden Thalmut Goßlar bey Johann Vogten.8.

Gespräch zwischen Graff Bernharden vnd einem studioso Chemiæ von der waren Præparation deß grossen Steins/2c.durch Conradt Schulern.Studtgart bey Gebhardt Grieb. 8.

Biblia sacra vulgatæ editioni Sixti v.Pont fi.M.iussa recognita. apud Joan.Moretum 8.

Annalium Ecclesiasticorum Cæsaris Baronii S.R.E.Cardinalis.Tom.xi.apud eund.fol.

M.Joannis Schroderi Theses de Communicatione proprii Swinfurti apud Casparum Kemlin.8.

M.Hippolyti Hubmeyeri Oratio de Aristotele & Ramo,Jenæ.

Viaticum extremi itineris Natanis Ghytræi,verteutscht durch HenricumRappæum.Stein-furt.apud Cæsarem.8.

Index librorum expurgandorum Fr.Joan Mariæ Braschellen. Colon. apud Cholin.8.

Sphæra Joannis de Sacro Bosco emendata:aucta cum scholiis Vineti,& Commentari-is Clauii apud eund.8.

Catechismus & declaratio Symboli Apostoli,Rob.Bellarmini ,&c. cum tractatu Joan. Rosseuini S.J.de Sacramento Altaris apud eund.8.

Tobiæ Knoblochii Phil.& Med.D.xxii. disquisitiones Physicæ , generales & speciales Witebergæ.apud Bergerum.4.

Heinrici Boceri JC tractatus de crimine læsæ maiestatis.Tubing.apud Cellium.8.

Joannis Harprechti U J.D.tractatus de legatis apud eund.8.

Heinrici Kitzschii L. Symbologia Heroica hexaglottos ex optimis quibusque auctori-bus collecta,&c.apud Henningium Groß seniorem. Eiusdem de magistratibus Romanis eorumque criteriis disertatio aphoristica numeralis.apud eund.

Eiusdem Onomasticum Symbolicum.

Centuria Symbolorum Chronologicorum.

Symbola priuata Illustriss.principis Anhaltinorum.

Tractatus de disciplina eiusque causis.&

Cæsare

Fig. 4 A page from the section of the Autumn Catalogue of 1607 listing books due to appear at the next Fair, one of which is a reprint of the edition of the Sphere by Cholinus that first appeared in 1601. The reprint eventually appeared in 1610. From: (*Catalogus…autumnalibus* 1607). Courtesy of the Herzog August Bibliothek Wolfenbüttel. http://diglib.hab.de/drucke/254-4-quod-3s/start.htm?image=00042

The widespread practice of unauthorized reprinting was engaged in also by his son, Pierre, who was born in 1572. He matriculated at the University of Heidelberg on August 24, 1591, where he acquired a sophisticated grasp of Latin to which his later prefaces bear witness, and succeeded his father in due course as a university book-seller (Toepke 1886, 2, 154 (no. 126: August 24th, 1591)). From 1591 to 1598, he published a number of works in the same subject areas as those chosen by his father. Two of his publications, both polemical works by the Marburg reformed philoso-pher Rudolphus Goclenius the Elder, attacked the gnesiolutheran Daniel Hoffmann (ca. 1538–1611), the Helmstedt scholar who argued that philosophy should not be divorced from theology: The relevance of this will emerge below.[26] After 1598, Mareschal's presses fell silent, but he continued to act as a bookseller in Heidelberg until 1622, when the Palatinate city was conquered after a siege by Bavarian and Spanish forces which ended on September 16. This led to the looting of the city's cultural heritage, including the magnificent Palatine library which was sent by the Duke of Bavaria as a bribe and a trophy to the Pope in Rome.[27]

Mareschal was linked to a number of prominent scholars and publishers who fled from the city as a consequence of this event: the Calvinist theologians Paul Tossanus (1572–1634) Abraham Scultetus (1566–1625), David (1548–1622) and Philip Paraeus (1576–1648), Janus Gruter (1560–1627), the Elector Palatinate's Librarian, as well as the distinguished scholarly printer-publishers Isaac (1598–1676) and Abraham Commelin (b. 1597), Gotthard Vögelin (b. 1597) (both of whose printing enterprises were already internationally based and afforded an easy transi-tion of the presses to Leiden and Leipzig, respectively), and Johann Ammon (1623–1656), whom he joined on his flight to Frankfurt.[28] Ammon was a son-in-law of Theodor de Bry (1561–1623), the prominent publisher of illustrated books, and had connections through him to the cartographer Matthaeus Merian (1593–1650), the engraver Paul de Zetter of Hanau (1600–1667), and his brothers Jacob (b. 1609) and Peter de Zetter (fl. 1629–1635), who was to join the consortium of printer-publishers known as the Wecheliani, the foremost Calvinist printing house in Frankfurt and nearby Hanau. All of these figures could be said to belong to the by then beleaguered "Calvinist international" (Grell 2011; Murdock 2011). Paul de Zetter was later to engrave a portrait of Pierre Mareschal, which was clearly intended for inclusion in an unpublished recension of the *Bibliotheca Chalcographica, Illustrium Virtute atque Eruditione in tota Europa, Clarissimorum Virorum Theologorum, Iurisconsul-torum, Medicorum, Historicorum, Geographicorum, Politicorum, Philosophorum, Poetarum, Musicorum, Aliorumque...*, a continuation of Jean-Jacques Boissard's (1528–1602) collection of images of prominent European scholars that appeared in 1650. Mareschal's portrait is one of the groups of Heidelberg worthies listed

---

[26] The Goclenius titles are *Defensio philosophica...adversus Danielem Hoffmannum* and *Quaes-tiones philosophicae adversus Danielem Hoffmannum*, both published in 1597 (Goclenius 1597a, b). On this debate, see (Friedrich 2004).

[27] The victorious Duke of Bavaria was hoping to have the Electorate of the Holy Roman Emperor, held by the defeated Frederick of the Pfalz, transferred to him by the Pope, in return for, inter alia, the Palatine Library (Thomas 2010, 295–297).

[28] On the Commelin house, see (Lüthy 2012, 30); on Gotthard Vögelin, see (Dyroff 1963).

above who left that city around the same time as Mareschal.[29] It describes him as enduring exile "for the love of religion" and dying in Strasbourg in 1622.[30] This cannot be the case, as he dates the dedications in his Frankfurt books after that year. Nothing is heard of Mareschal after 1625, which may well be the year of his death.[31] Shortly after his arrival in Frankfurt, he set up his bookshop and began printing and publishing, often in collaboration with his fellow refugee Johann Ammon, who remained active as a producer of books for several decades thereafter.[32] The titles that they produced jointly and severally reflect Ammon's connection to the de Bry family (maps and emblems), and the interests represented in the portfolio of publications of Mareschal's father: Calvinist pastoral literature, speculatively produced reprintings in various fields, and works in the higher faculties and philosophy. Like his father, Mareschal declared most of his publications at the Frankfurt Fair.[33]

Mareschal's most surprising titles—which are not reprints—are in my view those written by a scholar originally from Derry in Ireland called Bernardus Morisanus. He was a member of the Irish Catholic diaspora, a refugee like Mareschal, but of a very different religious persuasion.[34] Mareschal wrote prefaces to both of the previously unpublished textbooks of Morisanus that he had acquired (*In Aristotelis Logicam, Physicam, Ethicam, Apoletesma* and *In Sphaeram Joannis de S. Bosco commentarius...nunc primum publicae utilitati donatus*), which were intended to be sold together as a pedagogical package (Morisanus 1625; Sacrobosco and Morisanus 1625). The full title of the *Apoletesma* (*Commentariis luculentissimis, ad mentem*

---

[29] The others are David and Philippus Paraeus, Abraham Scultetus, Janus Gruter, and Daniel Tossanus.

[30] The verse under the portrait reads: "Hanc Heidelbergae Petrus Mareschallus opimae/Corporis ac vultus rettulit effigiem/Ast patria cedens ob relligionis amorem/Argentorati grandior exul obit;" see also (Müller 2017).

[31] It is worthy of note that when Morisanus'*Apotelesma* was readvertised in S1625, its bibliographical address was given as that of Johann Ammon.

[32] Mareschal's preface to Morisanus' *In sphaeram* printed in Frankfurt, "impensis Petri Mareschallus" (Sacrobosco and Morisanus 1625, 5), refers to "typis nostris." There are book ornaments that seem to be identical to those used by him in Heidelberg in the 1590s.

[33] The emblem book with De Bry provenance is the *Emblematum ethico-politicorum centuria* by Zincgref (Zincgref 1624). Two speculative reprintings, both appearing in 1625, were Ludovico Melzo's *Kriegsregeln* (Melzo 1625) and Niccolò Passeri [Janua]'s *De scriptura privata tractatus* (Passeri 1625) (earlier editions in 1611 and 1621, respectively). In S1625, there is a declaration of Hermann Nicephorus'*De analysi logica* (containing works by Cornelius Martini and Amandus Polanus) "apud Petrum Mareschallum;" the only surviving copy is recorded as being printed by Egenolphus Emmelius, who printed works for Mareschal, and it is possible that "apud" means in this case no more than "available at the shop of." The Calvinist works are by Josua Zevelius'*Christliche Practick* (Zevelius 1624), Daniel Tossanus'*Betbüchlein* (Tossanus 1624), Abraham Scultetus'*Kurzter Underricht* (Scultetus 1624) and the anonymous *Die Fünff Hauptstück Christlicher Religion* (*Fünff Hauptstück* 1624). Mareschal also published a translation of Thomas a Kempis'*Imitation of Jesus Christ* in the same year. All were declared in either S1624 or A1624 jointly with Johann Ammon.

[34] The name Morisanus could be the Latinisation of Mor[r]ison, or a rendering of the Gaelic O'Muirgheasain.

*Magni Magistri [*viz. *Aristotelis] penitus accommodatis, ut et Disputationibus inge-niosissimis, quibus cum veteres tum recentiores Controversiae solide pertractantur atque deciduntur, universum Peripateticae Philosophiae Corpus absolvens*) reveals Mareschal's targeted market: the purchasers of complete pedagogical courses for schools, colleges, and universities. In the preface to the work, Mareschal does not disclose any more about the deceased author than that he was "a most acute philosopher and a very subtle disputant," that his work is remarkable for its brevity and innovative discussions of philosophical issues arising from the texts, and that it should not be ignored because its author lacked the reputation of more famous writers of compendious philosophical texts (Zabarella, Francisco Suárez (1548–1617), Francesco Piccolomini (1520–1604), Benito Perera (1536–1610), Franciscus Toletus (1532–1596), Bartholomaeus Keckermann, Goclenius the Elder, and the Coimbra Fathers are here named by Mareschal).[35] The most significant name in this list may well be Keckermann, who set out to provide a complete philosophy course for the Reformed community of Europe to challenge the Jesuits' teaching program. Mareschal's interest in these works by Morisanus lay principally therefore in the fact that they belong to the genre of *cursus philosophicus.* There are indications that the preface was not without effect, for Morisanus' books are found in many European Catholic libraries; a Cambridge don called Richard Holdsworth also recommended them to a Cambridge undergraduate of the 1640s (Trentman 1982, esp.: 836–837).

Mareschal's act of publication follows what I believe to be the opportunistic acquisition of Morisanus' *Nachlass.*[36] Through Mareschal's own university training, the precedent publications of his father (Zabarella) and his own production of Goclenius' work, he was in a good place to judge the quality of Morisanus' work. Mareschal chose not to make anything of Morisanus' explicit statement in the *Apotelesma* of his membership of the Society of Jesus, where he associates himself with a group of Jesuits from the Spanish Netherlands.[37] This may suggest that Morisanus was resident for some time in that province (there were Jesuit Colleges in Louvain, Douai, and Antwerp, and a Jesuit house in Brussels) (Fraesen and Kenis 2012). Morisanus' *Apotelesma* contains summaries, commentaries, and disputations on parts of the Aristotelian corpus that are consistent with the program of the Coimbra Jesuit Fathers,

---

[35] The date of the preface is September 1, 1624. *Apotelesma* might be translated in this context as "Scholia" or "Definitive commentaries." The word appears in a title by Rudolphus Goclenius (*Apotelesma philosophicum sive conciliator…*) (Goclenius 1618), which may be why Mareschal chose it as a title for Morisanus' work (and might also be implying that Morisanus seeks to reconcile different philosophical positions in the same way). The poetry of the paratext is signed J[ohann] L[udwig] Gottfried, a client of the de Bry family. The various *cursus* of Keckermann, Zabarella, Piccolomini, Perera, and Goclenius appeared in 1614, 1607, 1600, 1576 and 1618–1619, respectively.

[36] The preface to the *Apotelesma* (Morisanus 1625, sig. 3v) makes it clear that nothing else survives by Morisanus.

[37] (Morisanus 1625, 34), where it refers to a *sententia* held by "praecipui Philosophi et Theologi, ex Societate nostra, quibus aliquando opposita sententia arrisit, Soarez, Leonardus Lesceus, Bonaert, imo et Soarem." The Jesuits in question are Cipriano Soarez (1524–1593), Leonardus Lessius (1554–1623), and Nicolaus Bonaert (1563–1610).

whose Aristotelian pedagogical works, published between 1591 and 1606, constituted part of what is now known as the second scholastic. The commentary on Sacrobosco's *De sphaera*, which also formed part of the curriculum of Portuguese Jesuits,
pays special tribute to that of Christophorus Clavius (Sacrobosco and Morisanus
1625, 13, 18; Carvalho 2018, 85).[38] The ensemble of texts does not correspond
to the mathematical component of the Jesuit *Ratio studiorum*, but to the genre of
cursus philosophicus, as Mareschal's preface makes clear.[39] This may suggest that
the market for the *De sphaera* in Jesuit Colleges was saturated by the mid-1620s;
if money was to be made out of pedagogical texts, then a new *corpus* needed to be
put in place. An unusual feature of Morisanus' text is its references to the works of
Galen (medicine forming no part of the Jesuit curriculum).[40] It was also unusual for
expatriate Irish scholars to be committed Thomists, as Morisanus was; most of them
at this time were Franciscans and Scotists (Binasco 2020).

In his preface to the *De Sphaera*, Mareschal declares that he was moved to publish
it for the benefit of learned youth and "of the Church and Commonwealth."[41] As the
Apotelesma and the *De Sphaera* are explicitly and even polemically Catholic in tone,
this is difficult to reconcile either with Mareschal's previous secularist publications
by Goclenius or his other Frankfurt imprints of the years 1624–1625 that are very
explicitly Calvinist. This would suggest that the edition of Morisanus' texts was
a speculative enterprise aimed at all those with an interest in scholastic pedagogy
(hence the silence over Morisanus' membership of the Society of Jesus, which would
not have been a recommendation to all potential purchasers).

Taken together, the two works constitute a not inconsiderable investment of money
by Mareschal. This implies that he had arrived in Frankfurt with some funds at his
disposal: an implication that can be drawn also from his financial support of another
publication at this time.[42] In this case therefore, Sacrobosco's *De sphaera* achieved
publication through its association with an Aristotelian philosophical corpus, and
the opportunism of an exiled learned publisher newly arrived on the Frankfurt scene
who suppressed its Jesuit provenance in an attempt to gain access to a broad market
for pedagogical material of a traditional kind.

---

[38] For further information, see also the Conimbricenses Encyclopedia: http://www.conimbricenses.
org/contents. Accessed 8 June 2021.

[39] Morisanus' work has also been associated with the philosophical *cursus* of the Parisian Cistercian
Eustache de Saint Paul (Kraye 2008, esp.: 1283–1284); Saint Paul asserted the superiority of the
intellect over the will. His *cursus* was published in 1609.

[40] (Morisanus 1625, 183). On logic: (Morisanus 1625, 580, 618), probably from a reference in the
Conimbricenses' commentary, as are also the references to Vesalius and Fernel in (Morisanus 1625,
604). See also (Sander 2014).

[41] The original text reads: "Quem commentarium etiam typis nostris et sumptibus studiosi iuvenuti
et quidem seorsim a caeteris eiusdem Philosophici studii commentationibus propinare placuit, qui
cum primi Authoris, aliorumque quibus idem quod nobis studium in hoc enucleando fuit, coniungi
facilius commodiusque posset. Tu [sci. lector] nostris studiis fave, sumptibusque et curis nostris ad
tui commodum Ecclesiae, reique publ[icae] emolumentum utere-fruere et vale."

[42] Daniel Meisner, the author of the emblem book entitled *Thesauri philo-politici pars*, printed
by Eberhard Kieser in 1623–1624, thanks Mareschal for his financial support for the publication
(Meisner 1623–1624, sig. A3r).

7 Some Conclusions

This investigation has been dependent on evidence from material bibliography, that is, the description of books as physical objects, and the use of contextual data to explain the circumstances of their publication. In a number of cases, it has been possible to show that editions with different title pages and dates are in fact the results of the same printing event (Ciotti 1601; Basa 1601; Ciotti 1603; Gabiano 1602; Crespin 1602; Gelli 1606, 1607; Gabiano 1593, 1594). This means that the seventeen editions of Clavius' commentary between 1570 and 1608 recorded in bibliographies represent only twelve printing events. It has also been possible to establish by visual inspection whether editions with the same pagination and text composition are reprints or not, and to form the hypothesis that there will always be a final reprint in a series that will not sell well and could well lead to a reissue. Thus, in the case of Wittenberg, the reprint of 1601 that was reissued in 1629 can be said to show that there was a clear decline in pedagogical use of the *De sphaera* at some point between these dates for the relevant market zone. But the Scoto edition of the *De sphaera* which appeared in Venice in 1620 (Sacrobosco et al. 1620) is a reprint, not a reissue, of the previous edition of 1586 (Sacrobosco et al. 1586), and probably indicates that the Scoto presses had some confidence in the saleability of the 1620 edition. The same would apply to the Cholinus 1610 reprint of the 1601 edition (Sacrobosco et al. 1601). Arguments about the "popularity" and the "impact" of a given text derived from the number of "editions" can be supported or undermined by determining whether a given printing event is a reprint or a reissue. This can throw light on the speculative commercial activity of publishers, who in some cases (reprintings) were predicting the continued existence of a given market, and in others (reissues), were simply trying to get any return whatsoever from dead stock.

In the case of illustrations, it has not been possible to create a stemma, or show which sets of plates and diagrams have been deployed in different editions. This is due to the remarkable skills of woodcutters, who were able to create near-exact reproductions of existing images and could also subtly adapt existing ones. If a stemma could be established, then more could be deduced about the relationships between publishers and editions. In the case of the images found in the Clavius commentaries, it would be reasonable to suppose that there were at least three sets: The 1570 set that was used subsequently by Basa, Gelli, and eventually Hierat, who asked for them to be sent to him as he was preparing the edition of the complete works of Clavius in the Spring of 1609 (Clavius 1992, VI, letter no. 305);[43] the set that Ciotti claims to have produced in 1591; and the de Gabiano set of 1593–1594, that was used by Crespin and bought by Michel of Tournon in 1599, but more work is needed to establish all of this.

The strong connection with the Society of Jesus is evident from the involvement of their chosen printer-publishers (Eliano, Gelli, Basa, Ciotti, Gabiano, Cholinus), and the effect of the *Ratio Studiorum*'s appearance on marketing (in 1601, notably). The 1607 cluster of declarations is linked more specifically to the latest edition of Clavius.

---

[43] The publisher Joannes Albinus of Mainz was the go-between.

The lull in publication after 1610 could be explained by market saturation.[44] The confessional rivalry which found expression in competing comprehensive philosophy courses is a feature of the 1610s, and may help to explain the last declared commentary of 1625 by Morisanus, which does not flag up its Jesuit connection, but has a potential place in a complete *cursus philosophicus* of a pre-Copernican kind.

Abbreviations

Digital Repositories

Sphaera Corpus*Tracer*   Max Planck Institute for the History of Science. https://db. sphaera.mpiwg-berlin.mpg.de/. Accessed 7 June 2021

Archives and Special Collections

MPM   Museum Plantin-Moretus, Antwerp

References

Primary Literature

Alsted, Johann Heinrich. 1620. *Cursus Philosophici Encyclopaedia...* Herborn: Christoph Corvinus.
Catalogus Universalis Pro Nundinis Francofurtensibus autumnalibus de anno 1601... 1601. Frankfurt am Main: Johannes Saur.
Catalogus Universalis Pro Nundinis Francofurtensibus Autumnalibus de Anno 1604... 1604. Leipzig: Abraham Lamberg.
Catalogus Universalis Pro Nundinis Francofurtensibus Autumnalibus De Anno M.DC.XXIV... 1624. Frankfurt am Main: Sigismundus Latomus.
Catalogus Universalis Pro Nundinis Francofurtensibus autumnalibus, de anno 1607... Frankfurt am Main: Johannes Saur.
Catalogus Universalis Pro Nundinis Francofurtensibus vernalibus de anno 1601... 1601. Frankfurt am Main: Johannes Saur.
Catalogus Universalis Pro Nundinis Francofurtensibus vernalibus de anno 1602... 1602. Frankfurt am Main: Johannes Saur.
Catalogus Universalis Pro Nundinis Francofurtensibus vernalibus de anno 1603... 1603. Frankfurt am Main: Johannes Saur.

---

[44] See Giovanni Giacomo Staserio to Christophorus Clavius, Naples, January 13, 1606, for evidence that the market for Clavius' commentary was near or at its peak in 1606 (Clavius 1992, VI, letter no. 257).

Catalogus Universalis Pro Nundinis Francofurtensibus Vernalibus de Anno 1608... 1608. Leipzig: Abraham Lamberg.

Catalogus Universalis Pro Nundinis Francofurtensibus Vernalibus, De Anno M.DC.X... 1610. Frankfurt am Main: Sigismundus Latomus.

Catalogus Universalis Pro Nundinis Francofurtensibus Vernalibus, De Anno M.DC.XII... 1612. Frankfurt am Main: Sigismundus Latomus.

Clavius, Christophorus. 1611–1613. *Opera mathematica.* 5 vols. fol. Mainz: Antonius Hierat.

Clavius, Christophorus. 1992. *Christoph Clavius. Corrispondenza.* 7 vols., ed. Ugo Baldini and Pier Daniele Napolitani. Pisa: Dipartimento di Matematica. Università di Pisa.

Clessius, Joannes. 1602. *Unius seculi; eiusque virorum literatorum monumentis...ab anno dom. 1500 ad 1602 nundinarum autumnalium inclusive elenchus.* Frankfurt am Main: Joannes Saur for Peter Kopf.

Collectio in unum corpus omnium librorum...qui in nundinis Francofurtensibus ab anno 1564 usque ad nundinas autumnales ann[i] 1592. Frankfurt am Main: Nicolaus Bassaeus.

Corpus omnium veterum poetarum.... 1603. Lyon: Hugues de la Porte, Samuel Crespin, and Jean de Gabiano.

Die Fünff Hauptstück Christlicher Religion. 1624. Frankfurt am Main: Johann Ammon and Peter Mareschal.

Draudius, Georgius. 1611. *Bibliotheca classica, sive Catalogus officinalis in quo singuli singularum facultatum ac professionum libri, qui in quavis fere lingua extant quique intra hominum fere memoriam in publicum prodierunt secundum artes et disciplinas, earumque titulos et locos communes, authorumque cognomina...recensentur.* Frankfurt: Nicolaus Hoffmannus.

Draudius, Georgius. 1625. *Bibliotheca classica, sive Catalogus officinalis, in quo singuli singularum facultatum ac professionum libri, qui in quauis fere lingua extant: quique intra hominum propemodum memoriam in publicum prodierunt, secundum artes et disciplinas, earumq[ue] titulos et locos communes, autorumque cognomina singulis classibus et rubricis subnexa, ordine alphabetico recensentur.* Frankfurt: Balthasar Oster.

Favre, Antoine. 1607. *Coniecturarum iuris civilis libri.* Lyon: Samuel Crespin.

Goclenius, Rudolph. 1618. *Apotelesma philosophicum sive conciliator.* Frankfurt am Main: Johann Carl Unckel.

Goclenius, Rudolph the Elder. 1597a. *Defensio philosophica aduersus Danielem Hofmannum. A quibusdam eius discipulis collecta.* Ed. Conradus Gumpellius. Heidelberg: Pierre Marsechal.

Goclenius, Rudolph the Elder. 1597b. *Explicatio quaestionis philosophicae, an et quatenus de Deo notiones logicae usurpandae sint? Adversus Danielem Hoffmannum.* Heidelberg: Pierre Mareschal.

Gonzalez, Hieronymus. 1610. *Dilucidum ac perutile glossema, seu commentatio ad regulam octavam cancellariae, de reservatione mensium et alternativa Episcoporum.* Frankfurt am Main: "in officina Samuelis Crispini."

Lutz, Thobias. 1592. *Catalogus novus nundinarum vernalium Francoforti ad Moenum anno M.D.LXXXXII celebratarum...apud Thobiam Lutz.* [Frankfurt am Main]: Peter Schmidt.

Meisner, Daniel. 1623–1624. *Thesauri philo-politici pars altera. Hoc est: Emblemata sive moralia politica....* Frankfurt am Main: Eberhard Kieser.

Melzo, Lodovico. 1625. *KriegsRegeln Desz Ritters Ludwig Meltzo, Malteser Ordens.* Frankfurt am Main: Kaspar Rötel for Pierre Mareschal.

Morisanus, Bernardus. 1625. *In Aristotelis Logicam, Physicam, Ethicam, Apotelesma: Commentariis luculentißimis, ad mentem Magni Magistri penitus accomodatis, ut [et] Disputationibus ingeniosissimis, quibus cum veteres tum recentiores Controversiae solide pertractantur atque deciduntur, universum Peripateticae Philosophiae Corpus absolvens.* Frankfurt am Main: Johann Friedrich Weiß for Pierre Mareschal.

Osorio, Juan. 1596. *Concionum Ioannis Osorii, Societatis Iesu, de sanctis. Tomus tertius.* Tournon: Claude Michel for Jean and David de Gabiano.

Passeri, Nicolò. 1625. *De Scriptura Privata Tractatus Novus pleniβimus...Ab eodem Authore, Quinta hac editione, summa diligentia revisus.* Frankfurt am Main: Egenolff Emmel for Pierre Mareschal.

Ramus, Petrus and Friedrich Beurhaus. 1583. *Ad P[etri] Rami Dialiecticam variorum et maxime illustrium exemplorum, naturali artis progressu, inductio...Auctore Frederico Beurhusio Menertzhagensi...* Cologne: Maternus Cholinus.

Ramus, Petrus and Friedrich Beurhaus. 1596. *Ad P[etri] Rami Dialiecticae proxin generalis introductio, et specialis illustrium exemplorum, naturali artis progressu, inductio... Auctore Frederico Beurhusio Menertzhagensi...* Cologne: Goswin Cholinus.

Sacrobosco, Johannes de and Franco Burgersdijk. 1626. *Sphaera Iohannis de Sacro-Bosco, decreto Illustr. & Potent. D D. Ordinum Hollandiae & West-Frisiae, in usum Scholarum ejusdem provinciae, sic recensita, Ut & Latinitas, & methodus emendata sit, multaque addita, qua ad huius doctrina illustrationem requirebantur. Operâ & studio Franconis Burgersdicii.* Leiden: Bonaventura & Abraham Elsevier. https://hdl.handle.net/21.11103/sphaera.101041.

Sacrobosco, Johannes de and Christoph Clavius. 1570. *Christophori Clavii Bambergensis, ex Societate Iesu, in Sphaeram Ioannis de Sacro Bosco commentarius.* Rome: Vittorio Eliano. https://hdl.handle.net/21.11103/sphaera.100365.

Sacrobosco, Johannes de and Christoph Clavius. 1581. *Christophori Clavii Bambergensis ex Societate Iesu in Sphaeram Ioannis de Sacro Bosco commentarius Nunc iterum ab ipso Auctore recognitus, & multis ac varijs locis locupletatus.* Rome: Domenico Basa and Francesco Zanetti. https://hdl.handle.net/21.11103/sphaera.101117.

Sacrobosco, Johannes de and Christoph Clavius. 1585. *Christophori Clavii Bambergensis ex Societate Iesu in Sphaeram Ioannis de Sacro Bosco commentarius Nunc tertio ab ipso Auctore recognitus, & plerisque in locis locupletatus. Permissu superiorem.* Rome: Domenico Basa. https://hdl.handle.net/21.11103/sphaera.101120.

Sacrobosco, Johannes de and Christoph Clavius. 1591. *Christophori Clavii Bambergensis ex Societate Iesu In Sphaeram Ioannis de Sacro Bosco commentarius, Nunc tertio ab ipso Auctore recognitus, & plerisque in locis locupletatus. Permissu Superiorum.* Venice: Giovanni Battista Ciotti. https://hdl.handle.net/21.11103/sphaera.100360.

Sacrobosco, Johannes de and Christoph Clavius. 1593. *Christophori Clavii Bambergensis ex Societate Iesu, in Sphaeram Ioannis de Sacro Bosco. Commentarius. Nunc quarto ab ipso Auctore recognitus, & plerisque in locis locupletatus.* Lyon: Brothers Gabiano. https://hdl.handle.net/21.11103/sphaera.100363.

Sacrobosco, Johannes de and Christoph Clavius. 1594. *Christophori Clavii Bambergensis Ex Societate Iesu, in Sphaeram Ioannis de Sacro Bosco. Commentarius. Nunc quarto ab ipso Auctore recognitus, & plerisque in locis locupletatus.* Lyon: Brothers Gabiano. https://hdl.handle.net/21.11103/sphaera.100676.

Sacrobosco, Johannes de and Christoph Clavius. 1596. *Christophori Clavii Bambergensis ex Societate Iesu in sphæram Ioannis de Sacro Bosco commentarius. Nunc tertio ab ipso Auctore recognitus, & plerisque in locis locupletatus. Maiori item cura correctus. Permissu superiorum.* Venice: Bernardo Basa. https://hdl.handle.net/21.11103/sphaera.100367.

Sacrobosco, Johannes de and Christoph Clavius. 1601a. *Christophori Clavii Bambergensis ex Societate Iesu in sphaeram Ioannis de Sacro Bosco commentarius. Nunc tertio ab ipso auctore recognitus, & plerisque in locis locupletatus. Maiori item cura correctus.* Venice: Giovanni Battista Ciotti. https://hdl.handle.net/21.11103/sphaera.101390.

Sacrobosco, Johannes de and Christoph Clavius. 1601b. *Christophori Clavii Bambergensis ex Societate Iesu In Sphæram Ioannis de Sacro Bosco Commentarius. Nunc tertio ab ipso Auctore recognitus, & plerisque in locis locupletatus. Maiori item cura correctus. Permissu Superiorum.* Venice: Isabetta Basa. https://hdl.handle.net/21.11103/sphaera.100673.

Sacrobosco, Johannes de and Christoph Clavius. 1602a. *Christophori Clavii Bambergensis ex societate Iesu, in Sphaeram Ioannis de Sacro Bosco, commentarius. Nunc quarto ab ipso Auctore recognitus.* Saint Gervais: Samuel Crispin. https://hdl.handle.net/21.11103/sphaera.100369.

Sacrobosco, Johannes de and Christoph Clavius. 1602b. *Christophori Clavii Bambergensis ex Societate Iesu, In Sphaeram Ioannis de Sacro Bosco, Commentarius. Nunc quarto ab ipso Auctore recognitus, & plerisque in locis locupletatus.* Lyon: Jean de Gabiano. https://hdl.handle.net/21.11103/sphaera.100368.

Sacrobosco, Johannes de and Christoph Clavius. 1603. *Christophori Clavii Bambergensis ex Societate Iesu, In Sphaeram Ioannis de Sacro Bosco Commentarius. Nunc septimò ab ipso Auctore recognitus, & plerisque in locis locupletatus. Maiori item cura correctus. Permissu superiorum.* Venice: Giovanni Battista Ciotti. https://hdl.handle.net/21.11103/sphaera.100370.

Sacrobosco, Johannes de and Christoph Clavius. 1606. *Christophori Clavii Bambergensis ex Societate Iesu, in Sphaeram Ioannis de Sacro Bosco. Commentarius. Nunc quinto ab ipso Auctore hoc anno 1606. recognitus, & plerisque in locis locupletatus. Accessit Geometrica, atque Uberrima de Crepusculis Tractatio.* Rome: Giovanni Paolo Gelli. https://hdl.handle.net/21.11103/sphaera.100665.

Sacrobosco, Johannes de and Christoph Clavius. 1607a. *Christophori Clavii Bambergensis ex societate Iesu, in Sphaeram Ioannis de Sacro Bosco, commentarius. Nunc quarto ab ipso Auctore recognitus.* Saint Gervais: Samuel Crispin. https://hdl.handle.net/21.11103/sphaera.100377.

Sacrobosco, Johannes de and Christoph Clavius. 1607b. *Christophori Clavii Bambergensis ex Societate Iesu, in Sphaeram Ioannis de Sacro Bosco, commentarius. Nunc quarto ab ipso Auctore recognitus, & plerisque in locis locupletatus.* Geneva: Samuel Crispin. https://hdl.handle.net/21.11103/sphaera.100380.

Sacrobosco, Johannes de and Christoph Clavius. 1607c. *Christophori Clavii Bambergensis ex Societate Iesu, In Sphaeram Ioannis de Sacro Bosco. Commentarius. Nunc quinto ab ipso Auctore hoc anno 1606. recognitus, & plerisque in locis locupletatus. Accessit Geometrica, atque Uberrima de Crepusculis Tractatio.* Lyon: Jacqes de Creux a.k.a. Molliard for Jean de Gabiano. https://hdl.handle.net/21.11103/sphaera.100378.

Sacrobosco, Johannes de and Christoph Clavius. 1607d. *Christophori Clavii Bambergensis ex Societate Iesu, in Sphaeram Ioannis de Sacro Bosco. Commentarius. Nunc quinto ab ipso Auctore hoc anno 1606. recognitus, plerisque in locis locupletatus. Accessit Geometrica, atque Uberrima de Crepusculis Tractatio.* Rome: Giovanni Paolo Gelli. https://hdl.handle.net/21.11103/sphaera.101015.

Sacrobosco, Johannes de and Christoph Clavius. 1608. *Christophori Clavii Bamb. ex societate Iesu, in Sphaeram Ioannis de Sacro Bosco, commentarius. Nunc postremo ab ipso Auctore recognitus, & plerisque in locis locupletatus. Accessit Geometrica atque uberrima de Crepusculis Tractatio.* Saint Gervais: Samuel Crispin. https://hdl.handle.net/21.11103/sphaera.100381.

Sacrobosco, Johannes de, Christoph Clavius, Francesco Giuntini, Albertus Hero, Élie Vinet, Pierio Valeriano and Pedro Nunes. 1610. *Sphaera Ioannis a Sacrobosco emendata, aucta et illustrata. Eliae Vineti Santonis, & Francisci Iunctini Florentini scholia in eandem Sphaeram restituta. Quibus praeter Alberti Heronis scholia nunc accessere R. P. Christoph. Clavij breves commentarij. Adiunximus huic libro compendium in Sphaeram per Pierium Valerianum Bellunensem: et Petri Nonii Salaciensis Demonstrationem eorum, quae in extremo capite de climatibus Sacroboscus scribit de inaequali climatum latitudine, eodem Vineto interprete.* Cologne: Peter Cholinus. https://hdl.handle.net/21.11103/sphaera.100357.

Sacrobosco, Johannes de, Christoph Clavius, Albertus Hero, Élie Vinet, Pierio Valeriano and Pedro Nunes. 1601. *Sphaera Ioannis de Sacrobosco emendata, aucta et illustrata. Eliae Vineti Santonis scholia in eandem Sphaeram ab ipso auctore restituta. Quibus praeter Alberti Heronis scholia nunc accessere R. P. Christoph. Clavij breves commentarij. Adiunximus huic libro compendium in Sphaeram per Pierium Valerianum Bellunensem: et Petri Nonii Salaciensis demonstrationem eorum, quae in extremo capite de climatibus Sacroboscus scribit de inaequali climatum latitudine, eodem Vineto interprete.* Cologne: Goswin Cholinus. https://hdl.handle.net/21.11103/sphaera.100327.

Sacrobosco, Johannes de and Francesco Giuntini. 1578. *Fr. Iunctini Florentini, sacrae theologiae doctoris, Commentaria in Sphæram Ioannis de Sacro Bosco accuratissima. Omnia iudicio*

S.R. Ecclesiae submissa sunto. Lyon: Philippe Tinghi. https://hdl.handle.net/21.11103/sphaera. 100393.

Sacrobosco, Johannes de, Francesco Giuntini, Albertus Hero and Élie Vinet. 1582. *Sphaera Ioannis de Sacro Bosco, emendata. In eandem Francisci Iunctini Florentini, Eliae Vineti Santonis, & Alberti Heronis Scholia.* Antwerp: Pierre Bellère. https://hdl.handle.net/21.11103/sphaera.100898.

Sacrobosco, Johannes de, Francesco Giuntini, Élie Vinet, Pierio Valeriano and Pedro Nunes. 1564. *Sphaera Ioannis de Sacro Bosco emendata. Cum additionibus in margine, et Indice rerum et locorum memorabilium, et familiarissimis scholijs, nunc recenter compertis, et collectis à Francesco Iunctino Florentino sacra Theologia Doctore. Interserta etiam sunt Eliae Vineti Santonis egregia Scholia in eandem Sphaeram. Adiunximus huic libro compendium in Sphaeram, per Pierium Valerianum Bellunensem. Et Petri Nonij Salaciensis demonstrationem eorum, quae in extremo capite de climatibus Sacroboscius scribit de inaequali climatum latitudine: eodem Vineto interprete.* Lyon: Symphorien Barbier for the Heirs of Jacopo Giunta. https://hdl.handle. net/21.11103/sphaera.101071.

Sacrobosco, Johannes de, Albertus Hero, Élie Vinet, Pierio Valeriano and Pedro Nunes. 1581. *Sphaera Ioannis de Sacro Bosco emendata. Eliae Vineti Santonis scholia in eandem Sphaeram, ab ipso authore restituta. Quibus nunc accessere scholia Heronis. Adiunximus huic libro compendium in Sphaeram per Pierium Valerianum Bellunensem, Et Petri Nonii Salaciensis demonstrationem eorum, quae in extremo capite declimatibus Sacroboscius scribit de inaequali climatum latitudine, eodem Vineto interprete.* Cologne: Maternus Cholinus. https://hdl.handle.net/21.11103/sphaera. 100655.

Sacrobosco, Johannes de, Albertus Hero, Élie Vinet, Pierio Valeriano and Pedro Nunes. 1591. *Sphaera Ioannis de Sacrobosco emendata. Eliae Vineti Santonis scholia in eandem Sphaeram, ab ipso authore restituta. Quibus nunc accessere scholia Heronis. Adiunximus huic libro compendium in Sphaeram per Pierium Valerianum Bellunensem, Et Petri Nonii Salaciensis demonstrationem eorum, quae in extremo capite de climatibus Sacroboscius scribit de inaequali climatum latitudine, eodem Vineto interprete.* Cologne: Goswin Cholinus. https://hdl.handle.net/21.11103/sph aera.100326.

Sacrobosco, Johannes de and Philipp Melanchthon. 1601. *Libellus de Sphaera Iohannis de Sacro Busto. Accessit eiusdem Autoris Computus Ecclesiasticus, Et alia quaedam, in studiosorum gratiam edita. Cum Praefatione Philippi Melanchthonis.* Wittenberg: Johann Krafft for Zacharias Schürer & Partners. https://hdl.handle.net/21.11103/sphaera.100646.

Sacrobosco, Johannes de and Bernardus Morisanus. 1625. *Bernardi Morisani Hiberni Derensis in Sphaeram Ioannis de S. Bosco commentarius; In quo praeter authoris explanationem facilem, iucundißimae & utilißimae Quaestiones ab aliis Interpretibus praetermissae, elegantißime & solidißime resolvuntur, & explicantur. Nunc primum publicae utilitati donatus.* Frankfurt am Main: Peter Marschall. https://hdl.handle.net/21.11103/sphaera.100584.

Sacrobosco, Johannes de and Francesco Pifferi. 1604. *Sfera di Gio. Sacro Bosco tradotta, e dichiarata da don Francesco Pifferi San Savino Monaco Camaldolense, e Matematico nello Studio de Siena. Misurato Intronato. Al Serensissimo don Cosimo Medici Gran Principe di Toscana. Con nuove aggiunte di molte cose notabili, e varie demostrazioni utili, e dilettevoli. Come nella seguente Tavola si vede.* Siena: Silvestro Marchetti. https://hdl.handle.net/21.11103/ sphaera.100591.

Sacrobosco, Johannes de, Élie Vinet, Pierio Valeriano and Pedro Nunes. 1586. *Sphaera Ioannis de Sacro Bosco emendata. Eliae Vineti Santonis scholia in eandem Sphaeram, ab ipso Authore restituta. Adiunximus huic Libro compendium in Sphaeram, per Pierium Valerianum Bellunensem: Et Petri Nonij Salaciensis demonstrationem eorum, quae in extremo capite de climatibus Sacroboscus scribit de inaequali climatum latitudine: eodem Vineto interprete. Ex postrema Impressione Lutetiae.* Venice: Heirs of Girolamo Scotto. https://hdl.handle.net/21.11103/sphaera. 100325.

Sacrobosco, Johannes de, Élie Vinet, Pierio Valeriano and Pedro Nunes. 1606. *Sphaera Ioannis de Sacrobosco, emendata. Eliae Vineti Santonis scholia in eandem Sphaeram, ab ipso auctore restituta. Adiunximus huic libro compendium in Sphaeram, per Pierium Valerianum Bellunensem: Et Petri Nonii Salaciensis demonstrationem eorum, quae in extremo capite de climatibus Sacroboscius scribit de inaequali climatum latitudine, eodem Vineto interprete. Post omnes omnium editiones, auctior & locupletior*. Lyon: Hugues Gazeau. https://hdl.handle.net/21.11103/sphaera. 100354.

Sacrobosco, Johannes de, Élie Vinet, Pierio Valeriano and Pedro Nunes. 1620. *Sphaera Ioannis de Sacro Bosco emendata. Eliae Vineti Santonis scholia in eandem sphaeram, ab ipso authore restituta. Adiunximus huic libro compendium in sphaeram, per Pierium Valerianum Bellunensem: et Petri Nonij Salaciensis demonstrationem eorum, quae in extremo capite de climatibus Sacroboscus scribit de inaequali climatum latitudine: eodem Vineto interprete. Ex postremo Impressione Lutetiae*. Venice: Heirs of Girolamo Scoto. https://hdl.handle.net/21.11103/sphaera. 100358.

Scultetus, Abraham. 1624. *Kurtzer Underricht von den vornem[m]sten Hauptpuncten Christlicher Religion für die jugend in der Chur-Pfalz…*. Frankfurt: Johann Ammon and Peter Mareschal.

Tossanus, Daniel. 1624. *Betbüchlein Oder Ubung der Christlichen Seel*. Frankfurt am Main: Pierre Mareschal.

Zabarella, Jacopo. 1590a. *Iacobi Zabarellae Patavini De rebus naturalibus libri XXX. Quibus questiones, quae ad Asristotelis interpretibus hodie tractari solent, accurate dicutiuntur*. Cologne: Giovanni Battista Ciotti.

Zabarella, Jacopo. 1590b. *Opera logica*. 2 vols. [Heidelberg]: Jean Mareschal.

Zevelius, Josua. 1624. *Christliche Practick, Wie jhm ein einfeltiger Christ die fünff Hauptstück der Christlicher Religion im leben und sterben solle zu nutz machen, in Frag und Antwort verfasset*. Frankfurt am Main: Pierre Mareschal.

Zincgref, Julius Wilhelm. 1624. *Emblematum Ethico-Politicorum Centuria Iulii Guilielmi Zincgrefii Coelo Matth. Meriani*. Frankfurt am Main: Pierre Mareschal.

Secondary Literature

Arbour, Roméo. 1979–1980. *L'Ere baroque en France: répertoire chronologique des éditions de textes littéraires*. 3 vols. Geneva: Droz.

Ascarelli, Fernanda and Marco Menato. 1989. *La tipografia del '500 in Italia*. Firenze: Olschki.

Baudrier, Henri-Louis. 1964–1965. *Bibliographie lyonnaise: recherches sur les imprimeurs, libraires, relieurs et fondeurs de lettres de Lyon au XVIe siècle, publiées et continuées par J. Baudrier*. 12 vols. Paris: F. de Nobele.

Benzing, Josef. 1977. *Die deutschen Verleger des 16. und 17. Jahrhunderts*. Frankfurt am Main: Buchhändler-Vereinigung.

Binasco, Matteo, ed. 2020. *Luke Wadding, the Irish Franciscans, and Global Catholicism*. Abingdon: Routledge.

Bonnant, Georges. 1999. *Le livre genevois sous l'Ancien Régime*. Geneva: Droz.

Brevaglieri, Sabina. 2008. Editoria e cultura a Roma nei primi tre decenni del Seicento: Lo spazio della scienza. In *Rome et la science moderne entre Renaissance et Lumières*, ed. Elisa Andretta, Irene Baldriga, Francesco Beretta, Jean-Marc Besse, and Elena Brambilla, 257–319. Rome: Ecole Française de Rome.

Casalini, Cristiano. 2012. *Aristotele a Coimbra: Il Cursus Conimbricensis e l'educazione nel Collegium Artium*. Rome: Anicia.

Cioni, Alfredo. 1970. Basa, Domenico. In *Dizionario biografico degli italiani* 7: 45–49.

Carvalho, Mário Santiago de. 2018. *The Coimbra Jesuit Aristotelian course*. Coimbra: Coimbra University Press.

De Lalande, Jérôme. 1803. *Bibliographie astronomique; avec l'histoire de l'astronomie depuis 1781 jusqu'á 1802*. Paris: De l'Imprimerie de la République. Reprint: Amsterdam: Meridian Publishing, 1970.

Dingel, Irene. 1996. *Concordia controversa: die öffentlichen Diskussionen um das lutherische Konkordienwerk am Ende des 16. Jahrhunderts*. Gütersloh: Gütersloher Verl.-Haus.

Dyroff, Hans-Dieter. 1963. Gotthard Vögelin Verleger, Drucker, Buchhändler 1597–1631. *Archiv für Geschichte des Buchwesens* 4: cols. 1131–1423.

Fraesen, Rob and Leo Kenis. 2012. *The Jesuits of the Low Countries: identity and impact (1540–1773): Proceedings of the International Congress at the Faculty of Theology and Religious Studies, KU Leuven*, 3–5 December 2009. Leuven/Walpole, MA: Peeters.

Friedrich, Markus. 2004. *Die Grenzen der Vernunft: Theologie, Philosophie und gelehrte Konflikte am Beispiel des Helmstedter Hofmannstreits und seiner Wirkungen auf das Luthertum um 1600*. Göttingen: Vandenhoeck and Ruprecht.

Grell, Ole P. 2011. *Brethren in Christ: A calvinist network in reformation Europe*. Cambridge: Cambridge University Press.

Hellyer, Marcus. 2005. *Catholic physics: Jesuit natural philosophy in early modern Germany*. Notre Dame: University of Notre Dame Press.

Hotson, Howard. 2007. *Commonplace learning: Ramism and its German ramifications 1543–1630*. Oxford: Oxford University Press.

Houzeau, Jean-Charles and Albert Benoit Marie Lancaster. 1882–1889. *Bibliographie générale de l'astronomie*. Vol. 2. Bruxelles: F. Hayez. Reprint: London: The Holland Press, 1964.

Kraye, Jill. 2008. Conceptions of Moral Philosophy. In *Cambridge history of seventeenth-century philosophy*, ed. Daniel Garber and Michael Ayers, 1279–1316. Cambridge: Cambridge University Press.

Lattis, James M. 1994. *Between Copernicus and Galileo. Christoph Clavius and the collapse of Ptolemaic cosmology*. Chicago: University of Chicago Press.

Lines, David A. and Eugenio Rufini, eds. 2015. *'Aristotele fatto volgare:' Tradizione aristotelica e cultura volgare nel Rinascimento*. Pisa: ETS.

Lüthy, Christoph. 2012. *David Gorlaeus (1591–1612): An enigmatic figure in the history of philosophy and science*. Amsterdam: Amsterdam University Press.

Maclean, Ian. 2009a. André Wechel at Frankfurt, 1572–1581. In *Learning in the marketplace; essays in the history of early modern books*, ed. Ian Maclean, 163–225. Leiden: Brill.

Maclean, Ian. 2009b. Melanchthon at the book fairs, 1560–1601: Editors, markets and religious strifc. In *Learning in the marketplace: Essays in the history of early modern books*, ed. Ian Maclean, 107–130. Leiden: Brill.

Maclean, Ian. 2021. *Episodes in the life of the early modern learned book*. Leiden: Brill.

Maranini, Anna. 2000. La biblioteca di un erudito: Francesco Pifferi da Siena. *Boletín de la Real Academia de Buenas Literas de Barcelona* 47: 127–196.

Marinheiro, Cristóvão S. 2012. The Conimbricenses: the last scholastics, the first moderns or something in between? The impact of geographical discoveries on late 16th century Jesuit Aristotelianism. In *Portuguese humanism and the republic of letters*, ed. Maria Louro Berbara and Karl A. E. Enenkel, 395–424. Leiden: Brill.

Müller, Johannes. 2017. Transmigrant literature: Translating, publishing, and printing in seventeenth century Frankfurt's migrant circles. *German Studies Review* 40: 1–21.

Murdock, Graeme. 2011. *Calvinism on the frontier, 1600–1660: International Calvinism and the reformed church in Hungary and Transylvania*. Oxford: Clarendon Press.

Ong, Walter J. 1958. *Ramus and Talon inventory*. Cambridge, MA: Harvard University Press.

Price, Audrey. 2014. Mathematics and mission: Deciding the role of mathematics in the Jesuit curriculum. *Jefferson Journal of Science and Culture* 1: 29–40.

Reske, Christoph. 2015. *Die Buchdrucker des 16. und 17. Jahrhunderts im deutschen Sprachgebiet, auf der Grundlage des gleichnamigen Werks von Josef Benzing*. Wiesbaden: Harrassowitz.

Rhodes, Dennis E. 2013. *Giovanni Battista Ciotti (1562–1627?): Publisher extraordinary at Venice*. Venice: Marcianum Press.

Sander, Christoph. 2014. Medical topics in the *De Anima* commentary of Coimbra (1598) and the Jesuits' attitude towards medicine in education and natural philosophy. *Early Science and Medicine* 19: 76–101.

Sander, Christoph. 2018. Johannes de Sacrobosco und die *Sphaera*-Tradition in der katholischen Zensur der Frühen Neuzeit. *NTM Zeitschrift für Geschichte der Wissenschaften, Technik und Medizin* 26 (4): 437–474. https://doi.org/10.1007/s00048-018-0199-6.

Schrörs, Heinrich. 1808. Der Kölner Buchdrucker Maternus Cholinus. *Annalen des Historischen Vereins für den Niederrhein* 85: 147–165.

Schwetschke, Gustav. 1850–1877. *Codex nundinarius Germaniae literatae bisecularis*. Halle: Selbstverlag.

Sigismondi, Costantino. 2012. Christopher Clavius astronomer and mathematician. ArXiv:1203.0476.

Thomas, Andrew L. 2010. *A house divided: Wittelsbacher confessional court cultures in the Holy Roman Empire, c. 1550–1650*. Leiden/Boston: Brill.

Toepke, Gustav. 1886. *Matrikel der Universität Heidelberg*. 8 vols. Heidelberg: Selbstverlag des Herausgebers.

Trentman, John A. 1982. Scholasticism in the seventeenth century. In *Cambridge history of later medieval philosophy*, ed. Norman Kretzmann, Anthony Kenny, Jan Pinborg, and Eleonore Stump, 818–837. Cambridge: Cambridge University Press.

Willer, Georg. 1972–2001. *Die Messkataloge Georg Willers*. ed. Bernhard Fabian. Hildesheim: Georg Olms Verlag.

Zannini, Masetti, and Gian Ludovico. 1980. *Stampatori e librai a Roma nella seconda metà del Cinquecento: Documenti inediti*. Rome: Fratelli Palombi.

Ian Maclean is an Emeritus Professor of Renaissance Studies of the University of Oxford, Emeritus Fellow, All Souls College, Oxford, and Honorary Professor of History, University of St Andrews. He is a Fellow of the British Academy and Member of the European Academy, and has held visiting positions in Australia, Canada, Germany, France, and the United States. He has published widely on intellectual history (including *The Renaissance notion of woman*, 1980, *Meaning and interpretation in the Renaissance: the case of law*, 1992, *Montaigne philosophe*, 1996, *Logic, signs and nature in the Renaissance: the case of learned medicine*, 2001) has edited Cardano's *De libris propriis*, and has produced three substantial works on book history: *Learning and the market place* (2009), *Scholarship, commerce, religion: the learned book in the age of Confessions, 1560-1630* (2021), and *Episodes in the life of the early modern learned book* (2021).

Chapter 7
The Iberian and New World Circulation of Sacrobosco's *Sphaera* in the Early Modern Period

Alejandra Ulla Lorenzo

Abstract The aim of this paper is to reconstruct the circulation of Sacrobosco's *Tractatus de sphaera* in early modern Iberian Peninsula and New World printing. We will present a survey on the locally active printers and publishers who contributed to the circulation of the *Sphaera* thanks to the information now offered by the *Iberian Books* database. This will be followed by a general discussion about the professional profile held by the printers and publishers who took part in the publication and circulation of the text in the Iberian Peninsula and America. Both markets were probably related through Seville. With this group of printers and publishers in mind, we will analyze what can be inferred from their production in terms of their approach to publishing, what audience were they generally targeting, their commercial scope, and how the *Sphaera* fit into their general production and commercial plan.

Keywords Early Modern printing · Iberian peninsula · New World · *Iberian Books* · *Tractatus de sphaera*

1 Introduction

The printing press arrived in the Iberian Peninsula in 1472 thanks to the printer Juan Párix (Johannes Parix) (1430–1501) and the patronage of the educational institution the Estudio General de Segovia. Shortly thereafter, in 1475, the printing of scientific books began with the publication of a medical tract in Barcelona, which was a Catalan translation of a Portuguese text. This first example demonstrates one of the main characteristics of the scientific publications of the time: the predominance of vernacular languages over Latin, as shall be examined below. Despite its early appearance, the scientific sector of the publishing industry was of little importance in the Iberian Peninsula during the early modern period and was mainly linked, albeit not

A. Ulla Lorenzo (✉)
Universidad de Santiago de Compostela, Santiago, Spain
e-mail: alejandra.ulla@usc.es

© The Author(s) 2022
M. Valleriani and A. Ottone (eds.), *Publishing Sacrobosco's De sphaera in Early Modern Europe*, https://doi.org/10.1007/978-3-030-86600-6_7

exclusively, to the printing industry based around the universities, in which, particularly during the sixteenth century, Johannes de Sacrobosco's (fl. 1256) *Tractatus de sphaera* became a manual for the study of the theoretical and practical sciences.

The aim of this article is to reconstruct the circulation of the *Tractatus de sphaera* in the context of the printing press and the publishing industry of the Iberian Peninsula and the New World during the early modern period. In this context, we shall present, first of all, the results of our research on local printers who published Sacrobosco's work in Latin or in a vernacular language and who contributed in such a way to its circulation on local, Peninsular, and extra-Peninsular levels. Furthermore, we shall examine the professional profile of the printers who participated in the publication and circulation of the text both in the Iberian Peninsula and in America. Then we shall analyze the work they produced, the target audience, the scope of their commercial activity, and how the tract fit in with their production as a whole and with their commercial plan. The data provided by *Iberian Books*[1] and the examination of the analyzed editions of Sacrobosco's text will help us to determine where and when the editions were published, the frequency with which they were reedited, and how popular the text was according to the number of editions. However, neither this bibliographic tool nor the editions themselves can tell us with any degree of precision where the books ended up, at which markets they were aimed, or in which markets they had greater success. To answer some of these questions we can rely on the invaluable postmortem inventories that list the estate of the deceased and, therefore, make it possible for us to establish the belongings of booksellers and owners of private libraries during the early modern period, among which, of course, their books can be found. As Bartolomé Bennassar has pointed out, these are "the most valuable documents when exploring intelligent, written culture, in order to know which books were owned and read in a certain age and by certain people" (Bennassar 1984, 141). In this way, these inventories will help us to establish a more accurate perspective of the success of Sacrobosco's text in the Iberian Peninsula. Last of all, we shall deal with the issue of the circulation of the Peninsular editions of Sacrobosco in America, thus enabling us to speak about international markets.

2 On the Distribution of the *Tractatus de sphaera* in the Iberian Peninsula (1472–1650)

There is much evidence that the *Tractatus de sphaera* was a fundamental work in Spanish education in the Golden Age for the teaching of astronomy in the *quadrivium*. However, it was also important in international expeditions, and in the private sphere as a reference manual held in private libraries for learning about the sky and the stars.

---

[1] The objective of the digital repository *Iberian Books* (https://iberian.ucd.ie. Accessed June 8, 2021) is to produce a foundational listing of all books published in Spain, Portugal, and the New World or printed elsewhere in Spanish or Portuguese during the Golden Age, 1472–1700, as well as to create a suite of digital search tools to permit their investigation.

This can be explained by the simplified presentation of the Ptolemaic system offered by Sacrobosco with the aim of presenting his students with an introductory astronomy text (Gómez Martínez 2013, 40–41). In the context of universities in the Iberian Peninsula, it is possible to note an increase in the demand for astronomy throughout the sixteenth century—in both its theoretical and practical aspects—perhaps due to the influence of navigation (Bonmatí Sánchez 2002, 1410).[2] It is interesting to contemplate, in this sense, the controversy that arose around the professorship of astrology of the master Juan de Aguilera of the University of Salamanca on January 9, 1552, between supporters of the reading of the *Sphaera* of Sacrobosco and those who supported the *Theorica planetarum* (Hurtado Torres 1982, 50). The controversy concluded with a vote before the vice chancellor; the work of Sacrobosco was chosen, as is stated in the text of the document:

> while lecturing at his chair of astrology with a great number of pupils, the mentioned doctor and treasurer Juan de Aguilera, recognized differences among them. Some asked him to read the *Sphaera*; others asked him to read the *theorica*. Because of this difference, the mentioned bachelor Cristobal de Perea, vice chancellor…, assigned him what he had to read [on the basis of a] *vota audientium*, and most of the said generals and listeners voted that he should read about the sphere…And the said doctor accepted it and said that accomplish the task in that way.[3] (Hurtado Torres 1982, 50 n. 6)

Also of note is the information offered by the *Lecturas de la Cátedra de Astrología* of the University of Salamanca for the years 1560–1641, which mentions several authors, one of whom is Johannes de Sacrobosco (Hurtado Torres 1982, 50). The influence of Sacrobosco's text in the field of education in the Iberian Peninsula is also shown by the fact that the humanist Antonio de Nebrija (1444–1522) wrote a dedication to him in his book *Introductorium Cosmographiae* (Nebrija 1498), a book in which he pointed out the former's mistake in calculating the total circumference of the Earth, despite noting him as a source of authority in many other parts of the text (Bonmatí Sánchez 1998, 513).

Also worthy of note is the mention made of Sacrobosco's text in the *Recopilación de leyes de los reinos de las Indias mandadas imprimir y publicar por la magestad católica del rey Carlos II nuestro señor*, originally published in 1681, in relation to works that should be read by cosmographers and mathematics professors of the Council of the Indies:

> To the cosmographer, who, as professor, teaches mathematics, we order that he lectures in that location that was indicated or that will be indicated in our house and palace, and following the Council of the Indies, a whole hour in the morning every day in winter from nine o'clock to ten o'clock, and in summer from eight

---

[2] As Kathleen Crowther has explained, a "major reason for interest in the *Sphaera* in Spain and Portugal was that the basic astronomical and geographical knowledge contained in this text could be used for navigation" (Crowther 2020, 162).

[3] Author's translation here and in the following.

o'clock to nine o'clock, changing the hours when the aforementioned Council changes them, and taking holidays for two months in July and August, and at Easter, given by the Council, with no others to be taken. Regarding the readings, the following order is to be kept: in the first year, which shall begin in September, from the beginning of the month until Christmas, Sacrobosco's *Sphaera* is to be read.[4]

Therefore, Sacrobosco's work was employed as a textbook and work of reference for various audiences from the thirteenth century up to the seventeenth. For this reason, it was printed both with and without commentaries on at least seventeen occasions in the Iberian Peninsula. It is interesting to consider, in this sense, the taxonomy established by the members of the research project *The Sphere: Knowledge System Evolution and the Shared Scientific Identify of Europe*, which distinguishes five types of books among the different editions preserved in their database: (a) original treatises, (b) annotated originals, (c) compilation of texts, (d) compilation of texts and annotated originals, and (e) adaptions (Valleriani 2020). The Iberian editions belong to the second, third, fourth, and fifth categories. Five of these editions were published in Latin between 1472 and 1650:

Sacrobosco, Johannes de, Pedro Ciruelo and Pierre d'Ailly. 1526. *Opusculum de sphera mundi*. Alcalá de Henares: Miguel de Eguía (Sacrobosco et al. 1526).

Aristotle and Pedro de Espinosa. 1535. *Philosophia naturalis Petri a Spinosa artium magistri*. Salamanca: Rodrigo de Castañeda (Aristotle and Espinosa 1535).

Sacrobosco, Johannes de and Pedro Espinosa. 1550. *Sphera Ioannis de Sacro Busto cum commentariis Petri a Spinosa*. Salamanca: Juan de Junta (Sacrobosco and Espinosa 1550).

Avelar, André do. 1593. *Sphaerae utriusque tabella, ad Sphaerae huius mundi faciliorem enucleationem*. Coimbra: Antonio de Barreira (Avelar 1593).

While twelve editions were published in vernacular languages (either Portuguese or Spanish), also between 1472 and 1650, thereby indicating a preference for the vernacular languages (Fig. 1):

---

[4] The quote belongs to a law enacted in 1636 during the reign of Philip the Fourth (1621–1665). In 1681 Charles II, who was King of Spain between 1665 and 1700, sent to publish the *Recopilación de leyes de los reinos de las Indias mandadas imprimir y publicar por la magestad católica del rey Carlos II nuestro señor*, which contained laws enacted from the sixteenth century until his reign in relation to the Indies. I followed an edition from 1841 for the quote. The original text reads: "El cosmógrafo, que como catedrático leyere la cátedra de matemáticas: Mandamos que lea en la parte que le fuere señalada o señalare nuestra casa y palacio, y cerca del consejo de las Indias todos los días que le hubiere, una hora entera a la mañana en invierno desde nueve a diez, y en verano de ocho a nueve, mudando las horas cuando el dicho consejo las mudare, y gozando de vacaciones los dos meses de julio y agosto, y las de las pascuas que gozare el consejo, y no pueda tener ni tenga otra mas; y en lo que toca a las lecturas guarden el orden siguiente. El primer año, que comenzará por setiembre, desde principìo de él hasta la Navidad, ha de leer la esfera de Sacrobosco" (Recopilación de leyes 1841, 207).

Fig. 1 Languages in which Sacrobosco's text was published in the Iberian Peninsula, 1472–1650. Author's plot based on data provided by the repository *Iberian Books*

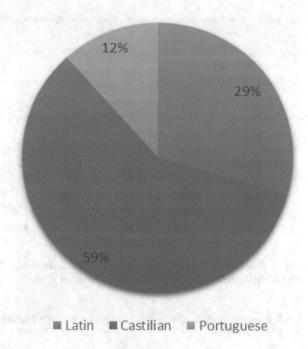

■ Latin   ■ Castilian   ■ Portuguese

Sacrobosco, Johannes de. [1510–1512]. *Tractado da Spera do mundo tirada de latim em liguoagem portugues.* [Lisbon: Germam Galharde] (Sacrobosco [1510–1512]).

Faleiro, Francisco. 1535. *Tratado del Esphera y del arte del marear.* Sevilla: Juan Cromberger (Faleiro 1535).

Sacrobosco, Johannes de and Pedro Nunes. 1537. *Tratado da sphera com a Theorica do Sol et da Lua.* Lisbon: Germam Galharde (Sacrobosco and Nunes 1537).

Sacrobosco, Johannes de and Jerónimo de Chaves. 1545. *Sphera del mundo.* Sevilla: Juan de León (Sacrobosco and Chaves 1545).

Sacrobosco, Johannes de and Jerónimo de Chaves. 1548. *Tractado de la Sphera.* Sevilla: Juan de León (Sacrobosco and Chaves 1548).

Cortés, Martin. 1551. *Breve compendio de la sphera y de la arte de navegar.* Sevilla: António Álvarez (Cortés 1551).

Cortés, Martin. 1556. *Breve compendio de la sphera y de la arte de navegar.* Sevilla: António Álvarez (Cortés 1556).

Sacrobosco, Johannes de and Rodrigo Sáenz de Santayana y Espinosa. 1567. *La sphera de Iuã de Sacro Bosco nueva y fielmente traduzida de Latin en Romance.* Valladolid: Adrián Ghemart (Sacrobosco and Santayana y Espinosa 1567).

Sacrobosco, Johannes de and Rodrigo Sáenz de Santayana y Espinosa. 1568. *La Sphera de Iuan de Sacrobosco Nueva y fielmente traduzida de Latin en Romance.* Valladolid: Adrián Ghemart for Pedro de Corcuera (Sacrobosco and Santayana y Espinosa 1568).

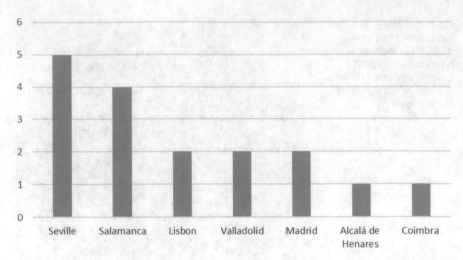

Fig. 2 Places of publication of Sacrobosco's work in the Iberian Peninsula, 1472–1650. Author's plot based on data provided by the repository *Iberian Books*

Rocamora y Torrano, Ginés. 1599. *Sphera del universo*. Madrid: Juan de Herrera (Rocamora y Torrano 1599).

Sacrobosco, Johannes de, Christoph Clavius, Francesco Giuntini and Élie Vinet. 1629. *Exposicion de la Esfera de Iuan de Sacrobosco*. Salamanca: Jacinto Taberniel (Sacrobosco et al. 1629).

Sacrobosco, Johannes de, Cosme Gómez Tejada de los Reyes and Aristotle. 1650. *El filosofo. Ocupacion de nobles, y discretos contra la cortesana ociosidad*. Madrid: Domingo García Morrás for Santiago Martín Redondo (Sacrobosco et al. 1650).

Among the places of publication, Seville and Salamanca stand out (Fig. 2). The former was linked to the Casa de Contratación and trading with America, whereas the latter was a seat of university education.

2.1 Printers and Publishers of the Tractatus de sphaera in the Iberian Peninsula (1472–1650): Latin Editions

In the context of the Iberian publishing industry, Sacrobosco's text was published in 1526 for the first time in the Peninsula in Latin by Miguel de Eguía (ca. 1495–1544), a printer from Alcalá de Henares (Sacrobosco et al. 1526). Later, two further editions, also published in Salamanca, can be found: in 1535 by Rodrigo de Castañeda (fl. 1533–1537) (Aristotle and Espinosa 1535) and in 1550 by Juan de Junta (fl. 1526–1558) (Sacrobosco and Espinosa 1550). Only at the end of the century, in 1593, was it printed in Coimbra (Avelar 1593). It is important to consider that the three cities in which the Latin text was published during the sixteenth century had been active university cities since the previous century. Indeed, if we examine the editions that came out of these publishing houses, we observe that all four cases are printers whose production, although not exclusively specialized in academic texts, was characterized by texts mainly aimed at the students and lecturers of the universities of their towns.

As stated above, the first edition of the Latin text was published in Miguel de Eguía's print shop in Alcalá de Henares (Sacrobosco et al. 1526), one of the foremost printing centers and the seat of the prestigious university founded by Francisco Jiménez de Cisneros (1436–1517). Miguel de Eguía was not only a prolific printer but also a "cultured man and a notable humanist who was a follower and advocate of the doctrines of Erasmus" (Fuente Arranz 2018, "Miguel de Eguía"). He produced his first printed works in Logroño in the workshop of Arnao Guillén de Brocar (fl. 1490–1524), with whom he formed a partnership in 1518 after marrying Brocar's daughter, María Brocar, who died at a young age. In 1511, he moved with his father-in-law to Alcalá de Henares and, in 1523, inherited Brocar's business after they had collaborated on such important projects as the printing of the *Biblia políglota complutense* (Brocar and Cisneros 1514). The *Iberian Books* database counts 197 editions of Eguía. Based on the same source, we can confirm that although he cannot be classified as a specialized printer, and books on religious matters occupy—as was typical in this period—the first position in terms of production (eighty-eight editions listed under Religious and eleven under Bibles), we may observe that his production, according to the *Iberian Books* database, was relevant to the common reading matter of the university at that time. In this respect, we cannot forget that the database classifies editions under these subjects: Classical Authors (nineteen editions), Educational (eleven editions), Sciences (ten editions), Philosophy and Morality (nine editions), Histories (eight editions), Linguistics and Philology (six editions), Literature (four editions), Dictionaries/Language Instruction (three editions), Poetry (three editions), Dialectics and Rhetoric (three editions), Astrology and Cosmography (two editions), Music (two editions), Agriculture/Veterinary (one edition), and Medical (one edition). As concrete examples, we note the many editions of the works of Antonio de Nebrija (1441–1522), the publication of classical Latin texts, and the first grammar reference of the Hebrew language (*Introductiones artis grammaticae hebraice*) by Alfonso de Zamora (1476–1544), printed in 1526 (Zamora 1526). Therefore, it can be stated that textbooks were of prime importance in Eguía's output (Martín Abad 1991, Vol. I, 79), although he also punctually published editions belonging to other subjects, such as Ordinances/Edicts (nine editions), News (five editions), Funeral Orations (one edition), and Military Handbooks (one edition). It should not be forgotten, however, that this edition of the *Sphaera* was undertaken by Pedro Ciruelo (1470–1554),[5] a Spanish mathematician and theologian who was a tutor of King Philip II (1527–1598) (Chap. 13). Therefore, in this case as well, Sacrobosco's text must be understood as an academic one, probably offered by Eguía to the students of Alcalá. It must not be forgotten, however, that he divided his time between printing in Alcalá and doing typographical work in Logroño, Toledo, and Burgos (Delgado Casado 1996, Vol. I,

---

[5] See (Lanuza Navarro 2020) for more information about Ciruelos's commentary on Sacrobosco's *Tractatus*.

199). It is possible that this circumstance may have favored him when establishing relationships with businesses outside of Alcalá for the sale of Sacrobosco's text.

It is also necessary to highlight the typographical perfection and innovative nature of his work, achieved through the innovation of types and decorative elements (Martín Abad 1991, Vol. I, 80). However, there was a clear influence of works printed abroad on some of Eguía's editions. One good example in this regard is, without doubt, the titlepage of his edition of Sacrobosco's *Opusculum de sphera mundi* (Fig. 3), on which the use of a typographical tabernacle of foreign origin can be observed. This may be attributed to the fact that he had been employed in Paris by Simon Vostre (fl. 1486–1521) many years before (Martín Abad 1991, Vol. I, 81).

In 1535, a new edition appears to have been published by the printer Rodrigo de Castañeda from Salamanca (Aristotle and Espinosa 1535). He was an extremely unprolific printer (with just seventeen editions listed in the *Iberian Books* database) despite having been active from 1533 to 1551. Although, as in the previous two cases, books of a religious nature are situated at the head of his output (five editions listed under Religious), certain works related to university reading matter, such as the edition of Sacrobosco, also came off his printing presses. Among them, the database counts editions on the next subjects: Jurisprudence (two editions), Philosophy and Morality (two editions), Sciences (two editions), Ordinances/Edicts (one edition), Literature (one edition), Educational (one edition), Classical Authors (one edition), Bibles (one edition), and Astrology and Cosmography (one edition). Thus, his workshop was often chosen by "the authors of rank (university lecturers, ecclesiastic authorities, or the university itself) in order to disseminate their writings" (Mano González 1998, 72) because of his way of working, which was in the Renaissance style (Ruiz Fidalgo 1991, 59). He worked closely with Pedro de Espinosa (fl. 1551), a sixteenth-century mathematician and astronomer, who prepared the edition of Sacrobosco.

The final Latin edition came off the presses of Juan de Junta in 1550 (Chap. 8). This printer, of Florentine origin, belonged to a family of printers and booksellers. It is believed that he arrived in Spain in 1514 and set himself up as a bookseller in Salamanca. However, he moved to Burgos shortly thereafter to assist the printer Isabel de Basilea (fl. 1517–1525), whom he later married. Around 1532, they moved to Salamanca and set up a second print shop (Fuente Arranz 2018, "Juan de Junta"), which produced, according to the *Iberian Books* database, at least 106 editions until 1552.[6] It should be noted that, after setting up the print shop in Salamanca, Junta left the business in the hands of Alejandro de Cánova (fl. 1569–1573) for the almost twenty years that he spent outside the Iberian Peninsula. Therefore, although Junta's is the name on the colophons, Cánova would have been responsible for selecting the books to be printed and for "taking on apprentices under his responsibility, agreeing

---

[6] It is important to consider that we are referring to those editions in whose imprints the name "Juan de Junta" appears as printer. It is also important to consider the importance of his very large library, well studied by William Pettas through his inventory, which includes up to 15,827 volumes (Pettas 1995, 9).

Fig. 3 Titlepage of (Sacrobosco et al. 1526). Universidad de Sevilla, Callnumber: 1294. Courtesy of HathiTrust. https://hdl.handle.net/2027/ucm.5320774244

and signing printing contracts and for the organization and supervision of the tasks carried out by the different members of the workshop" (Mano González 1998, 71).

The workshop went through different stages: the first had a moderate level of production, the second saw an increase in output, and finally there was a period of decline (Mano González 1998, 71). Its output is related with the publication of the works of new writers (five editions under Literature listed in *Iberian Books*), missals, breviaries, manuals for the local church (thirty-three editions under Religious and four under Bibles), pamphlets to commemorate historical events, and booklets of laws (thirty-one editions under the subject Ordinances/Edicts and four under Jurisprudence).

This set of works was, therefore, "in general of little scope, with a low cost and rapid and assured distribution, which, although it did not make it possible to obtain large profits, at least covered the costs of the investment made in terms of time, staff and material" (Mano González 1998, 73). Only occasionally did certain figures related to the university request the services of this print shop. In this respect, we found editions in Sciences (four), Philosophy and Morality (three), Linguistics and Philology (three), Classical Authors (three), Dialectics and Rhetoric (two), Economics (two), Educational (two), Agriculture/Veterinary (one), and Astrology and Cosmography (one) listed in *Iberian Books*. It is perhaps in this context, and as an exception to its common tasks, that we can situate the edition of Sacrobosco's *Sphaera*. It should also be recalled that the work of printing was only a complement to the selling of books, which was the fundamental commercial activity for the Junta firm, not only in Salamanca but also in Burgos, as mentioned above. This would have considerably broadened his center of sales and, therefore, of the distribution of the editions he printed, a circumstance that might have encouraged sales of his edition of Sacrobosco outside of Salamanca.

Only at the end of the century, in 1593, was a version of Sacrobosco's text, entitled *Sphaerae utriusque tabella ad sphaerae huius mundi faciliorem enucleationem*, printed in Portugal (Avelar 1593). This task was carried out by the printer Antonio de Barreira (fl. 1579–1597) in Coimbra, where the university had been established in 1537. This printer was only active between 1579 and 1597 and had a low output (seven editions) according to *Iberian Books*. In his production, we find mention of editions that belong to the most successful subjects in the market of the time—books of Religion (two editions) and those related to the Law (two editions under the category Ordinances/Edicts). It is known that he had connections in Salamanca, given that he financed at least the publication of one book in the city, and that he was related to the University of Coimbra, for which he published its charter. Part of its production may therefore have been limited to this area, which is why we find editions under subjects related with this field in the *Iberian Books* database: Sciences (one), Astrology and Cosmography (one), Dialectics and Rhetoric (one), Medical (one), Music (one), News (one), and Poetry (one). His edition of Sacrobosco's text could be related to his shop's work for the university.

As Gómez Martínez has explained, the inventories preserved from the old Spanish libraries demonstrate that in the first quarter of the sixteenth-century Venetian printers and booksellers—and then the French, particularly from Paris and Lyon—supplied the majority of Latin texts that students in the Iberian Peninsula needed for their education; only occasionally, as mentioned above, can Spanish printers be found to have produced Latin texts for students and lecturers (Gómez Martínez 2006, 197). In this regard, it should be recalled that "Spain was peripheral in the geography of the printing press, which could never compete with the great pioneering centers of the rest of Europe (Germany, France, and Italy) and was always relegated to the task of supplying, almost exclusively, the local markets" (González Sánchez and Maillard Álvarez 2003, 20).

This began to change around the end of the fifteenth century, although it would not be until the sixteenth century that, in Europe as a whole, a greater importance would begin to be assigned to the vernacular languages. There can be no doubt that the Spanish humanist Antonio de Nebrija (who considered the Spanish language to be a unifying factor for the various territories of the Catholic monarchs) contributed, with many of his works, to this paradigm shift in the use of the Spanish language. This circumstance may have contributed to the appearance of many translations of Sacrobosco's text produced by Spanish workshops for the local market. To this it must be added, as detailed below, that the Portuguese and Spanish editions of Sacrobosco's text were not only aimed at university students, but were also printed for the training of cosmographers and navigators who worked for the courts and who did not always have an in-depth knowledge of Latin.

The existence of translations of the text in different parts of the Peninsula raises the question of which language was used in university teaching in the sixteenth century. As explained by Gómez Martínez, "it is commonly held that the teaching of astronomy and cosmography during the sixteenth century was mainly carried out in Latin in the universities and in Spanish in other educational institutions focused on a more practical type of education, such as the Casa de Contratación in Seville and the Academia de Matemáticas in Madrid" (Gómez Martínez 2006, 206). However, it should be taken into consideration, on the one hand, that there are translations into Spanish dating from the end of the fifteenth century, of which at least one was made by a lecturer from the University of Salamanca, Diego de Torres, and, on the other hand, that, in accordance with the charter of the University of Salamanca, in the sixteenth century the use of Spanish was preferred for the teaching of certain laws and in the subjects of music and astrology—in other words, in *applied* subjects (Gómez Martínez 2006, 206). To this circumstance must be added the gradual increase in production of translations of works in other Romance languages, which was a common phenomenon throughout Europe and a demonstration of the progressive acceptance of the vernacular languages in the teaching of the sciences.

Some authors have stated that versions of the text in vernacular languages were few and late, given that it was an academic text (Ortiz Gómez and Menéndez Navarro 2004, 141). However, the figures indicate that the publication of the text in the vernacular languages was more common than in Latin; there were twelve editions in Portuguese and Spanish published in the sixteenth and seventeenth centuries.

2.2   Printers and Publishers of the Tractatus de sphaera in the Iberian Peninsula (1472–1650): Editions in the Vernacular Languages

The first Sacrobosco translation into a vernacular language of the Iberian Peninsula (Portuguese) was published in 1516 and came off the printing press of Germão Galherde (fl. 1519–1561) in Lisbon (Sacrobosco [1510–1512]). The same workshop printed the text again in 1537; however, the two texts are actually quite different in content (Sacrobosco and Nunes 1537).[7] The translator of both editions was Pedro Nunes (1502–1575), a cosmographer of King John III of Portugal (1502–1557). This would indicate, perhaps, that the primary aim of this translation was not academic, as was the case in previous examples, but rather administrative and connected with Portugal's international relations. It should be mentioned here that new lands were discovered in Asia and that the colonization of Brazil began during the reign of King John III.

The first information about Nunes dates from 1519. He probably continued working until 1557. Compared with the other cases analyzed above, astrology and cosmography played a greater role in the workshop of this printer (seven editions listed in the *Iberian Books* database); nevertheless, Religion and Ordinances remain the most important subjects (sixty-three editions listed under Religious, one mentioned under Funeral Orations, one mentioned under Bibles, thirty-eight mentioned under Ordinances/Edicts, and two under Jurisprudence). However, he also paid attention to other fields such as News (two editions), the Book Market (one edition), Texts for Education (eight editions under Histories, six editions under Educational, five editions under Sciences, four editions under Classical texts, three editions under Music, two editions under Medical), and Literary Texts from different genres (seven editions under Poetry, three editions under Drama, and three editions under Literature).

The first printed translation into Spanish would not arrive until the middle of the sixteenth century. During this period, four editions were printed in Spanish. The first two, published in Seville in 1545 and 1548, were made by Juan de León (Lyon) (fl. 1545–1555), a printer of French origin who was active between 1545 and 1555 (Sacrobosco and Chaves 1545, 1548).[8] His workshop was located in Calle Real and he used new typesets and ornamental elements of high quality. It is interesting to note that in 1549 he was named printer of the University of Osuna by the Count of Ureña, Juan Téllez de Girón (1494–1558), who also employed him as his personal printer (Ruiz Jiménez 2015). The translation was made by Jerónimo de Chaves (1523–1574), professor of the Art of Navigation and Cosmography in the Casa de Contratación in Seville. Therefore, it is possible that these editions were not destined for academic institutions such as the Casa de Contratación alone, but also for the count. Indeed,

---

[7] A description of this edition, see (Crowther 2020, 171–173).

[8] For a description of the 1545 edition, see (Crowther 2020, 173–175). The 1548 edition is a reissue of the one printed in 1545.

we shall later refer to the presence of Sacrobosco's text in the book collections of the Spanish nobility.

In the timeframe mentioned above, Juan de León published only sixteen editions divided between the subjects indicated hereafter. Perhaps the restricted nature of his production is due to the fact that he published for very specific audiences: it is possible that those editions classified in *Iberian Books* as Astrology and Cosmography (three) were destined for the Casa de Contratación, while perhaps the other editions he published could have been commissioned by the count himself and destined for him or the University of Osuna. Among them we find the following according to *Iberian Books*: News (three), Philosophy and morality (two), Poetry (two), Religious (two), Bibles (one), Histories (one), Literature (one), and Music (one).

Twenty years later, in 1567, Rodrigo Sáenz de Santaya y Espinosa made a new translation, which was printed by Adrian Ghemart in Valladolid and reissued in 1568 (Sacrobosco and Santayana y Espinosa 1567, 1568).[9] Adrian Ghemart (fl. 1550–1573) was a bookseller and publisher based in Medina del Campo, another significant printing center in the Iberian Peninsula during the sixteenth century. Later, around 1562, it is believed that he set up a shop as a printer in Valladolid, where he stayed, according to the colophons, until 1573 (Pérez Pastor 1895, 493), albeit without abandoning his main activities as a bookseller and publisher (Delgado Casado 1996, 273). He also appears in relation to the publication of certain editions in Alcalá de Henares (Delgado Casado 1996, 274). In his case, editions on religious and legal matters were also at the head of his output (eleven editions under Religious and two under Jurisprudence), while scientific works were merely incidental (two editions under Astrology and cosmography), given that the only examples are two editions of Sacrobosco's text. We must not lose sight of the fact that, as is typical in the Iberian publishing market at this time, this printer also welcomed the publication of works linked to the educational sphere (two editions under Government/Political theory, one edition under Dialectics and rhetoric, one edition under Educational), as well as literary texts (one edition under Literature and one under Poetry). The broad network of relationships created by this bookseller-printer working in both Valladolid and Medina del Campo should be noted, as it enabled him to distribute his work more widely.

At the end of the sixteenth century, in 1599, Ginés Rocamora y Torrano (1550–1614), alderman of Murcia and deputy in the court, offered a new translation of the text, which he included in his book entitled *Esfera del Universo*, which provided a summary of the explanations he offered during his stay in the court (Rocamora y Torrano 1599). It was published in Madrid by Juan de Herrera (fl. 1599–1614). We know of this printer's activity in the capital between 1599 and 1614. His output was extremely limited and was dominated by religious topics (five editions listed in *Iberian Books*); however, he also contributed to Jurisprudence (one edition), Literature (one edition), Military handbooks (one edition), and obviously to Astrology and cosmography (two editions). Yet again, the printing of scientific texts is limited—restricted in this case to the edition of Sacrobosco's texts.

---

[9] For a description of this edition, see (Crowther 2020, 177–179).

The editions reviewed up to this point were published, as explained by Gómez Martínez, in relevant cities at the time,

> due to having institutions dedicated to the teaching of astronomy and cosmography: in Salamanca there was the university, in Seville the Casa de Contratación, functioning since 1503, and in Madrid the Academia de Matemáticas, inaugurated in 1582. Furthermore, we know that in these educational institutions, not only were astronomy classes given, but the *Sphaera Mundi* was also used. For example, in the University of Salamanca, the subject of astrology, which was the main subject in the arts faculty of the university, included the teaching of both theoretical and practical astronomy, as did mathematics and geography over a three-year period, according to the charters of the University of Salamanca dating from both 1529 and 1538. However, the texts to be read by the professor are not specified. (Gómez Martínez 2006, 203)

However, we do have the university statutes of *Covarrubias*, dating from 1561, which offers a full list of the texts read on the subject of astronomy at the University of Salamanca during the sixteenth century. If we focus our attention only on those used for the teaching of astronomy, it can be stated that at least in 1577, 1580, 1588, and 1592 the *Sphaera* was read. Likewise, we know that Sacrobosco's text was used in the training of navigators and cosmographers in the Casa de Contratación in Seville (Gómez Martínez 2006, 204).

A good example in this regard is the edition by Francisco Faleiro (d. ca. 1574) from Portugal, published by Juan Cromberger (fl. 1502–1541) in 1535, in which part of the text is translated (Faleiro 1535).[10] It is known that Faleiro arrived in Spain to join Ferdinand Magellan's (1480–1521) expedition, but he ended up staying in Seville to serve the Crown of Castile as a cosmographer in the Casa de Contratación. The significance of his edition lies in the fact that it is the first in a series of tracts on cosmography published in connection with the Casa de Contratación. It is of interest to note an idea presented by Faleiro in the prologue to his translation: he points out that his work is eminently practical and is aimed at navigators who, perhaps due to the fact that they did not have a high level of education, needed a translation in a Romance language, as they did not understand Latin.

Juan Cromberger was one of the most important printers in the Iberian Peninsula during the sixteenth century, both in terms of the span of his career and his great publishing output (205 editions listed in the *Iberian Books* database). The first mention of his activity was in 1525 and the last in 1540. During this period, he printed editions aimed at a very wide audience. His production too was led by books linked to religion (eighty-two editions under Religious and three editions under Bibles); these are followed by literature (forty-seven editions under Literature, five editions under Poetry, and four editions under Drama), an extraordinarily important area of his production. He also dealt with works that were perhaps intended for the educational sphere (sixteen editions under Histories, nine editions under Classical authors, eight editions under Philosophy and morality, four editions under Adages/proverbs,

---

[10] For a description of this edition, see (Crowther 2020, 168–171).

two editions under Educational, two editions under Dialectics and rhetoric, and one edition under Music); he was also interested in science (seven editions under Medical, three editions under Astrology and Cosmography, two editions under Sciences) and, finally, Ordinances/Edicts (seven editions) and Culinary Arts (one edition).

A second example of the same circumstance is offered by the translation of Sacrobosco's text included with the text on the art of navigation written by Martín Cortés (1510–1582) and printed in Seville in 1551, then reprinted in 1556 (*Breve compendio de la esfera y de la arte de navegar*) by Antonio Álvarez (fl. 1544–1556) (Cortés 1551, 1556).[11] The latter was active as a printer from 1544 to 1556, during which time he mainly printed religious texts (eight editions listed in *Iberian Books*); he was however also interested in other topics, such as Astrology and Cosmography (two editions), News (one edition), Literature (one edition), Histories (one edition), Poetry (one edition), and Travel (one edition).

In the seventeenth century, we are only aware of one complete translation of Sacrobosco's work. It was made by Luis de Miranda (1600–1650) and published in Salamanca in 1629 (Sacrobosco et al. 1629) by Jacinto Taberniel (fl. 1628–1640).[12] Taberniel's output was mainly dedicated to religious books (sixteen editions listed under Religious and two editions under Bibles) and legal texts (eight editions under Jurisprudence and one edition under Ordinances/Edicts). Once again, the publication of an edition relating to cosmography (the *Sphaera*) is an isolated case (under Astrology and cosmography). Other subjects he was interested in were News (two editions), Histories (two editions), Panegyric (two editions), Educational (one edition), Military handbooks (one edition), Ordinances/Edicts (one edition), and Poetry (one edition).

Cosme Gómez de Tejada's seventeenth-century edition *El filosofo ocupacion de nobles y discretos contra la cortesana ociosidad sobre los libros de cielo y mundo, meteoros, parnos naturales, ethica, economica, politica de Aristoteles y esfera de sacro Bosco* should also be mentioned here (Sacrobosco et al. 1650). It was published in Madrid by Domingo García Morrás (fl. 1646–1699) in 1650 and includes an almost complete translation of Sacrobosco's text. García Morrás was one of the most prolific printers in the city of Madrid in the seventeenth century (332 editions listed in the *Iberian Books* database). He was active from 1643 to 1699, during which time he printed texts on a wide range of subjects. First of all, we must highlight religious works (141 editions, plus six editions under Funeral orations and one edition under Bibles are listed in *Iberian Books*) and other law-related matters (thirty-three editions under Ordinances/Edicts and twenty-eight under Jurisprudence); both subjects were the most successful at the time. Secondly, we must draw attention to the publication of news, imprints of great importance and easy to publish because of their customary

---

[11] For a description of these editions, see (Crowther 2020, 175–177).

[12] It can be noted that the various translations of the book progressively added additional explanations with a purely educational aim. In this regard, Luis de Miranda's edition is important as it significantly adds to and broadens the definitions of the terms included in Sacrobosco's text in instances when he considered a term particularly complicated. There is also a table of the cited terms (Gómez Martínez 2012, 97–98). In relation to the same topic, see (Gómez Martínez 2013, 39–58).

brevity (forty-nine editions under News). We cannot forget, however, that he also stood out for his publication of literary and historical texts (fourteen editions under Drama, nineteen editions under Histories, and seven editions under Literature listed in *Iberian Books*). He also dealt with texts intended, perhaps, for the educational field (three editions under Classical authors, two editions under Educational, and one edition under Dialectics and rhetoric). Texts belonging to the field of science are not particularly significant (three editions under Medical and one edition under Agriculture/Veterinary); neither are some occasional editions belonging to other areas that he published on an ad hoc basis. According to the *Iberian Books* database, these subjects are Memorial (ten editions), Adages/proverbs (three editions), Heraldic (three editions), Government (three editions), Military handbooks (two editions), Political tracts (two editions), Calendars and almanacs (one edition), and Economics (one edition).

3 On the Circulation of Sacrobosco's Text in the Iberian Peninsula via Inventories of Bookshops and Libraries (1472–1650)

Apart from the success enjoyed by Sacrobosco's text in the Iberian printing industry, which we are aware of through the printed editions we have reviewed, it is also necessary to make use of other bibliographic tools that enable us to examine in more depth the circulation, reading, and possession of the text in Spain, distinguishing, where possible, between different editions. In order to do so, we can first make use of the inventories of bookshops of the period. This is an area of research that, though it has been looked into (Dadson 1998), still requires a research project to organize the material in a systematic way, given that the information available is fragmented, therefore leading to fragmented conclusions. No mention has been found of Sacrobosco's text in bookshops prior to 1571. The first reference we can cite is the 1571 inventory of Martín de Salvatierra's bookshop, located in the city of Granada, reproduced in the 2001 study by Osorio Pérez, Moreno Trujillo, and Obra Sierra on the city's bookshops in the sixteenth century. Five copies of a Latin edition were found, along with five more copies of an edition in Castilian and one other edition of which no more details are offered (Pérez et al. 2001, 271–272). Secondly, we can cite the example of the bookseller Francisco García, the 1583 inventory of whose bookshop included a copy of an edition of "Sfera de Sacrobusto con comento" (Pérez et al. 2001, 362). The third example is that of the bookshop of the French bookseller and humanist Guillaume Rouillé (1518–1589) in Medina del Campo.[13] Here, there

---

[13] For the inventory of the bookseller Guillaume Rouillé as well as for the inventories of the private libraries of Pedro Díez Barruelo, Sebastián de Salinas, Pedro Enríquez, León de Castro, Antonio de Hormaza, Pedro Simón Abril, Mateo de Vargas, Lorenzo Ramírez de Prado, Pedro Gutiérrez Ramírez, Juan López de Fuentesdaño, Francisco López, Juan Flores Torrecilla, and Cristóbal Salas de León, it is not possible to specify exact bibliographic information. Such inventories were

are three generic mentions of the work, which could refer to three different editions in octavo. There were forty-two copies of the first, one of the second, and four of the third: "42 sphera de sacrobosco 8° 5 r; 1 sphera de sacrobosco 8° en pergamino 5 r;[14] 4 sphera de sacro bosco 8° 5 r."

The final bookshop we can cite is that of Francisco de Aguilar (d. 1582) in Seville, whose inventory has been transcribed and studied by González Sánchez and Maillard Álvarez (2003, 92). In this shop, scientific imprints occupied nine percent of the total, with a total number of 389 editions recorded. Among them, medical editions stand out, a key discipline in the science of the early modern period. The remaining editions in the scientific-technical section consist of a set of works by significant authors, among which Sacrobosco's text is mentioned with one copy being cited in a Spanish edition and four possibly in Latin: "una sfera de Sacrobosco en romance" (González Sánchez and Maillard Álvarez 2003, 177) and "quatro esfera de Sacrobosco" (González Sánchez and Maillard Álvarez 2003, 182). It is not known which specific editions are referred to here, and given the date of the inventory the edition in Castilian could refer to up to nine different editions. It should be recalled, however, that by that time four editions had been published in Seville. Perhaps one of these was the edition on sale. Furthermore, it must not be forgotten that Francisco Aguilar's bookshop was located on Calle Génova, the same street as Juan de León's print shop, which had printed two editions, in 1545 and 1548. However, we must bear in mind that Aguilar did not only buy stock in Castile (with his connections with printers in Medina and Salamanca); to some extent, he also depended on foreign suppliers, mainly from the Netherlands (González Sánchez and Maillard Álvarez 2003, 43), whence it is possible that these four *Sphaerae* came (again, no information is given about their language). Aguilar also had connections in Portugal, more specifically in Lisbon.

Apart from the bookshop inventories, which merely serve to confirm the sale of Sacrobosco's text (at least in the sixteenth century), and the coexistence of editions in Latin and in Spanish, it is also of interest to review the inventories of private collections, which can offer an idea of the kind of people who owned, and perhaps read, the text. Sacrobosco's text is mentioned in inventories of private libraries belonging to members of the nobility and professionals, among whom can be found medical practitioners, university lecturers, humanists in close proximity to the field of education in the sixteenth and seventeenth centuries, and occasionally other members of the administration of the crown, such as members of the Council of the Inquisition.

---

published online by the historian Anastasio Rojo Vega, who passed away in 2017 (https://invest igadoresrb.patrimonionacional.es. Accessed June 8, 2021). Rojo Vega published hundreds of such inventories in transcribed form, often together with the electronic copies of manuscript folios from which he was retrieving the information. Unfortunately, however, none of these historical documents is accompanied by the bibliographic metadata and no one was apparently able to complete his work. We consider the website, hosted by the Real Biblioteca of Madrid, as well as its content to be trustworthy.

[14] It is possible that this is a manuscript and not a printed edition. The presence of manuscripts in individual collections, or in this case a bookshop, may indicate a collection built on the sedimentation of previous collections.

We shall begin by looking at the private collections of noblemen of different kinds. First, we will mention the inventory of Don Alonso de Osorio Seventh Marquis of Astorga (d. 1582), dating from 1573, in which mention is made of "La espera de Juan Sacrobosco, traduçida de latín en castellano por Rodrigo Sanz de Santaparra (sic) y Espinosa," which he most probably kept in his palace in Valladolid or in his fortress in Astorga (Cátedra 2002, 333). This is the first explicit mention of a specific edition beyond invocation of its language, as we saw in the inventories of the bookshops.

The inventory of the Count of Gondomar, Diego Acuña y Sarmiento (1567–1626) (Manso Porto 1996), dating from 1626, includes mention of some editions and commentaries of Sacrobosco's book published outside the Iberian Peninsula (Sacrobosco et al. 1490; Sacrobosco 1538, 1546, 1551; Sacrobosco and Beyer 1552; Sacrobosco and Clavius 1585), though it also mentions those published in Alcalá de Henares in Latin in 1526 (Sacrobosco et al. 1526), one from Seville published in Spanish in 1545 (Sacrobosco and Chaves 1545), and one from Valladolid from 1568 (Sacrobosco and Santayana y Espinosa 1568). One octavo edition in Spanish, though it is not known which one, is mentioned in the inventory of the seventh Duke of Medinaceli, Antonio Juan Luis de la Cerda (1607–1671).

Without a doubt, the most significant book collection in this group is that of the Torre Alta del Alcázar of King Philip IV (1605–1665), the reconstruction of which we have thanks to Fernando Bouza Álvarez (Bouza Álvarez 2005) by way of the inventory carried out in 1637 by Francisco de Rioja (1583–1659), the King's librarian. The inventory is divided by subjects that the king had the duty to know in order to rule well. One such area of knowledge was the *Sphaera*, which deals with "the celestial and, particularly, with its applications in navigation, although its twenty-four entries for twenty-six bodies bear a close relationship with the cosmographic works" (Bouza Álvarez 2005, 110). Sacrobosco's *Sphaera* "appears alone…and in the adaptation by Jerónimo de Chaves…allowing its influence to be felt in the Hispano-Portuguese tracts, though always with a practical bias, by Pedro Nunes…, Martín Cortés…, Ginés de Rocamora…." (Bouza Álvarez 2005, 110). An edition in Castilian is mentioned— perhaps the one from Valladolid (1568)—along with a copy of Juan de León's edition (1545), another from Lisbon (1537), one from Seville (Antonio Álvarez), and one from the 1599 edition from Madrid (Bouza Álvarez 2005, 110). In other words, this book collection held copies of the editions of the 1545 and 1551 Seville editions, the 1537 Lisbon edition, and the 1599 Madrid edition (Sacrobosco and Nunes 1537; Sacrobosco and Chaves 1545; Cortés 1551; Rocamora y Torrano 1599).

It should be pointed out that, unlike the inventories of the libraries of the nobility, it is not possible to find detailed references on specific editions in professional libraries. First, we shall look at libraries belonging to university professors and lecturers. Chronologically, the first mention of a copy of an edition of Sacrobosco's text, most likely in Spanish, is found in the 1558 inventory of books belonging to Pedro Díez Barruelo, professor of logic at the University of Valladolid: "esphera de sacrobusto yn folio en quatro reales." In a later inventory—of 1572—of the professor of grammar and rhetoric of the same university, Sebastián de Salinas, one edition with commentaries by Ciruelo and another in Castilian are mentioned: "491. Sphera de sanobosto cum comento Ciruelo; 511. Sphera en romanze en 4." Two editions are also mentioned in the 1584 inventory of the book collection of Pedro Enríquez, a professor from Valladolid. One is the edition published in Seville in 1551 (Cortés 1551) and the other is probably a Spanish edition: "583. brebe compendio de la sphera y arte de navegar fo.; 594. sphera de sacrobosco." Last of all, the 1585 inventory of a professor from the University of Salamanca, León de Castro (1505–1585), mentions an edition of the *Sphaera* with comments and another with Greek and Latin annotations: "203. sphera con comento; 254. sphera con anotationes greçe et latine."

Sacrobosco's text also sat on the shelves of the libraries of illustrious humanists. Antonio de Hormaza, humanist and archdean of the Bierzo, owned (according to his 1575 will) a Castilian edition of the text: "455. esphera en romance." Another is mentioned in the inventory of Pedro Simón Abril's (1530–1595) book collection, dating from 1595. In the seventeenth century, there is mention of an edition in Castilian in Mateo de Vargas's 1623 book collection: "58. esfera del mundo en romance seis reales." Finally, a copy of a commentary of Sacrobosco's text has been found in the postmortem 1662 inventory of the book collection of Lorenzo Ramírez de Prado (1583–1658), a Spanish humanist and bibliophile (Entrambasaguas 1954). More specifically, it is the *Commentaria in Sphaeram Sacro Bosco* written by Francisci Iunctini and published in two volumes in Lyon in 1577 (Sacrobosco and Giuntini 1577a, b) (Chaps. 6 and 8). It is also necessary to add the reference of Joaquín Entrambasaguas to the same book collection in which a copy of the edition from 1545 was preserved (Entrambasaguas 1954, Vol. I, 92): "La Esphera de Sacrobosco en romance por Gerónimo Chaves con el Alonso de Fuentes Philosophía—natural en Romance, Madrid 1545."

This group is followed by the libraries of various members of the administration of the Crown of Castile, for example Pedro Gutiérrez Ramírez, a supplier of royal works, in whose 1617 inventory an edition in Castilian is mentioned ("113. otro libro intitulado esfera de sacrobera"). We can also mention the 1631 book collection of the inquisitor of Valladolid, Juan López de Fuentesdaño, which held a copy of the 1567 edition from that city (Sacrobosco and Santayana y Espinosa 1567).

Last of all, we can examine inventories of the libraries belonging to medical practitioners and surgeons. Those of Francisco López (1557), Juan Flores Torrecilla (1590), and Cristóbal Salas de León (1616) mention, with no further details, a copy of an edition of Sacrobosco's text, probably in Spanish: "13. Esfera; 92. otro libro esfera de sacriobosco; 188. sfera de sacrobosco." The latter, however, quotes an edition published in Rome in 1581: "403. iten una sphera de sacrobosco 4 perg roma año 1581" (Sacrobosco and Clavius 1581).

The few inventories that offer us data specific to the editions present in these libraries allow us to surmise that the two most popular editions were those by Antonio Álvarez, published in Seville in 1551 (Cortés 1551), and the edition printed in Valladolid by Adrian Ghemart in 1567 (Sacrobosco and Santayana y Espinosa 1567). It does not appear to be a coincidence that the first one was reprinted in 1556 (Cortés 1556), while there is a reissue of the second published in 1568 (Sacrobosco and Santayana y Espinosa 1568). These are followed by the 1537 Lisbon, 1545 Seville, and 1599 Madrid editions (Sacrobosco and Nunes 1537; Sacrobosco and Chaves 1545; Rocamora y Torrano 1599).

4   On the Circulation of the *Tractatus de sphaera* in America (1472–1650)

Not a great deal of information is available relating to the publication and dissemination of Sacrobosco's text in America. One important work on this topic is Pedro Rueda's monographic study of the book trade between the Iberian Peninsula and America (Rueda 2005). In this regard, Seville's monopoly on human trafficking and trade with the New World must not be forgotten. This situation "supposed an exceptional circumstance which was felt throughout the whole Ancien Régime, thanks, in part, to its geographically strategic position and a wide experience and tradition in long-distance trading" (González Sánchez and Maillard Álvarez 2003, 18). In 1503, the Catholic monarchs established the only (and obligatory) means of navigation and trade with America in Seville—the Casa de Contratación. From that moment on, the city underwent a deep transformation in every regard (González Sánchez and Maillard Álvarez 2003, 19). In 1550, faced with the advance of Protestant doctrine and in an attempt to block its penetration into the New World, King Charles V (1550–1558) gave the order to the officials of the Casa de Contratación in Seville that "when you have to take permitted books to the Indies, register them one by one, declaring the subject matter of each book, and do not register them wholesale" (González Sánchez and Maillard Álvarez 2003, 23). Therefore, merchants sending or taking books to the Indies were obliged to present a written and signed declaration with the specific number and title of the books to the officials of the Casa de Contratación who controlled intercontinental trading (González Sánchez and Maillard Álvarez 2003, 24). This strategy continued under later kings, which makes it possible for us to study, at least in part, the book trade between Seville and America. A significant part of Rueda's study mentioned above deals with the genre of books shipped to America and who sent them from Seville. Chapter XII deals specifically with practical and scientific literature (Rueda 2005, 411). With regard to this type of book, Rueda states that "the observations of the celestial phenomena of astronomy, which focus on the movement of the stars and planetary theories, have a lesser presence in the lists; it is easier to find works such as nautical books which make a practical use of the knowledge and techniques of this discipline" (Rueda 2005, 426–427). Rueda only located two shipments by the booksellers Ana Vernagli and Nicolás Antonio, in 1603 and 1609, respectively, which included two editions of Clavius's version of the *Sphaera* published in Rome in 1570 (Sacrobosco and Clavius 1570) and in Venice in 1601 (Chaps. 6 and 11).[15] This enables us to confirm that, apart from the editions and translations of the works published in the Peninsula, there were also editions circulating in its overseas territories that came from international circuits, with Seville serving as a distributor of these editions to the American market. However, Ginés Rocamora's Spanish translation, published in Madrid in 1599 by Juan de Herrera, was also sent from Seville to Mexico (Rocamora y Torrano 1599). We know that this edition was

---

[15] Two printings of the same edition of Clavius's commentary on *De sphaera* were put on the market in Venice in 1601 (Chap. 6): one by Isabetta Basa (Sacrobosco and Clavius 1601b) and one by Giovanni Battista Ciotti (Sacrobosco and Clavius 1601a).

sent in 1604 by the bookseller Fernando Mexía and in 1605 by the merchant Diego Correa. There is also evidence from 1640 indicating that Duarte Álvarez de Osorio took "2 Esfera de Rocamora" (Rueda 2005, 427, n. 44). It is therefore evident that the 1599 edition was still in circulation in Seville in 1640.

As these shipments show, Sacrobosco's *Sphaera* was known, and perhaps also read, in America. However, we have no proof that new editions were printed in the New World apart from the edition entitled *De sphera* by the Italian mathematician Francesco Maurolico (1494–1575). Some authors have pointed out that although universities were established early in colonial America during the sixteenth century, certain teaching, such as in the faculty of astronomy, did not begin to function until the seventeenth century (Chang-Rodriguez 2002, 16). It would not be until this century that a new current would be felt in mathematical and astronomical studies, although in a very timid way (Mazin 2008, 71). On the other hand, one should not lose sight, in considering the teaching of the trivium and the quadrivium in New Spain, of the fact that both parts of the liberal arts had been well established in the cathedral schools and in the European universities since the late Middle Ages; however, "to think that these teachings could be taught completely in the schools of the mendicant friars is almost utopian" (Cuesta Hernández 2018, 108). We may imagine that the subject matter of Sacrobosco's treatise was less than prominent in the educational contexts of America, which meant that the work was less than successful at its printing presses.

In any case, we must not lose sight of Maurolico's work, mentioned earlier, which was printed in Mexico by Antonio Ricardo (d. 1606) in 1578 (Maurolico 1578). It was financed, according to the colophon, by "Petri Nunnesij a prado" and published at the request of the Italian Jesuit Vicenzo Le Noci: "Rogatu R. P. Vincêtij Nutij societates Iesu, and Rectoris D. Petri and Pauli Collegialum." The edition is an adaption of Sacrobosco's treatise; this means that it is a work that significantly resembles to Sacrobosco's treatise in terms of content and structure, but does not include the original text. In this regard, Antonella Romano's work on the first scientific books published in New Spain is of great interest, as it considers the reasons why Sacrobosco's original, or annotated original, text was not published in America, with Maurolico's 1578 edition being preferred instead. In her research, Romano demonstrates that the publication of this text was carried out in the first decades of educational advances by the Jesuits in New Spain, thanks to the impetus of the Jesuit lecturer Vincenzo Le Noci. He was sent to Mexico in 1574 and had trained in Messina—the hometown Maurolico. It seems plausible, then, that Le Noci was a key factor in the printing of Maurolico's text in Mexico (Romano 2005, 115–116).

5   Conclusions

At the beginning of this study, we posed a series of questions regarding the circulation of Sacrobosco's text in the Iberian Peninsula, based on which it is possible to draw certain general conclusions.

First of all, it is interesting to note the peninsular preference for the vernacular languages when printing Sacrobosco's text. This circumstance can, perhaps, be attributed to the advancement of humanism in various European nations. However, we must not ignore the fact that printers offered editions in vernacular languages for those collectors and readers who may have needed them at least in part because the demand for Latin editions was covered by editions published in the Peninsula and in other countries with circulation in the Peninsula, as we have seen in reviewing the inventories of bookshops and libraries. Furthermore, the aesthetic influence of some of these foreign editions on those published in the Peninsula must be considered; foreign editions must have reached the hands of Spanish printers in some way. The connection between Miguel de Eguía's edition and that of Simon Vostre from Paris should be kept in mind.

It is also of interest to consider the main centers of the printing industry, Salamanca and Seville, and their respective connections with the university and the Casa de Contratación—the former a scholarly setting and the latter an administrative center. This provides us with a sense of the readership and collectors of Sacrobosco's text: students and members of the university administration, mainly navigators and cosmographers. As we have stated in this study, it was an obligatory text both for students and for navigators and cosmographers involved in the expeditions to America. In general, all of the editions studied were linked with these characteristics. We can recall here the examples of the Portuguese printer Germao Galherde, linked to the University of Coimbra, and Juan de León, who was the printer of the University of Osuna, although León also printed the translation made by the professor of the art of navigation and cosmography of the Casa de Contratación in Seville. We must bear in mind the debate between cosmographers and pilots in the central years of the century on the relative importance of theory versus experience—a debate in which cosmographers were important defenders of theory, of general rules, of universal truths, and of the systematic knowledge documented in the treaties of the period. This circumstance shortens, perhaps, the distance between two of the mentioned large groups of receptors of the work of Sacrobosco and the distance between students and cosmographers was perhaps not so significant after all. Maybe these two groups did not constitute two separate sets of receivers but a single one, if we take into consideration their interest in the theory expressed in treatises such as that of Sacrobosco.[16]

The dates of publication of the text in the Iberian Peninsula must be highlighted, given that, although the text appears to have had an extraordinary degree of success in the sixteenth century, only two editions are preserved from the seventeenth. It is possible that it was supplanted by other texts, which resulted in a lack of public interest. It should also not be ruled out that perhaps the market was saturated and the number of existing editions satisfied what academic interest might still have existed. Unfortunately, we have no reliable evidence of either circumstance. On the other hand, during the seventeenth century the Iberian Peninsula entered a period of

---

[16] We cannot forget that the Casa de Contratación in Seville enjoyed, like other educational centers of the time, an intellectual atmosphere typical of the humanist period. For more information, see (Gulizia 2016, 131).

greater decadence, during which the era of great international discoveries was left behind.

We find ourselves faced with a group of printers who were not specialized in the publication of scientific books—books that, as we have mentioned, did not achieve great success in the Iberian Peninsula. However, in many cases the printers were familiar with the printing of academic texts of different kinds, such as grammar references, historical books, scientific texts, and literary texts by classical authors. The publication of Sacrobosco's text must be considered within this context.

It is important to take into account the mixed profile of some of the printers who published Sacrobosco's text. All of them were printers, but some also worked as publishers or booksellers and had access to a much broader commercial network. Some—such as the case of Juan Junta, who was part of a very large European family network—had international connections, while others had businesses in different cities, thus enabling them to sell in more places. Such was the case of Miguel de Eguía, who had businesses in Alcalá de Henares, Logroño, Toledo, and Burgos; Juan Junta, who had a print shop and traded books in Burgos and in Salamanca; and Adrian Ghemart, who worked in Valladolid and in Medina del Campo, where there was an important book fair. This event gathered and distributed books printed in Lyon, Paris, Antwerp, and Venice, as well as those published in different places in the Iberian Peninsula.

We should remember that, in some cases, Sacrobosco's text must have been one from which its printers made a profit, as it was a successful book. In this regard, we must recall the editions that were reprinted within a short space of time. The most striking is, without a doubt, that of Adrian Ghemart, published in Valladolid in 1567 and reissued only a year later. Also worthy of note is Antonio Álvarez's edition, printed in Seville in 1551 and reprinted in the same city in 1556.

The examination of certain inventories of Iberian bookshops in the sixteenth century has enabled us to observe that, although editions in Romance languages began to be published quite early, they did not substitute the Latin editions. Rather, they appear to have coexisted on the shelves of these establishments. Perhaps of more interest are the inventories of the private libraries of the sixteenth and seventeenth centuries. Their examination allows us to broaden the portrait of the readership and collectors of the text, given that copies were found in the libraries of the nobility and of professionals, such as university lecturers and humanists in the field of education, medical practitioners, and other members of the crown's administration. The sparse clear data available regarding the specific editions found in the inventories indicate that the most successful editions were those by Adrian Ghemart and Antonio Álvarez—the editions that were rapidly reissued and reprinted as mentioned above. Furthermore, it is possible to highlight, thanks to the libraries of the nobility, the circulation in the Peninsula of editions of Sacrobosco's text printed in other parts of Europe.

Finally, it is of interest to highlight that it seems that the original, or annotated original, text was not printed in America; however, we do have an adaption of Sacrobosco's book published in Mexico in the sixteenth century thanks to the intervention of the Italian Jesuit Vincenzo Le Noci. It is interesting to point out that it was Italy and not the Iberian Peninsula that promoted the publication of Sacrobosco's work in America, even though, as we have shown, the work traveled from the Peninsula to America on several occasions. We do know that Seville was the point of departure for editions published in the Peninsula, specifically the 1599 Madrid edition. Likewise, Seville also launched the shipment of foreign editions of the text, specifically those published in Rome (1570) and Venice (1601), which were sent to the New World by two Spanish booksellers.

This information highlights, yet again, the significance of editions imported for sale in the Iberian Peninsula from important European cities such as Lyon, Paris, Venice, and Antwerp. Thanks to its links with America, the trade from Iberia became a flourishing market that attracted the interest of both local and foreign printers. This is evident in the shipments through which Sacrobosco's *Sphaera* reached the New World in both Latin and Spanish.

Abbreviations

Digital Repositories

Iberian Books | Wilkinson, Alexander S., Ulla Lorenzo, Alejandra, Cruz Redondo, Alba de la, eds. Dublin: University College, Dublin. Library https://iberian.ucd.ie/index.php. Accessed 07 June 2021.

Sphaera Corpus*Tracer* | Max Planck Institute for the History of Science. https://db.sphaera.mpiwg-berlin.mpg.de/resource/Start. Accessed 07 June 2021.

References

Primary Literature

Aristotle and Pedro de Espinosa. 1535. *Philosophia naturalis Petri a Spinosa artium magistri: opus inquam tripartitum: quo continet tres partes. Prima pars erit Emporium refertissimum bone philosophie: currens per omnes textus philosophi cum aptis questionibus ibidemque proprijs. Secunda pars erit Calculatoria: quam appello Roseam. Tertia pars erit Flos campi, Lilium agri, continens omnes naturales questiones ordine alphabetico. Nil optabis quod hec philosophia non clare tibi ostendat: si textum ibidem habes expositionem lucidissimam. Si questiones ad idem. Si calculationes, habes eas in secunda parte. Si denique probleumata, habes omnia ordine alphabetico: quo sit tibi minor labor inveniendi quod vellis.* Salamanca: Rodrigo de Castañeda. https://hdl.handle.net/21.11103/sphaera.101007. *Iberian Books*: http://n2t.net/ark:/87925/drs1.iberian.17454.

Avelar, André do. 1593. *Sphaerae utriusque tabella, ad Sphaerae huius mundi faciliorem enucleationem. Autore Andrea D'Avellar Olysiponensi, Artium, ac Philosophiae Magistro, and publico in Conimbricensi Academia Mathematum professore.* Coimbra: Antonio de Barreira. https:// hdl.handle.net/21.11103/sphaera.100534. *Iberian Books:* http://n2t.net/ark:/87925/drs1.iberian. 13578.

Brocar, Arnoldo Guillén de, Cisneros, Francisco Jiménez de. 1514. *Libri veteris et novi Testamenti multiplici lingua impressi.* Alcala de Henares: Arnaldo Guillén de Brocar.

Cortés, Martin. 1551. *Breve compendio de la sphera y de la arte de navegar, con nuevos instrumentos y reglas, exemplificado con muy subtiles demonstraciones: compuesto por Martin Cortes natural de burjalaroz en el reyno de Aragon y de presente vezino de la ciudad de Cadiz: dirigido al invictissimo Monarcha Carlo Quinto Rey de las Hespañas etc. Señor Nuestro.* Seville: António Alvares. https://hdl.handle.net/21.11103/sphaera.101044. *Iberian Books:* http://n2t.net/ ark:/87925/drs1.iberian.11223.

Cortés, Martin. 1556. *Breve compendio de la sphera y de la arte de navegar, con nueuos instrumentos y reglas, exemplificado con muy subtiles demonstraciones: compuesto por Martin Cortes natural de burjalaros en el reyno de Aragon y de presente vezino de la ciudad de Cadiz: dirigido al invictissimo Monarcha Carlo Quinto Rey de las Hespanas etc. Senor Nuestro.* Seville: António Alvares. https://hdl.handle.net/21.11103/sphaera.101394. *Iberian Books:* http://n2t.net/ ark:/87925/drs1.iberian.4190.

Faleiro, Francisco. 1535. *Tratado del Esphera y del arte del marear: con el regimiento de las alturas: con algunas reglas nuevamente escritas muy necessarias.* Seville: Juan Cromberger. https:// hdl.handle.net/21.11103/sphaera.101182. *Iberian Books:* http://n2t.net/ark:/87925/drs1.iberian. 4614.

Maurolico, Francesco. 1578. *Reverendi Do. Francisci Maurolyci, Abbatis Messanensis, atque mathematici celeberrimi. De sphaera. Liber unus.* Mexico City: Antonio Ricardo. https://hdl.handle. net/21.11103/sphaera.101292.

Nebrija, Antonio de. [1498]. *Introductorium in cosmographiam Pomponii Melae.* [Salamanca]: s.n.

Rocamora y Torrano, Ginés. 1599. *Sphera del universo. Por Don Gines Rocamora y Torrano, Regidor de la Ciudad de Murcia, y Procurador de Cortes por ella, y su Reyno. Dirigida a Don Luis Faxardo, Marques de los Velez y de Molina, Adelantado mayor y Capitan general del Reyno de Murcia, y Marquesado de Villena, etc.* Madrid: Juan de Herrera. https://hdl.handle.net/21. 11103/sphaera.100672. *Iberian Books:* http://n2t.net/ark:/87925/drs1.iberian.18222.

Sacrobosco, Johannes de. [1510–1512]. *Tractado da Spera do mundo tirada de latim em liguoagem portugues Com huma carta que hunum gramde doutor Alemam mandou a eli'Rey de Portugall dom Ioum ho segundo.* Lisbon: German Gaillard. https://hdl.handle.net/21.11103/sphaera. 100048.

Sacrobosco, Johannes de. 1538. *Ioannis de Sacro Bosco Sphaera mundialis, summa diligentia nuper correcta atque emendata, adiectis insuper figuris nonnullis, atque annotatiunculis marginibus adpositis.* Paris: Regnault Chaudière. https://hdl.handle.net/21.11103/sphaera.101024.

Sacrobosco, Johannes de. 1546. *La sphere de Iehan de Sacrobosco, traduicte de Latin en langue Francoyse, augmentée de nouvelles figures: avec une Preface contenant arguments evidents par lesquels est prouvée l'utilité d'Astrologie, et qu'i celle ne doibt estre mesprisee de l'homme Chrestien.* Paris: Jean Loys. https://hdl.handle.net/21.11103/sphaera.101030.

Sacrobosco, Johannes de. 1551. *Sphaera Ioannis de Sacrobosco.* Antwerp: Jean Richard. https:// hdl.handle.net/21.11103/sphaera.100161.

Sacrobosco, Johannes de and Hartmann Beyer. 1552. *Quaestiones in libellum De sphaera Ioannis de Sacro Busto, in gratiam studiosae iuuentutis collectae ab Hartmanno Beyer and nunc denuo recognitae.* Frankfurt am Main: Peter Braubach. https://hdl.handle.net/21.11103/sphaera.100150. *Iberian Books:* http://n2t.net/ark:/87925/drs1.iberian.19742.

Sacrobosco, Johannes de and Jerónimo de Chaves. 1545. *Sphera del mundo. Tractado de la sphera que compuso el doctor Ioannes de Sacrobusto con muchas adiciones, Agora nuevamente traduzido de Latin en lengua Castellana por el Bachiller Hieronymo de Chaves: el qual annidio muchas figuras tablas, y claras demonstrationes: juntamente con unos breves Scholios, necessarios á mayor illucidation, ornato y perfection del dicho tractado*. Seville: Juan de León. https://hdl.han dle.net/21.11103/sphaera.101051. *Iberian Books*: http://n2t.net/ark:/87925/drs1.iberian.7763.

Sacrobosco, Johannes de and Jerónimo de Chaves. 1548. *Tractado de la Sphera que compuso el doctor Ioannes de Sacrobusto con muchas additiones. Agora nuevamente traduzido del Latin en lengua Castellana por el Bachiller Hieronymo de Chaves: el qual añidio muchas figuras, tablas, y claras demostrationes: junctamente con unos breves Scholios, necessarios á mayor illucidation, ornato y perfectio del dicho tractado*. Trans. Jerónimo de Chaves. Seville: Juan de Léon. https:// hdl.handle.net/21.11103/sphaera.101033. *Iberian Books*: http://n2t.net/ark:/87925/drs1.iberian. 7764.

Sacrobosco, Johannes de, Pedro Ciruelo, and Pierre d'Ailly. 1526. *Opusculum de sphera mundi Joannis de sacro busto: cum additionibus: et familiarissimo commentario Petri Ciruelli Daro- censis: nunc recenter correctis a suo autore: intersertis etiam egregijs questionibus domini Petri de Aliaco*. Alcalá de Henares: Miguel de Eguia. https://hdl.handle.net/21.11103/sphaera.100884. *Iberian Books*: http://n2t.net/ark:/87925/drs1.iberian.7761.

Sacrobosco, Johannes de and Christoph Clavius. 1570. *Christophori Clavii Bambergensis, ex Soci- etate Iesu, in Sphaeram Ioannis de Sacro Bosco commentarius*. Rome: Vittorio Eliano. https:// hdl.handle.net/21.11103/sphaera.100365.

Sacrobosco, Johannes de and Christoph Clavius. 1581. *Christophori Clavii Bambergensis ex Soci- etate Iesu in Sphaeram Ioannis de Sacro Bosco commentarius Nunc iterum ab ipso Auctore recognitus, and multis ac varijs locis locupletatus*. Rome: Domenico Basa and Francesco Zanetti. https://hdl.handle.net/21.11103/sphaera.101117.

Sacrobosco, Johannes de and Christoph Clavius. 1585. *Christophori Clavii Bambergensis ex Soci- etate Iesu in Sphaeram Ioannis de Sacro Bosco commentarius Nunc tertio ab ipso Auctore recog- nitus, and plerisque in locis locupletatus. Permissu superiorem*. Rome: Domenico Basa. https:// hdl.handle.net/21.11103/sphaera.101120.

Sacrobosco, Johannes de and Christoph Clavius. 1601a. *Christophori Clavii Bambergensis ex Soci- etate Iesu in sphaeram Ioannis de Sacro Bosco commentarius. Nunc tertio ab ipso auctore recog- nitus, and plerisque in locis locupletatus. Maiori item cura correctus*. Venice: Giovanni Battista Ciotti. https://hdl.handle.net/21.11103/sphaera.101390.

Sacrobosco, Johannes de and Christoph Clavius. 1601b. *Christophori Clavii Bambergensis ex Soci- etate Iesu In Sphæram Ioannis de Sacro Bosco Commentarius. Nunc tertio ab ipso Auctore recognitus, and plerisque in locis locupletatus. Maiori item cura correctus. Permissu Superiorum*. Venice: Isabetta Basa. https://hdl.handle.net/21.11103/sphaera.100673.

Sacrobosco, Johannes de, Christoph Clavius, Francesco Giuntini, and Élie Vinet. 1629. *Exposicion de la Esfera de Iuan de Sacrobosco Doctor Parisiense. Traduzida de Latin en lengua vulgar, augmentada y enriquecida, con lo que della dixeron Francisco Iuntino, Elias Veneto, Christoforo Clavio, y otrossus expositores, y comentadores. Por F. Luys de Miranda de la Orden de san Francisco, Lector jubilado, y Provincial que ha sido, de la Provincia de Santiago, Consultor del supremo Consejode la santa general Inquisicion. Dirigida Al serenissimo señor Cardenal Infante D Fernando Arcobispo de Toledo, y Primado de las Españas*. Trans. Luis de Miranda. Salamanca: Jacinto Taberniel. https://hdl.handle.net/21.11103/sphaera.100555. *Iberian Books*: http://n2t.net/ ark:/87925/drs1.iberian.28640.

Sacrobosco, Johannes de and Pedro Espinosa. 1550. *Sphera Ioannis de Sacro Busto cum commen- tariis Petri a Spinosa Artium Magistri, celeberrimique praeceptoris Salmanticensis gymnasij, aeditis*. Salamanca: Juan de Junta. https://hdl.handle.net/21.11103/sphaera.100199. *Iberian Books*: http://n2t.net/ark:/87925/drs1.iberian.7762.

Sacrobosco, Johannes de and Francesco Giuntini. 1577a. *Fr. Iunctini Florentini, sacrae theolo- giae doctoris, Commentaria in Sphaeram Ioannis de Sacro Bosco accuratissima. Omnia iudicio S.R. Ecclesiae submissa sunto*. Lyon: Philippe Tinghi. https://hdl.handle.net/21.11103/sphaera. 100921.

Sacrobosco, Johannes de and Francesco Giuntini. 1577b. *Fr. Iunctini Florentini, sacrae theologiae doctoris, Commentaria in tertium et quartum capitulum Sphaerae Io. de Sacro Bosco. Ad nobilem virum D. Marcum Bonavoltam Florentinum.* Lyon: Jean de Tournes II for Philippe Tinghi. https://hdl.handle.net/21.11103/sphaera.100674.

Sacrobosco, Johannes de and Pedro Nunes. 1537. *Tratado da sphera com a Theorica do Sol et da Lua. E ho primeiro livro da Geographia de Claudio Ptolomeo Alexandrino. Tirados novamente de Latim em lingoagem pello Doutor Pero Nunez Cosmographo del Rey Don João ho terceyro deste nome nosso Senhor. E acrecentatos de muitas annotaçiones et figuras per que mays facilmente se podem entender. Item dous tratados quo mesmo Doutor fez sobre a carta de marear. Em os quaes se decrarano todas as principaes du vidas da navegação. Con as tavoas do movimento do sol: et sua declinação. Eo regimento da altura assi ao meyo dia: como nos outros tempos. Com previlegio real.* Lisbon: Galharde. https://hdl.handle.net/21.11103/sphaera.101010. *Iberian Books*: http://n2t.net/ark:/87925/drs1.iberian.12878.

Sacrobosco, Johannes de, Joannes Regiomontanus, and Georg von Peurbach. 1490. *Spaerae mundi compendium foeliciter inchoat. Noviciis adolescentibus: ad astronomicam rem publicam capessendam aditum impetrantibus: pro brevi rectoque tramite a vulgari vestigio semoto: Ioannis de Sacro busto sphaericum opusculum una cum additionibus nonnullis littera A sparsim ubi intersertae sint signatis: Contraque cremonensia in planetarum theoricas delyramenta Ioannis de monte regio disputationes tam acuratiss. atque utills. Nec non Georgii purbachii in erundem motus planetarum acuratiss. theoricae: dicatum opus: utili serie contextum: fausto sidere inchoat.* Venice: Ottaviano Scoto I. https://hdl.handle.net/21.11103/sphaera.100885.

Sacrobosco, Johannes de and Rodrigo Sáenz de Santayana y Espinosa. 1567. *La sphera de Iuã de Sacro Bosco nueva y fielmente traduzida de Latin en Romance, por Rodrigo Saenz de Santayana y Spinosa. Con una Exposicion del mismo. Dirigida al Serenissimo y Excellentissimo Infante Don Iuan de Austria, Hijo del Invictissimo Cesar Carlo Quinto.* Valladolid: Adrián Ghemart. https://hdl.handle.net/21.11103/sphaera.101077. *Iberian Books*: http://n2t.net/ark:/87925/drs1.iberian.7759.

Sacrobosco, Johannes de and Rodrigo Sáenz de Santayana y Espinosa. 1568. *La Sphera de Iuan de Sacrobosco Nueva y fielmente traduzida de Latin en Romance, por Rodrigo Saenz de Santayana y Spinosa. Con una Exposicion del mismo. Dirigida al Serenißimo y Excellentißimo Principe Don Iuan de Austria, Hijo del Invictißimo Caesar Carlo Quinto.* Valladolis: Adrián Ghemart for Pedro de Corcuera. https://hdl.handle.net/21.11103/sphaera.101079.

Sacrobosco, Johannes de, Cosme Gómez Tejada de los Reyes, and Aristotle. 1650. *El filosofo. Ocupacion de nobles, y discretos contra la cortesana ociosidad. Sobre los libros de cielo, y Mundo, Meteoros, Parvos Naturales, Ethica, Economica, Politica de Aristoteles, y Esfera de Sacro Bosco. Epitome claro, y curioso. Tratanse estas materias con rigor escolastico: y dividense en dos libros, Filosofo Natural, y Filosofo Moral. Por el licenciado Cosme Gomez Texada de los Reyes, Capellan mayor de las Bernardas Descalças, y Patronazgo en S. Ilefonso de Talavera. Al licenciado Frey Martin Rodriguez de Corrales, del Abito de San Iuan, Prior de San Christoval de Salamanca, y Vicario General de dicha Sacra Religion, etc.* Madrid: Domingo Garcia Morrás for Santiago Martín Redondo. https://hdl.handle.net/21.11103/sphaera.100659. *Iberian Books*: http://n2t.net/ark:/87925/drs1.iberian.7760.

Zamora, Alfonso de. 1526. *Introductiones artis grammaticae hebraice nunc recenter edite.* Alcala de Henares: Miguel de Eguía.

Secondary Literature

Bennassar, Bartolomé. 1984. Los inventarios postmortem y la historia de las mentalidades. In *La documentación notarial y la Historia. Actas del II Coloquio de Metodología Histórica Aplicada*, ed. Antonio Eiras Roel, 139–146. Santiago de Compostela: Universidad de Santiago de Compostela.

Bonmatí Sánchez, Virginia. 1998. El Tratado de la Esfera (1250) de Juan de Sacrobosco en el Cosmographiae de Antonio de Nebrija (c.1498). *Cuadernos de filología clásica: Estudios latinos* 15: 509–513.

Bonmatí Sánchez, Virginia. 2002. La revolución científica del siglo XVI: de la "Sphaera Mundi" de Juan de Sacrobosco al "De Revolutionibus" de Nicolás Copérnico (1543). In *Humanismo y pervivencia del mundo clásico: homenaje al profesor Antonio Fontán*, eds. José María Maestre Maestre, Luis Charlo Brea, Joaquín Pascual Barea, and Antonio Fontán Pérez, 1407–1412. Madrid: CSIC.

Bouza Álvarez, Fernando. 2005. *El libro y el cetro: la biblioteca de Felipe IV en la Torre Alta del Alcázar de Madrid.* Salamanca: Instituto de Historia del Libro y de la Lectura.

Cátedra, Pedro M. 2002. *Nobleza y lectura en tiempos de Felipe II: la biblioteca de don Alonso Osorio Marqués de Astorga.* Valladolid: Consejería de Educación y Cultura.

Crowther, Kathleen M. 2020. Sacrobosco's *Sphaera* in Spain and Portugal. In *De sphaera of Johannes de Sacrobosco in the early modern period: The authors of the commentaries*, ed. Matteo Valleriani, 161–184. Cham: Springer. https://doi.org/10.1007/978-3-030-30833-9_7.

Cuesta Hernández, Javier. 2018. La educación indígena y la memoria en Nueva España en el siglo XVI. *Boletín de Antropología. Universidad de Antioquia* 33: 103–116.

Chang–Rodríguez, Raquel. 2002. *Historia de la literatura mexicana. 2. La cultura letrada en la Nueva España del siglo XVII*. México: Universidad Nacional Autónoma de México / Siglo XXI Editores.

Dadson, Trevor J. 1998. *Libros, lectores y lecturas.* Madrid: Arco/Libros.

Delgado Casado, Juan. 1996. *Diccionario de Impresores Españoles (Siglos XV–XVII)* (2 vols.). Madrid: Arco Libros.

Entrambasaguas, Joaquín de. 1954. *La biblioteca de Ramírez de Prado.* 2 Vol. Madrid: CSIC.

Fuente Arranz, Fernando. 2018. *Diccionario Biográfico Español.* Madrid: Real Academia de la Historia. http://dbe.rah.es. Accessed 07 June 2021.

Gulizia, Stefano. 2016. Printing and Instrument Making in the Early Modern Atlantic, 1520–1600. *Nuncius* 31: 129–162.

Gómez Martínez, Marta. 2006. *Sacrobosco en castellano.* Salamanca: Ediciones Universidad de Salamanca.

Gómez Martínez, Marta. 2012. Un glosario de astronomía escondido en las páginas de un manual traducido en el siglo XVII. *Quaderns de filología. Estudis lingüístics* 17: 97–110.

Gómez Martínez, Marta. 2013. Claves didácticas en un manual de astronomía: De Sphaera Mundi de Sacrobosco. *Relaciones: Estudios de historia y sociedad* 34 (135): 39–58.

González Sánchez, Carlos Alberto y Natalia Maillard Álvarez. 2003. *Orbe tipográfico. El mercado del libro en la Sevilla de la segunda mitad del siglo XVI*. Gijón: Trea.

Hurtado Torres, Antonio. 1982. La "Esfera" de Sacrobosco en la España de los siglos XVI y XVII. Difusión bibliográfica. *Cuadernos bibliográficos* 44: 49–58.

Lanuza Navarro, Tayra M. C. 2020. Pedro Sánchez Ciruelo. A commentary on Sacrobosco's Tractatus de sphaera with a defense of astrology. In *De sphaera of Johannes de Sacrobosco in the early modern period: The authors of the commentaries*, ed. Matteo Valleriani, 53–89. Cham: Springer. https://doi.org/10.1007/978-3-030-30833-9_3.

Leitão, Henrique. 2008. *A Ciência na "Aula da Esfera" no Colégio de Santo Antão, 1590–1759*. Lisboa: Comissariado Geral das comemorações do V centenario do nascimento de São Francisco Xavier.

López Piñeiro, José María and Francesc Bujosa Homar. 1981. *Los impresos científicos españoles de los siglos XV y XVI. Inventario, bibliometría y thesaurus, Volumen I: Introducción. Inventario A–C*. Valencia: Cátedra de Historia de la Medicina / Universidad de Valencia.

Mano González, Marta de la. 1998. *Mercaderes e impresores de libros en la Salamanca del siglo XVI*. Salamanca: Ediciones de la Universidad de Salamanca.

Manso Porto, Carmen. 1996. *Don Diego Sarmiento de Acuña, conde de Gondomar (1567–1626). Erudito, mecenas y bibliófilo*. Santiago de Compostela: Xunta de Galicia.

Martín Abad, Julián. 1991. *La Imprenta en Alcalá de Henares (1502–1600)* (3 vols.). Madrid: Arco/Libros.

Mazín, Óscar. (2008). Gente de saber en los virreinatos de Hispanoamérica (siglos XVI-XVIII). In *Historia de los intelectuales en América Latina*, ed. Carlos Altamirano, 53–78. Buenos Aires: Katz.

Ortiz Gómez, Teresa and Alfredo Menéndez Navarro. 2004. Sphera mundi cum commentis…: de Johannes de Sacrobosco. In *Domus sapientiae: fondos bibliográficos de la Universidad de Granada de la época de Isabel la Católica*, ed. María Amparo Moreno Trujillo, 138–141. Granada: Universidad de Granada.

Osorio Pérez, María José, María Amparo Moreno Trujillo, and José María de la Obra Sierra. 2001. *Trastiendas de la cultura: Librerías y libreros en la Granada del siglo XVI*. Granada: Universidad de Granada.

Pérez Pastor, Cristóbal. 1895. *La imprenta en Medina del Campo*. Madrid: Establecimiento tipográfico "Sucesores de Rivadeneyra.".

Pettas, William. 1995. *A sixteenth-century Spanish bookstore: The inventory of Juan de Junta*. Philadelphia: American Philosophical Society Independence Square.

Recopilación de leyes de los reinos de las Indias, mandadas imprimir y publicar por la magestad católica del rey Carlos II nuestro señor. 1841. Madrid: Librería Española.

Rojo Vega, Anastasio. s.d. Real Biblioteca / Investigadores, *Historia del Libro*. https://investigador esrb.patrimonionacional.es/. Accessed 07 June 2021.

Romano, Antonella. 2005. Las primeras enseñanzas científicas en Nueva España: México entre Alcalá, Messina y Roma. *Takwá* 8: 93–118.

Rueda Ramírez, Pedro J. 2005. *Negocio e intercambio cultural: El comercio de libros con América en la Carrera de Indias (siglo XVII)*. Sevilla: Diputación de Sevilla, Universidad de Sevilla, Consejo Superior de Investigaciones Científicas.

Ruiz Fidalgo, Lorenzo. 1991. *La imprenta en Salamanca (1501–1600)*. Madrid: Arco Libros.

Ruiz Jiménez, Juan. 2015. *Impresión de un libro de música en cifras para vihuela (1546). Historical soundscape*. http://www.historicalsoundscapes.com/evento/398/sevilla/es. Accessed 07 June 2021.

Valleriani, Matteo. 2020. Prolegomena to the study of early modern commentators on Johannes de Sacrobosco's Tractatus de sphaera. In *De sphaera of Johannes de Sacrobosco in the early modern period: The authors of the commentaries*, ed. Matteo Valleriani, 1–23. Cham: Springer Nature. https://doi.org/10.1007/978-3-030-30833-9_1.

Alejandra Ulla Lorenzo is a Lecturer in Spanish Literature at Universidad de Santiago de Compostela. Her current research interests include the Golden Age Spanish theatre and the Spanish and Portuguese book trade. She is co-editor, with Alexander S. Wilkinson, of *Iberian Books* Volumes II & III (2015) and with Fernando Rodríguez-Gallego, of *Un fondo desconocido de comedias españolas impresas conservado en la biblioteca pública de Évora (con estudio detallado de las de Calderón de la Barca)* (2016). She is currently researching the involvement of women in the early-modern Spanish and Portuguese book trade.

Chapter 8
The Giunta's Publishing and Distributing Network and Their Supply to the European Academic Market

Andrea Ottone

Abstract This essay presents the Giunta publishing firm as a transnational network, highlighting its ideal center and peripheries. It describes the construction of a business model in conjunction with marketing channels and a consequent publishing plan intended to enhance the firm's reputation in a specific slice of the book market: the clergy and the high professions. At the center of this narrative are several instances of the Giunta endeavoring to commercialize Sacrobosco's *Sphaera*. I argue that, regardless of the eight known instances in which the Giunta family published Sacrobosco, the *Tractatus de sphaera* remained of marginal interest in the general publishing plan laid down by the firm.

Keywords Giunta publishing house · Johannes de Sacrobosco · Astronomical books · History of science · European book market

1 Introduction

Giunta publishing rapidly rose in the ranks of the late Renaissance European book market. The firm's strength was mainly based on the ability of its leaders to build a transnational network of production and distribution with branches in some of the most prominent hubs of the book trade in Catholic Europe.

The synergy between the various branches of the firm is represented in their shared use of the lily as a common trademark. The lily, a proud assertion of their Florentine origins, eventually became a statement of quality standards recognized by customers around Europe. This chapter will attempt to reconstruct the steps through which the Giunta built their organic network and the reasoning behind their choices. Further, it will describe the development of a common business model and a shared publishing strategy. This will elucidate the Giunta's approach to the publishing of Johannes de Sacrobosco's *Tractatus de sphaera* in the context of their business vision. Ultimately,

A. Ottone (✉)
Department of Economics, Management, Quantitative, University of Milan, Milan, Italy
e-mail: andrea.ottone@unimi.it

Max Planck Institute for the History of Science, Berlin, Germany

© The Author(s) 2022
M. Valleriani and A. Ottone (eds.), *Publishing Sacrobosco's* De sphaera *in Early Modern Europe*, https://doi.org/10.1007/978-3-030-86600-6_8

the aim is to reconstruct a small fragment of the integrated infrastructure that granted the *Sphaera* a wide circulation during the early stages of print culture.

2 Building an International Network

The family's firm was started mainly under the initiative of Lucantonio (1457–1538) and Filippo Giunta (1456–1517). The first steps they took in the late fifteenth century reveal the non-local aspirations of their enterprise. The mastermind of the business strategy seems to have been Lucantonio, the younger of the two brothers, who, from the start, took on a leading role. He is the one who moved to Venice in 1477 with the intention of book dealing.[1] As late as 1485, Lucantonio's older brother Filippo would pursue the same profession in Florence (Pettas 2013, 4), thus creating the premises for a multi-centered project.

Lucantonio's relocation in Venice brought him in contact with an emerging industry and placed him in a propitious commercial position. This may be the reason why he often proved to be one step ahead of his brother. Lucantonio's first known publications are dated 1489 (Camerini 1962–1963, I, 59–62), whereas Filippo's first signed editions that we know of are dated 1497 (Pettas 2013, 223–224). The same year Lucantonio started publishing, his brother Filippo opened a stationery shop in Florence (Pettas 2013, 4–5), a synchronized move that should not be overlooked. Two years later, the two brothers signed a partnership consolidating what seems to have been a common project already (Camerini 1962–1963, I, 34–37). This laid the foundation of a polycentric firm with the Venetian branch progressively taking the lead. Lucantonio's swifter and greater professional achievements brought him to demand a larger share of the revenues, thus asserting a de facto leadership (Camerini 1962–1963, I, 35). This imbalance would continue in the decades to follow; with multiple branches flourishing below and across the Alps, Venice would remain the natural barycenter of the Giunta's transnational network.

Lucantonio's entrepreneurial talent, along with the initial vision, developed in the years to follow, as the firm experienced at least three visible stages of expansion in the continental market. The first instance of this thoroughly planned process came with the partnership signed in 1517 between Giuntino di Biagio Giunta (1477–1521) and his uncle, Lucantonio.[2] The four-year contract between the two secured a

---

[1] His elder brother, Bernardo, accompanied Lucantonio to Venice, where he also entered the profession of bookdealer. However, his career would not take the same momentum as that of young Lucantonio (Camerini 1962–1963, I, 32). I would like to thank Carolin Strecker and Diane Booton for reading and commenting the first drafts of my paper. I would also like to share my gratitude to Gudrun and Reiner Strecker for the help they provided during challenging 2020 and beyond, when this work was still in the making.

[2] Giuntino's partnership with Lucantonio was preceded by a long stay in Venice, which is attested from at least 1507 by a small set of publications that carry his name (EDIT16, CNCT 1219). Giuntino may have trained in the profession under the supervision of his uncle, whose reputation was by then well established in the sector.

capital of 32,153 Venetian ducats. Giuntino contributed only 5,000 leaving the rest to Lucantionio, who clearly asserted his weight in the company. The stated purpose of the company was to have Giuntino "exercise in any approved merchandise…in Venice and any other place."[3] A few elements are worth mentioning. Giuntino's role in the partnership consisted in brokering business for Lucantonio, acting as proxy agent in an undefined commercial space. The merchandise of interest, one may assume, consisted mainly in books, but the loose definition of the commercial objective allows for the possibility that the company's trade may have also comprised other merchandise if it proved profitable. Commercial diversification is a feature of the Giunta's business model that would emerge more clearly and systematically on the eve of the sixteenth century (Tenenti 1957), as the Venetian book industry felt the bite of northern competition. However, the partnership with Giuntino suggests that this was a strategic vision already in place when Lucantonio first ventured into the publishing business. Networking and commercial expansion were also visible features of his vision. As far as this partnership goes, the geographic scope of the company was still limited to the Italian-speaking territories. Venice and Florence were already established hubs for the family. Iacopo di Francesco Giunta (1486–1547) settled in Rome in 1504, providing his family an important presence in the Papal State.[4] Giuntino Giunta, who originally had no solid settlement outside Venice, finally set up his base in Sicily, where he opened a bookshop in Palermo in 1517 (Camerini 1962–1963, I, 44), thus allowing Lucantonio and Filippo to stretch their peninsular network further south.

By moving deep in the periphery of the Viceroyalty of Naples, a state entangled with the Spanish crown, the Giunta were likely tightening their relationship with the Iberian market, a commercial area that had already fallen under the family's interest. The presence of a Giunta in Spain can be traced back to 1514 when Giovanni (1494–1557)—later known by his Spanish name Juan de Junta—was active in Salamanca (Pettas 2004, 18) (Chap. 7). Juan's relocation to Spain was followed by that of his brother Iacopo (1486–1547)—later known by his French name Jacques—who established himself in Lyon in 1520. Interestingly, Jacques's move abroad also involved Lucantonio, who signed an *accomandita* contract with Iacopo.[5] The stipulated contract, as in the case of Giuntino, granted much freedom to Jacques in conducting his undertaking in Lyon. As in the case of Giuntino, Jacques had only loose obligations in determining the direction of his enterprise, which was explicitly oriented toward printed books but comprised "any other merchandise that would be

---

[3] "…per exercitarsi in qualunche…mercantia venisse approbata…così in Vinegia come in ogni altro luogo dove detta compagnia distendesse…" (Pettas 1980, 304–308).

[4] Iacopo di Francesco Giunta is attested in Rome until 1531. The remaining known editions suggest that his publishing activity in Rome may have not been impressive (EDIT16, CNCT 1923). It is likely that Iacopo's role in Rome was that of agent for Lucantonio and Filippo (Pettas 1974, 340). The Giunta quickly filled the gap left by Iacopo in Rome with Benedetto Giunta, who was active there from 1531 until 1548 (EDIT16, CNCT 746). In later times the family mainly resorted to contracted proxy agents (Tenenti 1957, 1034).

[5] On the *accomandita* system, see (Carmona 1964).

held adequate."[6] The prime interest of the company was to conduct trade in Lyon, but it was explicitly stated in the contract that Jacques's operational area could comprise the whole of France.[7] Some interesting analogies emerge with the case of Juan. Both brothers chose not to set their base close to the court of Spain or that of France. They instead established themselves at the center of key commercial networks, thus choosing mercantile resourcefulness over the comfort of serving modern state bureaucracies. In fact, Lyon, not Paris, was the seat of a prominent book fair serving France and beyond. Likewise, Salamanca and Burgos were a safe distance from Medina del Campo, site of a prominent national fair.[8]

From these initial moves of Lucantonio, it emerges that the firm's ambition was to reach a wide market, albeit carefully confined to Catholic lands. What has been accounted thus far is the network that was built by securing the presence of a family member on site. A less visible network is that which employed occasional proxy agents. We know for a fact that the third-generation leader of the Venetian branch, Lucantonio junior (ca. 1535–1602), counted on a rather impressive web of representatives (Tenenti 1957, 136–139). These were mainly concentrated in northern Italy, but were also present in the rest of the peninsula and in at least one case across the Alps (Fig. 1).

Much emphasis so far has been put on how this commercial infrastructure could have benefited the Venetian branch, which appears to have been largely responsible for structuring and indirectly financing it. However, it is likewise true that the availability of such an integrated structure was a valuable asset for each node of the network. From this perspective, even in the absence of a formal contract of partnership, the cooperation between the branches of the Giunta would be granted by mutual convenience. One revelatory example of the clan-like mentality underlying the business held by the Giunta is linked to the papal privilege that Lucantonio senior earned in August 1530 to cover three works of theologian Tommaso de Vio (1469–1534).[9] These expensive editions were protected by a ten-year book privilege encompassing all of the Italian states, Germany, and France. Infringements of the standing privilege would have caused an automatic excommunication and a fine of 1,000 ducats. Sanctions for reprints or unauthorized commercialization, it was stated, would have applied to everyone except those who carried Lucantonio's family name (Ginsburg 2013, 383).[10] Vatican privileges were costly instruments and strategic assets capable of regulating competition over a vast portion of the European market. In the interest of smoothing the circulation of his own imprints, Lucantonio Giunta considered the sharing of a papal privilege a matter of common interests.

---

[6] "…et in ogni altra mercantia come parra a decto Iacobo…" (Pettas 1980, 298).

[7] For a comprehensive account of Jacques de Giunta's enterprise in Lyon, see (Pettas 1997).

[8] On Juan's attendance to the fair of Medina del Campo, see (Pettas 1995, 3).

[9] The works in question were, (*Psalmi* 1530) and (De Vio 1530, 1531). Papal privileges were a luxury legal protection for transnational firms like the Giunta; their legal stipulations were valid everywhere in Catholic Europe due to the fact that they could be enforced, among other means, by excommunication.

[10] I am grateful to Professor Jane Ginsburg for sharing her data on papal privileges in this and several other occasions.

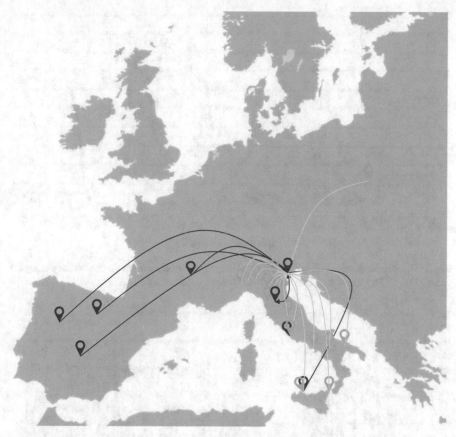

Fig. 1 Diagram of the Giunta's transnational network. The red pins represent cities in which a Giunta family member would be present in place (years 1489–1602). Green pins mark out the presence of a proxy agent working for Lucantonio Giunta the Younger, third generation leader of the Venetian branch. Author's plot

3   Sorting Out a Publishing Strategy

An operative business strategy for Lucantonio senior went hand in hand with his publishing plan. The former would not have worked without the latter and vice versa. Lucantonio's interest in a larger market is revealed in the progression of his output in both vernacular Italian and Latin (Fig. 2). During his first ten years of publishing activity, tighter contact with the local market would have been more of a necessity than a choice. However, his vocation toward a transnational market emerged rapidly, as the crossing of the two lines shows as early as 1493. These were the years when Lucantonio was in partnership with his brother Filippo. After this date, vernacular publishing became largely episodic for him, with a significant gap between 1513 (around when Juan de Junta moved to Spain, and 1528). The same

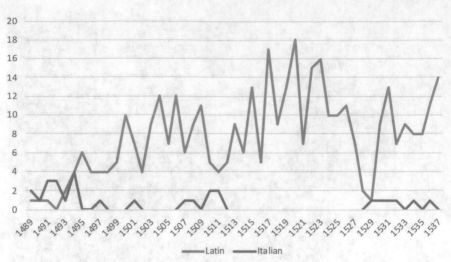

Fig. 2 Diachronic distribution of the output by language for Lucantonio Giunta senior during the years 1489–1537. Data source (Camerini 1962–1963, I). Author's plot

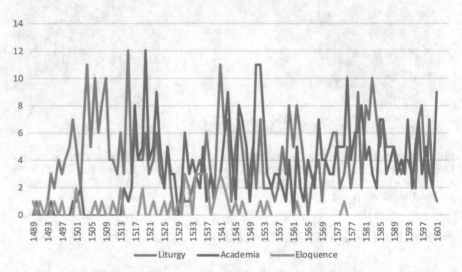

Fig. 3 Diachronic distribution of the output by the three main literary genres for the Giunta of Venice (years 1489–1601). Data source (Camerini 1962–1963, I–II). Author's plot

correlation between the widening of the Giunta's commercial network and a realignment of their publishing strategy is visible when dissecting Lucantonio's output by literary genre, with particular reference to the three main categories of his publishing portfolio: liturgical literature, academic literature, and eloquence (Fig. 3).[11] These

---

[11] The taxonomy used in Figs. 3, 4, 5, 6, 7, 8 and 10 follows a categorization of literary genres in use by the Giunta firm itself, as it will be detailed later. For this purpose, sales catalogues and

categories have been singled out to better represent three of the main commercial targets that a publisher of the time may have had in mind when drafting a publishing plan: clergy, high professions, and grammar-schools students or classic literature enthusiasts. From 1516 onward, the higher professions became a steady target of Lucantonio's publishing strategy, whereas in the previous decades they were virtually disregarded. By 1516, Juan de Junta was at least in his second year in Spain, Giuntino was on his way to Palermo, and Jacques would have opened the Lyon branch in 1520. Understandably, Lucantonio felt that in order to approach the high professions market he was required to build an adequate distribution network to make the project financially sustainable. This was due to the higher costs of production for academic editions, their slower sale, and the expectation of higher and less predictable transnational competition. From this perspective, the choice of both Juan and Jacques to follow the commercial routes of national and international fairs acquires a clearer meaning. Assuring a steady presence at fairs opened up the network to an even wider market.

Lucantonio's publishing plan settled into a stable pattern soon after the 1520s (Fig. 3). Liturgical works went hand in hand with academic ones, one taking the lead over the other alternatively, roughly every decade. The pattern remained steady for the two generations to follow. Liturgical texts, a category that Lucantonio chose as his signature product from the beginning of his career, granted safe revenue. This was

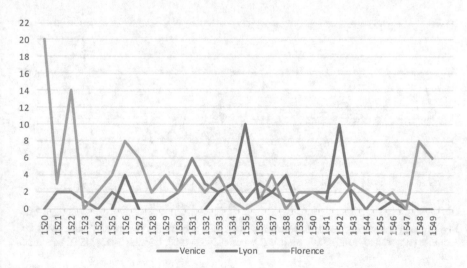

Fig. 4 Comparative diachronic distribution of humanities works for the Giunta of Venice, Florence, and Lyon (years 1520–1549). Data sources (Camerini 1962–1963, I; Pettas 2013; USTC). Author's plot

other commercial documents have been used to retrieve the nomenclature in use at the time. This conservative approach relies on the idea that said literary categories corresponded to adequate commercial targets and well-identified readership typologies. In Fig. 3, the category of *academic literature* aggregates canon and civil law, medicine, philosophy, and scholastic theology.

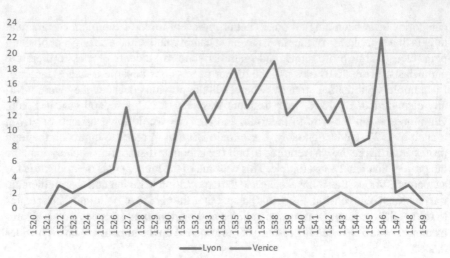

Fig. 5 Comparative diachronic distribution of works of jurisprudence for the Giunta of Venice and Lyon (years 1520–1549). Data sources (Camerini 1962–1963, I; USTC). Author's plot

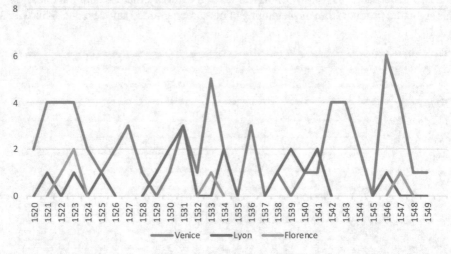

Fig. 6 Comparative diachronic distribution of medical works for the Giunta of Venice, Lyon, and Florence (years 1520–1549). Data sources (Camerini 1962–1963, I; Pettas 2013; USTC). Author's plot

the benefit of serving a fairly predictable audience, reachable in the urban space of Venice and at short and medium distances in the rest of Italy. Liturgical texts granted a steady flow of income, making it easier to cope with the higher risks of academic publishing in the wide-open transnational market.[12]

---

[12] On the role played by liturgical works in early modern publishing, see (Grendler 1977, 170).

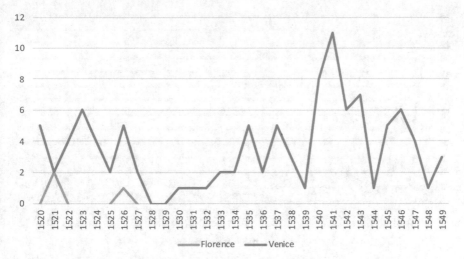

Fig. 7 Comparative diachronic distribution of liturgical works for the Giunta of Venice and Florence (years 1520–1549). Data sources (Camerini 1962–1963, I; Pettas 2013). Author's plot

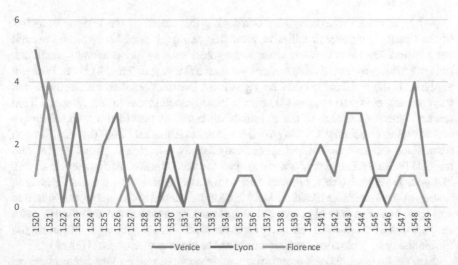

Fig. 8 Comparative diachronic distribution of philosophical works for the Giunta of Venice, Florence, and Lyon (years 1520–1549). Data sources (Camerini 1962–1963, I; Pettas 2013; USTC). Author's plot

Works of eloquence remained a secondary interest for Lucantonio. This remained true when his heirs, Tommaso (1494–1566) and Giovanni Maria Giunta (d. 1569), led the Venetian branch between 1538 and 1566. The category virtually fades away during the tenure of Lucantonio junior between 1566 and 1601. Keeping up with the audience interested in Greek and Roman classics or contemporary humanists was, in fact, mainly the craft of the Giunta branch of Florence (Fig. 4).

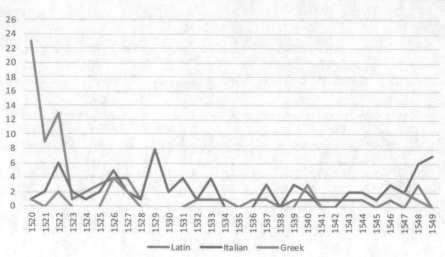

Fig. 9 Diachronic distribution of published works by language for the Giunta of Florence (years 1520–1549). Data source (Pettas 2013). Author's plot

A key element that emerges by comparing the output of the various branches of the Giunta is an overall cohesive publishing strategy aimed at avoiding mutual competition. Overlaps between macro-categories such as jurisprudence, medicine, liturgy, philosophy, and the humanities were rare (Figs. 5, 6, 7 and 8).[13] An in-depth analysis of the overlaps reveals no significant intersections between authors. On the contrary, evidence suggests that each local branch chose to feed a specialized market, whereas the sum of the output of all branches provided a comprehensive and diverse commercial offering to the continental market. The Venetian branch maintained a more varied output specializing mainly in liturgical, philosophical, and medical literature. Legal works were instead the specialization of Jacques de Giunta in Lyon. Latin and Greek classics and vernacular works were the distinguishing feature of the Florentine branch. Lastly, looking at output by language, the Venetian and Lyon branches proved successful in approaching a transnational audience (with Jacques de Giunta showing virtually no interest in national languages), whereas the Florentine branch remained mainly anchored to a peninsular market (Fig. 9).

Juan de Junta's publishing portfolio was in contrast much more comprehensive (Fig. 10), showing significant overlaps with the literary genres explored by the other branches of the Giunta. In this case, however, competition was systematically avoided by publishing the vast majority of the editions in vernacular Spanish, thus restricting the market of reference mainly to Spain and, eventually, its colonies (Fig. 11). Aside

---

[13] Figures 4, 5, 6, 7 and 8 only account for the output of the branches active in Venice, France, and Lyon. The Spanish branch is not accounted for, as its non-competition policy is indisputably proven by Juan de Junta's output being mainly in vernacular Spanish (see Fig. 11). In Figs. 4, 5, 6, 7 and 8, whenever a branch is not shown it means that said branch did not visibly engage in publishing the literary genre in question.

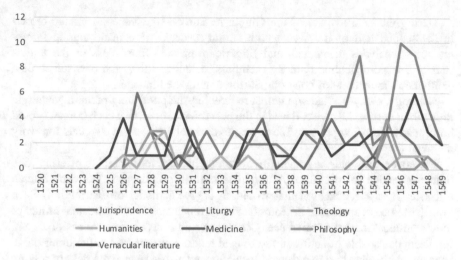

Fig. 10 Diachronic distribution of the output by main literary genres for Juan de Junta in Spain (years 1525–1549). Data source (Pettas 2004, 184–367). Author's plot

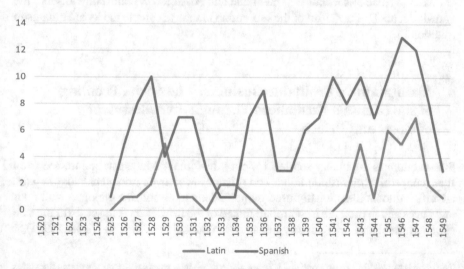

Fig. 11 Diachronic distribution of published works by language for Juan de Giunta (years 1525–1549). Data source (Pettas 2004, 184–367). Author's plot

from serving the Spanish-speaking market, Juan de Junta also operated as an outlet for the Giunta's network in the Iberian Peninsula, particularly for the Lyon branch.[14]

---

[14] Identifiable editions of Latin texts inventoried in 1556 in Juan de Junta's store in Burgos (Pettas 1995, 37–103) show that—aside from a justified 39% of acquisitions from Spanish publishers, and an expected relevance of Venetian editions (20%)—a large quantity of imprints came from France (16% from Paris and 5% from Lyon, whereas the growing market of Antwerp accounted for the 10% of identified provenances leaving a tiny 1% to Florence and 5% to other minor printing centers).

During the long activity of the Giunta in Europe (the Venetian branch closed in the second half of the seventeenth century), deals were made among family members, societies disbanded, and litigations arose.[15] Regardless of the understandable legal differences among members, evidence shows the persistence of a gentlemen's agreement of non-competition among the branches.

Naturally, everyone had an interest in keeping the publishing planning synergic rather than hostile. Likewise, it was in the best interest of all that each branch should hold steady for as long as possible to ensure optimal distribution channels for every member.

Moreover, in order to maintain a comprehensive commercial offering, the branches tended to specialize. A hypothesis worth proposing is that each branch felt safer in a specialization that would fit a *glocal* model of distribution. Each branch seemed to specialize in sectors that best represented the intellectual milieu of their own local market of reference. Competitive editions needed professionals and intellectuals capable of acting in the role of authors or editors. On the other hand, specializing in what best represented local demand also assuaged the risks of relying too much on a wide and competitive market. Hence, for example, the choice of the Giunta of Venice to specialize in medicine and Aristotelian philosophy to serve the Patavine school, or the effort of the Florentine branch to cater to the local humanistic tradition.

4 Maintaining a Profitable Business: The Social Profiling of the Giunta's Customers Through an Assessment of Costs and Prices

Sale catalogues are luxury sources for book historians. Much can be inferred from them concerning publishing trends and prices.[16] Moreover, printed catalogues were advertising tools intentionally used by publishers to establish a dialogue with their audiences and to promote a controlled image of the firm. Such is the case for the sale catalogue published in Venice by Lucantonio Giunta junior in 1591 (*Index* 1591a),[17]

The choice of isolating Latin imprints, leaving aside those in vernacular Spanish, reflects the status of transnational competition from the perspective of Juan de Junta's own book trade.

[15] The partnership between Lucantonio and Filippo Giunta ended in 1509 with an arbitration to reevaluate fair shares of the profits (Camerini 1962–1963, I, 37–43); likewise, arbitration was necessary to dissolve the contract between Lucantonio and Giuntino in 1521 (Pettas 1980, 37–43). A power struggle occurred in 1560 between the heirs of Bernardo Giunta to settle which of the five sons would lead the Florentine branch (Pettas 2013, 86). Patrimonial disputes arose as late as 1604 between various members (Santoro 2013, xxix, 205–207, 252).

[16] For a survey of the topic see (Coppens 2008; Ammannati and Nuovo 2017; Coppens and Nuovo 2018). On the applied methodologies, see (Ammannati 2018).

[17] A digitized copy of this earlier catalogue has been published in (Fratoni 2018, 99), which provides an example of how printed catalogues were used by individual collectors to orient their own acquisitions through the case of humanist scholar Prospero Podiani from Perugia.

which he reprinted in 1595 with marginal variation (*Index* 1595).[18] The dating of
these sources is late compared to the data considered thus far, but in light of a
substantial continuity in the publishing strategy of each branch of the firm, they may
be considered equally representative, although only for the activity of the Venetian
branch, which was, however, the epicenter of the Giunta's network.

Both catalogues group their listings in the following categories: humanities,
philosophy, theology, medicine, astronomy, Greek works, civil law, canon law, ecclesiastical works, and vernacular works.[19] The sequence mirrors the *cursus studiorum*
of the time quite beautifully, from bottom up. From grammar studies to the disciplines
worth a doctorate, from works suitable to magistrates to those necessary to low and
high clergy, the academic ranks and the social orders are all paid the proper tribute.
Specialist-allied disciplines like astronomy and Greek are conveniently placed next
to medicine. Liturgical texts stand out from the sequence of academic disciplines, but
they literally occupy the center of the page as they are largely listed in the second of
the three columns composing each broadsheet. Vernacular works for non-specialized
collectors close the catalogue in a marginalized position, entirely compatible with
the interest that the publisher shows toward this commercial target overall. Playing
with hierarchies of arts and professions was part of the commonplace communicative strategies in a time when scholars such as Conrad Gesner (1516–1565) and
Antonio Possevino (1533–1611) were redrafting the tree of knowledge. Commercial
and scholarly taxonomies followed very different agendas, and the publishers interested in flirting with their audiences of reference knew how to use these taxonomies
accordingly. They could even become erudite divertissements for Venetian booksellers like Bernardo di Bernardo Giunta (ca. 1550–ca. 1527), who noted on the
opening page of his personal work book a common motto of the time: "Theology
queen, philosophy lady, medicine servant."[20]

One way to shift these categories, breaking the ideal order based on academic and
social hierarchies is by taking into account the number of listings by category (see
Table 1).

With an eye to quantities, it emerges clearly that ecclesiastical works were the chief
interest of the firm. Nonetheless, products related to higher education (philosophy,
theology, medicine, and law) made up 121 listings in the 1595 catalogue, thus proving
to be an equally relevant focus for the company. Works of eloquence, astronomy, and
Greek are a marginal digression in the Giunta's catalogue. Vernacular works are
a notable presence, but not the strongest category advertised. It is worth noting the
impressive presence of canon and civil law editions, which were scarcely represented
in the output of the Venetian branch for the years 1520–1549. This is the characteristic
that distinguished Lucantonio junior from all other leaders of the Venetian branch.

---

[18] A known copy of said catalogue is preserved at the YRL, Z233.G44G 448i 1595.

[19] *Libri humanitatis, Libri philosophiae, Libri theologiae, Libri medicinae, Libri de re astronomica,
Libri Graeci, Libri iure canonico Libri in iure ciuili, Libri ecclesiastici nigri ac rubei, Libri volgari*
(*Index* 1591a, 1595).

[20] "La theologia regina, la filosofia donzella, la medicina serva." Bernardo di Bernardo Giunta, held
a large bibliographic repertoire, now catalogued as the "Giunta publishing house stockbook" (YRL,
Collection 170/622, f. 1r).

Table 1 Number of listings per literary category in the Giunta 1591 and 1595 sale catalogues

Literary genre	Listings	
	Index (1591a)	*Index* (1595)
Ecclesiastical works	180	176
Medicine	34	37
Theology	25	23
Philosophy	21	20
Civil law	18	22
Vernacular works	19	20
Canon law	18	19
Humanities	5	5
Astronomy	2	2
Greek	2	2

Data source (*Index* 1591a, 1595)

The probable cause of this innovation can be found in the progressive decline of the Lyon branch of the firm, which was chiefly specialized in publications of interest to legal practitioners. With the Lyon branch declining in the late sixteenth century, Lucantonio took the initiative to fill the disciplinary gap by publishing law books in Venice (Ottone 2003, 72).

Prices provide invaluable information for reconstructing the ideal link between publishers and their audiences. In a standard sale catalogue, prices would be associated with a short but clear description of the product. An example from the Giunta 1595 catalogue might read as follows: "Roman Breviary. With Saint Laurence's insignia and copper made illustrations. In 8°, ducats 1, grossi 12."[21] The purchaser, either a wholesaler, an individual, or an institution, would know that the advertised item corresponded to an in 8° edition of the reformed Breviary with a special insignia on the titlepage, and copper-plate illustrations. All this would justify the price of 1 venetian *ducato* and 12 *grossi*. Seemingly, the exact same item was available with woodcut illustrations. This would have reduced the price by almost 1 *ducato*, bringing the total price to 18 *grossi* (*Index* 1595, f.1rc, no. 51). One was given the opportunity to negotiate between convenience and quality and get either the cheap product or the deluxe model. All prices in the Giunta 1591 and 1595 catalogues are displayed in *ducati*, whereas other Venetian catalogues of around the same period more often used *lire*. Most likely this was due to the fact that the firm privileged that currency in its own accounting. *Ducati* had the advantage of flattening big prices into small figures (one Venetian *ducato* at the time was worth six *lire* and four *soldi*). Whatever the cause may have been, it is fair to say that a knowledgeable customer who approached Lucantonio Giunta's catalogue would have realized at a glance that the advertised merchandise was on average expensive.

---

[21] "Breu. Ro. Cum signo S. Laur. ac figuris in aere. In 8, D. 1, G. 12" (*Index* 1595, f.1rc, n. 50).

Table 2 Average price per printing sheet per edition in *denari* (i.e., sub unit of Venetian lire)

Literary genre	Average price per printing sheet	
	Index (1591a)	Index (1595)
Greek	20.61	20.31
Ecclesiastical works	18.70	18.54
Canon law	15.23	18.49
Astronomy	13	17.22
Humanities	12.92	10.91
Medicine	12.81	12.35
Vernacular	12.53	13.24
Civil law	12.52	12.95
Philosophy	10.30	10.66
Theology	10.20	10.11

Data source (*Index* 1591a, 1595)

Links between product, value, and price were familiar to those who normally approached book sale catalogues at the time. Their awareness was based on routine contact with merchandise and a commercial commonsense that is now lost. In order to recreate such links at least in part, prices will be handled with a mechanical artifice of price per printing sheet. Printing sheets were the basic unit that both publishers and printers used to measure the extent and material investment of imprints; they calculated labor and wages on the basis of printing sheets per print run. Similarly, they estimated costs linked to the consumption of raw materials when planning a publishing endeavor. Ultimately, piled groups of unfolded printed sheets were also the raw product that customers saw stacked on display for sale (Nuovo 2013, 389–392). Thus, reducing prices to the unit of printing sheets not only allows for the leveling down of variegated commodities to a common denominator, but it also approximates the outlook that was most familiar to producers and sellers alike.[22] Table 2 proposes the breakdown of average prices per printing sheet of items listed in the Giunta 1591 and 1595 catalogues by literary genre. To ease readers' understanding, prices have all been reconverted to Venetian *denari*.

"Theology queen, philosophy lady, medicine servant" was a motto of the time. Yet the catalogue under scrutiny reveals an opposite hierarchy when parameters cross-reference prices and costs. Medicine took the lead over philosophy, which passed theology by an inch, but none of them could compete with all other categories, especially canon law and liturgical works. Higher prices per printing sheet were understandably determined primarily by the higher costs of production. These impacted the price for technical reasons. Hence, the high price of works in Greek, which was not only directed to a niche market but which also required specialized philologists, uncommon types, competent compositors, and proofreaders to produce them. In the

---

[22] Using price per printing sheet has become a common method within the EMoBookTrade project that most of the data in this section originates from.

case of astronomical texts, copious illustrations and diagrams were likely the reason behind the higher price per printing sheet. Liturgical texts often carried evocative and illustrative images and comprised music, which also implied higher costs. Moreover, liturgical texts, often referred to as "red and black imprints" in that they carried a main text in black and the rubrics in red, required an additional passage under the press. This doubled the effort and increased the risk of misprints, which resulted in waste that, ultimately, translated into additional costs. Furthermore, liturgical texts and canon law works, following the Council of Trent (1545–1563), came under tight quality control by Vatican institutions, which made them very cost-sensitive products. Moreover, in consequence of Rome's policy of allocating papal privileges on reformed canonical texts (both canon law and liturgies), prices for these products toward the end of the sixteenth century became largely inflated.[23] Below medical books, whose cost was linked to illustrations, it is not surprising to find purely speculative literature, which would be richer in text than images and could thus drop the cost-price balance. Thus, on average, these occupy the lower positions.

A comparison between the Giunta 1591 catalogue and a sale catalogue printed in Venice by the Giolito in 1592 (*Indice* 1592) may assist us in understanding the way in which the Giunta catalogue stood out.[24] The average price in the Giunta catalogue is 18.86 *denari*, whereas the average price found in the Giolito catalogue published the year after is 13.87 *denari*. The Giolito 1592 catalogue has been selected not only due to its chronological proximity to the Giunta's catalogue, but also because, unlike the latter, the former was primarily aimed at a localized market, as is revealed by its very heading in vernacular Italian and the imbalance between the twenty-six advertised Latin editions versus the 176 in vernacular.[25] Interestingly, the average price of the Giolito catalogue gets surprisingly close to the 12.53 average price per sheet that characterizes the vernacular section of the Giunta catalogue of 1591 (13.24 in the Giunta 1595 catalogue). The Giolito and the Giunta had different audiences, different geographic scopes, and different infrastructures of distribution, and the average prices advertised by the two firms reflect these structural differences. A structurally closer competitor of the Giunta at the time were the heirs of Girolamo Scoto (1505–1572). A comparison between the average prices of the products advertised by the Scoto around the same time is somewhat surprising. In their 1591 multidisciplinary catalogue the average price per printing sheet amounts to no more than 10.01 Venetian *denari*

---

[23] Papal privileges, enforceable in all Catholic lands via automatic excommunication, created a regime of large monopolies that allowed grantees to set high prices for products that were legally sheltered from competition. The result was a general increase of prices for this literary genre (Mercati 1937; Grendler 1977, 169–181).

[24] It is worth recalling that the Giunta 1595 catalogue is a mere reprint of a 1591 catalogue, and that the advertised publications and related prices are substantially the same.

[25] "Copious index of all books printed in Venice by the Giolito up to the year 1592." ("Indice copioso e particolare di tutti li libri stampati dalli Gioliti in Venetia fino all'anno M.D.XCII"), (*Indice* 1592). A known copy of this catalogue is held at the Biblioteca Nazionale Marciana under call number 193. D.443/1. The average price per printing sheet is based on data analysis made by Dr. Giliola Barbero in the context of the EMoBookTrade project. The catalogue is the object of an essay that she authored and to which I direct the reader for further details. See (Barbero 2018).

(*Index* 1591b),[26] some 4 *denari* less than the average price set by Lucantonio Giunta in his catalogue published the same year.

One aspect that does not easily emerge from the Giunta 1591 and 1595 catalogues is their internal chronology. By matching the listed editions with surviving copies, it is possible to reconstruct a chronological morphology of the catalogue. This allows speculation on the consistency of the Giunta's stock at the end of the sixteenth century. The aim is to highlight how fast the Giunta expected to exhaust their print runs. To better illustrate this aspect, it is useful to focus on scholarly literature alone, which emerged as one of the signature products of the Giunta—one that they chased with greater effort when designing their business model. In the 1591 catalogue (*Index* 1591a), fifty out of 118 listings ascribable to an academic target (medicine, philosophy, theology, Greek or Hebrew grammar, civil and canon law) would match editions that were thirty years old or older.[27] A similar figure, fifty-two out of 121 listings, emerges from the 1595 catalogue. The Giunta seemed overall able to cope with slow sales. The profit, based upon some of the signature products of the firm, was in fact expected to come within a considerably long timespan. The Giunta being conscious of the slow sales of most of their products seemed capable of measuring their profits even on a very long run. The capacity that the Giunta had in handling slow profits could be measured in their formidable access to credit during periods of severe financial difficulty.[28]

Prices per printing sheet derived from publishers' sale catalogues are especially beneficial in retrieving the perspectives of the book market professionals: publishers, printers, and sellers. Retail prices instead tell us the same story from a slightly different perspective; they reveal how much collectors or consumers—either individuals or institutions—were willing to take out of their wallets in order to access the product that mattered to them. An imperfect way to access this standpoint is by comparing average total prices (Table 3).

With some sensible exceptions, this view of the Giunta sale catalogue reestablishes in part the ideal hierarchy of literary genres and speculative disciplines. Theology goes above philosophy, which is still surpassed by medicine due to the design and technical features mentioned above. If theology is queen again, the true *imperatrix* is law, with civil law giving right of way to canon law. The higher professions take back the lead and, overall, the academic ranking seems to be reestablished.

---

[26] A known copy of this catalogue is held at Milan's Biblioteca Ambrosiana under the call number S.M.I.VII.3/6. The average price per printing sheet is based on data analysis made by Dr. Giliola Barbero.

[27] The theology section comprises an edition of Bernardus Claraevallensis' (ca. 1090–1153) *Flores* dated 1503 (Claraevallensis 1503).

[28] In 1553, the Venetian branch went through financial difficulties that led to a default. The already challenging situation was aggravated by a second incident in 1557 when a fire damaged the Giunta's print shop with a probable loss of part of their stock. The two joint incidents are accounted as hurtful memories in the will of Tommaso Giunta. For Tommaso Giunta's will, see (Camerini 1927). Both adversities resulted in a visible drop in the output of the Venetian branch, which nonetheless never hit zero and had resumed its normal course by 1560 (Ottone 2003, 69, Fig. 2). The quick recovery shows that, regardless of adversities, the Giunta were still considered fully viable through their longstanding cosmopolitan reputation. On the bankruptcy of 1553, see also (Pettas 1980, 92).

Table 3 Average total price per edition in *denari* (i.e., sub unit of Venetian lire)

Literary genre	Total price	
	Index (1591a)	*Index* (1595)
Canon law	5545.55	5290.66
Civil law	3368.66	3129.65
Vernacular	2288.53	2078.55
Theology	1996.4	1951.65
Greek	1860	1860
Medicine	1503.97	1461.43
Philosophy	1174.9	1230.7
Humanities	1159.4	1159.4
Astronomy	1116	1116
Ecclesiastical works	1057.93	944.24

Data source (*Index* 1591a, 1595)

5 A Network of Information

The transnational infrastructure built by the Giunta over the decades granted them adequate opportunities to circulate their books, but it also exposed them to wider and wilder competition. In this respect, timely information on where the European market was heading was vital. Naturally, a capable commercial network as that available to the Giunta was fit to circulate information as well as merchandise. In this respect the Giunta's preference for cosmopolitan commercial hubs would allow them to feel the pulse of the European book market. International book fairs were places where dealers boasted their merchandise, made deals, consolidated alliances, and shared intelligence.[29] Having someone on your payroll in sensitive marketplaces who could browse stacks, acquire catalogues, and glean updated knowledge of what other European publishers were up to was a vital asset for entrepreneurs with transnational aspirations, such as the Giunta. We know for a fact that Venetian printers visiting the Frankfurt fair in the early seventeenth century would head back home carrying more than just merchandise. In fact, they would carry one or more copies of the fair's catalogue to hold on to or share according to convenience.[30] Likewise, proxy

---

[29] On the presence of Italian publishers at international European book fairs, see (Nuovo 2013, 281–314).

[30] In their pursuit of censorial policies, the Roman Congregation for the Index was often eager to acquire copies of the latest Frankfurt catalogue for investigative purposes. The Roman Curia would primarily refer to Venetian publishers knowing their equal interest in catalogues in their pursuit of commercial inquiries. In several instances the Congregation would trade sensitive material, such as certified emended texts or special dispensations, in exchange for recent catalogues shipped from Frankfurt. For instance, in July 1601, in exchange for a catalogue from the fair, the Congregation for the Index offered Venetian printers the certified copy of an integrative text of Martín Alfonso Vivaldo's *Candelabrum Aureum* alongside the authorization to emend and commercialize suspended texts by Giovanni Zabarella and Scipione Manzano (ACDF, Index, V.1, f. 140v). A similar give-and-take dynamic emerges in a letter dated November 1601 in which the Congregation promises

agents disseminated in sensitive spots were themselves part of an echelon of valuable information. For instance, given their activism in producing canonical texts, the Giunta of Venice were keen to maintain a steady presence in Rome and to pull the right strings in a space that merged commercial and political interests. It also helped them cope with the turbulence of post-Tridentine policies.[31] This form of soft insider trading was easy to gather even in a world in which communication was fairly limited. Furthermore, cosmopolitan publishing firms like that of the Giunta kept a tight epistolary connection among branches to coordinate strategies. Most of this is only visible through secondary evidence as surviving correspondence is very rare for Venetian printers. One of the rarest exceptions regards the Gabiano family from Monferrato, who, like the Giunta, were in control of a transnational network. One single year of epistolary correspondence in and out of Lyon in 1522 is sufficient to represent the level of detailed information that traveled among publishers around Europe.[32]

Ultimately, the productive capacity and the quasi-standardized mode of production introduced by the printing press opened the sector to a new level of competition, but it also provided useful tools to cope with it. Books were the only commodity of the time that carried durable information on producers, financers, time, and site of production. If, on the one hand, the information carried on titlepages and colophons amplified the perception of the ongoing transnational competition, on the other, it also offered possible remedies.

Publishers' response to competition could be political; they could in fact seek institutional protection in the form of book privileges.[33] Response to competition

---

to make progress in expurgating suspended Venetian editions of the *Roman Missal* only under the condition of receiving a newer copy of the catalogue (ACDF, Index, V.1. f. 144v and ACDF, Index, III.6, f. 298r). The censorial purposes that led the Congregation for the Index to acquire copies of the Frankfurt fair catalogues emerges clearly from an instance dated July 1602 (ACDF, Index, V.1, f. 163v). On their part, the Venetian printers had little to no interest in aiding the Congregation, as it could indirectly inhibit their traffic with Germany. They likely had even less interest in parting with such a good source of information; thus, they seemed to do that only when they could earn a sufficiently high favor from the Roman Congregation. Furthermore, at the back of their mind the Venetian printers must have had the thought that leaking such information could have benefitted their Roman competitors, who they had close to no interest helping.

[31] This emerges quite clearly from an incident that took place in 1601. Following a scandal regarding a number of corrupted editions of the *Roman Missal* printed in Venice from 1597 onward, Vatican authorities factually blockaded the commercialization of two vital products for the Venetian book industry: the *Roman Missal* and the *Roman Breviary*. Among Venetians, anxiety grew that behind the doctrinal reasons propelling the blockade could have been the hidden intent of favoring Roman publishers by granting them a commercial advantage on the production of key liturgical texts. In consequence, the Venetian guild kept their Roman competitors under tight watch, sending well-informed complaints to the Roman authorities (Ottone 2019, 312). When times were ripe, this detailed intelligence was used to build a case with the Venetian Senate and move the action at a diplomatic level (Grendler 1977, 247–250).

[32] Said correspondence is currently being published by Professor Angela Nuovo in a forthcoming volume to which I refer for further details.

[33] Privileges granted temporary monopolies on specific products and provided a commercial advantage to holders. They could be local, as in the case of privileges granted by most secular authorities,

could be tactical and manifest in the form of temporary partnerships.[34] Lastly, and most interestingly for the purpose of this section, competition could be approached and resolved strategically. Since the high monetary investment of publishing and growing competition resulted in high financial risks for publishers, the profession required more than just good instincts, it required planning and sound methods of market assessments. The book market was a dynamic and complex environment that needed systematization. The question of how early modern printers oriented their market strategies in a world of limited communication remains open. Evidence is limited.

A revealing source, however, emerges within the network of the Giunta family. Looking at the periphery of the Giunta's network, at an advanced stage of the firm's history one can find an early seventeenth-century manuscript once owned and, for the most part, compiled by Bernardo di Bernardo Giunta. He was a fourth-generation member of the Florentine branch. In the 1570s, he moved to Venice and remained active in the Serenissima until the late 1620s. His achievements in the publishing business were not impressive compared to the standards held by the leaders of the Venetian branch of his family. The golden age of his career coincided with a partnership he initiated in 1600 with Giovanni Battista Ciotti (ca. 1564–ca. 1635), an expat from Siena who established himself quite successfully in the publishing community of Venice.[35] Aside from his marginal publishing career, we know of Bernardo di Bernardo Giunta mainly due to a manuscript he began compiling in March 1600 and that he, and at least two other unknown compilators, continued augmenting for the following forty years.[36] The codex is now preserved at UCLA's Department of Special Collections under the call number *Collection 170/622*. The incipit states the identity of the owner and the year of creation of the manuscript, but reveals nothing of its nature and purpose. The manuscript consists of an extensive sequence

---

or they could be transnational (or universal, at least theoretically), as in the case of privileges granted by the emperor, or even more so by the Pope. On book privileges in general, see (Nuovo 2013, 194–257); on a valid example of secular privileges, the Venetian system is paradigmatic—see (Squassina 2019); on universal papal privileges, see (Ginsburg 2013).

[34] Multiple publishers could team up to pursue joint editions. In doing so not only did they share the risks linked to the commercial venture, but they also limited local competition on specific products. An example worth mentioning is the Venetian *Societas Aquilae renovantis* (EDIT16, CNCT 90). This was formed in 1571 and lasted at least until 1608. Throughout the years it included some of the most prominent families of publishers active in the Serenissima. In the year 1584 alone, the society counted some fourteen members: Lucantonio Giunta junior, Filippo Giunta junior, Bernardo Giunta junior, heirs of Bernardino Magiorino, Francesco De Franceschi, Francesco Ziletti, Giovanni and Andrea Zenaro, Girolamo Zenaro, Damiano Zenaro, Felice Valgrisi, the heirs of Girolamo Scoto, Giovanni Varisco, and the heirs of Melchiorre Sessa senior. The society was devoted to financing large, expensive, and slow-selling editions of law books. On the *Societas Aquilae renovantis*, see (Nuovo 2013, 64–67).

[35] For a comprehensive account of Giovanni Battista Ciotti's activity, see (Rhodes 2013).

[36] Accounts of the relevance of this manuscript for book history and especially to topics pertaining the economics behind early modern publishing can be found in (Lowry 1991; Pettas 2004, 105–106; Nuovo and Ammannati 2017; Bruni 2018).

of bibliographic records: a first estimation counts approximately 11,555 entries scattered throughout 313 leaves (Nuovo and Ammannati 2017, 12). Data are organized into different logical categories. A large section gathers data following a taxonomy based on literary genres which, aside from a few exceptions, tightly mimics that of the Giunta catalogues of 1591 and 1595 discussed above (*Index* 1591a, 1595). A second section organizes largely the same data according to provenance and by publisher. The bulk of the data, I contend, was compiled between 1600 and 1604. A third section hosts, for the most part, later entries—mainly from 1608 onward. These entries can be largely attributed to the publishing output of Giovanni Battista Ciotti, in or out of his partnership with Bernardo Giunta. The various sections are conceived and organized to ease targeted searches and repeated browsing in accordance with different investigative aims. For the most part, the bibliographic records listed carry information on authors, titles, formats, and numbers of printing sheets. This standard is more or less consistent, but records are occasionally incomplete and carry, for instance, only sparse information (e.g., author and title, only author, or only title). For roughly half of the entries a corresponding price is provided. To facilitate quick data retrieval, leather tags were placed at the right margin of the leaves to single out macro-categories, such as literary genre, provenance, or publisher.[37] Additionally, sections of greater relevance carry letter tags to speed up alphabetical searches by author or title. Within each letter section, records are grouped by format (folio, 4°, 8°, or smaller). Clear signs of consumption are visible in the lateral tags, thus providing tangible evidence of frequent use during the active life of the manuscript (as expected for a tool that required considerable effort to be compiled). On the other hand, its extraordinarily good state of preservation and its fairly clean handwriting are evidence of the value that this tool had for its users.

To this day the source is catalogued by its holding institution under the label "Giunta publishing house stockbook" and so it is known to field scholarship. A systematic discussion of the inner features of the manuscript that conflict with the idea that this was a catalogue of books in stock is beyond the purpose of the present essay.[38] What is more pressing is to assert that, aside from other possible purposes that this manuscript may have served, evidence suggests that it was chiefly valuable in conducting empirical market research. This was likely aimed at catching profitable publishing endeavors or checking the viability of conspicuous acquisitions by assessing market risks or opportunities and avoiding the hazard of oversaturation.

---

[37] Digital images of the manuscript with examples of such search tags are visible in (Ammannati and Nuovo 2017, 14–18).

[38] I am currently working on a comprehensive account of the evidence supporting a reassessment of the nature of the manuscript *Collection 170/60*, which is the object of an ongoing publication (Ottone forthcoming). First results and tentative hypotheses have been presented at the conference *Merchants, Artisans and Literati: The Book Market in Renaissance Europe* (University of California, Los Angeles, 1–2 March 2019) and during the cycles of annual colloquia at the Institut für Philosophie, Literatur-, Wissenschafts- & Technikgeschichte (Technische Universität Berlin, December 2, 2019). The aforementioned evidence has been gathered during two years of systematic examination of the empirical data recorded in the manuscript in the context of the EMoBookTrade project. Results of the ongoing data retrieval process conducted on the manuscript are being published in the online database EMoBookPrices.

The manuscript was in fact a tidy directory of information on the publishing portfolio of those who qualified as direct competitors (i.e., Italian publishers, mainly Venetian, or foreign publishers with greater influence on the Venetian market).

We have no evidence to assess how widespread the use of similar devices was among early modern book dealers. It is, however, hard to believe that Bernardo di Bernardo Giunta was the only publisher of his time compiling and using such a tool. In fact, it is rather difficult to argue that he was the first. A somewhat similar device is known to have been in use by the Plantine press in the years 1555–1593.[39] Sources of this kind are very rare in the already scanty remains of the private archives of early modern publishing firms. However, the two instances represented by the Plantin and the Giunta sources indicate some continuity. If one accepts the hypothesis that tools of this kind were largely in use among early modern publishers, a hypothesis may be that Bernardo junior learned this practice within the circle of the Giunta family, where he conducted his apprenticeship and made his early professional steps (Decia and Delfiol 1978, 6; Camerini 1962–1963, II, 447–448).

Arguably, large-scale publishing houses such as that of Plantin and Giunta could hardly keep themselves afloat in a growingly competitive market unless they had an effective method to predict its complex fluctuations. This level of awareness of the difficult harmony between demand and supply among early modern publishers shall not be overlooked. This is especially true when analyzing how publishers dealt with a popular work like Sacrobosco's *Tractatus de sphaera*.

6   The Giunta as Publishers of the *Sphaera*

Taking into account the prices habitually assigned to Sacrobosco's *Sphaera* may be helpful in placing it in the larger context of the book market of the time. A commercial profile of the text positions it within the general scope of the Giunta's publishing portfolio. Having already taken into account general prices per literary genre set by the Giunta of Venice between 1591 and 1595, it will be fruitful to compare them with prices of the *Sphaera* set elsewhere at around the same time. In 1591, the heirs of Girolamo Scoto had set a price of 120 *denari* in Venetian lira for their 1586, 8° edition (Sacrobosco et al. 1586).[40] The 1601, 4° edition of Clavius' commentaries to the *Sphaera* credited to Giovanni Battista Ciotti (Sacrobosco and Clavius 1601–1603) would instead go for 720 *denari* around the same year of its publication.[41] The price per printing sheet of these two editions amounted to 11.43

---

[39] MPM, M296. The manuscript is currently being investigated by Renaud Milazzo in the context of the EMoEuropeBookPrices. To his forthcoming publications I address for further details.

[40] For the price see (EMoBookPrices 9772).

[41] The price originates from YRL, Collection 170/600. The manuscript displays a dynamic internal chronology that makes dating each price rather complicated. My own conclusion is that the indicated price was set between 1603 and 1608 (EMoBookPrices 15272). The declared price was for a 1603 edition; however, evidence shows that this was in fact a reissue of a 1601 edition that carried shared credits for both Basa and Ciotti (Chap. 6)

denari for the former and 11.16 for the latter, thus justifying the higher total cost of Clavius' commentaries due to a higher consumption of raw materials and labor. A comparison with the average price per literary category found in the Giunta 1591 catalogue (Tables 2 and 3) shows that pricewise the *Sphaera Mundi* would sit in the lower ranks of the Giunta's publishing portfolio. With regard to the total price alone, the *Sphaera* falls below the average price of astronomy books (undoubtedly its category of reference), which in Table 3 occupies the bottom line. Taking into account the price per printing sheet, the two editions would fit the space between medicine on one side and humanities and philosophy books on the other. Much of this is probably due to material features. However, it is worth noting that in both cases Sacrobosco's text is assimilated with rather coherent epistemological categories. A possible explanation is that publishers, in the act of setting prices for specific products, among other variables, also took the social profile of the targeted audience into serious consideration.

Being a formative book useful to students entering higher education, the *Sphaera mundi* connected publishers to a large pool of users and collectors. This opened up wider opportunities for publishers in pursuit of yet unspecialized readers.[42] This, however, also exposed them to higher competition. Measuring competition is extremely difficult when lacking information on print runs and, therefore, on how many copies entered the market at a given time. One imperfect solution is to observe the behavior of publishers in regard to a specific work, with particular reference to the chronology of their reprints. For reasons functional to the argument being pursued, it will be useful to momentarily shift attention from the Giunta firm to one of their direct competitors in Venice, the Scoto family.

The Scoto, active from the late fifteenth to the first half of the seventeenth century, occupied a leading position within the Venetian community of publishers. However, their commercial infrastructure could not compete with that of the Giunta and their publishing portfolio was carved around this fact (Chap. 6). Over the years, the Scoto family had placed their name on at least eight editions linked to the Sacrobosco tradition. Ottaviano Scoto (fl. 1479–1498) placed an abridged version of the text on the market in 1490 (Sacrobosco et al. 1490) (Chap. 3). It was a 4° edition of the text curated by Georg von Peuerbach (1423–1461) and Johannes Regiomontanus (1436–1476). Four years later, he sent out a commentary to the text by Gasparino Borro (1430–1498), also in 4° (Sacrobosco and Borro 1494). This ended Ottaviano's pursuit of readers and collectors interested in Sacrobosco. After his death in 1495 his heirs chose not to pick up this pursuit until 1518, when they proposed an in-folio edition of the full text boasting a plethora of commentators: Campano da Novara (1220–1296), Pierre d'Ailly (1351–1420), Cecco d'Ascoli (1260–1327), Theodosius of Bithynia (ca. 160 BC–ca. 100 BC), Francesco Capuano di Manfredonia (1450–1490), Jacques Lefèvre d'Étaples (1450–1536), Michael Scot (ca. 1175–ca. 1234), Robert Grosseteste (ca. 1175–1253), Johannes Regiomontanus (d'Ascoli et al. 1518a). Then nothing until 1562 when Girolamo Scoto seems to have found a new formula for

---

[42] The Sphaera Database (Sphaera Corpus*Tracer*) counts fifty-four recurrences of works related the Sacrobosco tradition in Venice in the sixteenth century.

the market: an allegedly revised text with notes by Élie Vinet (1509–1587) and contributions by two more contemporary authors: Pierio Valeriano (1477–1560) and Pedro Nuñes (1502–1578). For this edition, Girolamo Scoto chose the 8° format (Sacrobosco et al. 1562a). This formula seems to have worked well. In fact, he proposed it again in 1569 (Sacrobosco 1569). One may assume that in the span of six years he had exhausted the 1562 print run and believed that a new one might have given satisfactory results. He was not wrong—in 1574 the heirs of Girolamo Scoto decided to reprint the same formula a third time (Sacrobosco et al. 1574). A fourth had to wait a much longer time (Sacrobosco et al. 1586), illustrating that the market for this product was in fact slowing. The 1574 reprint took some eleven years to exhaust; the 1586 reprint was still on the market five years later, as the Scoto family was advertising it in their sale catalogue in 1591 (*Index* 1591b). This would justify the family's choice to stop dealing the product for well over thirty years.[43] One piece of evidence worth mentioning in understanding the Scoto's attitude toward the publishing of the *Sphaera* is that none of the editions mentioned above claimed the coverage of a book privilege in or out of Venice. In their pursuit of the market for Sacrobosco's text, the Scoto family would have been engaged in open competition with nothing more than their reputation and their commercial channels as safeguards. On the other hand, none of the published editions may have met the criteria of undisputedly novel content, which was a prerequisite to earn a book privilege, at least in Venice.[44]

The experience that the Scoto family had with the *Sphaera* shows what may be a general pattern in attempting to commercialize a very popular product: it was necessary to test the market, build a reputation, then find the right formula and use it until it proved profitable. Girolamo Scoto may have been the one who found the right one in 1562; his successors, however, failed to recognize when the market was no longer willing to welcome the same formula years later. Endeavors in popular imprints had their upside, but they could also quickly show their limits. The Giunta must have come to this conclusion much earlier than the Scoto did. This is reflected by the publishing history of the Giunta in relationship to Sacrobosco's text:

> *Textvs Sphaerae Ioannis De Sacro Bvsto.* (Impressio Veneta: per Ioannem Rubeum & Bernardinum fratres Vercellenses: ad instantiam Iunctae de Iunctis florentini, 1508 die VI mensis Maii).
> *Sphera mundi nouiter recognita cum commentarijs et authoribus in hoc volumine contentis.* (Venetijs: impensis nobilis viri domini Luceantonij de Giunta Florentini, die vltimo Iunij 1518).
> *Spherae tractatuus Ioannis de Sacro Busto.* (Impressum fuit volume istud in urbe Veneta orbis & vrbium regina: calcographica Luce Antonii Iuntae Florentini officina aere proprio ac typis excussum, 1531. Labente mense Martio).

---

[43] The last known edition of this kind that carries the family's name is dated 1620: see (Sacrobosco et al. 1620).

[44] On the treatment of "ordinary works" (*opere comunali*) in Venice's privileges system, see (Nuovo 2013, 213).

Sphera Ioannis de Sacro Busto cum commentariis Petri a Spinosa Artium Magistri, celeberrimique praeceptoris Salmanticensis gymnasij, aeditis. Salmanticæ: excudebat Ioannes Iunta, 1550.
Sphaera Ioannis de Sacro Bosco emendata. Lugduni: apud haeredes Iacobi Iunctae, 1564 (Lugduni: excudebat Symphorianus Barbier).
Sphaera Ioannis de Sacro Bosco emendata. Lugduni: apud haeredes Iacobi Iunctae, 1567 (Lugduni: excudebat Symphorianus Barbier).
La sfera di messer Giovanni Sacrobosco. In Fiorenza: nella stamperia de Giunta, 1571 (In Firenze: appresso i Giunta, 1572).
La sfera di messer Giovanni Sacrobosco. In Firenze: nella stamperia de' Giunta, 1579 (In Firenze: appresso i Giunta, 1579).[45]

One general observation is that in the Giunta's publishing history of the *Sphaera*, chronology follows a dynamic geography. Within the Giunta's network, interest in the Sacobosco tradition migrates between different branches over the years. The first account is the 1508 Venetian edition. Although this is the epicenter of the transnational firm, one might actually be compelled to see this initiative as coming from the periphery of the network. The one primarily responsible for this publishing endeavor was the same Giuntino Giunta who we saw partnering with Lucantonio senior no earlier than 1517.

The two subsequent editions (d'Ascoli et al. 1518b; Sacrobosco et al. 1531) came instead as a direct initiative of Lucantonio, mastermind of the Giunta network. Two observations on the 1518 edition are worth considering. First, this edition arrived in the period when the Giunta's transnational network was still in the making, and therefore when their commercial scope was still limited but already comprised Florence, Rome, and Palermo as steady commercial hubs, and when Juan de Junta was consolidating his presence in Spain. Lucantonio was not yet targeting scholars and high professionals. Primarily he was still a medium-sized publisher concerned with a wide local market. Clues suggest that the 1518 Sacrobosco edition was primarily conceived to settle local unresolved issues. This in-folio gothic types edition was basically a specular copy of the one published by Scoto just five months earlier (d'Ascoli et al. 1518a), but of a slightly better quality.[46] The year of publication shall

---

[45] For the editions enlisted above see, respectively (d'Ascoli 1518b; Sacrobosco et al. 1508, 1531, 1564; Sacrobosco and Espinosa 1550; Sacrobosco and Danti 1571–1572, 1579).

[46] The Scoto edition displays the date January 19, 1518, whereas the Giunta claims the date of 30 June of the same year. The better quality of Lucantonio Giunta's product is particularly appreciable in the composition work with punctuation systematically followed by a fair blank space, which the Scoto edition does not provide with equal consistency. Moreover, the Giunta edition boasted Gerard of Cremona among its featured commentators. Indications of an ongoing competition between the two publishing houses may be visible in Lucantonio Giunta's choice to publish the commentaries of Ugo Benzi (ca. 1360–1439) to the fourth *Fen* of the first *Canon* of Avicenna in December 1517 (Benzi 1517b), after the heirs of Ottaviano Scoto placed Benzi's commentaries to the first *Fen* of the fourth *Canon* on the market in August of the same year (Benzi 1517a). Whatever issue they may have had, it must have been resolved by 1539 when the heirs of Lucantonio Giunta senior partnered with the heir of Ottaviano Scoto in the *Compagnia delli libri della Corona* set to publish costly law books (Nuovo 2013, 59; Nuovo and Coppens 2005, 86–91).

not go unnoticed. This aggressive move by Lucantonio Giunta toward the heirs of Ottaviano Scoto came after the Venetian Senate passed a law in 1517 that formally suspended all standing book privileges and imposed the criterion of absolute novelty to grant any in the future (Squassina 2019, 342). Thus, the time was right to play tricks on one's competitors and reposition yourself in the market. All considered, Lucantonio's reprint of Scoto's edition seems more a crude retaliation against a competitor than a genuine entrepreneurial or cultural choice, especially considering that this was an author he had never shown interest in, and an audience from whom he was disengaging in those very years. Lucantonio was hurting the Scoto family right where they had substantial interest (as their general publishing history of the *Sphaera* shows). He knew he could distract a good number of the Scotos' potential purchasers around the Italian states and beyond by proposing a better option.

The next time Lucantonio signed his name to a Sacrobosco edition seems to have been a more genuine choice. The 1531 edition was published when he had nearly completed restructuring his publishing strategy toward a more specialized professional audience, in light of the availability of an expanded commercial network. Lucantonio found motivation for feeding the market a Sacrobosco edition in the fact that, thanks to the editorial work of Luca Gaurico (1476–1558), he could cover the edition with a ten-year privilege granted by the Senate.[47] The 1531 edition is the last known Sacrobosco edition published by the Giunta of Venice. From that moment on, the Venetian branch, having settled its publishing strategy elsewhere, would no longer enter the competitive orbit of the *Sphaera mundi*. Instead, the Venetian branch choose to compete in a rather more specialized academic market.

Some twenty years passed before the Giunta chose to offer Sacrobosco to their network of users. This time the offer came from the Spanish hub. Juan de Junta, typically more inclined to publish in vernacular Spanish, proposed a Latin version of the text with the commentaries of the local scholar Pedro Espinosa (1485–1536) (Sacrobosco and Espinosa 1550), thus aiming at a *glocal* market. Relying on the fact that the Iberian market would rather count on foreign imports than local imprints (Chap. 7), Juan did not feel the urge to display a privilege on his edition.

Fourteen years passed after Juan de Junta's edition; then, in 1564, the *Sphaera* appeared again as an initiative of the heirs of Jacques de Giunta in Lyon (Sacrobosco et al. 1564), and namely of the then regent Florentine expat Filippo Tinghi (1523–1580). As far as the whole Giunta network was concerned, the Venetian marketplace must have felt saturated with the 1562 edition by Girolamo Scoto, whose path the heir of Lucantonio Giunta did not want to cross. France must have felt like a safer spot to commercialize the *Sphaera*. Again, the edition was sheltered by a book privilege, issued by the French king in 1563. The privilege was valid for seven years in the whole of France and protected the commentaries and the textual emendations of Carmelite theologian and astronomer Francesco Giuntini (1523–1590).[48] The editorial work

---

[47] The privilege was in fact granted to Luca Gaurico, who had already declared his willingness to have his work published by Lucantonio Giunta in the petition; the privilege was approved by 144 senators with the contrary vote of only ten (EMoBookPrivileges 850).

[48] The privilege can be found on the back of the titlepage of the 1564 edition. Francesco Giuntini was a recent member of the Florentine community in Lyon, where he landed in 1561 to escape

of the fellow Florentine scholar (also an expat in Lyon) was the Trojan horse by which Tinghi hoped to enter the high competition surrounding Sacroboco. The royal privilege granted to the heir of Jaques de Giunta secured them an advantageous position in the wide French market, where the edition lured buyers with the newly revised version of a foreign scholar. It is unlikely that the edition was expected to reach the Italian peninsula due to the rumors surrounding Giuntini's heretical religious inclinations.[49] For the exact same reason, however, there must have been hope for a greater sympathy among the readers and collectors of central Europe—and Tinghi was not wrong about that. Giuntini's rendition of the *Sphaera* in fact raised the immediate attention of Antwerp publisher Jean Richard, as well as the heirs of Arnold Birckmann, who both published Giuntini's work in 1566 (Sacrobosco et al. 1566a, b) while the French privilege awarded to the Giunta edition still stood.

Tinghi's initiative to publish Giuntini's *Sphaera* was thoroughly planned, and was aimed at maximizing sales while maintaining an advantageous position in France. This is shown by the fact that, in the context of the first and only known print run of the 1564 edition, Tinghi thought to print a separate batch of copies with a postponed date of 1567. The purpose of doing so was to offer an alleged fresh reprint four years later, while the edition was still shielded for three additional years by the royal privilege. To prove profitable, this marketing strategy required some advance planning. Tinghi, or his advisers, felt they could measure the size of the two batches in order to have the first batch fully or adequately sold before the second was set to enter the market.[50] Once the privilege expired, the Giunta of Lyon did not experiment with the *Sphaera* again, although Francesco Giuntini's contribution to the debate on Sacrobosco would not prove marginal in the years to follow.[51]

The challenge of commercializing the *Sphaera* in the Giunta's network was lastly taken by the heirs of Bernardo Giunta in Florence. Their motivation was to attract a more popular audience by offering a new vernacular Italian translation of Sacrobosco's text. Interestingly, making the *Sphaera* more accessible to a general audience seems to have been a progressive trend in the Giunta's approach to the *Tractatus*. Hence, Lucantonio senior's 1518 in-folio, Gothic types edition (large formats and Gothic types were the standard layout of scholastic works) evolves to a more agile 8°

---

persecution due to his alleged inclination toward religious heterodoxy. On Francesco Giuntini see (Ernst 2001).

[49] On the treatment of Giuntini's work by censorial authorities, see Sander (2018).

[50] This astute marketing strategy was not at all an invention of Tinghi; as far as Italian printers go, this technique has been largely documented for the Giolito (Nuovo and Coppens 2005).

[51] Francesco Giuntini is in fact the attested author of several sixteenth-century publications of the *Sphaera* issued in Antwerp and Lyon between 1566 and 1583; he would experience a brief seventeenth-century revival with two instances in 1610 and 1629, in Cologne and Salamanca, respectively (Sphaera CorpusTracer https://hdl.handle.net/21.11103/sphaera.100357). The idea of bringing Giuntini's work into the publishing portfolio of the Giunta of Lyon was likely Tinghi's personal initiative, which he replicated in several instances over the years (Rozzo 2007, 240). The disengagement between the Giunta of Lyon and the fortunate rendition of the *Sphaera* by Giuntini may have to do with the disengagement between the heirs of Jaque de Giunta and Filippo Tinghi from 1572 (Rozzo 2007, 247). The transnational aspiration of Giuntini's redaction of Sacrobosco is testified to by its circulation at the Frankfurt fair of 1578 (Chap. 6).

for a less erudite readership. As far as vernacular translations go, the Giunta initiative (Sacrobosco et al. 1571–1572) came some twenty years after that of Valerio da Meda and brothers, printed in Milan circa 1550 (Sacrobosco 1550). The text had experienced two previous Venetian imprints in vernacular Italian: a 1537 edition by Bartolomeo Zanetti (1487–1550), translated by Mauro da Firenze (1493–1556) (Mauro da Firenze 1537), and a 1543 edition by Francesco Brucioli (fl. 1541–1545) and brothers, featuring the translation of future apostate Antonio Brucioli (1487–1566) (Sacrobosco and Brucioli 1543). The short timespan between the two Venetian editions shows that there was a market for a vernacular edition of Sacrobosco which the Milanese brothers da Meda served again ten years later. No one, however, followed their example for the next twenty years, knowing that the market was saturated.

With a gap of two decades and a probable void created by the disgraced edition carrying the unfavorable name of Antonio Brucioli on the frontispiece, Iacopo and Filippo Giunta considered the time suitable for a new vernacular edition. On this occasion, they revived an old but unpublished translation by Piervincenzo Danti (1460–1512). The translation earned a privilege from the Grand Duke of Tuscany, Cosimo I de' Medici (1519–1574), likely held by Danti's family.[52] A short sequence of editions, one printed in Perugia in 1574 and another by the Giunta in 1579, proves that the intuition of Iacopo and Filippo was not wrong. The fact that these editions were solely covered by the Grand Duke's privilege would suggest that Tuscany was the prime market for this work, whereas the rest of Italy was still considered something of a secondary market.

Following the 1579 reprint of Danti's translation, none of the branches of the Giunta found sufficient reason to engage in the competition surrounding the *Tractatus de sphaera*.

7   Conclusions

This essay attempted to offer a professional profiling of the Giunta firm, in an effort to illustrate their publishing style and commercial sensibility. It argues that the dynamic definition of the firm's publishing strategy cannot be easily disjointed from its general business plan. More specifically, I have described the Giunta's key vision of an organic network of distribution as the necessary infrastructure to sustain a consistent effort in specializing their publishing offering toward the higher professions.

These being the premises, the publishing history of the *Sphaera* was folded into the general scope of the Giunta's publishing vision. I contend that the Giunta had an episodic interest in engaging in the fierce transnational competition that emerged around the longstanding tradition of Sacrobosco's text during the late Renaissance.

---

[52] The promoter of the edition was Piervincenzo Danti's nephew, Egnazio (Fiore 1986). The fact that a Perugian edition of 1574 also displays the same privilege (Sacrobosco and Danti 1574, 2) is consistent with the tenure of the privilege by the Danti family rather than by the Giunta.

Lucantonio Giunta senior, who has been described as the mind behind much of the Giunta's business plan and publishing strategy, proved interested in the *Sphaera* only after his associate Giuntino first experimented with the potential of this product in the Giunta's target market.

Lucantonio senior's first dealing with Sacrobosco falls under suspicion of having been more an unfair play toward the Scoto house (in the context of market warfare) than a genuine editorial initiative. This may limit sincere instances of interest from the Venetian Giunta to one edition (Sacrobosco et al. 1531)—and this choice originated principally in the possibility of earning a Venetian privilege to protect the investment. Later, the initiative came only from the periphery of the network, and almost only when motivated by textual innovations that might not only captivate the market but also secure the issuing of a book privilege and the consequent commercial advantage.

The categories *center* and *periphery* of the Giunta's wide network have been used to maintain that Sacrobosco quickly escaped the radius of the Giunta's primary interest. The firm, chiefly invested in building a reputation with the high professions and the clergy, did not prioritize the publishing of the *Sphaera*, finding the high competition that surrounded this product too risky, and the revenues too marginal to fit the general plan of the firm.

Abbreviations

Digital Repositories

EDIT16	Istituto Centrale per il Catalogo Unico delle biblioteche italiane e per le informazioni bibliografiche, Censimento nazionale delle edizioni italiane del XVI secolo. https://edit16.iccu.sbn.it/web/edit-16. Accessed 7 June 2021
EMoBookPrices	Early Modern Book Prices. Università degli Studi di Milano. https://emobooktrade.unimi.it/prices. Accessed 7 June 2021
EMoBookPrivileges	Early Modern Book Privileges in Venice. Università degli Studi di Milano. https://emobooktrade.unimi.it/privileges. Accessed 7 June 2021
Sphaera Corpus*Tracer*	Max Planck Institute for the History of Science. https://db.sphaera.mpiwg-berlin.mpg.de/resource/Start. Accessed 7 June 2021
USTC	Universal Short Title Catalogue. University of St. Andrews. https://www.ustc.ac.uk. Accessed 07 June 2021

Archives and Special Collections

ACDF   Archivio della Congregazione per la Dottrina della Fede,
                        Vatican City
MPM   Museum Plantin-Moretus, Antwerp
YRL   University of California, Los Angeles, Charles E. Young
                        Research Library, Department of Special Collections

References

Primary Literature

Benzi, Ugo. 1517a. *Expositio Ugonis senensis in primam fen primi canonis Auicenne cum questionibus eiusdem. Item questionem de febre Antonij Fauentini.* Venice: Heirs of Ottaviano Scoto I.

Benzi, Ugo. 1517b. *Super quarta primi. Super quarta fen primi Aui. preclara expositio cum annotationibus Jacobi de Partibus.* Venice: Lucantonio Giunta I.

Claraevallensis, Bernardus. 1503. *Flores.* Impressum Venetijs, per nobilem virum Luceantonium de Gionta Florentinum.

d'Ascoli, Cecco, Francesco Capuano, Jacques Lefèvre d'Étaples, Johannes Regiomontanus, Michael Scot, Pierre d'Ailly, Robert Grosseteste, Ptolemy (Ps.), and Theodosius of Bithynia. 1518a. *Sphera cum commentis.* Venice: Heirs of Ottaviano Scoto I. https://hdl.handle.net/21.11103/sphaera.101057.

d'Ascoli, Cecco, Francesco Capuano di Manfredonia, Jacques Lefèvre d'Étaples, Johannes Regiomontanus, Michael Scot, Pierre d'Ailly, Robert Grosseteste, Campano da Novara, Gerard of Cremona, Ptolemy, and Theodosius of Bithynia. 1518b. *Sphera mundi.* Venice: Lucantonio Giunta. https://hdl.handle.net/21.11103/sphaera.100047.

De Vio, Tommaso. 1531. *Epistolae Pauli et aliorum apostolorum.* Venice: Lucantonio Giunta I.

De Vio, Tommaso. 1530. *Euangelia.* Venice: Lucantonio Giunta I.

Index librorum omnium qui Venetijs in nobilissima Iuntarum typographia usque ad annum MDXCI. 1591a. Venice: Lucantonio Giunta II.

Index librorum omnium tam ad principales scientias, nempe theologiam, philosophiam et iuris utriusque peritiam, Index librorum omnium tam ad principales scientias, nempe theologiam, philosophiam et iuris utriusque peritiam, quam ad quascunque alias artes et facultates cuiuscunque generis pertinentium, qui ad annum usque praesentem MDXCI editi extant atque habentur penes haeredem Hieronymi Scottii. 1591b. Venice: Heirs of Girolamo Scoto.

Indice copioso e particolare di tutti li libri stampati dalli Gioliti in Venetia fino all'anno M.D.XCII. 1592. Venice: Giovanni Paolo Giolito De Ferrari & Nephews.

Index librorum omnium qui Venetijs in nobilissima Iuntarum typographia usque ad annum MDXCV. 1595. Venice: Lucantonio Giunta II.

Mauro da Firenze. 1537. *Sphera.* Venice: Bartolomeo Zanetti. https://hdl.handle.net/21.11103/sphaera.101009.

Psalmi Dauidici. 1530. Venice: Lucantonio Giunta I.

Sacrobosco, Johannes de. 1550. *Trattato della sphera.* Milan: Valerio Meda & Brothers. https://hdl.handle.net/21.11103/sphaera.101037.

Sacrobosco, Johannes de, and Antonio Brucioli. 1543. *Trattato della Sphera.* Venice: Francesco Brucioli & Brothers. https://hdl.handle.net/21.11103/sphaera.101028.

Sacrobosco, Johannes de, and Pedro Espinosa. 1550. *Sphera*. Salamanca: Juan de Junta. https://hdl. handle.net/21.11103/sphaera.100199.

Sacrobosco, Johannes de, and Gasparino Borro. 1494. *Commentum super tractatum sphaerae mundi*. Venice: Ottaviano Scoto I. https://hdl.handle.net/21.11103/sphaera.100272.

Sacrobosco, Johannes de, and Christoph Clavius. 1601–1603. *In sphaeram Ioannis de Sacro Bosco commentarius*. Venice: Giovanni Battista Ciotti. https://hdl.handle.net/21.11103/sphaera.100370.

Sacrobosco, Johannes de, and Piervincenzo Danti. 1571–1572. *La sfera*. Florence: Filippo & Jacopo Giunta. https://hdl.handle.net/21.11103/sphaera.101084.

Sacrobosco, Johannes de, and Piervincenzo Danti. 1574. *La sfera*. Perugia: Giovanni Bernardino Rastelli. https://hdl.handle.net/21.11103/sphaera.100491.

Sacrobosco, Johannes de, and Piervincenzo Danti. 1579. *La sfera*. Florence: Jacopo Giunta. https:// hdl.handle.net/21.11103/sphaera.101098.

Sacrobosco, Johannes de, Pierre d'Ailly, Francesco Capuano, Jacques Lefèvre d'Étaples, Robert Grosseteste, Johannes Regiomontanus, and Bartolomeo Vespucci. 1508. *Oratio de laudibus astrologiae. Textus Sphaerae. Expositio sphaerae. Annotationes nonnullae. Commentarii in eandem sphaeram. In eandem quaestiones subtilissimae numero 14. Sphaerae compendium. Disputationes contra cremonensia deliramenta. Theoricarum nouarum textus cum expositione*. Venice: Giuntino Giunta. https://hdl.handle.net/21.11103/sphaera.100915.

Sacrobosco, Johannes de, Pierre d'Ailly, Nūr al-Dīn Abū Isḥāq al- Bitrūǧi, Prosdocimo Beldomandi, Francesco Capuano, Gerard of Cremona, Jacques Lefèvre d'Étaples, Luca Gaurico, Robert Grosseteste, Georg von Peuerbach, Johannes Regiomontanus, Michael Scot, and Bartolomeo Vespucci. 1531. *Spherae tractatus*. Venice: Lucantonio Giunta I. https://hdl.handle.net/21.11103/sphaera. 100999.

Sacrobosco, Johannes de, Francesco Giuntini, Pedro Nunes, Élie Vinet. 1566a. *Sphaera*. Antwerp: Jean Richard. https://hdl.handle.net/21.11103/sphaera.101114.

Sacrobosco, Johannes de, Francesco Giuntini, Pedro Nunes, Élie Vinet. 1566b. *Sphaera*. Antwerp: Heirs of Arnold Birckmann. https://hdl.handle.net/21.11103/sphaera.101115.

Sacrobosco, Johannes de, Francesco Giuntini, Pedro Nunes, Pierio Valeriano, and Élie Vinet. 1564. *Sphaera*. Lyon: Heirs of Jacopo Giunta. https://hdl.handle.net/21.11103/sphaera.101071.

Sacrobosco, Johannes de, Francesco Giuntini, Pedro Nunes, Pierio Valeriano, and Élie Vinet. 1567. *Sphaera*. Lyon: Heirs of Jacopo Giunta. https://hdl.handle.net/21.11103/sphaera.101111.

Sacrobosco, Johannes de, Pedro Nunes, Pierio Valeriano, and Élie Vinet. 1562. *Sphaera*. Venice: Girolamo Scoto. https://hdl.handle.net/21.11103/sphaera.101094.

Sacrobosco, Johannes de, Pedro Nunes, Pierio Valeriano, and Élie Vinet. 1569. *Sphaera*. Venice: Girolamo Scoto. https://hdl.handle.net/21.11103/sphaera.101040.

Sacrobosco, Johannes de, Pedro Nunes, Pierio Valeriano, and Élie Vinet. 1574. *Sphaera*. Venice: Heirs of Girolamo Scotto. https://hdl.handle.net/21.11103/sphaera.100309.

Sacrobosco, Johannes de, Pedro Nunes, Pierio Valeriano, and Élie Vinet. 1586. *Sphaera*. Venice: Heirs of Girolamo Scotto. https://hdl.handle.net/21.11103/sphaera.100325.

Sacrobosco, Johannes de, Pedro Nunes, Pierio Valeriano, and Élie Vinet. 1620. *Sphaera*. Venice: Heirs of Girolamo Scotto https://hdl.handle.net/21.11103/sphaera.100358.

Sacrobosco, Johannes de, Georg von Peuerbach, and Johannes Regiomontanus. 1490. *Sphaerae mundi compendium*. Venice: Ottaviano Scoto I. https://hdl.handle.net/21.11103/sphaera.100885.

Secondary Literature

Ammannati, Francesco. 2018. Book prices and monetary issues in Renaissance Europe. *JLIS.it* 9: 179–191. DOI: https://doi.org/10.4403/jlis.it-12454.

Ammannati, Francesco and Angela Nuovo, 2017. Investigating book prices in early modern Europe: questions and sources. *JLIS.it* 8: 1–25. DOI: https://doi.org/10.4403/jlis.it-12365.

Barbero, Giliola. 2018. Ordinary and extraordinary prices in the Giolito *Libri Spirituali* sales list. *JLIS.it* 9: 222–264. DOI: https://doi.org/10.4403/jlis.it-12462.

Bruni, Flavia. 2018. Peace at the Lily. The De Franceschi section in the stockbook of Bernardino Giunti', *JLIS.it* 9: 265–279. DOI: https://doi.org/10.4403/jlis.it-12468.

Camerini, Paolo. 1927. Il testamento di Tommaso Giunti. *Atti e memorie dell'Accademia di Scienze e Lettere in Padova* 43: 191–210.

Camerini, Paolo. 1962–1963. *Annali dei Giunti*. 2 vols. Firenze: Sansoni.

Carmona, Maurice. 1964. Aspects du capitalisme toscan aux XVIe et XVIIe siècles: Les sociétés en commandite à Florence et à Lucques. *Revue d'histoire moderne et contemporaine* 11: 81–108.

Coppens, Christian. 2008. I cataloghi degli editori e dei librai in Italia (secoli XV–XVI). *Bibliologia* 3: 107–124.

Coppens, Christian, and Angela Nuovo. 2018. Printed catalogues of booksellers as a source for the history of the book trade. *JLIS.it* 9: 166–178. DOI: https://doi.org/10.4403/jlis.it-12465.

Decia, Decio, and Renato Delfiol. 1978. *I Giunti, tipografi, editori di Firenze*. Vol. 2. Firenze: Giunti Barbera.

Ernst, Germana. 2001. Francesco Giuntini. In *Dizionario biografico degli italiani*, ed. Mario Caravale, vol. 57: Giulini—Gonzaga. Rome: Treccani.

Fiore, Francesco P. 1986. Piervincenzo Danti. In *Dizionario biografico degli italiani*, ed. Mario Caravale, vol. 32: Dall'Anconata—Da Ronco. Rome: Treccani.

Ginsburg, Jane C. 2013. Proto-property in literary and artistic works: Sixteenth-century papal printing privileges. *The Columbia Journal of Law and the Art* 36: 345–458.

Grendler, Paul F. 1977. *The Roman inquisition and Venetian press, 1540–1605*. Princeton (N.J.): Princeton University Press

Lowry, Martin. 1991. *Book prices in Renaissance Venice: The stockbook of Bernardo Giunti*. Los Angeles: Department of Special Collections, University Research Library, University of California, Los Angeles.

Mercati, Michele. 1937. Vecchi lamenti contro il monopolio de' libri ecclesiastici, specie liturgici. In *Opere minori: raccolte in occasione del settantesimo natalizio sotto gli auspici di S.S. Pio 11*, 482–489. Città del Vaticano: Biblioteca Apostolica Vaticana.

Nuovo, Angela. 2013. *The book trade in the Italian Renaissance*. Leiden/Boston: Brill.

Nuovo, Angela. 2017. The price of books in Italy (XV–XVI Centuries). *I prezzi delle cose nell'età preindustriale. The prices of things in pre-industrial times*, ed. Istituto internazionale di storia economica F. Datini, 107–127. Firenze: Firenze University Press.

Nuovo, Angela, and Christian Coppens. 2005. *I Giolito e la stampa nell'Italia del XVI secolo.* Genève: Droz.

Ottone, Andrea. 2003. L'attività editoriale dei Giunti nella Venezia del Cinquecento. *Dimensioni e problemi della ricerca storica* 2: 43–80.

Ottone, Andrea. 2019. Il privilegio del Messale riformato. Roma e Venezia fra censura espurgatoria e tensioni commerciali. In *Privilegi librari nell'Italia del Rinascimento*, eds. Erika Squassina and Andrea Ottone, 289–329. Milano: Franco Angeli.

Ottone, Andrea. Forthcoming. Market Risks and Empiric Methods of Assessment: A Revised Interpretation of Bernardo di Bernardo Giunti's Stockbook.

Panzanelli Fratoni, Maria Alessandra. 2018. Building an up-to-date library. Prospero Podiani's use of booksellers's catalogues, with special reference to law books. *JLIS.it* 9: 74–113.

Pettas, William. 1974. An international Renaissance publishing family: The Giunti. *The library quarterly: Information, community, policy* 44: 334–349.

Pettas, William. 1980. *The Giunti of Florence: Merchant publishers of 16th century, with a checklist of all the books and documents published by the Giunta in Florence from 1497 to 1570, and with the texts of twenty-nine documents, from 1427 to the eighteenth century.* San Francisco: Bernard M. Rosenthal.

Pettas, William. 1995. *A Sixteenth-century Spanish bookstore: The inventory of Juan de Junta.* Philadelphia: The American Philosophical Society.

Pettas, William. 1997. The Giunti and the book trade in Lyon. *Libri tipografi biblioteche: Ricerche storiche dedicate a Luigi Balsamo*, eds. Arnaldo Ganda, Elisa Grignani, and Alberto Petrucciani, vol. 1. Parma/Florence: Olschki: 169–192.

Pettas, William. 2004. *A history & bibliography of the Giunti (Junta) printing family in Spain 1526–1628.* New Castle: Oak Knoll Press.

Pettas, William. 2013. *The Giunti of Florence: A Renaissance printing and publishing family. A history of the Florentine firm and a catalogue of the editions.* Newcastle: Oak Knoll.

Rhodes, Dennis Everard. 2013. *Giovanni Battista Ciotti (1562–1627): Publisher extraordinary at Venice.* Venezia: Marcianum Press.

Rozzo, Ugo. 2007. Filippo Tinghi editore tipografo e libraio tra Firenze, Lione e Ginevra. *La bibliofilía* 109: 239–270.

Sander, Christoph. 2018. Johannes de Sacrobosco und die Sphaera-Tradition in der katholischen Zensur der Frühen Neuzeit. *N.T.M.* 26: 437–474. https://doi.org/10.1007/s00048-018-0199-6.

Santoro, Marco. 2013. *I Giunta a Madrid: vicende e documenti.* Pisa: Fabrizio Serra editore

Squassina, Erika. 2019. I privilegi librari a Venezia (1469–1545). In *Privilegi librari nell'Italia del Rinascimento*, eds. Erika Squassina and Andrea Ottone, 331–399. Milano: Franco Angeli.

Tenenti, Alberto. 1957. Luc'Antonio Giunti il giovane, stampatore e mercante. In *Studi in onore di Armando Sapori*, vol. II, 1023–1060. Milano: Istituto editoriale cisalpino.

Andrea Ottone earned a doctorate in history at the University of Naples. He is currently a post-doctoral research fellow at the University of Milan's Department of economics, management, and quantitative methods in the context of the ERC founded EMoBookTrade project (Grant Agreement no 694476). He is also a fellow at Berlin's Max Planck Institute for the History of Science. He teaches classes in European history at the Technische Universität Berlin.

Chapter 9
Mathematical Books in Paris (1531–1563): The Development of Publishing Strategies in a Competitive International Market

Isabelle Pantin

Abstract In the first half of the sixteenth century, the printing of mathematical books in Paris developed rapidly. Parisian printers were able to sell their productions to a sufficient local market, and even to achieve such quality as to become attractive at a European level. They achieved this quality with printing houses that could afford to be equipped with all necessary material, but their chief asset was collaboration with skilled teachers and mathematicians who had a talent for innovation. This article first analyzes the Parisian book market in the European context and examines ways we can evaluate the circulation of Parisian mathematical books. Then it focuses on the career and practices of Guillaume Cavellat (ca. 1500–1576), who devoted the main part of his activity to the publication of mathematical books. The network of his scientific collaborators, both inside and outside the University of Paris, was a crucial factor in his prolonged success. Lastly, the essay discusses the impact of the quality of the layout and illustration on the capacity of mathematical books to circulate, and to attract attention and customers, across a wide area: Did the emergence around 1540 of a "Parisian style," with unique features from the beauty of the books as material objects to their efficiency as learning and thinking tools, contribute to the visibility and attractiveness of Parisian mathematical books in the European market?

Keywords Guillaume Cavellat · Parisian printers · Mathematical books · Book history · Book layout

1 Introduction

As Alissar Levy and Richard Oosterhoff show in the present book (Chaps. 2 and 13), the printing of mathematical books in Paris developed rapidly from the late 1480s up to the first decades of the sixteenth century, due to the prominent status of the University of Paris and the heft of its faculty of arts, the affluence of students from

I. Pantin (✉)
Institut d'Histoire Moderne et Contemporaine (IHMC), Ecole Normale Superieure, Paris, France
e-mail: isabelle.pantin@ens.fr

© The Author(s) 2022
M. Valleriani and A. Ottone (eds.), *Publishing Sacrobosco's* De sphaera *in Early Modern Europe*, https://doi.org/10.1007/978-3-030-86600-6_9

different French and foreign nations, and the successful efforts of reformers, who strove for establishing a "mathematical culture" in the bosom of the *Alma mater Parisiensis* (Oosterhoff 2018).

During this period, Parisian printers of mathematical books were able to sell their production to a sufficiently vibrant local market, and even to achieve such quality as to become attractive at a European level. They achieved this quality with printing houses that could afford to be equipped with all necessary material, from fonts for special characters and numerals to woodcuts for diagrams (Levy 2020, 199–204). But their chief asset was their collaboration with skilled teachers and mathematicians who had a talent for innovation.

Richard Oosterhoff has shown that Jacques Lefèvre d'Étaples (1450–1526) and his disciples had influence in Spain and Germany (Oosterhoff 2018, 16–7), but with Oronce Finé (1494–1555), the first royal lecturer in mathematics at the Collège Royal (appointed in 1531) and some of his successors, especially Petrus Ramus (1515–1572), the possibilities of international reach were scaled up. Indeed, though King Francis I (1494–1547) had not drawn up a chart of all the tasks he entrusted to his royal lecturers, he expected them, in any case, to promote disciplines that were not taught (or that were badly taught, according to the humanists) in the very conservative University of Paris, and to teach these disciplines to as wide an audience as possible (Pantin 2006). Finé brilliantly fulfilled this task, mainly through his books. Paradoxically, Pedro Nuñez (1502–1578) acknowledged this fact when he published, in Coimbra, his *De erratis Orontii Finaei*, a devastating critique of the royal lecturer's work (Nuñez 1546; Leitão 2009).

From 1530 onward, the networks of the international book trade had become denser and wider, but the way this impacted the printing of mathematical books in Paris is far from clear. This is the first question I shall investigate.

In any case, a strong indication that the publishers of mathematical books were able to count on a sufficiently stable and large customer base is the appearance of publishers that specialized in this field. Guillaume Cavellat (ca. 1500–1576) is the best example. He was a Parisian bookseller and publisher. From 1549 to 1563, before entering into an association with another bookseller, Jérôme de Marnef (ca. 1515–1595), which led him to diversify his production, he devoted the main part of his activity to the publication of mathematical books. I shall analyze his production and inquire into the reasons why it could be maintained over such a period of time. In particular, the network of his scientific collaborators, either inside or outside the University of Paris was a crucial factor of Cavellat's prolonged success.

The third and last issue I shall address is that of the emergence of a "Parisian style" (notably in layout), which may have contributed to the visibility and attractiveness of Parisian mathematical books in the European market.

2 The Parisian Mathematical Books in an International Context

2.1 The Parisian Book Market and Its International Openness

Without too much exaggeration, it can be argued that the book market of the early modern period had always been an international market: even before the advent of the printing press, books were circulated through well-organized commercial and scholarly networks, which were already highly intensified at the end of the incunabula era, notably through the activity of major printing firms in different European countries, such as Aldus Manutius (ca. 1450–1515), Lucantonio Giunti (1457–1538), and Giovan Battista Sessa (d. ca. 1509) in Venice, Anton Koberger (ca. 1442–1513) in Nuremberg, Johann Amerbach (1440–1513) in Basel, Antoine Vérard in Paris (d. ca. 1512), and others (Harris 2009; Hellinga 2018; Nuovo 2013, 21–96). Moreover, books were sold at seasonal fairs, which were the very nodes of the international trade network, notably those of Frankfurt in Germany (Thompson 1911) (Chap. 6), Lyons in France (Gascon 1971, 237–262; Cassandro 1979; Matringe 2016), and Medina del Campo in Castile (Lapeyre 1955; Casado 2018). During the first half of the sixteenth century, the traffic of books between different towns and countries of Europe increased steadily (Febvre and Martin 1997, 216–247; Pettegree 2011b, 65–90).

Paris, one of the largest printing centers in Europe, participated in this trade, mainly through the activity of important booksellers whose commercial success was supported by extensive networks in France and abroad, like Jean Petit II (fl. 1518–1540), Oudin Petit (fl. 1540–1572), Chrestien Wechel (1495–1554), Jacques du Puys (fl. 1540–1589), and Michel de Vascosan (1500–1576). Moreover, numerous foreign publishers sent agents or correspondents (*facteurs*) to Paris; some of them settled there permanently and opened a bookshop. For instance, the celebrated *Ecu de Bâle* ("Arms of Basel," or *Scutum Basiliense*) was run by Conrad Resch (d. 1552) from 1516 to 1526 to sell books printed in Basel, and then was transferred to Chrestien Wechel (Parent-Charon 1974, 159–160; Bietenholz 1971, 33–34, 171–172 and *passim*). Agents who worked for foreign publishers also bought Parisian books that they sent to the home branch.

Paris attracted foreign booksellers, among many foreign merchants, for it offered an exceptionally large pool of potential customers. It was then the most populated city in northern Europe, with probably about two hundred thousand inhabitants around 1550—more than twice as many as London (Braudel 1976, 83; Chaunu 1978, 198)— and it contained a numerous, wealthy, and educated elite. Paris was the administrative heart of France and the principal residence of the royal court. Its parliament was the highest sovereign court of the kingdom and played a political as well as a judicial role, which brought about bustling activity. Moreover, in the period from 1520 to

1550, the University of Paris was still the most frequented in Europe, with 10,000–11,000 enrolled students, before its rapid decline during the religious wars (Brockliss 1989). This had always been crucial to the prosperity of the book trade in the town: a large community of printers, engravers, bookbinders, and booksellers was able to live on the local market.

Yet this was detrimental to Paris's active engagement in the international trade. Compared with Lyons, Venice, or Antwerp, where book production was export-oriented (Adam 2017; Coornaert 1961; Gascon 1971; Nuovo 2013, 21–87, 298–301; Pettegree 2011a; Voet 1973), Paris imported more books than it exported, to satisfy the needs and curiosities of its students and professors, not to mention its lettered and wealthy citizens, the lawyers of its parliament, the courtiers, and the officers of the royal administration (Parent-Charon 1974, 154–166; Parent-Charon 2000). In this regard, it remained an exception in Europe until at least the first decades of the seventeenth century (Maclean 2012, 207–208).

In any case, mathematical publishing was a niche. In Paris, during the whole period of 1480–1550, the average production of mathematical books to the whole production of books was about one percent (Levy 2020, 138–141). The commercial traffic of this product could thus never affect the external trade balance of any city or country. This essay discusses the possibility of small quantities of mathematical Parisian books sold in the foreign market—it being understood that the booksellers obtained their main profit from the trade of more largely saleable books, like religious books.

2.2   The Circulation of Parisian Mathematical Books: A Few Clues

We have only a few clues to clarify this issue, for the main indicators and data sources ordinarily used to evaluate the commercial circulation of books are incomplete or absent. Registers of account and other documents from notarial records and bank archives do not specify the kind of books that were sold or bought, except in the case of the account books of the booksellers themselves, which are extremely rare in the archives, notably in Paris. Also rare are the postmortem inventories of booksellers and wholesalers, which could give some idea of the breadth of their trade.

The inventories of Renaissance libraries, private and public, provide precious but sparse information, as they remain largely unexplored.[1] The collective catalogues of the Frankfurt fairs are the most detailed and complete sources of information on the international book trade, but they were not published before 1564, so we must resort to more diffuse indicators.

---

[1] The inventory of Renaissance libraries is a field of research in itself and goes beyond the scope of this study. For useful pointers, see (Mandelbrote 2000); for a provisional list of scientist's private libraries, see (Wells 1983).

First of all, as already mentioned, the publishers of mathematical books in Paris had above-average means for the community of printers and booksellers, and they had also a wider expanse of trade.[2] For sixteenth-century Paris, only 25 booksellers' catalogues with prices are known. Drawing up and printing a catalogue was still a rare practice that indicated that the author of the catalogue was actively involved in trade with other booksellers and wholesalers, either inside or outside of Paris. The 25 Parisian catalogues were all published between around 1540 and 1562, by only seven booksellers: Simon de Colines (fl. 1520–1546), Robert Estienne (1503–1559), Chrestien Wechel, Regnault Chaudière (fl. 1509–1554), Jean Loys (fl. 1535–1547), Mathieu David (fl. 1544–1562), and Guillaume Morel (fl. 1548–1564) (Coppens 1992; Proot 2018). At least four of these seven booksellers—Colines, Chaudière, Wechel and Loys—were particularly involved in the publication of mathematical books.

At a more general level, mathematical books were not, as a rule, local-interest products, even in the case of textbooks published at the request of some lecturer: elementary formation—based on a small set of standard manuals—was the same in all European universities. Moreover, the rise of interest in the discipline, due to the development of so-called mathematical humanism, nourished an appetite for pedagogical and editorial novelties in all European countries, wherever those novelties came from. An even stronger appetite was aroused by technical development in many fields, notably in the arts and in tactics of war, and increased the demand for books to teach various aspects of practical mathematics applied to the needs of artisans, engineers, and merchants (Bennett 2006; DeVries 2006; Hall 1997; Long 2011; Oosterhoff 2014; Büttner 2017; Valleriani 2013, 35–39, 2017).

One obvious sign of the circulation of mathematical books is the fact that the publishers of these textbooks frequently borrowed texts, commentaries, diagrams, and layout patterns from foreign editions. In particular, from the beginning of the 1530s, Paris began to play a growing role in this kind of exchange "bank." I have already observed this phenomenon in the case of the two main astronomical textbooks: the *Sphaera* of Johannes de Sacrobosco (d. ca. 1256) and the *Theoricae novae planetarum* of Georg von Peuerbach (1423–1461). In both treatises, several innovative diagrams, conceived by Oronce Finé and first printed in Paris, were borrowed in Wittenberg editions of the same textbooks, and through this intermediary were in turn copied in later Italian, French, and Dutch editions (Pantin 2012b, 2020). Oronce Finé probably also had a key role in the diffusion of the practice of printing the diagrams with the accompaniment of a detailed legend—a practice soon adopted and perfected in the Wittenberg editions of the *Theoricae novae planetarum* (Pantin 2012b, 2013).[3]

---

[2] For the period we consider here, Simon de Colines and Michel de Vascosan, who published mathematical books, were among the wealthiest booksellers in Paris (Parent-Charon 1974, 200–202). Jean Loys was also quite well off; his daughter, Madeleine, married after his death with a dowry of 2,000 *livres tournois* (Renouard and Beaud 1995, 21–22). On Cavellat's fortune, see infra 3.1.

[3] Finé was not the original inventor of this practice. A few preceding examples can be found, notably in the astronomical part of (Reisch 1503).

2.3 Toward a European Regulation of Production?

More important still, when we catalogue the European production of different kinds of mathematical books (scholarly editions of fundamental sources like Euclid and Ptolemy; university textbooks; manuals for practitioners; books of instruments) year by year, a sort of topo-chronological logic emerges.[4] Without ever having a monopoly, different centers of production successively exercised a kind of leadership role in the production of certain books: they printed the largest number of copies through frequent editions (which indicates a wide circulation), and they launched innovations that were imitated elsewhere.

2.3.1 Astronomical Textbooks

Thus, in the case of the pocket treatises of the *Sphaera*, Paris had its moment of leadership during the 1540s and 1550s (Pantin 2020). The *Theoricae novae planetarum* does not provide so clear an example, as it was less frequently printed, but it is worth noting that after the death of Erasmus Reinhold (1511–1553), though his commented edition of Peuerbach was not printed again in Wittenberg (or elsewhere in Germany) before 1580,[5] this remarkable commentary in pocket format was circulated in Catholic countries in editions published in Paris by Charles Perier (fl. 1550–1572) in 1553 and 1556—with many copies also bearing the date 1557 or 1558.[6] Whereas the surviving copies of the 1542 and 1553 Wittenberg editions are in majority kept in the Germanic area, those of the Perier editions are kept in France, in Spain, and, in a significant proportion, in Italy.

Surviving Copies of the First Editions of Reinhold's Commentary on the *Theoricae novae planetarum*:

Edition: (Peuerbach and Reinhold 1542), Wittenberg:

Germanic area: Aschaffenburg, Stiftsbibliothek; Bremen, Staats- und Universitätsbibliothek; Budapest, Országos Széchényi Könyvtár; Freiburg im Breisgau, Universitätsbibliothek; Gotha, Forschungsbibliothek; Jena, Thüringer Universitäts- und Landesbibliothek; München, Bayerische Staatsbibliothek; Nürnberg, Germanisches Nationalmuseum Bibliothek; Strasbourg, Univ.;

---

[4] For important remarks on this phenomenon, and also on the existence of so-called market zones concerning learned books in general, see (Maclean 2012, 194–200).

[5] (Peuerbach and Reinhold 1542); (Peuerbach and Reinhold 1553b): augmented edition; (Peuerbach and Reinhold 1580).

[6] (Peuerbach and Reinhold 1553a), copied from (Peuerbach and Reinhold 1542); (Peuerbach and Reinhold [1556–1558]), copied from the augmented edition: (Peuerbach and Reinhold 1553b). It is worth mentioning that in his augmented 1556 edition Perier used several sheets from his 1553 edition. Thus, this new edition was partly a reprint of the old one.

Wolfenbüttel, Herzog August Bibliothek; Wien, Österreichische Nationalbibliothek; Wien, Universitätsbibliothek.
Other countries: Avignon, BM; Grenoble, BM; Marseille, BM; Oxford, Bodleian; Paris, Observatoire.

Edition: (Peuerbach and Reinhold 1553b), Wittenberg:

Germanic area: Erfurt, Stadt- und Regionalbibliothek; Freiburg im Breisgau, Universitätsbibliothek; Göttingen, Niedersächsische Staats- und Universitätsbibliothek; Halle, Universitäts- und Landesbibliothek; Jena, Universitätsbibliothek; Mainz, Stadtbibliothek; München, Bayerische Staatsbibliothek; Wolfenbüttel, Herzog August Bibliothek; Zürich, Zentralbibliothek; Zwickau, Ratschulbibliothek.

Other countries: Toronto, Univ., Thomas Fisher Rare Book Library.

Edition: (Peuerbach and Reinhold 1553a), Paris:

France: Bibliothèques municipales of Besançon, Bourg-en-Bresse, Lyon, Nancy, Poitiers, and Rouen; Paris, Bibliothèque nationale de France.

Italy: Bologna, Biblioteca universitaria; Firenze, Riccardiana; Lecce, Biblioteca provinciale Nicola Bernardini; Pisa, Biblioteca universitaria; Piacenza, Biblioteca comunale Passerini-Landi; Roma, Biblioteca nazionale centrale; Roma, Biblioteca dell'Osservatorio astronomico.

England: London, Wellcome Library; St. Andrews, University Library.

Edition: (Peuerbach and Reinhold 1556 (A)/1557 (B)/1558 (C), Paris:

France: Bibliothèques municipales of La Rochelle (A), Nancy (B and C), Saint-Mihiel (C), Toulouse (A), Troyes (C), and Versailles (B); Nice, Bibliothèque patrimoniale Romain Gary (B); Paris, Arsenal (C); Paris, Bibliothèque nationale de France (C); Paris, Observatoire (A and B).

Italy: Alessandria, Biblioteca civica Francesca Calvo (B); Bologna, Univ., Dipartimento di Fisica ed Astronomia (A and C); Cremona, Biblioteca statale (B); Fermo, Biblioteca civica Romolo Spezioli (inc., B); Nardò, Biblioteca comunale Achille Vergari (C); Roma, Biblioteca nazionale centrale (B); Roma, Biblioteca Lancisiana (C); Roma, Biblioteca universitaria Alessandrina (B); Roma, Osservatorio astronomico (B); Roma, Vallicelliana (B); Serra Sant'Abbondio, Biblioteca del venerabile Eremo di Fonte Avellana (C).

Spain: Madrid, Compiutense (B); Sevilla, Univ. (B).

Other countries: Dillingen, Studienbibliothek (B); Munich, BSB (A); Salisbury, Cathedral Library (C).

2.3.2 Treatises of the Astrolabe

The books on the astrolabe provide an even more telling example.[7] During the 1540s and 1550s, Paris, and secondarily Lyons, became the first centers in Europe for their production, in spite of the fact that Lyonese and Parisian mathematicians had no outstanding expertise in this field. According to Anthony Turner, mathematical instrument making was "a relatively new trade in Renaissance Paris" where there was "no commercial manufacture and no craft organization" inherited from the Middle Ages (Turner 1998, 63–64). Only the clockmakers were united in a corporation, whose statutes were granted by King Francis I in 1544 at the request of "seven masters clockmakers resede in our city of Paris" (*7 maistres orlogeurs demourans en nostre ville de Paris*) (Lespinasse 1897, 549–552; Bourcerie-Savary and Marc 2019, 132).[8] Despite the efforts of Finé, who strove for the development of practical mathematics, and conceived, described, explained, and sometimes made several instruments (Eagleton 2009), their production in Paris and in Lyon remained limited compared to that in Nuremberg, Antwerp, or Louvain: the French collections bear witness to this fact (Chapiro and Turner 1989; Turner 2018).[9] As concerns, the astrolabes, between Johannes Fusoris (ca. 1365–1436) and Philippe Danfrie (ca. 1532–1608), no Parisian maker left a mark in history (Poulle 1963; Turner 1989), and French mathematicians gave only modest contributions to the literature of the astrolabe.

Anyway, the topo-chronological scheme mentioned above also operates in this case. In the beginning, the books on the astrolabe were only printed in Northern Italy (in Perugia, Ferrara, and Venice). They began as a few medieval Latin treatises written by Robert of Ketton (fl. 1141–1157), Henry Bate of Malines (Heinrich von Mecheln) (1246–1310) (Gunther 1932, II, 368–376), Andalò del Negro (1260–1334), and Prosdocimo Beldomandi (ca. 1370–1428), and an adaptation by Marcantonio Cadamosto (1476–1556) of the *Compendium astrolabii*, then attributed to the Persian Jewish astronomer Messahala (ca. 740–815). To these were added some Byzantine sources, thanks to Giorgio Valla (1447–1500), who included several texts on the astrolabe in his miscellaneous collections of translations (Segonds 1981, 105–111; Raschieri 2012): works by Proclus (412–485), Johannes Philoponus (ca. 490–ca. 570), and

---

[7] Diverse Renaissance astronomical instruments were labeled "astrolabium," in order to exploit the metaphorical meaning of the term (taker of stars), but, unless otherwise specified, I refer to the planispheric astrolabe. See (D'Hollander 2002; Gunther 1932; Webster 1998).

[8] The "maistres orlogeurs" were Fleurant Valleran, Jehan de Presles, Jehan Patin, Michel Pothier, Antoine de Beauvais, Nicolas Miret, and Nicolas Le Contançois.

[9] On the production of mathematical instruments in Lyon in the 1540s and 1550s, see (Virassamynaïken 2015, 104–109): astrolabes and clocks made by Pierre de Fobis, Noël Dauville, and Jean Naze. In Paris, we know of two astrolabes made by Finé, and two made by an itinerant Spanish instrument maker, Michael Picquet or Piquer, in 1542 (Turner 1998, 65, 90). Anthony J. Turner lists Thirty-two names of mathematical instrument makers in Paris from 1500 to 1650. Among them, only four were active between 1540 and 1560: Pierre Le Compassier, "Hilarius," Piquer, and Jehan Quenif (Turner 1998). Philippe Mestrel can be added. Philippe Danfrie, a type-founder and cutter established in Paris from 1556, and later "tailleur général des monnaies," probably made no mathematical instrument before the 1570s (Turner 1989).

Nicephorus Gregoras (1295–1360), and the *Explicatio altera*, an anonymous treatise falsely attributed to Nicephorus (Delatte 1939). Only one contemporaneous work appeared in this period: the *Annulus astronomicus* of Bonetus de Latis (ca. 1450–ca. 1510/1515), which describes an astrolabe finger ring (Gunther 1932, II 326–327; Rodríguez-Arribas 2017).

Treatises on the Astrolabe Printed in Italy (1475–1507)

(Listed in the order of their first publication)

Edition: (Negro 1475).
Edition: (Beldomandi ca. 1477).
Edition: (Bate of Malines 1484).
Edition: (Latis [1492–1493]).
Edition: ([Ketton] ca 1497). Repr. 1502, 1512.
Edition: (Valla and Proclus, Nicephorus Gregoras, Pseudo-Nicephorus 1498).
Edition: (Valla and [Johannes Philoponus] 1501).
Edition: (Cadamosto 1507).

Then, from 1510 to 1545, the more interesting publications were printed in the Germanic area. There were new editions of translations of Arabic and Greek sources (published in Nuremberg and in Basel)—notably the explanation of the *Saphea*, or universal astrolabe, by Azarchel (1028–1087) with corrections by Johannes Schöner (1477–1547), and an edition by Jacob Ziegler (1470–1549). But the main feature of the period was the appearance of modern treatises. Johann Copp (1487–ca. 1558) even published his work in German, in two very different versions: one printed in Augsburg, the other in Bamberg. The *Astrolabium imperatorium* of Johann Stabius (ca. 1450–1522) was not a planispheric astrolabe, but a rectangular instrument devised for making horoscopes and printed on a single sheet (Kremer 2016). Two treatises would be particularly successful: the *Declaratio* of Jacob Köbel (1462–1533) and the *Elucidatio* of Johannes Stöffler (1452–1531).

Treatises on the Astrolabe Printed in the Germanic Area (1510–1545)

(Listed in the order of their first publication)

Edition: (Valla and Pseudo-Nicephorus. 1510).
Edition: (Messahala 1512). In (Reisch 1512); repr. in (Reisch 1515), (Reisch 1535).
Edition: (Stöffler 1513); repr. 1524, 1534, 1535.
Edition: (Stabius [1515]).
Edition: (Copp 1525a).
Edition: (Copp 1525b).
Edition: (Boemus 1529).

Edition: (Colb 1532).
Edition: (Köbel 1532); repr. 1535.
Edition: (Bronkhorst 1533).
Edition: (Arzachel and Schöner 1534).
Edition: (Jordanus Nemorarius 1536).
Edition: (Latis 1537).
Edition: (Dryander 1538).
Edition: (Sophianos 1545).

Over the next two decades, the production of treatises of the astrolabe dwindled in Germany and remained dormant in Italy. Its primary site of production ought to have moved to the Netherlands where the work of Gemma Frisius (1508–1555) was giving a strong impulse to instrument making, cartography, and surveying. In fact, while the astrolabe was being supplanted in some of its traditional functions by more efficient tools (notably astronomical ephemerids), it remained useful in surveying and also in the determination of longitudes (as reliable and exact clocks were still rare). Gemma Frisius himself conceived a new model of astrolabe, described in a treatise posthumously published in Antwerp, in 1556.[10] However, at that date, Paris (followed by Lyons) had already established itself as the first place for the printing of such books.

After a small group of early publications, launched in Paris by Henri Estienne (fl. 1502–1520), Nicolas Savetier (fl. 1524–1532), and Simon de Colines between 1519 and 1534, the number of editions of treatises on the astrolabe increased significantly from 1545, and this production was maintained until the end of the 1550s. In the next decade, the Parisian book trade was to be disorganized by the impact of the wars of religion.

Treatises on the Astrolabe Printed in France up to 1560

(Listed in the order of their first publication)

(Poblacion 1519); repr.: (Poblacion 1527). Both printed in Paris.

(Fernel 1527): Description of an astrolabe invented by Fernel. Dedicated to Diego de Gouvea. Printed in Paris.

(Finé 1527): Description of an astrolabe quadrant, an improved version of the one devised by Profatius Judaeus (ca. 1236–ca. 1305). Printed in Paris (Savetier).

(Finé 1534): Revised and augmented version of (Finé 1527), dedicated to Louis Lasseré (fl. 1498–1546), head of the Collège de Navarre. Printed in Paris (Colines).

---

[10] Gemma's *astrolabus catholicus* was a variant of Arzachel's *Saphaea*: a universal astrolabe using a stereographic projection whose center of projection is the vernal equinox.

(Köbel 1545a): Printed in Paris (Loys and Richard); repr.: (Köbel [1551–1552]). Printed in Paris (Cavellat).

(Jacquinot 1545), in French. Dedicated to Catherine de Medicis. Printed in Paris (Barbé).

(Poblacion 1545): new edition with (Regiomontanus 1545). Printed in Paris (Corbon).

(Focard 1546). Dedicated to Noël Alibert, valet de chambre of Marguerite de Navarre. Printed in Lyons (De Tournes).

(Poblacion 1546): new edition with (Pseudo-Nicephorus 1546; Proclus 1546): Printed in Paris (Barbé and Gazeau); repr.: (Poblacion [1553–1554], [1556–1557]). All in Paris (Cavellat).

(Rojas Sarmiento 1550). Dedicated to Charles the Fifth; repr.: (Rojas Sarmiento 1551). Both printed in Paris (Vascosan).

(Stöffler 1553); repr.: (Stöffler 1564). Both printed in Paris (Cavellat).

(Bassantin 1555) with (Focard 1555). Printed in Lyons (De Tournes).

(Bassantin 1558) with (Jacquinot 1558). Printed in Paris (Cavellat).

(Battink [1557–1558]). Printed in Paris (Du Puys).

(Stöffler 1560). French translation with notes. Printed in Paris (Cavellat).

The treatises printed in Paris and in Lyons were all modern with only one exception: the texts of Pseudo-Nicephorus and Proclus translated by Giorgio Valla (Valla 1498) were joined to Juan Martinez Poblacion's (fl. 1517–1535) *De usu astrolabi* in three successive editions (Poblacion 1546, [1553–1554], [1556–1557]). Two of these treatises had already been printed in Germany (Köbel 1545b; Stöffler 1553), but, for the most part, they were original treatises, some of which explained the construction of new instruments (Fernel 1527; Finé 1527, 1534; Rojas Sarmiento 1550). The authors taught mathematics, but they also wished to attract patrons. Jean Fernel (1497–1558), then a regent at the college of Navarre, dedicated his *Monalosphaerium* (Fernel 1527) to the head of this college, Diego de Gouvea (ca. 1470–1558), who had ties with King João III of Portugal (1502–1557).[11] Dominique Jacquinot offered a manuscript copy of his treatise to Catherine de Médicis (1519–1589), then dauphine of France,[12] two years before dedicating the printed version to her (Jacquinot 1545).

These authors were either French natives (Fernel, Finé, Jacquinot, Focard) or foreigners who stayed in France for various lengths of time: Juan Martinez Poblacion, a Spaniard, who was to be the physician of Eleonora of Habsburg (1498–1558),

---

[11] Fernel's next work, Cosmotheoria (Fernel 1528) was to be dedicated to King João III himself.

[12] This manuscript, dated 1543 and bearing the arms of Catherine, is in Chantilly, Musée Condé, Ms 323.

queen of France (Levy 2020, 95–98); James Bassantin (d. 1568), a Scottish astrologue who remained in France until 1562; and Rudolf Battink (1542–1622), a young mathematician and physician from Groningen, who explained, in his profuse dedication letter to the "Count of Emden," Edzard II, count of East Frisia (ca. 1533–1599), dated from Paris, that he had lived the three previous years in Brabant, and that he hoped he would find in Frisia, his homeland, the patron for his work that he could not find in either Brabant or in France (Battink 1557–1558, A6r-v).

More significant still, though puzzling, is the Paris edition of the treatise of Juan de Rojas Sarmiento (fl. 1545–1550), the most remarkable contribution to the subject since Stöffler's *Elucidatio*. In fact, there is no evidence that Rojas, the younger son of Juan de Rojas y Rojas, first Marquis of Poza, who lived between Spain and Flanders, ever stayed in Paris (the dedication letter to Charles the Fifth bears no date or address). Though it made some impact in Paris,[13] the edition represented, above all, an accomplishment of the Flemish school of instrument making: it published, for the first time, an orthogonic projection conceived by Hugo Helt (ca. 1525–after 1570), a pupil of Gemma Frisius, who was in the service of the Marquis of Poza in Palencia between 1546 and 1549.[14] In 1545–1546, Rojas had also been a pupil of Gemma Frisius in Louvain (Maddison 1966; Turner 1985, 161–165; Esteban Piñeiro and Vicente Maroto 1991, 266–280; Pantin 2009). The fact that this beautiful and prestigious book was printed in Paris, and not in Louvain or Antwerp, is probably indicative of the high reputation of the Parisian production of mathematical books in the mid-sixteenth century. Part of this reputation was certainly due to the quality and elegance of the layout and illustration of these books, but the soundness of the editorial and commercial organization of the booksellers involved in their edition played a key role.

3   The Case of Guillaume Cavellat: A Publishing Strategy Centered on Mathematical Books

The examples provided above show that the production of mathematical books in Paris enjoyed a golden age in the 1540s and 1550s and that the possible causes for this phenomenon are complex. Among them, we find the growing place of mathematics in university syllabuses, the diffusion of mathematical culture in the larger public, encouraged by the cultural politics of King Francis I and King Henry II (1519–1559), and the existence of a sufficient group of collectors and patrons of the arts interested in instrument making. However, the involvement of several printers and booksellers was a prerequisite.

---

[13] The Museum of the History of Science of Oxford has a Rojas astrolabe, made in Paris, by Anthoine Mestrel, in 1551 (inv. 32,378).

[14] In his dedication to the marquis of Poza (dated from Salamanca, 27 September 1549) of a treatise on a kind of astrolabe, Helt thanked the marquis for having received him in his house during three years and having treated him with extreme kindness (Helt 1549, A2r).

For reasons already explained, the printers and publishers of mathematical books had to be specialized and required particular means and capabilities; consequently, they remained a minority. Moreover, even when they succeeded in that field, mathematical books occupied only a limited place in their production, as, in the book trade, economic equilibrium was generally based on the production and circulation of a diversity of books. Henri I Estienne, Jean II Petit, and Simon de Colines are examples of this (Renouard 1843, 1–23; Renouard 1894; Armstrong 1952; Schreiber 1982; Schreiber and Veyrin-Forrer 1995; Amert 2012; Levy 2020). Thus, the case of Guillaume Cavellat, who based the success of his business on the production of mathematical books, is worth noting. Cavellat developed and maintained this policy from 1549 to the beginning of 1563. I described and analyzed his career many years ago (Renouard and Pantin 1986, 1988), and here I shall only highlight some factors that contributed to the success of his project.

3.1 Cavellat's Status Within the Parisian Book Trade

3.1.1 *Libraire Juré* of the University of Paris

Guillaume Cavellat was the first of his family to possess a bookshop in Paris. We know nothing about the social status or occupation of his parents—a likely hypothesis would be that they belonged to the merchant class.[15] In February 1547, Cavellat was in the service of Jean II Petit, one of the wealthiest booksellers in Paris, engaged in the national and international trade (Parent-Charon 1974, 201). Later in the year, he opened his own bookshop and began to publish some books in association with other booksellers (Renouard and Pantin 1986, 15–17). On April 15, 1549, he was named *libraire juré* of the University of Paris. He succeeded his brother-in-law Conrad Badius (d. 1562), son of the celebrated Jodocus Badius (ca. 1462–1535), who had been forced to resign his office after being convicted of Calvinism and fleeing to Geneva (Pallier 2002, 59–60).

The *libraires jurés* were a group of twenty-four booksellers licensed by the University of Paris, and forming part of its *suppôts* (servants). Their number did not change between the middle of the fifteenth century and 1618, when the institution of the "Communauté des libraires et imprimeurs de Paris" made their function obsolete. The university had created this office (well before the advent of the printing press) to gain better control of the book industry in Paris. The *libraires jurés*, who bought their office, were supposed to be good Catholics. They took oaths to observe the rules imposed by the university upon the printers and booksellers (and to report breaches of these rules) and to manage their trade in an honest way (by producing correct

---

[15] A document, dated 6 November 1556 (AnMc, LXXIII, 50), mentions "Jehan Cavellart" (with no detail about his profession), dead before the majority of his sons, Jehan and Guillaume. The document concerns a lawsuit filed against Jehan and Guillaume's former guardian, Guillaume Cardinal, master at arms, about the guardianship accounts (Renouard and Pantin 1986, 13).

editions and charging fair prices). They were granted some honors and privileges, mainly tax exemption, and constituted a kind of aristocracy.

More than three hundred printers and booksellers were settled in Paris in the mid-sixteenth century, and the *libraires jurés* formed an elite in terms of education, wealth, and professional status (Pallier 2002). They were entrusted with missions in collaboration with representatives of the university, like control of the books imported to Paris, when the legislation against the propagation of heretical ideas was reinforced (Higman 1979, 50–52). The benefits received for their trade were many. They have accredited suppliers for doctors, regents, and students, and, in return, their privileged contacts with university members enabled them to find any kind of collaborators they wanted, be they authors, annotators, copy editors, or proofreaders (Shaw 2011, 339–341).

The mere fact that Cavellat was named *libraire juré* at the beginning of his career, taking possession of the office once held by Jodocus Badius, indicates that he had means and ambition, that he was probably well educated, and that he had a good reputation among printers and booksellers, as well as useful connections.

3.1.2 Family Connections[16]

Cavellat first married, probably in 1547, Marie Aleaume (d. ca. 1558), the widow of Guillaume Richard, or Rikart (d. 1545), a native of Leuven who had had a bookshop in Paris from 1540 until his death. From 1547, Cavellat ran Richard's bookshop, situated in rue Saint Jean de Latran, within the precincts of the Commanderie de saint Jean de Latran, in front of the College of Cambray (*ex adverso Collegii Cameracensis*), where the royal lecturers gave their public lectures. He also adopted Richard's emblem and mark, the fat hen (*in pingui gallina*, 'à l'enseigne de la poule grasse').

Among Marie Aleaume's brothers and sisters were Jean Aleaume (d. 1573), a *Doctor Regent* in theology at the University of Paris, Jérôme Aleaume, hosier merchant, husband of Madeleine, daughter of Jodocus Badius (meaning Cavellat became Conrad Badius's brother-in-law), and Pierrette Aleaume—who married first Jean Loys, from Thielt in Flanders (*Tiletanus*), printer and bookseller in Paris from 1535 to his death in 1547, and, second, Thomas Richard (d. 1568), brother of Guillaume Richard.[17] Thomas Richard then inherited Loys's bookshop and printing house.

Some of these family ties certainly influenced Cavellat's editorial program. The Belgian connection was an important point. Guillaume Richard was probably related to Jean Richard, bookseller in Antwerp, who notably published (in 1543 and 1547)

---

[16] If not otherwise mentioned, the references to the sources used in this section are in (Renouard and Pantin 1986, 1–15, 170–179, 393–394, 414–417, 453–454). See also (Renouard 1901, 39–42).

[17] The family ties of Marie Aleaume are mentioned in the contract of marriage of Jehanne Richard, daughter of Marie Aleaume and Guillaume Richard (Paris, Archives nationales, Minutier central, LXXIII, 51), see (Renouard and Pantin 1986, 13).

editions of Sacrobosco that had some similarities with Guillaume Richard's own Parisian editions (Pantin 2020, 284–285).

The emblem and mark *In pingui gallina* had its origins in Cologne and in Antwerp. It had been chosen by Franz Birckmann (fl. 1503–1529), who ran bookshops in Cologne, London, and Antwerp, and had many of his editions printed in Paris (Renouard and Beaud 1986, 48–56). Franz Birckmann published mainly religious works, and this printer device was linked to Matthew 23:37:

> Jerusalem, Jerusalem, you who kill the prophets and stone those sent to you, how often I have longed to gather your children together, as a hen gathers her chicks under her wings, and you were not willing.[18]

Then the mark was used by Franz's brother, Arnold I Birckmann (d. 1542), who in turn had bookshops in Cologne, London, and Antwerp until 1542. Peter Apian (1495–1552) and Gemma Frisius were then published "at the fat hen" long before entering Cavellat's catalog (Van Ortroy 1902, no. 27, 30, 31). Arnold I's widow, Agnès van Gennep, who had bookshops in Paris (1547–1549) and in Cologne (1547–1550), and who settled in Antwerp in 1549, also adopted this printer device, as did her son, Arnold II Birckmann (d. 1576), bookseller in Antwerp (Renouard and Beaud 1986, 58–68).

To get back to more precise data, Guillaume Richard and Jean Loys had published some mathematical books in association (Richard was only a bookseller, whereas Loys was also a printer): Gemma Frisius's *Arithmetica*, Sacrobosco's *De sphaera* and *De anni ratione*, Köbel's *Astrolabii declaratio*, and Heinrich Loriti's (1488–1563) *De geographia*. These books were in turn published by Cavellat, either unchanged or with some improvements.

Arithmetica

(Gemma Frisius 1545a) with (Peletier du Mans 1545): published by Loys and Richard.

(Gemma Frisius 1549) with (Peletier du Mans 1549): copied from (Gemma Frisius 1545b), and published by Guillaume Cavellat and Thomas Richard.

De sphaera (Pantin 2020, 284–285)

(Sacrobosco and Melanchthon [1542–1543]): copied from an edition printed by Josef Klug in Wittenberg in 1540 and published by Loys and Richard.

(Sacrobosco and Melanchthon 1545): new edition with additions and anonymous *scholia*, published by Loys and Richard.

---

[18] "Hierusalem Hierusalem quae occidis prophetas et lapidas eos qui ad te missi sunt quotiens volui congregare filios tuos quemadmodum gallina congregat pullos suos sub alas et noluisti."

(Sacrobosco and Melanchthon 1549): follows (Sacrobosco and Melanchthon 1545) with changes and additions borrowed from an edition printed in Wittenberg by Peter Seitz in 1543; published by Cavellat and Thomas Richard.

De anni ratione

(Sacrobosco 1543): copied from an edition printed in Wittenberg, by Peter Seitz, in 1543, and published by Loys and Richard.

(Sacrobosco 1550): follows (Sacrobosco 1543); published by Cavellat.

Astrolabii declaratio (supra 2.3.2)

(Köbel 1545a): copied from an edition printed in Mainz, by Petrus Jordan, in 1535, and published by Loys and Richard.

(Köbel 1550): follows (Köbel 1545b); published by Guillaume Cavellat.

De geographia [a treatise of mathematical cosmography]

(Loriti 1540) and (Loriti 1542): copied from an edition printed in Freiburg by Johannes Faber in 1530, and published by Loys and Richard.

(Loriti 1550): follows (Loriti 1542); published by Cavellat.

These editions, which played a significant role in the constitution of Cavellat's early catalog, obviously belonged to the "Richard legacy"—all the more so as the first of these editions (dated 1549) were published in association with Thomas Richard, the younger brother of Guillaume Richard, who inherited Loys's bookshop when he married Pierrette Aleaume, Loys's widow and Cavellat's sister-in-law. It is worth noting that the few mathematical books printed by Loys alone were not republished by Cavellat. These are Juán Martínez Siliceo's *Arithmetica* (1540); a Latin anonymous translation, with notes, of the *Sphaera*, falsely attributed to Proclus, with an edition of the Greek text (1543), and the French translation, by Martin de Perer, of Sacrobosco's *Sphaera* (1546). Cavellat was to publish other translations of the treatises on the *Sphere* by Pseudo-Proclus and Sacrobosco.

Anyway, Cavellat did not use the Richard legacy indiscriminately. Among the books published by Loys and Richard, he left aside all those which were not mathematical,[19] which shows that his program had been conceived quite early.

Cavellat's second marriage occurred in June 1559. He married Denise Girault (d. ca. 1616), the daughter of Denise de Marnef (d. 1555), and the bookseller Ambroise Girault (d. 1546). After the death of Girault, an association had been formed, from 1548 to 1555, between Denise de Marnef and her brother, Jérôme de Marnef. At his sister's death, Jérôme became the tutor of Denise Girault, then a child, and arranged her marriage with Cavellat. Cavellat thus allied himself with a family who had played an important role in the history of printing from the end of the fifteenth century, not only in Paris but also in Poitiers, Angers, and Bourges. The size of the dowry shows

---

[19] (Renouard and Beaud 1995, no. 99, 134, 145, 177, 178, 210, 214, 216, 219, 220, 224, 325, 343).

that the newly-wed couple was rather wealthy, according to the standards of the Parisian community of booksellers and printers.[20]

However, this marriage presaged a deep change in the practices of Cavellat's bookshop. On May 1, 1563, Guillaume Cavellat and Jérôme de Marnef signed a contract of permanent association. On the same day, Jérôme de Marnef, who was not married and had no children, made a donation of all his property, after his death, to his two nieces, Guillemette and Denise Girault, and to their husbands (Renouard and Pantin 1986, 170, 177). Therefore, Cavellat abandoned his address and his mark, and the mathematical works he continued to publish were immersed in the diversity of books bearing the mark of the Pelican.[21]

3.2 The Constitution of a Catalogue

Guillaume Cavellat, who was a bookseller, not a printer, deployed all the capabilities and activities of a publisher. This is documented by some publishing contracts that remain in the *Minutier central des notaires de Paris* (Renouard and Pantin 1986, 5, 11–15); by the "notes to the reader" in Latin or in French, which he repeatedly inserted in his books to highlight his efforts for producing useful, well illustrated, and correctly printed books in every branch of mathematics; and, above all, by his remarkably coherent catalog.

Between 1549 and 1563, Cavellat published 76 editions of mathematical books of a total of 150 (the numerous reissue editions are excluded from this count). A few of these editions were luxurious in-quartos, produced in association with other booksellers or with the aid of noble or rich patrons, but the large majority were in-octavos intended for a public of students, teachers, practitioners, and moderately cultured readers. Only the books reserved to specialists, like scholarly editions of Greek sources or astronomical tables, were excluded.

All these publications covered a large part of the field of mathematical knowledge: from the disciplines taught at the university (arithmetic, geometry, music, astronomy, cosmology) to a diversity of applied mathematics (cosmography, surveying, architecture, the making and use of instruments). As already seen in the case of the treatises on the astrolabe, Cavellat tried to publish a variety of works on each topic.

The following census has been drawn in order to better understand the role of Cavellat as a broker of mathematical knowledge. It lists only the first Cavellat editions of the works and, if need be, the first Cavellat editions of improved or revised editions of the same texts.

---

[20] Denise Girault's dowry amounted to 2,000 *livres tournois*. Apart from a few exceptional cases (the dowry of the widow of Jean II Petit amounted to 6,000 *livres tournois*.), the dowries in the wealthiest Parisian bookseller families did not exceed 3,000 *livres tournois*. As a comparison, the dowry of Perrette Bade, when she married Robert Estienne, was 1,000 *livres tournois*. (Renouard and Pantin 1986, 14; Parent-Charon 1974, 193–194).

[21] The Pelican, emblem of Christ's sacrifice, had been the mark of the Marnef family since the end of the fifteenth century.

A–Publication of Foreign Works (in chronological order)

A-a Cavellat's Editions that Follow Preceding French Editions
A–a–1 Reprints without modifications

(Gemma Frisius 1549). Follows (Gemma Frisius 1545a), see *supra* 3.1.2.

(Köbel 1550). Follows (Köbel 1545a), see *supra* 3.1.2.

(Loriti 1550). Follows (Loriti 1542), see *supra* 3.1.2.

(Sacrobosco 1550). Follows (Sacrobosco and Melanchthon 1545), see *supra* 3.1.2.

(Loriti 1551). Follows (Loriti 1543), published by Jacques Gazeau.

(Sagredo 1555). Follows (Sagredo 1550). On this anonymous translation, first published in Paris in 1536, of a Spanish treatise of architecture, see (Lemerle 2011).
A–a–2 With Additions, Corrections, or Other Modifications

(Sacrobosco and Melanchthon 1549), see *supra* 3.1.2.

(Sacrobosco et al. 1550): with new anonymous annotations and other changes. See (Pantin 2020, 285).

(Sacrobosco et al. 1551). Cumulates 1550 notes with scholia by Elie Vinet. See (Pantin 2020, 285).

(Poblacion 1553), see *supra* 2.3.2. Follows (Poblacion 1546), but the illustration is modified, as highlighted by Cavellat in a note to the reader.

(Sacrobosco et al. [1555–1556]): entirely revised by Elie Vinet, with new scholia, and the addition of (Nuñez [1555–1556]). See (Pantin 2020, 286).

(Annuli 1557). A collection built up from several sources, published in Antwerp (Beausard 1553; Gemma Frisius 1548), and Marburg (Annulorum, 1536; Mithobius 1536), to which was added a fragment from (Finé 1532). A note to the reader by Cavellat.

A–b First Editions in France that Follow Foreign Editions
A–b–1 Reprints without Modifications

(Apian 1550). Follows (Apian 1533), not later editions of the *Cosmographiae introductio* published in Venice and in Cologne.

(Valeriano 1550). Follows (Valeriano 1537), published in Rome. This edition was produced as a complement to (Sacrobosco et al. 1550).

(Sacrobosco and Beyer 1551). Follows either the first edition, printed in Frankfurt (Sacrobosco and Beyer 1549) or the second, published in Wittenberg (Sacrobosco and Beyer 1550).

(Ptolemy and Reinhold 1556). This translation and commentary of the first book of the *Almagest* follow the edition published the same year in Wittenberg (Ptolemy and Reinhold 1556).

(Maurolico 1558). Follows the first edition published in Venice (Maurolico 1543).
A–b–2 With Additions, Corrections, or other Modifications

(Borrhaus 1551). Follows the edition published in Strasbourg (Borrhaus 1536), with corrections, notes, and new diagrams by Oronce Finé. A note to the reader by Cavellat.

(Scheubel 1551). First self-contained edition of this introduction to algebra, previously published in Basel with a slightly different title as a prologue to an edition of the *Elements* (Euclid 1550). A note to the reader by Cavellat.

(Stöffler 1553). Follows the editions published in Oppenheim (Stöffler 1513, 1524). The diagrams have been reduced by Guillaume Des Bordes. A note to the reader by Cavellat.

(Gemma Frisius 1556a). Cavellat has combined two editions of the *De principiis* published in Antwerp: (Gemma Frisius 1548), which adds Schöner's *De usu globi*, and (Gemma Frisius 1553), which offers a revised text.

(Gemma Frisius 1557). Cavellat has added to the *De radio astronomico*, published in Antwerp and Leuven (Gemma Frisius 1545b), extracts on the Jacob's staff from (Spangenberg 1539) and (Münster 1551), published in Wittenberg and Basel. The diagrams are new.
A–c French Translations

(Piccolomini 1550). A translation by Jacques Goupyl, of *De la sfera del mundo* (Piccolomini 1540). Dedicated by Goupyl to Catherine de Medicis. The diagrams are new.

(Gemma Frisius 1556b). A translation (with several modifications) by Claude de Boissière of (Gemma Frisius 1556a, see *supra A–b–2*), with a description of Gemma Frisius's world map, published in Antwerp in 1540 (Boissière 1556c).

(Stöffler 1560). See *supra* 2.3.2. An annotated translation, by Guillaume Des Bordes and Jean-Pierre de Mesmes. Dedicated by Guillaume Des Bordes to Jean de Maynemares, seigneur de Bellegarde.

(Gemma Frisius 1561b). A translation with commentary by Pierre Forcadel, of the *Arithmeticae practicae methodus*, already published by Cavellat (Gemma Frisius 1549, 1559, 1561a), without the notes and opuscules by Jacques Peletier du Mans. Dedicated by Forcadel to Hierosme de la Rovere, bishop of Toulon, ambassador of the Duke of Savoie to the king of France. Note of Cavellat to the reader.

B – Publication of French Authors or Foreigners Staying in France

B-a Cavellat's Editions that Follow Preceding French Editions

B–a–1 Reprints without Modifications

(Peletier du Mans 1549). Two opuscules, *De fractionibus astronomicis. De cognoscendis per memoriam calendis*, in (Gemma Frisius 1549). Follows (Gemma Frisius 1545a), see *supra* 3.1.2.

(Bovelles 1555). This edition of the *Geometrie practique* follows one of the editions published in Paris by Regnault Chaudière (Bovelles 1547,1551).

(Finé 1556). This edition of *Les canons...touchant l'usage et practique des communs Almanachs* follows another edition published by Regnault Chaudière (Finé 1551).

(Vinet and Psellos 1557). This edition of translations of treatises by Psellos and Pseudo-Proclus follows an edition given in Bordeaux (Vinet and Psellos 1554).

(Toussain 1560). A commentary to Pseudo-Proclus's *De sphaera* in (Pseudo-Proclus 1560). Follows an edition published in Paris by Martin Le Jeune (Pseudo-Proclus 1556).

B–a–2 With Additions, Corrections, or other Modifications

(Lefèvre d'Etaples 1551). First self-contained edition of the treatise on music, published before in a collection of mathematical textbooks (Jordanus Nemorarius 1496; Jordanus Nemorarius 1514). A note to the reader by Cavellat.

(Boissière 1556b). Adds new chapters to the first edition of *Le...jeu Pythagorique* published in Paris by Anet Brière and Jehan Gentil (Boissière 1554b).

(Poblacion [1553–1554]). See *supra* 2.3.2. Follows (Poblacion 1546) but the illustration is different.

(Finé 1557). Follows the first edition (Peuerbach and Finé 1528). With an epitaph of Oronce Finé, and a dedication by Claude Finé (Oronce's son) to Claude Guiot, king's counselor.

(Jacquinot 1558). See *supra* 2.3.2. Thoroughly revised edition of the texts and figures of (Jacquinot 1545), with the addition of (Bassantin 1558), a revised edition of (Bassantin 1555).

(Finé 1560). First self-contained edition of this treatise on sundials that had been a part of (Finé 1532). Dedication by Jean Finé (Oronce's son) to Odet de Coligny, cardinal of Châtillon. The original diagrams have been reduced by Guillaume Des Bordes. Note to the reader by Cavellat.

(Boissière 1561), New edition, revised and augmented by Lucas Tremblay, of *L'art d'arithmetique*, published before in Paris, by Anet Brière (Boissière 1554a).

B–b First Editions

(Mizauld 1552). A treatise of the sphere in Latin verse. Dedicated by Mizauld to Marguerite de France, sister of Henri II.

(Postel 1552a, b, c). Three anopistographic broadsheets, corresponding to lectures in astronomy, arithmetic, and music.

(Berthot 1554). A Latin arithmetic in the form of a series of aphorisms. Dedicated to Claude d'Espence, doctor in theology.

(Baeza 1555). A Latin arithmetic.

(Boissière 1556d). A translation, by the author, of his *Jeu Pythagorique* (Boissière 1556b), see *supra* B–a–2. Dedicated by Boissière to Antoine Escalin des Aimars.

(Boissière 1556a). *La proprieté et usage des quadrans*. Dedicated to Antoinette de Luynes, wife of Jean de Morel.

(Boissière 1556c) in (Gemma Frisius 1556b). See *supra* A–c.

(Forcadel 1556a, b; Forcadel 1557). A treatise on arithmetic in French. Dedicated by Forcadel to Michel de L'Hospital.

(Ptolemy and Gracilis 1556b). A translation of the first book of the *Almagest*. Dedicated by Gracilis to Jean Magnien, royal lecturer in mathematics.

(Du Hamel 1557). A commentary on Archimedes's *De numero arenae*. Dedicated by the author, royal lecturer in mathematics, to Cardinal Charles of Lorraine.

(Euclid 1557). With a preface to the reader by Gracilis.

(Harambour 1557). Inaugural lecture of a newly appointed royal lecturer in mathematics.

(Forcadel 1558). On the method of calculation by counters. Dedicated by Forcadel to Robert Hurault, seigneur de Belesbat, Michel de L'Hospital's son-in-law.

(Peletier du Mans 1559). A treatise on the square and cubic roots added in (Gemma Frisius 1559).

(Peletier du Mans 1560). Latin translation, by the author, of his *Algebre* (Peletier du Mans 1554). Dedicated by Peletier to Jean Chapelain, first royal physician.

(Bullant 1562). A treatise on geometry for the clock makers.

This census shows that Cavellat drew on all possible means to set up and stock a mathematical bookstore where teachers, students, practitioners, and amateurs could find all they needed. On the one hand, he harnessed the legacy not only of his direct predecessor, Guillaume Richard, as we have seen but also of some great figures of the previous generation of mathematical printers, notably Regnault Chaudière, son-in-law of Jean Higman (d. 1500) and successor of Simon de Colines. Thus, he became the publisher of Jacques Lefèvre d'Étaples and Charles de Bovelles (1479–1566) emblematic authors of the first French mathematical school (Lefèvre d'Étaples 1551; Bovelles 1555).

On the other hand, he offered his customers up-to-date books, either by finding new titles in France or abroad, or by refreshing old ones with improved diagrams, new commentaries, or the addition of accompanying texts—for he liked thematic anthologies: his *Annuli astronomici* (*Annuli* 1557) is a prime example of it. His catalog, where Dutch, German, Swiss, Italian, and Spanish authors were represented, was a showcase of European mathematical knowledge and could attract foreign customers, but it also fitted perfectly into the trend of the "défense et illustration" of the French language and culture, promoted by the Valois kings, for he published French translations of Latin books, and books originally written in French.

3.3 Cavellat's Collaborators

In his "notes to the reader," Cavellat portrayed himself as the main actor in this remarkable achievement, but he also wished to suggest that he was surrounded and assisted by a network of advisors, authors, and collaborators, all experts in mathematics, who ensured the quality of his production. Some of these remained anonymous, like the annotators and revisers of (Sacrobosco and Melanchthon 1549; Sacrobosco et al. 1550) and of (Jacquinot 1558), but a majority signed their work or were named by Cavellat. As they belonged to different milieu, they permit us to glimpse the intricate and elusive topography of the world of mathematical knowledge in mid-sixteenth-century Paris.

3.3.1   The Collège Royal and the University

At the forefront, we find the royal lecturers in mathematics, connected both to the court and to the university (and also very often to the parliament of Paris). The royal lecturers were selected and appointed directly by the king, their salaries were paid from the royal treasury, and they were under the authority of the king's grand almoner, but they nevertheless belonged to the university: their complete title was "lecteur royal en l'Université de Paris" (Compère 2002; Farge 2006).

Cavellat had his shop "in front of the college of Cambray" where the royal lecturers gave their lectures, for want of a building of their own, and he published works by Oronce Finé[22] and his colleagues and successors: Pascal Du Hamel (d. 1565) who

---

[22] Cavellat settled too late to be the first publisher of Finé's main works, but he asked him to revise, annotate, and illustrate a cosmographical textbook (Borrhaus 1551), and he made arrangements with Finé's heirs, Jean and Claude Finé, to republish works unavailable for a long time (Finé 1557, 1560).

joined Finé in 1540 as a second royal lecturer in math; Jean Magnien (d. ca. 1556) who succeeded Finé in 1555; Augier d'Harambour (d. between 1557 and 1562), appointed in 1557 to succeed Magnien; and Pierre Forcadel (d. ca. 1572), appointed in 1563. The sole exception was Jean Pena (d. 1558), appointed in 1557, who had his books published by André Wechel (d. 1581), as did Ramus, Pena's master and protector (Pantin 2006).

Magnien was in touch with Cavellat before his appointment as royal lecturer. In 1551 lectures he referred to work on algebra by Johann Scheubel (1494–1570), previously printed in Basel by Johann Herwagen (ca. 1497–ca. 1558) as the first part of an edition of the *Elements*, and persuaded Cavellat to provide a self-contained edition (Euclid 1550; Scheubel 1551). Cavellat probably sent his edition of the work to Scheubel, a professor in Tübingen, and received a thank-you letter: he related this story in his edition of the *Elucidatio astrolabii* (Stöffler 1553)—also a Tübingen affair—as an illustration of his devotion to mathematics and of his efficiency. This self-promotional discourse was addressed to a foreign as well as a Parisian audience.

In addition to the holders of an "ordinary" chair, Cavellat gathered a larger group of university men into his circle of collaborators and authors: "extraordinary" royal lecturers,[23] ex-lecturers, future royal lecturers, or simply the pupils and friends of actual royal lecturers.

Guillaume Postel (1510–1581) had been a royal lecturer in oriental languages (and probably also, occasionally, in mathematics) from 1538 to 1542, before being dismissed from this charge.[24] Disgraced and condemned by the Sorbonne, he had been absent from Paris from 1544 to the beginning of 1552. Then, until the beginning of 1553, and without being a royal lecturer, he gave lectures at the Collège des Lombards, this time most probably in mathematics in addition to his usual orientalist and messianic themes. His *De universitate*, which reflects his complex concerns, was published in 1552 by the bookseller Jean Gueullart (d. 1554). But the "brevissima synopsis" of his lectures on astronomy, arithmetic, and music was printed on anopisthograph broadsheets that bore the address of Cavellat. According to the title of the astronomical synopsis, they were new augmented edition of sheets printed when Postel was royal lecturer (Postel 1552a, b, c).

Pierre Forcadel, a young mathematician from Béziers, was not a former but a future royal lecturer. Protected by Ramus, who had been appointed in 1551 as royal lecturer in eloquence and philosophy, he entrusted Cavellat with the publication of his books on arithmetic, all written in French (Forcadel 1556a, b, 1557, 1558), and provided him an annotated translation of Gemma Frisius's arithmetic (Gemma Frisius 1561b). When he became a royal lecturer in 1563, he kept Cavellat (then associated with Jérôme de Marnef) as his publisher.

---

[23] On the category of "extraordinary royal lecturers" and, more generally, on the somewhat nebulous signification of the title of "lecteur royal," see (Girot 1998).

[24] In his address to the "professors of the University of Paris," at the beginning of his edition of the *De motibus corporum coelestium*, Postel calls himself "Mathematum et pereg[rinarum] ling[uarum] regius interpres" (Amico 1540, A1r). See also (Postel 1992, 12–20).

Cavellat was acquainted with other pupils or protégés of the royal lecturers. He published a versified treatise of the sphere by Antoine Mizauld (ca. 1512–1578), Finé's closest disciple (Mizauld 1552), the *Numerandi doctrina* of Lodoico Baeza, a Spanish scholar recommended by Jean Magnien (Baeza 1555), and, thanks to Stephanus Gracilis (fl. 1556–1564), an otherwise unknown mathematician, he was able to have Magnien's final works completed and printed (Ptolemy and Gracilis 1556a; Euclid 1557).

3.3.2   Professors of Mathematics Outside the University of Paris

Other professors of mathematics entered Cavellat's circle—some obscure, like Lucas Tremblay (fl. 1561) (Boissière 1561), some famous, like Jacques Peletier du Mans (1517–1582). From 1548 to 1557, Peletier lived in Poitiers, Bordeaux, Torino, and mainly Lyons, where he published his mathematical books. But from the end of 1557–1560 he was in Paris, matriculated at the faculty of medicine to obtain his baccalaureate and license degrees. He entrusted Cavellat to publish a small work on the square and cubic roots, and his Latin translation of his *Algèbre* (Gemma Frisius 1559; Peletier du Mans 1560).

Provincial towns often lacked print shops capable of publishing scholarly books. Claude Berthot, doctor in theology, principal, and superintendent of the College of Dijon, asked Cavellat to publish an arithmetic reduced to a long sequence of aphorisms (Berthot 1554). Berthot had previously resided in Paris, first as one of the bursars of the Collège de Navarre, then, until 1551, as the principal of the Collège de La Marche,[25] and he certainly was well acquainted with Cavellat's bookshop. More importantly, Cavellat came into contact with Élie Vinet (1509–1587).

Vinet spent the greater part of his career at the Collège de Guyenne, in Bordeaux, where the university's imprimeur juré, François Morpain (fl. 1542–1562), had only limited means. Simon Millanges (1540–1623), a disciple of Vinet, the first to establish a large humanist printing firm in Bordeaux, would settle there in 1572. Vinet had thus the majority of his works printed in Lyons by Jean I de Tournes (1504–1564), and in Poitiers by his friends, the brothers De Marnef, Jean III, and Enguilbert II (both d. 1568). However, Vinet made several sojourns in Paris, notably in summer 1543 (when he had his edition of Theognis's *Sententiae elegiacae* printed by Jean Loys), and from July 1549 to the beginning of 1550, during a plague and an insurrection against the salt tax in Bordeaux. At his departure, Vinet left two works that would be printed by Guillaume Morel: corrections to Lucius Annaeus Florus's (70–140) Roman history (Vinet 1550) and an edition of Ausonius, the printing of which was carefully supervised by Jacques Goupyl (ca. 1525–ca. 1564), Vinet's friend and collaborator (Ausonius 1551). Vinet had also prepared scholia on Sacrobosco's

---

[25] In 1540, Berthot, still bachelor in theology, was bursar of the Collège de Navarre (Coyecque 1905, 298, no. 1536); in 1543, now a doctor in theology, he was principal of La Marche (Coyecque 1905, 499, no. 2708; 508, no. 2759); in 1551 he resigned his post at La Marche in favor of Jean Chollet. See the documents in AN, MC/ET/XI/22, MC/ET/XI/72, MC/RE/XI/1, MC/ET/XI/31.

Sphaera. As Loys, whom Vinet had much appreciated, was dead, Goupyl probably introduced his friend to the actual publisher of cosmological textbooks of all kinds, Guillaume Cavellat, to whom he had entrusted a translation of his own (Piccolomini 1550).

However, Goupyl did not go so far as to keep an eye on the whole operation, and Vinet fumed when he received the book (Sacrobosco et al. 1551): he discovered that Cavellat, who had a talent for mixtures, had combined his new scholia with anonymous annotations that had been added to the preceding edition of the *Sphaera* (Sacrobosco et al. 1550). In a letter to Guillaume Guérende, a colleague at the Collège de Guyenne, Vinet bitterly complained about his "shamefully defiled" scholia, "mixed with totally inept" material. This had taught him to be more careful when choosing a publisher:[26] the same letter, by contrast, warmly praises Enguilbert de Marnef. However, Vinet reconsidered his position—he probably realized that Cavellat was, with all his faults, the best publisher for his works on mathematics. An agreement was achieved; Vinet prepared, for Cavellat, a totally revised edition of the *Sphaera* (Sacrobosco 1555–1556) and entrusted him with his translations of Michael Psellos and Pseudo-Proclus (Vinet and Psellos 1557, 1573).

3.3.3 Connections with the Court and the Parisian Elite

Jacques Goupyl can be grouped with the collaborators of Cavellat who had links with the Collège Royal, but his other connections have greater importance. A fine Hellenist, he was named extraordinary royal lecturer in 1552 (Omont 1894) and ordinary lecturer in medicine in 1555 (he had obtained his doctorate in 1548). But he was also interested in poetry and in mathematics and was well introduced at the court and among the Parisian literary elite. He had participated in the anthologies published in 1550–1551 to honor the memory of Marguerite de Navarre (1492–1549) and had connections with the young poets who were to form the *Pléiade*. As mentioned above, Cavellat published a French translation by Goupyl, dedicated to Catherine de Médicis (Piccolomini 1550). In this early phase of his career as a bookseller, he also published the first works of Joachim Du Bellay (1522–1560) and Pierre de Ronsard (1524–1585) (Du Bellay 1549; Ronsard 1550a, b); obviously, he had contacts in this milieu.

These contacts did not bear visible fruit, as Cavellat totally abandoned the field of literature, but they probably provided him with useful acquaintances. For instance, Cavellat collaborated with Claude de Boissière (ca. 1530–ca. 1560), a protégé of Jean de Morel (ca. 1511–1581), who played a crucial role in the success of the *Pléiade* (Boissière 1556a, b, d; Boissière 1561; Magnien 2000; Ford 2004).

Cavellat also collaborated with Jean-Pierre de Mesmes (1516–ca. 1579), nephew of Jean-Jacques de Mesmes (1490–1569), *maître des requêtes* and king's counselor.

---

[26] "Mea nanque in Sacrobosco scholia, tam, foede contaminata, illisque immista longe ineptissima, Lutetiae meo nomine omnia nuper edita, me monuerunt, ut diligenter etiam atque etiam posthac videam, cui lucubratiunculas meas publicandas committam" (Desgraves 1977, 112).

Jean-Pierre de Mesmes, an Italianist and poet who committed himself to studying astronomy from 1552, was a prime collaborator in the Marguerite de Navarre anthologies and was welcomed by Ronsard among the group of the Pléiade from 1553 to 1560 (Pantin 1986; Bingen 2004). Jean-Pierre de Mesmes, an early reader of Copernicus and a disciple of Jofrancus Offusius (ca. 1505–ca. 1570), one of the early annotators of *De revolutionibus*, had a wide and thorough knowledge of recent German astronomical publications (Gingerich 2002, 35–36; Pantin 1986).

3.3.4   The Elusive Milieu of Mathematical Practitioners

The analysis of Cavellat's catalog also provides a glimpse into the world of practitioners who were not attached to an institution and have escaped notice, unless they had published work or signed their name on a notarial document: private teachers, engineers, architects, instrument makers, and instrument sellers.

Jean-Pierre de Mesmes, just mentioned above, was a nobleman and had not to earn his living as a professional mathematician, but he collaborated several times with his friend, Guillaume Des Bordes, "gentilhomme Bordelais," professor of mathematics (probably in private circles) and designer of instruments. From 1551 to 1556, Des Bordes accompanied Jean de Maynemares, Lord of Bellegarde, in military campaigns, as mathematical advisor (Stöffler 1560, â3r). He had a long-term relationship with Cavellat (Stöffler 1553; Finé 1560; Stöffler 1560; Des Bordes 1570); Cavellat called this excellent draughtsman "the Apelles of printers," *typographorum Apellis* (Finé 1560, â4r).

Jean Bullant (ca. 1515–1578), the architect of Anne de Montmorency (1493–1567), Constable of France, is better known. He probably came to a dip in his career around 1560, for he had been dismissed from his charge as Contrôleur des Bâtiments in 1559, and planned a program of publications. In January 1561, he was granted the privilege for printing—at his own expense [*à ses propres cousts et despens*]—a book on sundials (Bullant 1561, A2r). This *Recueil d'horlogiographie* was printed in Paris for Vincent Sertenas (d. 1562) in 1561. But on January 2, 1562, by a notarial act, Bullant sold Cavellat his privilege and six hundred copies of this *Recueil*—probably the entire remaining stock (Renouard and Pantin 1986, 14–15), and Cavellat soon published a sort of appendix to the first treatise, *Petit traicté de geometrie et d'horologiographie pratique* (Bullant 1562). Bullant's main book, *Reigle generalle d'architecture* (Bullant 1564) was also published by Cavellat (and Marnef).

As a publisher of books of instruments, Cavellat was acquainted with instrument makers and instrument sellers. On May 2, 1556, Jehan Gentil sold him the privilege he had obtained for the publication of Claude de Boissière's *Jeu Pythagorique* (Boissière 1556b), on the condition that Cavellat would give him 25 copies of the book, and that an advertisement for his shop would be inscribed on the first page: "The boards and instruments for the game of this book are made in Jehan Gentil's shop, in the courtyard of the courthouse," (Renouard and Pantin 1986, 11–12).[27] The same year, another book by Boissière, *La proprieté et usage des quadrans*, ended with an advertisement for "Jehan Quenif, sur les grans degres du Palais, à l'enseigne du Cylindre," who sold all kinds of dials and clocks (Boissière 1556a). Cavellat and Marnef made a similar deal with "Maistre Benoist Forfait compassier à Paris," concerning Guillaume Des Bordes's *Canomettre*. Benoît Forfait (fl. 1557), who lodged in the Louvre, like other skillful artists and craftsmen, made the instrument described by Des Bordes and provided some improvements to facilitate its use, as mentioned in the title page. He probably supervised the printing process, in the absence of the author, and wrote the dedication to Pierre de Picquet (d. 1579), treasurer of France in the generality of Champagne, to whom he presented the instrument and offered his services (Des Bordes 1570, A2).

This network of collaborators and advisors, which Cavellat visibly worked hard to develop, certainly helped him to keep well informed and to maintain the quality of his publications and his position in the market of mathematical books, not only in Paris but at a European level.

4 The Importance of Being Stylish

This last part is limited to a few remarks about the impact of the quality of the layout on the capacity of mathematical books to circulate and to attract attention and customers across a wide area. By "quality of the layout," I mean a variety of features, from the beauty of the books as material objects to their efficiency as learning and thinking tools.

We are facing an apparent paradox: on the one hand, the circulation of books across borders depended upon sharing a common visual language—all the more so in the case of mathematical texts where symbols, special characters, and other graphic conventions played a crucial role;[28] on the other hand, the ability to stand out in the market was important. To combine both conditions, it was necessary to produce books whose visual language could be appreciated everywhere, and that, at the same time, had the allure of innovation. Yet historians of typography generally agree that between 1530 and the mid-sixteenth century, Parisian printing craft achieved this twofold perfection.

The quality and abundance of the paper manufactured in France was the first asset, while German, English, and Dutch printers depended on importation for getting fine white paper and often had to make do with second-rate, cheaper products (Febvre and Martin 1997, 39–44; Bidwell 2002). Another factor was the cultural policy of King

---

[27] "On faict les jeux de ce present livre et instruments en la cour du pallays en la bouticque de Jehan Gentil." The courtyard and corridor of the Courthouse (the "Palais" on the Île de la Cité) were filled with shops.

[28] By leafing through Smith's *Rara arithmetica* (Smith 1908), one sees that at the beginning of the sixteenth century, there were print shops equipped with special types necessary to print mathematical textbooks in several towns in Italy, Germany, France, and Spain.

Francis I, who encouraged the renewal of the design of books. Under this impetus, Parisian printers progressively replaced their old blackletter fonts with new roman and italic fonts (Martin 2000, 162–209). At that time, Claude Garamont (d. 1561), among other punch-cutters, created typefaces that, in the second half of the century, would be sold or imitated in other European countries (Vervliet 2008, 149–214, 287–320; Updike 2001, 235–236). Until the end of the 1520s, Parisian printers had been strongly influenced by German typography, then Italian influences arrived, and a new style emerged, which stemmed from the fusion of several styles, but "with uniquely Parisian features" (Amert 2012, 47). At the same moment, French and Italian artists, commissioned by the king to decorate his several residences, created a French variant of the Italian style, the so-called Bellifontain style (of the *School of Fontainebleau*), whose motifs and decorative patterns also reached books (Fig. 1).

This was more than a question of aestheticism. In Kay Amert's words, Parisian printers shaped "a graphic style and a set of presentational practices" that would be accepted in all Europe and become an "international style." This "universal graphic apparatus," notably used in scholarly and scientific texts, enabled readers to "preview a text, read it rapidly, ascertain the relative significance of its components, understand how images fit in with it, and refer to things within and outside of it" (Amert 2012, 41, 49)—all facilities offered by the "modern book" (Martin 2000).

Among the features of this "modern book," some are of particular importance in mathematical works, notably the capacity to make complex texts easily readable by clear organization in the architecture of the pages. In the 1530s and 1540s, Simon de Colines was, notably, a master in this art, and his collaboration with Oronce Finé, who was a book designer as well as a mathematician, probably produced the best available balance (for the time) between legibility and richness of information (Pantin 2013). For instance, in a page of Finé's arithmetic, provided with a running title and a folio number, the use of paragraph numbering in the left margin, notes in the right margin, capital letters, aldine leaves, and pilcrows helps the reader to follow the lesson step by step. Examples of operations with fractions are neatly printed and embedded within the text blocks at exactly the right place, which is a measure of the precision of the layout (Fig. 2). In the following decades, other forms of *mise en page*, with fewer graphic markings and more blank spaces, became the standard, for instance in the books printed by Michel de Vascosan (Pantin 2012a).

We can compare two quasi-identical editions printed the same year, one in Wittenberg (Sacrobosco and Melanchthon 1550), the other in Paris (Sacrobosco et al. 1550). Both follow the same pattern and are illustrated by similar diagrams (Pantin 2020), but some features of the layout make a difference. The Wittenberg edition (Fig. 3) has no running titles or folio numbers, contrary to the Paris edition (Fig. 4), which also adds marginal notes. Moreover, there is some coarseness in the contrast between roman and italic used by the Wittenberg printer, while the elegant italic font of the Parisian edition creates a sense of unity; here, there is no need to alternate different typefaces to highlight textual divisions, for the Parisian edition divides each of the four sections of Sacrobosco's *De sphaera* into chapters.

Fig. 1 (Finé 1560, title page). Courtesy of the Library of the Max Planck Institute for the History of Science, Berlin

ARITH. PRACT. LIB. II. 28

14 ꝓPRAETEREA, SI PLVRES DVABVS FRACTIONES
simplices ad vnam simplicem proponátur conuertendæ:fiat primùm
duarum primarum,ad vnam simplicem & communem reductio , eo
quippe modo,quo præfatovndecimo tradidimus numero.Deinde ipsa
communis & simplex fractio,ad quam duæ primæ reductæ sunt , vnà
cum sequenti & in ordine tertia fractione(nec refert quam primam,
secundam,tertiámve feceris)ad vnam simplicem & communem fra=
ctionem simili via reducatur . Rursum eadem communis & simplex
fractio,ad quam tres primæ fractiones conuersæ sunt, vnà cum succe
denti quarta,ad vnam fractionem itidem vertatur simplicem. Idque
deinceps continuetur,pro datarum reducendarúmve fractionũ mul=
titudine:non secus acsi duæ solummodò fractiones simplices,ad vnã
simplicem fractionem continuò proponerentur reducendæ. ¶Placet

<Exemplum>

exemplum adijcere.Sint itaque $\frac{1}{2}$,& $\frac{3}{4}$,& $\frac{5}{6}$,cōuer
tenda ad vnam simplicem fractionem. Reducantur
igitur primùm,duæ primæ fractiones , vtpote $\frac{1}{2}$,&
$\frac{3}{4}$,ad vnam fractionē simplicem : & si præallegatũ
vndecimi numeri documentum 11ó prorsus ignora=
ueris,cóperies ipsas fractiones efficere $\frac{10}{8}$,veluti ob=
iecta numerorum indicat formula : ex quibus $\frac{10}{8}$,4 fiunt ab $\frac{1}{2}$,& 6 à
$\frac{3}{4}$.Per idem rursus documentum vndecimi numeri huiusce capitis,
reducito eadem $\frac{10}{8}$,vnà cum succedenti fractione,vtpote $\frac{5}{6}$,ad vnam
simplicem fractionem:& modò non erraueris, colli=
gentur ex hac vltima reductione $\frac{100}{48}$, quemadmo=
dùm ipsa descriptio numerorum , hic ad maiorem
elucidationem adiuncta, demonstrat. Concludendũ
igitur, $\frac{1}{2}$,& $\frac{3}{4}$,& $\frac{5}{6}$,integri, cóponere $\frac{100}{48}$:quæ 2 fa=
ciunt integra, & præterea $\frac{4}{48}$ siue $\frac{1}{12}$ eiusdem integri.

15 ꝓEODEM MODO, CVM PLVRES QVAM DVAE FRA=
ctionum fractiones,ad vnã simplicem sese offerent reducēdæ fractio=
nem,cócludas fore procedendum.Quælibet enim fractionis fractio,
ad vnam simplicem fractionē seorsum reducenda est:veluti sexto do
cuimus numero. Deinde fractiones ex qualibet singulari reductione
productæ,in vnã tãdem simplicem fractionem conuertantur: quem=
admodùm proximo numero sufficienter expressimus.¶Exemplí cau

<Exemplum>

sa,proponantur reducēda ad vnam fra
ctione simplicem $\frac{1}{2}$ $\frac{1}{3}$,& $\frac{4}{3}$ $\frac{1}{4}$,atque
$\frac{3}{4}$ $\frac{1}{3}$. Conuertes itaque primùm, per
regulã præallegati sexti numeri, quãli=
bet fractionis fractionē seorsum & per

D.iiij.

II.

Terram esse centrum mundi, hoc est, in medio uniuersi sitâ, & uelut punctum respectu firmamenti esse, immo= bilemq̃ consistere.

QVOD autem terra sit in medio
firmamenti sita sic patet. Existentibus in
superficie terræ, stellæ apparent eiusdem quantita
tis, siue sint in medio cœli, siue iuxta ortum, siue
iuxta occasum, & hoc ideo, quia æqualiter terra
distat ab eis.

    Si enim terra magis accederet ad firmamen=
tum in una parte quam in alia, sequeretur quod
aliquis existens in illa parte superficiei terræ, quæ

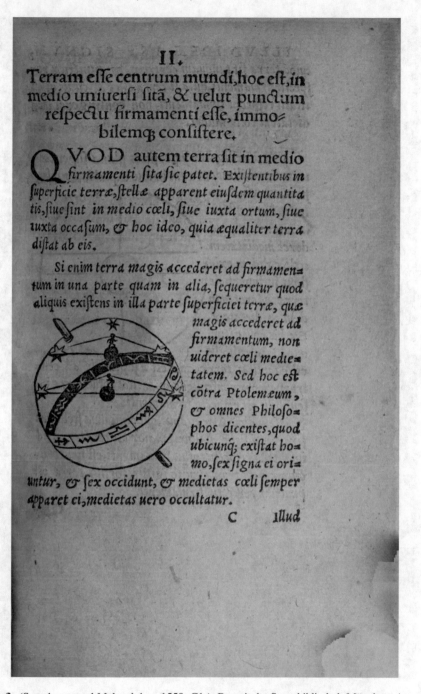

magis accederet ad
firmamentum, non
uideret cœli medie=
tatem. Sed hoc est
côtra Ptolemæum,
& omnes Philoso=
phos dicentes, quod
ubicunq; existat ho=
mo, sex signa ei ori=
untur, & sex occidunt, & medietas cœli semper
apparet ei, medietas uero occultatur.

                  C   Illud

Fig. 3   (Sacrobosco and Melanchthon 1550, C1r). Bayerische Staatsbibliothek München, Astr.u. 164. https://nbn-resolving.org/urn:nbn:de:bvb:12-bsb10998883-7

DE SACRO BOSCO. 19

TRRRAM ESSE CENTRVM
mundi, hoc eſt, in medio uniuerſi ſitam, & ue-
lut punctum reſpectu firmamenti eſſe,
immobilemq́ue conſiſtere.

Cap. 6.

Q̃V̇od autem terra ſit in medio firmamẽ-
ti ſita, ſic patet. Exiſtentibus in ſu-
perficie terræ, ſtellæ apparent eiuſdem quanti-
tatis, ſiue ſint in medio cœli, ſiue iuxta ortum,
ſiue iuxta occaſum: & hoc ideo, quia æqualiter terra diſtat ab eis.

Si enim terra magis accederet ad firmamen
tum in una parte quã in alia, ſequeretur quòd
aliquis exiſtens in illa parte ſuperficiei terræ,

quæ magis acce-
deret ad firma-
mẽtum, nõ uide- Lib. 1. cap. 5.
ret cœli medieta- mag.cõſtru.
tem. Sed hoc eſt
cõtra Ptolemæũ,
& omnes Phi-
loſophos, dicẽtes,
quòd ubicunque
exiſtat homo, ſex ſigna ei oriuntur, & ſex occidũt, & medietas cœli ſemper apparet ei, medietas uerò occultatur.

C iij

Fig. 4 (Sacrobosco et al. 1550, 19r). Universidad Complutense (Alcalá de Henares), BH DER 930(1). Public Domain

Illustration was another key issue, and numerous skilled artists and engravers worked in Paris—some of them, like Guillaume Des Bordes, specialized in mathematics. The vitality of this activity was at least partly due to the exceptional durability of the manuscript trade in the capital, where clients for luxury goods abounded (Orth 2015; Rouse 2019). Some printed mathematical books had been previously presented to their dedicatee in manuscript form. This was the case for (Jacquinot 1545) and for several treatises by Finé.[29]

Not only were the Parisian publishers able to produce books with precise and well-executed diagrams like those of the construction of Finé's sundials (Fig. 5) or Rojas's astrolabe (Rojas Sarmiento 1550), but their illustrations had a touch of modernity. For instance, in (Stöffler 1553), the geometrical figures of the original folio edition have been reduced to fit the in-octavo format, as we have seen, and the images showing the "usages" of the astrolabe have been changed. The typically German woodcuts (Fig. 6) are replaced by gracious pictures with elongated human forms and landscapes in the background in the mannerist style (Fig. 7).[30]

5   Conclusion

All the necessary conditions to produce high-quality mathematical books and to give them exposure in the international market were present in mid-sixteenth-century Paris: the commercial strength and dynamism of the town, the availability of talents of all sorts (mathematicians, copyeditors, drawers, and engravers), the existence of well-equipped print shops, and, more generally, the high standard and good name of Parisian book production. Moreover, ambitious publishers could give an important place to mathematical books in their publishing programs, for they were sure to sell a significant part of their output to the students and lecturers of the University of Paris and its numerous colleges; the possibility of living on the local market was, paradoxically, an aid to venturing securely, at a modest scale, into the international market.[31]

Even before the burgeoning development of bibliographic tools (Charon et al. 2016), Parisian mathematical books were known abroad, at least enough to be noticed by foreign mathematicians, imitated by foreign publishers, and collected in foreign libraries. This does not mean that they circulated in large quantities. We have to keep

---

[29] See BnF, Ms Fr. 1334 (dedication manuscript to King Francis I, dated 1538, of Finé's *Quarré géometrique*, printed in 1556); BnF, Ms Fr. 1337 (dedication manuscript, dated 1543, of the French version of the treatise on the *Metheoroscope*, printed in Latin in 1544); Harvard, Houghton Library, Ms Typ. 57 (dedication manuscript to King Henry II, dated 1551, of Finé's *Sphere du monde*, printed the same year).

[30] In the case of Fig. 6 the model is Lyonese. These figures are inspired by those in (Focard 1546), attributed to Bernard Salomon (Sharratt 2005, 77–8).

[31] On the difficulties of the publishers of learned books who depended mainly on the export of their production, see (Maclean 2009, 12–14, 2012, 211–234).

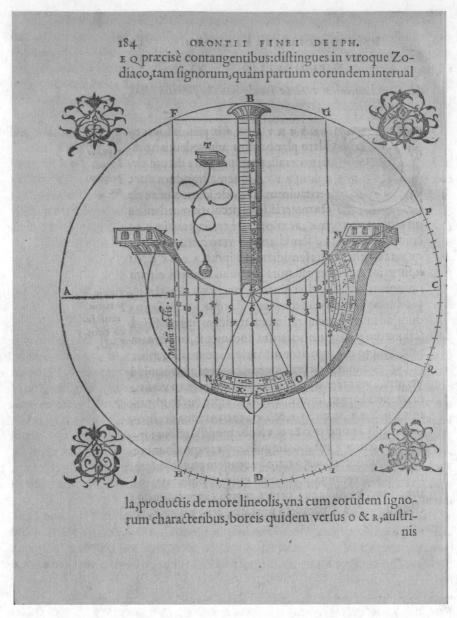

Fig. 5 (Finé 1560, 184). Courtesy of the Library of the Max Planck Institute for the History of Science, Berlin

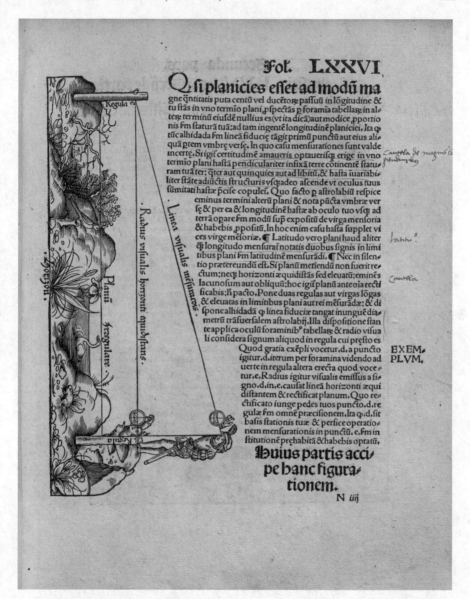

Fig. 6   (Stöffler 1513, 76r). Augsburg, Staats- und Stadtbibliothek, 2 Math 101, fol. LXXVI. https://nbn-resolving.org/urn:nbn:de:bvb:12-bsb11199838-6

DE GEOMETRICIS

videndo aduerte in regula altera erecta, quod vo-
cetur, e. Radius igitur visualis emissus à signo, d,
in, e, causat lineam horizonti æquidistantem : &
rectificat planum. Quo rectificato, iunge pedes tuos
puncto, d, regulæ secundum omnem præcisionem, ita
quòd, d, sit basis stationis tuæ & perfice operatio-
nem mensurationis in punctum, e, secundũ institu-
tionem præhabitam, & habebis optatum.

Huius partis accipe hac figurationem.

Fig. 7 (Stöffler 1553, 170v). Courtesy of the Library of the Max Planck Institute for the History of Science, Berlin

in mind that mathematical books were relatively rare products, even luxury goods targeting a limited clientele, which did not necessarily lessen their visibility.

Abbreviations

Digital Repositories

Sphaera Corpus*Tracer*	Max Planck Institute for the History of Science. https://Db.sphaera.mpiwg-berlin.mpg.de/resource/Start. Accessed 07 June 2021

Archives and Special Collections

An. Archives nationales	Paris
AnMc. Archives nationales	Minutier central, Paris
BnF. Bibliotheque nationale de France	Paris

References

Primary Literature

Amico, Giambattista. 1540. *De motibus corporum coelestium*. Paris: Jacques Kerver.

Annuli. 1557. *Annuli astronomici usus ex diversis authoribus*. Paris: Cavellat.

Annulorum. 1536. *Annulorum trium…ratio atque usus*. Marburg: Eucharius Cervicornus.

Apian, Peter. 1533. *Cosmographiae introductio*. Ingolstadt: Apian.

Apian, Peter. 1550. *Cosmographiae introductio*. Paris: Cavellat.

Ausonius. 1551. *Opera diligentius iterum castigata, et in meliorem ordinem restituta*, ed. Elie Vinet, and Jacques Goupyl. Paris: Guillaume Morel for Jacques Kerver.

Arzachel, and Johann Schöner, ed. 1534. *Sapheae recentiores doctrinae patris Abrusahk Azarchelis*. Nuremberg: [Schöner].

Baeza, Lodoico. 1555. *Numerandi doctrina*. Paris: Cavellat.

Bassantin, James. 1555. *Amplification de l'usage de l'astrolabe*. In *Paraphrase de l'astrolabe*, ed. Focard, Jacques. Lyons: Jean de Tournes.

Bassantin, James. 1558. *Amplification de l'usage de l'astrolabe*. In *L'usage de l'astrolabe*, ed. Dominique Jacquinot. Paris: Cavellat.

Bate of Malines, Henry. 1484. *Magistralis compositio astrolabii*. In *De Nativitatibus*, ed. Abraham ben Haya. Venice: Ratdolt.

Battink, Rudolf. [1557–1558]. *Nova quaedam et compendiosa usus Astrolabii methodus*. Paris: Jacques Du Puys.

Beausard, Pierre. 1553. *Annuli astronomici usus.* Anvers: Steelsius.

Beldomandi, Prosdocimo. [1477]. *Compositio astrolabii.* In *De astrolabio canones*, ed. [Robert of Ketton]. Perugia: Ulysses Lanciarinus.

Berthot, Claude. 1554. *De numerandi ratione aphorismi.* Paris: Cavellat.

Boemus, Matthias. 1529. *Canones astrolabii...de circulis arcubus ac lineis astrolabii.* Wittenberg: Schirlentz.

Boissière, Claude de. 1554a. *L'art d'arithmetique.* Paris: Anet Brière.

Boissière, Claude de. 1554b. *Le tresexcellent et ancien jeu Pythagorique.* Paris: Anet Brière and Jehan Gentil.

Boissière, Claude de. 1556a. *La proprieté et usage des quadrans.* Paris: Cavellat.

Boissière, Claude de. 1556b. *Le tresexcellent et ancien jeu Pythagorique.* Paris: Cavellat.

Boissière, Claude de. 1556c. *L'exposition de la Mappemonde.* See (Gemma Frisius 1556b).

Boissière, Claude de. 1556d. *Nobilissimus et antiquissimus ludus Pythagoreus.* Paris: Cavellat.

Boissière, Claude de. 1561. *L'art d'arithmetique...reveu et augmenté par Lucas Tremblay Parisien.* Paris: Cavellat.

Borrhaus, Martin. 1536. *Elementale cosmographicum*, ed. Achilles Pirmin Gasser. Strasbourg: Crato Mylius.

Borrhaus, Martin. 1551. *Elementale cosmographicum*, ed. Achilles Pirmin Gasser and Oronce Finé. Paris: Cavellat.

Bovelles, Charles de. 1547. *Geometrie practique.* Paris: Regnault Chaudière.

Bovelles, Charles de. 1551. *Geometrie practique.* Paris: Regnault Chaudière.

Bovelles, Charles de. 1555. *Geometrie practique.* Paris: Cavellat.

Bronkhorst, Jan van. 1533. *De astrolabii compositione.* Cologne.

Bullant, Jean. 1561. *Recueil d'horlogiographie, contenant la description, fabrication et usage des horloges solaires.* Paris: Sertenas.

Bullant, Jean. 1562. *Petit Traicté de géométrie et d'horologiographie pratique.* Paris: Cavellat.

Bullant, Jean. 1564. *Reigle géneralle d'architecture des cinq manieres de colonnes.* Paris: Marnef and Cavellat.

Cadamosto, Marcantonio. 1507. *Compendium in usum et operationes Astrolabii Mesahallach*, ed. Francesco Tanzi. Milan: Petrus Martyr and Brothers de Mantegatiis.

Colb, Caspar. 1532. *Astrolabii instrumenti geometricique tabulae auctiores.* Cologne: Hero Alopecius.

Copp, Johann. 1525a. *Erklärung unnd Gründtliche underweysung alles nutzes so in dem Edlen Instrument Astrolabium genannt.* Augsburg: Silvan Otmar.

Copp, Johann. 1525b. *Wie man dieses hochberühmte astronomischer und geometrischer Kunst Instrument Astrolabium brauchen soll.* Bamberg: Georg Erlinger.

Des Bordes, Guillaume. 1570. *La declaration et usage de l'instrument nommé Canomettre.* Paris: Marnef and Cavellat.

Dryander, Johannes. 1538. *Astrolabii canones brevissimi.* Marburg: Eucharius Cervicornus.

Du Bellay, Joachim. 1549. *Recueil de poesie.* Paris: Cavellat.

Du Hamel, Pascal. 1557. *Commentarius in Archimedis librum de numero arenae.* Paris: Cavellat.

Euclid. 1550. *Sex libri priores de geometricis principiis graeci et latini.* In *Brevis regularum algebrae descriptio*, ed. Johann Scheubel. Basel: Johann Herwagen.

Euclid. 1557. *Elementorum libri XV*, ed. and trans. Stephanus Gracilis, and Jean Magnien. Paris: Cavellat.

Fernel, Jean. 1527. *Monalosphaerium sive astrolabii genus generalis horarii structura et usus.* Paris: Simon de Colines.

Fernel, Jean. 1528. *Cosmotheoria.* Paris: Simon de Colines.

Finé, Oronce. 1527. *Descriptio partium et...utilitatum elucidatio quadrantis cujusdam universalis.* Paris: Nicolas Savetier.

Finé, Oronce. 1532. *Protomathesis.* Paris: Gérard Morrhy and Jean Pierre. https://hdl.handle.net/21.11103/sphaera.101190.

Finé, Oronce. 1534. *Quadrans astrolabicus, omnibus Europae regionibus inserviens*, Paris: S. de Colines.

Finé, Oronce. 1542. *Arithmetica practica...Aeditio tertia*. Paris: Simon de Colines.

Finé, Oronce. 1551. *Les canons et documens tres amples touchant l'usage et practique des communs Almanachs, que l'on nomme Ephemedides*. Paris: Regnault Chaudière.

Finé, Oronce. 1556. *Les canons et documens tres amples touchant l'usage et practique des communs Almanachs, que l'on nomme Ephemedides*. Paris: Cavellat.

Finé, Oronce. 1560. *De solaribus horologiis*. Paris: Cavellat.

Focard, Jacques, ed. 1546. *Paraphrase de l'astrolabe*. Lyons: Jean de Tournes.

Forcadel, Pierre. 1556a. *L'arithmeticque*. Paris: Cavellat.

Forcadel, Pierre. 1556b. *Le second livre de l'arithmeticque*. Paris: Cavellat.

Forcadel, Pierre. 1557. *Le troysiesme livre de l'arithmeticque*. Paris: Cavellat.

Forcadel, Pierre. 1558. *L'arithmetique par les gects*. Paris: Cavellat.

Gemma Frisius, Reiner. 1543. *Arithmeticae practicae methodus*. Paris: Jean Loys and Guillaume Richard.

Gemma Frisius, Reiner. 1545a. *Arithmeticae practicae methodus*. In *De fractionibus astronomicis. De cognoscendis per memoriam calendis*, ed. Jacques Peletier du Mans. Paris: Jean Loys and Guillaume Richard.

Gemma Frisius, Reiner. 1545b. *De radio astronomico*. Antwerp and Leuven: Gregorius Bontius and Petrus Phalesius.

Gemma Frisius, Reiner. 1548. *De principiis astronomiae. Annuli astronomici usus*. In *De usu globi astriferi*, Johann Schöner. Antwerp: Steelsius.

Gemma Frisius, Reiner. 1549. *Arithmeticae practicae methodus*. In *De fractionibus astronomicis. De cognoscendis per memoriam calendis*, ed. Jacques Peletier du Mans. Paris: Guillaume Cavellat and Thomas Richard.

Gemma Frisius, Reiner. 1553. *De principiis astronomiae...opus nunc demum ab ipso auctore multis in locis auctum*. Antwerp: Steelsius.

Gemma Frisius, Reiner. 1556a. *De principiis astronomiae et cosmographiae...opus nunc demum ab ipso auctore multis in locis auctum. Annuli astronomici usus*. Johann Schöner. *De usu globi astriferi*. Paris: Cavellat.

Gemma Frisius, Reiner. 1556b. *Les principes d'astronomie et cosmographie avec l'usage du Globe...Plus est adjousté l'usage de l'anneau astronomic....* In *L'exposition de la Mappemonde*, ed. and trans. Claude de Boissière. Paris: Cavellat.

Gemma Frisius, Reiner. 1557. *De radio astronomico*. Paris: Cavellat.

Gemma Frisius, Reiner. 1559. *Arithmeticae practicae methodus*. In *Radicis quadratae et cubicae demonstratio* [*et alia opera*], ed. Jacques Peletier du Mans. Paris: Cavellat.

Gemma Frisius, Reiner. 1561a. *Arithmeticae practicae methodus*. In *Radicis quadratae et cubicae demonstratio* [*et alia opera*], ed. Jacques Peletier du Mans. Paris: Cavellat.

Gemma Frisius, Reiner. 1561b. *L'arithmetique*, trans. and comment. Pierre Forcadel. Paris: Cavellat.

Harambour, Augier d'. 1557. *De instituti sui ratione oratio*. Paris: Cavellat.

Helt, Hugo. 1549. *Declaración y uso del relox español entretexido en las armas de la muy antigua y esclarecida casa de Roias*. Trans. Francisco Sánchez de las Brozas. Salamanca: Juan de Junta.

Jacquinot, Dominique (ed.). 1545. *L'Usage de l'astrolabe, avec un petit traicté de la sphère*. Paris: Barbé.

Jordanus Nemorarius. 1496. *Elementa arithmetica*. In *Elementa musicalia. Epitome in libros arithmeticos Severini Boetii. Rithmimachie*, ed. Jacques Lefèvre d'Étaples. Paris: Wolfgang Hopyl and Johannes Higman.

Jordanus Nemorarius. 1514. *Elementa arithmetica*. In *Elementa musicalia. Epitome in libros arithmeticos Severini Boetii. Rithmimachie*, ed. Jacques Lefèvre d'Étaples. Paris: Henri I Estienne.

Jordanus Nemorarius. 1536. Planisphaerium. In *Sphaerae atque astrorum coelestium ratio, natura et motu*, ed. Jacob Ziegler. Basel: J. Walder.

Köbel, Jacob. 1532. *Astrolabii declaratio*. Mainz: Petrus Jordan.

Köbel, Jacob. 1545a. *Astrolabii declaratio*. Mainz: Petrus Jordan.

Köbel, Jacob. 1545b. *Astrolabii declaratio*. Paris: Loys and Richard.

Köbel, Jacob. 1550. *Astrolabii declaratio*. Paris: Cavellat.

Köbel, Jacob. [1551–1552]. *Astrolabii declaratio*. Paris: Cavellat.

Latis, Bonetus de. [1492–1493]. *Annuli per eum compositi super astrologiam utilitates*. Rome: [Andreas Freitag].

Latis, Bonetus de. 1537. *Annuli astronomici utilitatum liber*. In *Annulorum trium diversi generis*, ed. Johannes Dryander. Marburg: Eucharius Cervicornus.

Lefèvre d'Étaples, Jacques. 1551. *Musica libri quatuor demonstrata*. Paris: Cavellat.

Loriti, Heinrich. 1540. *De geographia liber unus*. Paris: Loys and Richard.

Loriti, Heinrich. 1542. *De geographia liber unus*. Paris: Loys and Richard.

Loriti, Heinrich. 1543. *De sex arithmeticae practicae speciebus epitome*. Paris: Gazeau.

Loriti, Heinrich. 1550. *De geographia liber unus*. Paris: Cavellat.

Loriti, Heinrich. 1551. *De sex arithmeticae practicae speciebus epitome*. Paris: Cavellat.

Maurolico, Francesco. 1543. *Cosmographia in tres dialogos distincta*. Venice: heirs of Giunta.

Maurolico, Francesco. 1558. *Cosmographia in tres dialogos distincta*. Paris: Cavellat.

Messahala. 1512. De compositione astrolabii. In *Margarita philosophica*, ed. Gregor Reisch. Strasburg: J. Grüninger.

Mithobius, Burckard. 1536. *Annuli cum sphaerici tum mathematici usus et structura*. Marburg: Eucharius Cervicornus.

Mizauld, Antoine. 1552. *De mundi sphaera*. Paris: Cavellat.

Münster, Sebastian. 1551. *Rudimenta mathematica*. Basel: Heinrich Petri.

Negro, Andalò del. 1475. *Opus clarissimum astrolabii*. Ferrara: Johannes Picardus.

Nuñez, Pedro. 1546. *De erratis Orontii Finaei regii mathematicarum Lutetiae professoris*. Coimbra: Joannes Barrerius and Ioannes Alvarus.

Peletier du Mans, Jacques. 1554. *L'Algebre*. Lyons: Jean de Tournes.

Peletier du Mans, Jacques. 1560. *De occulta parte numerorum*. Paris: Cavellat.

Peuerbach, Georg von, and Oronce Finé. 1528. *La theorique des cieulx et sept planetes*. Paris: Simon Dubois.

Peuerbach, Georg von, and Oronce Finé. 1557. *La theorique des cieulx et sept planetes*. Paris: Cavellat. (Trans.)

Peuerbach, Georg von, and Erasmus Reinhold. 1542. *Theoricae novae planetarum Georgii Purbachii...ab Erasmo Reinholdo...pluribus figuris auctae et illustratae scholiis...Inserta item methodica tractatio de illuminatione lunae*. Wittenberg: Hans Lufft.

Peuerbach, Georg von, and Erasmus Reinhold. 1553a. *Theoricae novae planetarum auctore G. Purbachio ab Erasmo Reinholdo pluribus figuris auctae et illustratae scholiis*. Paris: Charles Perier.

Peuerbach, Georg von, and Erasmus Reinhold. 1553b. *Theoricae novae planetarum Georgii Purbachii...ab Erasmo Reinholdo...pluribus figuris auctae et illustratae scholiis...auctae novis scholiis in theoria solis ab ipso autore. Inserta item methodica tractatio de illuminatione Lunae*. Wittenberg: Hans Lufft.

Peuerbach, Georg von, and Erasmus Reinhold. [1556–1558]. *Theoricae novae planetarum...ab Erasmo Reinholdo...pluribus figuris auctae et illustratae scholiis...auctae novis scholiis in theoria solis ipso autore. Inserta item methodica tractatio de illuminatione lunae*. Paris: Charles Périer [the copies of this edition bear either the date 1556 (A), or 1557 (B), or 1558 (C)].

Peuerbach, Georg von, and Erasmus Reinhold. 1580. *Theoricae novae planetarum...recens editae et auctae novis scholiis in theoria solis ab ipso autore. Inserta item methodica tractatio de illuminatione lunae*. Wittenberg: Heirs of Johannes Kraft.

Piccolomini, Alessandro. 1540. *De la sfera del mundo*. Venice: [Andrea Arrivabene]. https://hdl.handle.net/21.11103/sphaera.101026.

Piccolomini, Alessandro. 1550. *La sphere du monde*. Trans. Jacques Goupyl. Paris: Cavellat. https://hdl.handle.net/21.11103/sphaera.100661.

Poblacion, Juan Martinez. 1519. *De usu astrolabi compendium*. Paris: Henri Estienne.

Poblacion, Juan Martinez. 1527. *De usu astrolabi compendium*. Paris: Nicolas Savetier.

Poblacion, Juan Martinez. 1545. *De usu astrolabi compendium*. Regiomontanus. *Super usu et constructione astrolabii armillaris Ptolemei*. Paris: Jean Corbon.

Poblacion, Juan Martinez. 1546. *De usu astrolabi compendium*. Pseudo-Nicephorus. *Astrolabus*. Trans. Giorgio Valla. Proclus. *Fabrica, ususque astrolabi*. Trans. Giorgio Valla. Paris: Barbé and Gazeau.

Poblacion, Juan Martinez. [1553–1554]. *De usu astrolabi compendium*. Pseudo-Nicephorus. *Astrolabus*. Trans. Giorgio Valla. Proclus. *Fabrica, ususque astrolabi*. Trans. Giorgio Valla. Paris: Cavellat.

Poblacion, Juan Martinez. [1556–1557]. *De usu astrolabi compendium*. [With] Pseudo-Nicephorus. *Astrolabus*. Proclus. *Fabrica, ususque astrolabi*. Trans. Giorgio Valla. Paris: Cavellat.

Postel, Guillaume. 1552a. *Astronomicae considerationis…brevissima synopsis*. Paris: Cavellat.

Postel, Guillaume. 1552b. *Musices ex theorica ad praxim aptatae compendium*. Paris: Cavellat.

Postel, Guillaume. 1552c. *Theoricae arithmetices compendium*. Paris: Cavellat.

Pseudo-Proclus. 1556. *Sphaera*. Trans. Thomas Linacre, annot. Jacques Toussain. Paris: Martin Le Jeune.

Pseudo-Proclus. 1560. *Sphaera*. Trans. Thomas Linacre, annot. Jacques Toussain. Paris: Cavellat.

Ptolemy, and Stephanus Gracilis. 1556a. *Mathematicae constructionis liber secundus*. Paris: Cavellat.

Ptolemy, and Erasmus Reinhold. 1556a. *Mathematicae constructionis liber primus*. Wittenberg: Hans Luft.

Ptolemy, and Stephanus Gracilis 1556b. *Mathematicae constructionis liber primus*. Paris: Cavellat.

Reisch, Gregor, ed. 1503. *Margarita philosophica*. Friburg: Schott.

Reisch, Gregor, ed. 1512. *Margarita philosophica*. Strasburg: J. Grüninger.

Reisch, Gregor, ed. 1515. *Margarita philosophica*. Strasburg: J. Grüninger.

Reisch, Gregor, ed. 1535. *Margarita philosophica*, ed. Oronce Finé. Basel: Heinrich Petri for Conrad Resch.

[Ketton, Robert of], ed. [1497]. *Astrolabii quo primi mobilis motus deprehendentur canones* [= *De astrolabio canones*]. [Anon.]. *De mensurationibus rerum tractatulus*. Venice: Paganinus de Paganinis.

[Ketton, Robert of], ed. 1502. *Astrolabii quo primi mobilis motus…*. Venice: Petrus Liechtenstein.

[Ketton, Robert of], ed. 1512. *Astrolabii quo primi mobilis motus…*. Venice: Petrus Liechtenstein.

Rojas Sarmiento, Juan de. 1550. *Commentaria in astrolabium quod planisphaerium vocant*. Paris: Vascosan.

Rojas Sarmiento, Juan de. 1551. *Commentaria in astrolabium quod planisphaerium vocant*. Paris: Vascosan.

Ronsard, Pierre de. 1550a. *Les quatre premiers livres des Odes de Pierre de Ronsard…Ensemble son Bocage*. Paris: Cavellat.

Ronsard, Pierre de. 1550b. *Ode de la paix*. Paris: Cavellat.

Sacrobosco, Johannes de. 1543. *Libellus de anni ratione seu computus ecclesiasticus*. Paris: Jean Loys and Guillaume Richard.

Sacrobosco, Johannes de. 1550. *Libellus de anni ratione seu computus ecclesiasticus*. Paris: Thomas Richard and Guillaume Cavellat.

Sacrobosco, Johannes de, and Guillaume Des Bordes. 1570. *La sphere augmentee de nouveaux commentaires et figures*. Paris: Marnef and Cavellat. https://hdl.handle.net/21.11103/sphaera. 100245.

Sacrobosco, Johannes de, and Hartmann Beyer. 1549. *Quaestiones novae in libellum de sphaera*. Frankfurt: Peter Brubach. https://hdl.handle.net/21.11103/sphaera.100645.

Sacrobosco, Johannes de, and Hartmann Beyer. 1550. *Quaestiones novae in libellum de sphaera*. Wittenberg: Peter Seitz. https://hdl.handle.net/21.11103/sphaera.101121.

Sacrobosco, Johannes de, and Hartmann Beyer. 1551. *Quaestiones novae in libellum de sphaera*. Paris: Cavellat. https://hdl.handle.net/21.11103/sphaera.101038.

Sacrobosco, Johannes de, and Philip Melanchthon. [1542–1543]. *De sphaera liber. Plurimis novis typis auctus*. Paris: Jean Loys. https://hdl.handle.net/21.11103/sphaera.101027.

Sacrobosco, Johannes de, and Philip Melanchthon. 1545. *Sphaera typis auctior, quam antehac, atque...castigatior*. Paris: Jean Loys and Guillaume Richard. https://hdl.handle.net/21.11103/ sphaera.101054.

Sacrobosco, Johannes de, and Philip Melanchthon. 1549. *Sphaera typis auctior, quam antehac....* Paris: Thomas Richard and Guillaume Cavellat. https://hdl.handle.net/21.11103/sphaera.100137.

Sacrobosco, Johannes de, and Philip Melanchthon. 1550. *Libellus de sphaera*. Wittenberg: Johann Krafft. https://hdl.handle.net/21.11103/sphaera.100157.

Sacrobosco, Johannes de, Philip Melanchthon, and Pierio Valeriano. 1550. *Sphaera typis auctior, quam antehac...cum annotationibus*. Paris: Cavellat. https://hdl.handle.net/21.11103/sphaera. 100952.

Sacrobosco, Johannes de, Philip Melanchthon, Élie Vinet, Pierio Valeriano. 1551. *Sphaera typis auctior, quam antehac...cum annotationibus, et scholiis...Eliae Vineti*. Paris: Cavellat. https:// hdl.handle.net/21.11103/sphaera.101039.

Sacrobosco, Johannes de, Élie Vinet, Pierio Valeriano, and Pedro Nuñez. [1555–1556]. *Sphaera emendata. Eliae Vineti Santonis scholia...ab ipso autore restituta*. Trans. Elie Vinet. In *Annotatio in extrema verba capitis de climatibus*, ed. Pedro Nuñez. Paris: Cavellat. https://hdl.handle.net/ 21.11103/sphaera.100198.

Sagredo, Diego de. 1550. *Raison d'architecture antique*. Paris: Claude and Regnault Chaudiere.

Sagredo, Diego de. 1555. *Raison d'architecture antique*. Paris: Cavellat.

Scheubel, Johann. 1551. *Algebrae compendiosa facilisque descriptio*. Paris: Cavellat.

Sophianos, Nikolaos. 1545. *De praeparatione et usu astrolabii annulario*. Basel.

Spangenberg, Johann. 1539. *Computus ecclesiasticus in pueriles quaestiones redactus*. Wittenberg: G. Rhaw.

Stabius, Johannes. [1515]. *Astrolabium imperatorium*. Nuremberg: Johann Stuchs.

Stöffler, Johannes. 1513. *Elucidatio fabricae ususque astrolabii*. Oppenheim: Jacob Köbel.

Stöffler, Johannes. 1524. *Elucidatio fabricae ususque astrolabii*. Oppenheim: Jacob Köbel.

Stöffler, Johannes. 1553. *Elucidatio fabricae ususque astrolabii*. Paris: Cavellat.

Stöffler, Johannes. 1564. *Elucidatio fabricae ususque astrolabii*. Paris: Marnef and Cavellat.

Stöffler, Johannes. 1560. *Traité de la composition et fabrique de l'Astrolabe et de son usage*. Trans. Jean-Pierre de Mesmes. Paris: Cavellat.

Valeriano, Giovanni Pietro. 1537. *Compendium in sphaeram*. Rome: Antonio Blado. https://hdl.han dle.net/21.11103/sphaera.101194.

Valeriano, Giovanni Pietro. 1550. *Compendium in sphaeram*. Paris: Cavellat.

Valla, Giorgio, 1498. *Giorgio Valla Placentino Interprete. Hoc in volumine hec continentur Nicephori logica....* Venice: Simon Papiensis Bevilacqua.

Valla, Giorgio. 1501. *De expetendis et fugiendis rebus*. Venice: Aldus.

Valla, Giorgio, and Pseudo-Nicephorus. 1510. *Ratio de compendiaria arte disserendi et de astrolabio*. [Basel?].

Vinet, Elie. 1550. *Castigationes in Luci Flori libros quatuor de historia Romana*. Paris: Guillaume Morel.

Vinet, Elie, and Psellos. 1554. *Ex mathematico Pselli breviario, arithmetica, musica, geometria: Sphaera vero ex Procli graeco*. Bordeaux: François Morpain.

Vinet, Elie, and Psellos. 1557. *Michael Psellus de arithmetica, musica, geometria: et Proclus de Sphaera*. Paris: Cavellat, 1557.

Vinet, Elie, and Pseudo-Proclus. 1573. *La Sphaire de Procle*. Paris: Marnef and Cavellat.

Secondary Literature

Adam, Renaud. 2017. Living and printing in Antwerp in the late fifteenth and early sixteenth centuries: A social enquiry. In *Netherlandish culture of the sixteenth century: Urban perspectives*, eds. Ethan Matt Kavaler and Anne-Laure Van Bruaene, 83–98. Turnhout: Brepols.

Amert, Kay. 2012. *The scythe and the rabbit. Simon de Colines and the culture of the book in Renaissance Paris*, ed. Robert Bringhurst. Rochester, NY: Cary Graphic Art Press.

Armstrong, Elizabeth. 1952. Jacques Lefèvre d'Etaples and Henri Estienne the Elder, 1502–1520. In *The French mind: Studies in honour of Gustave Rudler*, ed. W. Grayburn Moore, 17–33. Oxford: Sutherland and Starkis.

Bennett, James A. 2006. The mechanical arts. In *The Cambridge history of science. III: Early modern science*, eds. Lorraine Daston and Katharine Park, 673–695. Cambridge: Cambridge University Press.

Bidwell, John. 2002. French paper in English books. In *The Cambridge history of the book in Britain. IV. 1557–1695*, eds. John Barnard and D.F. McKenzie, 583–601. Cambridge: Cambridge University Press.

Bietenholz, Peter G. 1971. *Basle and France in the sixteenth century: The Basle humanists and printers in their contacts with francophone culture*. Geneva: Droz.

Bingen, Nicole. 2004. Jean-Pierre de Mesmes: à propos de deux contributions récentes. *Bibliothèque d'humanisme et Renaissance* 66: 331–357.

Bourcerie-Savary, Émeline., and Marc Bourcerie. 2019. *Le temps: Aspects scientifiques et historiques*. London: Iste.

Braudel, Fernand. 1976. Pre-modern towns. In *The early modern town: A reader*, ed. Peter Clark, 53–90. London: Longman.

Brockliss, Laurence W. 1989. Patterns of attendance at the University of Paris. In *Dominique Julia and Jacques Revel*, ed. Les Universités and européennes du XVIe au XVIIIe siècle. Histoire sociale des populations étudiantess, vol. 2: France, 1400–1800, 489–492. Paris: EHESS.

Büttner, Jochen. 2017. Shooting with ink. In *The structures of practical knowledge*, ed. Matteo Valleriani, 115–166. Dordrecht: Springer Nature.

Casado Alonso, Hilario. 2018. Comprar y vender en las ferias de Castilla durante los siglos XV y XVI. In *Faire son marché au Moyen Âge: Méditerranée occidentale, XIIIe-XVIe siècle*, eds. Judicaël Petrowiste and Mario Lafuente Gómez, 111–131. Madrid: Casa de Velázquez.

Cassandro, Michele. 1979. *Le fiere di Lione e gli uomini d'affari italiani nel Cinquecento*. Florence: Baccini & Chiappi.

Chapiro, Adolphe, Chantal Mestin-Perrier, and Anthony Turner. 1989. *Musée national de la Renaissance. Château d'Écouen. Catalogue de l'horlogerie et des instruments de précision du début du XVIe au milieu du XVIIe siècle*. Paris: Réunion des musées nationaux.

Charon, Annie, Sabine Juratic, and Isabelle Pantin. 2016. *L'Annonce faite au lecteur: La circulation de l'information sur les livres en Europe (16e–18e siècles)*. Louvain: Presses Universitaires de Louvain.

Chaunu, Pierre. 1978. *La Mort à Paris*. Paris: Fayard.

Compère, Marie-Madeleine. 2002. Collège royal. In *Les collèges français 16e–18e siècles. Répertoire 3: Paris*, 407–413. Paris: Institut national de recherche pédagogique.

Coornaert, Emile. 1961. *Les Français et le commerce international à Anvers: Fin du XVe-XVIe siècle*. Paris: Marcel Rivière.

Coppens, Christian. 1992. Sixteenth-century octavo publishers' catalogues mainly from the Omont collection. *De Gulden Passer* 70: 5–34.

Cousseau, Marie-Blanche. 2016. *Etienne Colaud et l'enluminure parisienne sous François 1er*. Tours: Presses universitaires François-Rabelais.

Coyecque, Ernest. 1905. *Recueil d'actes notariés relatifs à l'histoire de Paris et de ses environs au XVIe siècle, t. I, 1498–1545*. Paris: Imprimerie nationale.

Delatte, A. 1939. L'*Explicatio altera*, traité anonyme sur l'astrolabe. In *Anecdota Atheniensia et alia*, vol. 2, 254–262. Liège: Faculté de Philosophie et de Lettres de l'Université de Liège.

Desgraves, Louis. 1977. *Élie Vinet humaniste de Bordeaux (1509–1587): Vie, bibliographie, correspondance, bibliothèque*. Geneva: Droz.

DeVries, Kelly. 2006. Sites of military science and technology. In *The Cambridge history of science*, vol. 3: *Early modern science*, eds. Lorraine Daston and Katharine Park, 306–319. Cambridge: Cambridge University Press.

D'Hollander, Raymond. 2002. *L'Astrolabe, histoire, théorie et pratique*, 2nd ed. Paris: Eyrolles.

Eagleton, Catherine. 2009. Oronce Fine's sundials: the sources and influences of *De solaribus horologiis*. In *The worlds of Oronce Fine: Mathematics, instruments and print in Renaissance France*, ed. Alexander Marr, 83–99. Donington: Shaun Tyas.

Esteban Piñeiro, Mariano, and Maria Isabel Vicente Maroto. 1991. *Aspectos de la ciencia aplicada en la España del Siglo de Oro*. Salamanca: Junta de Castilla y León.

Farge, James K. 2006. Les lecteurs royaux et l'Université de Paris. In *Histoire du Collège de France*, vol. 1: *La création 1530–1560*, ed. André Tuilier, 209–228. Paris: Fayard.

Febvre, Lucien and Henri-Jean Martin. 1997. *The coming of the book: The impact of printing 1450–1800*. Trans. David Gerard. London: Verso.

Ford, Philip. 2004. An early French Renaissance salon: The Morel household. *Renaissance and Reformation* 28: 9–20.

Gascon, Richard. 1971. *Grand commerce et vie urbaine au 16e siècle: Lyon et ses marchands v. 1520-v. 1580*, 2 vols. La Haye: Mouton and SEVPEN.

Gingerich, Owen. 2002. *An annotated census of Copernicus' De revolutionibus (Nuremberg, 1543 and Basel, 1566)*. Leiden: Brill.

Girot, Jean-Eudes. 1998. La notion de lecteur royal: Le cas de René Guillon (1500–1570). In *Les Origines de Collège de France (1500–1560)*, ed. Marc Fumaroli, 43–108. Paris: Klicksieck.

Gunther, Robert T. 1932. *The astrolabes of the world, 2 vols*. Oxford: Oxford University Press.

Hall, Bert S. 1997. *Weapons and warfare in Renaissance Europe*. Baltimore: Johns Hopkins University Press.

Harris, Michael, Giles Mandelbrot. 2009. *Books for sale. The advertising and promotion of print since the fifteenth century*, ed. Robin Myers. London: The British Library.

Hellinga, Lotte. 2018. *Incunabula in transit: People and trade*. Leiden: Brill.

Higman, Francis M. 1979. *Censorship and the Sorbonne*. Geneva: Droz.

Kremer, Richard L. 2016. Playing with geometrical tools: Johannes Stabius's *astrolabium imperatorium* (1515) and its successors. *Centaurus* 58: 104–134.

Lapeyre, Henri. 1955. *Une famille de marchands: Les Ruiz*. Paris: Colin.

Leitão, Henrique. 2009. Pedro Nunes against Oronce Fine: Content and context of a refutation. In *The worlds of Oronce Fine. Mathematics, instruments and print in Renaissance France*, ed. Alexander Marr, 156–171. Donington: Shaun Tyas.

Lemerle, Frédérique. 2011. Sagredo, *Raison d'architecture antique*. In *Architectura: Books on Architecture published in France, written in French or translated into French (16th–17th centuries)*. Tours: CESR.

Lespinasse, René de. 1897. *Les métiers et corporations de la ville de Paris*, vol. 3: *XIVe-XVIIIe siècles. Tissus, étoffes, vêtements, cuirs et peaux, métiers divers*. Paris: Imprimerie Nationale.

Levy, Alissar. 2020. *Du quadrivium aux disciplinae mathematicae: histoire éditoriale d'un champ disciplinaire en mutation (1480–1550): Une recherche de bibliographie matérielle et d'histoire sociale du livre à Paris au XVIe siecle*. Thèse pour le diplôme d'archiviste paléographe. Paris: École nationale des Chartes.

Long, Pamela O. 2011. *Artisan/Practitioners and the rise of the new sciences, 1400–1600*. Corvallis: Oregon State University Press.

Maclean, Ian. 2009. *Learning and the market place: Essays in the history of the early modern book*. Leiden: Brill.

Maclean, Ian. 2012. *Scholarship, commerce, religion: The learned book in the age of confessions, 1560–1630*. Cambridge, MA: Harvard University Press.

Maddison, Francis. 1966. Helt and the Rojas astrolabe projection. *Revista do Faculdade de Ciencias da Universidade de Coimbra* 39: 195–251.

Magnien, Michel. 2000. "Ordre" et "méthode" dans l'*Art poetique reduict et abregé* de Claude de Boissière (1554). *Nouvelle Revue du XVIe Siècle* 18 (1): 113–130.

Mandelbrote, Giles. 2000. Scientific books and their owners: a survey to c. 1720. In *Thornton and Tully's scientific books, libraries and collectors*, ed. Andrew Hunter, 333–366. Aldershot: Ashgate.

Martin, Henri-Jean. 2000. *La Naissance du livre moderne. Mise en page et mise en texte du livre français (XIVe-XVIIe siècles)*. Paris: Editions du Cercle de la Librairie.

Matringe, Nadia. 2016. *La banque en Renaissance. Les Salviati et la place de Lyon au milieu du XVIe siècle*. Rennes: Presses Universitaires de Rennes.

Nuovo, Angela. 2013. *The Book Trade in the Italian Renaissance*. Trans. Lydia G. Cochrane. Leiden: Brill.

Omont, Henri. 1894. Jacques Goupyl, professeur extraordinaire au Collège Royal (1552). *Bulletin de la Société de l'Histoire de Paris et de l'Île de France* 21: 184–185.

Oosterhoff, Richard. 2014. *Idiotae*, mathematics, and artisans: The untutored mind and the discovery of nature in the Fabrist circle. *Intellectual History Review* 24: 1–19.

Oosterhoff, Richard. 2018. *Making mathematical culture: University and print in the circle of Lefèvre d'Étaples*. Oxford: Oxford University Press.

Orth, Myra. 2015. *Renaissance manuscripts: The sixteenth century. A survey of manuscripts illuminated in France*, 2 vols. London: Harvey Miller Publishers.

Pallier, Denis. 2002. L'office de libraire juré de l'Université de Paris pendant les guerres de religion. *Bulletin du Bibliophile* 1: 47–69.

Pantin, Isabelle. 1986. Jean-Pierre de Mesmes et ses Institutions astronomiques. *Revue de Pau et du Béarn* 13: 167–182.

Pantin, Isabelle. 1988. Les problèmes spécifiques de l'édition des livres scientifiques à la Renaissance: l'exemple de Guillaume Cavellat. In *Le Livre dans l'Europe de la Renaissance*, eds. Pierre Aquilon, Henri-Jean Martin, and François Dupuigrenet Desroussilles, 240–252. Paris: Promodis.

Pantin, Isabelle. 2006. Teaching mathematics and astronomy in France: The *Collège Royal* (1550–1650). *Science and Education* 15: 189–207.

Pantin, Isabelle. 2009. La projection de Rojas. In *Passeurs de Textes. Imprimeurs, éditeurs et lecteurs humanistes dans les collections de la Bibliothèque Sainte-Geneviève*, ed. Yann Sordet, 168–171. Turnhout: Brepols.

Pantin, Isabelle. 2012. Le style typographique des ouvrages scientifiques publiés par Michel de Vascosan In *Passeurs de Textes: Imprimeurs et libraires à l'âge de l'Humanisme*, ed. Christine Bénévent, Anne Charon, Isabelle Diu, and Magalie Vène, 167–184. Paris: Ecole nationale des Chartes.

Pantin, Isabelle. 2012. The first phases of the *Theoricae planetarum* printed tradition (1474–1535): The evolution of a genre observed through its images. *Journal for the History of Astronomy* 43: 3–26.

Pantin, Isabelle. 2013. Oronce Finé, mathématicien et homme du livre: la pratique éditoriale comme moteur d'évolution. In *Mise en forme des savoirs à la Renaissance. À la croisée des idées, des techniques et des publics*, eds. Isabelle Pantin and Gérald Péoux, 19–40. Paris: Armand Colin.

Pantin, Isabelle. 2020. Borrowers and innovators in the history of printing Sacrobosco: The case of the 'in-octavo tradition.' In *De sphaera of Johannes de Sacrobosco in the early modern period: The authors of the commentaries*, ed. Matteo Valleriani, 265–312. Cham: Springer. https://doi.org/10.1007/978-3-030-30833-9_9.

Parent-Charon, Annie. 1974. *Les métiers du livre à Paris au XVIe siècle (1535–1560)*. Geneva: Droz.

Parent-Charon, Annie. 2000. Le commerce du livre étranger à Paris au XVIe siècle. In *Le livre voyageur: Constitution et dissémination des collections livresques dans l'Europe moderne*, ed. Dominique Bougé-Grandon, 95–108. Paris: Klincksieck.

Pettegree, Andrew. 2011a. Printing in the Low Countries in the early sixteenth century. In *The book triumphant: Print in transition in the sixteenth and seventeenth centuries*, eds. Graeme Kemp and Malcolm Walsby, 3–25. Leiden: Brill.

Pettegree, Andrew. 2011b. *The book in the Renaissance*. New Haven: Yale University Press.

Postel, Claude. 1992. *Les écrits de Guillaume Postel publiés en France et leurs éditeurs: 1538–1579*. Geneva: Droz.

Poulle, Emmanuel. 1963. *Un constructeur d'instruments astronomiques au XVe siècle, Jean Fusoris*. Paris: Champion.

Proot, Goran. 2018. Prices in Robert Estienne's booksellers' catalogues (Paris 1541–1552): Statistical analysis. In *Selling and collecting: Printed book sale catalogues and private libraries in early modern Europe*, ed. Giovanna Granata and Angela Nuovo, 177–210. Macerata: Edizioni Università di Macerata.

Raschieri, Amedeo. 2012. Giorgio Valla editor and translator of ancient scientific texts. In *Greek science in the long Run: Essays on the Greek scientific tradition (4th c. BCE–16th c. CE)*, ed. Paula Olmos, 127–149. Newcastle upon Tyne: Cambridge Scholars Publishing.

Renouard, Antoine-Augustin. 1843. *Annales de l'imprimerie des Estienne*. Paris: Renouard.

Renouard, Philippe. 1894. *Bibliographie des éditions de Simon de Colines, 1520–1546*. Paris: E. Paul, L. Huard, and Guillemin.

Renouard, Philippe. 1901. *Documents sur les imprimeurs, libraires, cartiers, graveurs...ayant exercé à Paris, de 1450 à 1600*. Paris: Champion.

Renouard, Philippe, Marie-Josèphe Beaud, and Sylvie Postel. 1986. *Imprimeurs et libraires parisiens du XVIe siècle. Ouvrage publié d'après les manuscrits de Philippe Renouard*, vol. 4: *Binet-Blumenstock*. Paris: Service des Travaux historiques de la Ville de Paris.

Renouard, Philippe, Marie-Josèphe Beaud, and Sylvie Postel. 1995. *Imprimeurs et libraires parisiens du XVIe siècle. Ouvrage publié d'après les manuscrits de Philippe Renouard. Fascicule Jean Loys*. Paris: Paris Musées.

Renouard, Philippe, and Isabelle Pantin. 1986. *Imprimeurs et libraires parisiens du XVIe siècle. Ouvrage publié d'après les manuscrits de Philippe Renouard. Fascicule Cavellat–Marnef et Cavellat*. Paris: Bibliothèque nationale.

Rodríguez-Arribas, Josefina. 2017. The astrolabe finger ring of Bonetus de Latis: Study, Latin text, and English translation with commentary. In *Astrolabes in Medieval Cultures*, eds. Josefina Rodriguez-Arribas, Charles Burnett, and Silke Ackermann, 45–106. Leiden: Brill.

Rouse, Richard, and Mary Rouse. 2019. *Renaissance illuminators in Paris: Artists & artisans 1500–1715*. London: Harvey Miller Publishers.

Schreiber, Fred. 1982. *The Estiennes: An annotated catalogue of 300 highlights of their various presses*. New York: E. K. Schreiber.

Schreiber, Fred, and Jeanne Veyrin-Forrer. 1995. *Simon de Colines: An annotated catalogue of 230 examples of his press 1520–1546*. Provo: Friends of the Brigham Young University Library.

Segonds, Alain Philippe. 1981. Introduction. In *Jean Philopon, Traité de l'astrolabe*. Paris: A. Brieux.

Sharratt, Peter. 2005. *Bernard Salomon: Illustrateur lyonnais*. Geneva: Droz.

Shaw, David J. 2011. Book trade practices in early sixteenth-century Paris: Pierre Vidoue (1516–1543). In *The book triumphant: Print in transition in the sixteenth and seventeenth centuries*, eds. Malcolm Walsby, and Graeme Kemp, 334–346. Leiden: Brill.

Smith, David Eugene. 1908. *Rara arithmetica: A catalogue of the arithmetics written before the year MDCI*. Boston: Ginn and Company.

Thompson, James Westfall. 1911. *The Frankfort Book Fair: The Francofordiense Emporium of Henri Estienne edited, with historical introduction, original Latin text with English translation*, 1911. Chicago: The Caxton Club.

Thorndike, Lynn. 1943. Robertus Anglicus. *Isis* 34: 467–469.

Turner, Anthony J. 1985. *Astrolabes, astrolabe related instruments*. Vol. 1, part 1 of *The Time Museum: Time measuring instruments*, ed. Bruce Chandler. Rockford: The Time Museum.

Turner, Anthony J. 1989. Paper, Print and Mathematics: Philippe Danfrie and the making of mathematical instruments in late sixteenth-century Paris. In *Studies in the history of scientific instruments*, ed. Christine Blondel, Françoise Parot, Anthony Turner, and Mari Williams, 22–42. London: Rodgers Turner Books.

Turner, Anthony J. 1998. Mathematical instrument making in early modern Paris. In *Luxury trades and consumerism in Ancien Régime Paris. Studies in the history of the skilled workforce*, ed. Robert Fox, and Anthony Turner, 63–96. Aldershot: Ashgate.

Turner, Anthony J., Taha Arslan, and Silke Ackermann. 2018. *Mathematical Instruments in the collections of the Bibliothèque Nationale de France*. Paris/Turhout: BnF/Brepols.

Updike, Daniel Berkeley. 2001. *Printing types. Their history, forms, and use*, 2 vols. London/New Castle: The British Library/Oak Knoll Press.

Valleriani, Matteo. 2013. *Metallurgy, ballistics and epistemic instruments. The Nova scientia of Nicolò Tartaglia. A New Edition*. Berlin: Edition Open Access. https://edition-open-sources.org/sources/6/index.html.

Valleriani, Matteo. 2017. The epistemology of practical knowledge. In *Structures of practical knowledge*, ed. Matteo Valleriani, 1–19. Dordrecht: Springer.

Van Ortroy, Fernand. 1902. *Bibliographie de l'oeuvre de Pierre Apian*. Besançon: P. Jacquin.

Van Ortroy, Fernand. 1920. *Bio-bibliographie de Gemma Frisius, fondateur de l'école belge de géographie, de son fils, Corneille, et de ses neveux, les Arsenius*. Bruxelles: Lamertin and Hayez.

Vervliet, Hendrik D.L. 2008. *The paleography of the French Renaissance. Selected papers on sixteenth-century typefaces*, 2 vols. Leiden: Brill.

Virassamynaïken, Ludmila, ed. 2015. *Lyon Renaissance. Arts et humanisme*. Lyon: Musée des Beaux-Arts de Lyon.

Voet, Leon. 1973. *Antwerp the golden age: The rise and glory of the metropolis in the sixteenth century*. Antwerp: Mercatorfonds.

Webster, Roderick and Marjorie Webster. 1998. *Western astrolabes*. Vol. 1 of *Historic scientific instruments of the Adler Planetarium and Astronomy Museum*. Chicago: Adler Planetarium and Astronomy Museum.

Wells, Ellen B. 1983. Scientists' libraries: A handlist of printed sources. *Annals of Science* 40: 317–389.

Isabelle Pantin is Emerita Professor at the Ecole Normale Supérieure (PSL University Paris), and member of the Institut d'Histoire Moderne et Contemporaine (IHMC, UMR 8066). Her research interests concern notably the circulation of knowledge in printed books in the early modern period, and scientific illustrations.

Chapter 10
Paratexts, Printers, and Publishers: Book Production in Social Context

Matteo Valleriani and Christoph Sander

Abstract Paratexts, such as dedication letters or epigrams, in early modern printed books can be used by historians to situate a book's production in its institutional and social context. We depart from the general assumption that two publishers or printers were in a relation of awareness of each other if they printed and put on the market two different editions that contain at least one identical paratext. In this paper, we analyze the circulation of the paratexts among the 359 editions of the "*Sphaera* corpus." First, we discuss the available data, the conditions to build a social network, and the latter's characteristics. Second, we interpret the results—potential relationships among printers and publishers—from a historical point of view and, at the same time, discuss the sorts of potential relationships that this method can disclose. Third, we corroborate the historical results among different approaches, namely by using editions' fingerprints and by investigating the book production of those printers and publishers tangentially involved in relevant relationships, but who fall outside the "*Sphaera* corpus." Finally, we identify local communities of printers and publishers and, on a transregional level, printers, and publishers who were observing and influencing each other.

Keywords Paratext · *Tractatus de sphaera* · Johannes de Sacrobosco · Social network · Local market

1 Premise

In the context of the research project *The Sphere: Knowledge System Evolution and the Shared Scientific Identity of Europe* (https://sphaera.mpiwg-berlin.mpg.de), we

M. Valleriani (✉)
Max Planck Institute for the History of Science, Berlin, Germany
e-mail: valleriani@mpiwg-berlin.mpg.de

Technische Universität Berlin, Berlin, Germany

Tel Aviv University, Tel Aviv, Israel

C. Sander
Bibliotheca Hertziana, Max Planck Institute for Art History, Rome, Italy

© The Author(s) 2022
M. Valleriani and A. Ottone (eds.), *Publishing Sacrobosco's* De sphaera *in Early Modern Europe*, https://doi.org/10.1007/978-3-030-86600-6_10

337

investigate processes of evolution of knowledge. Specifically, we focus on basic astronomical knowledge in the period from the thirteenth to the seventeenth century. Our major historical goal is to reconstruct such processes of evolution by means of a large number of historical sources in reference to both their content and the context in which they were produced.

Concerning the early modern period, we have been able to collect a meaningful corpus of printed editions, as described below. Besides the fact that the corpus is sufficiently large and covers a certain subject systematically and completely, we were moreover able not only to apply digital humanities techniques but also to move forward toward a method for analyzing historical data by making use of mathematical means. Within such framework—which we call computational humanities—methods originally developed to analyze the physics of complex systems are applied to questions from the humanities. Therefore, we formalize data as multi-layer networks.

By way of the analysis of some of the data extracted from the textual aspects of the sources, we were able to build a relevant empirical network of five layers. We have also examined its structural and topological characteristics (Valleriani et al. 2019). It is our intention to expand that network by adding new layers, particularly layers that contain information about the relationships between the various actors related to the editions of this corpus. Authors, printers, and publishers are our main focus. Once such layers are in place, we will be able to examine correlations between data on the content of the treatises and data on the social aspects of the production and dissemination of the same books on the market.

Because of the formal and mathematical character of the investigations in the context of computational history, for each aspect we intend to investigate, we must find a systematic approach that is more than a simple accumulation of results from micro case studies. The present study was conceived while looking for ways to systematically detect relationships among printers and publishers involved in the corpus under our scrutiny.

2 The Research Question

Early modern printers, publishers, and booksellers undoubtedly had a strong impact on the development of scientific knowledge in their period, although their contribution to the history of science is rarely acknowledged.[1] When we think about scientific achievements, we often forget about those actors like printers and publishers—rather businessmen than scientists—who nevertheless provided the conditions for the publication of scholarly books. The role of these actors within the larger scientific milieu

---

[1] "Printers," "publishers," and "booksellers" are categories that denote different roles in the context of the production and distribution of printed books in the early modern period, but not necessarily different persons (Maclean 2012, 101–102). In this essay we will mostly focus on the first two categories and define them as "book producers."

can be investigated from a variety of different perspectives. The most obvious angle might be to approach their influence by looking for connections between publishers and the authors of the books they published. This line of investigation has been followed in scholarship to some extent and has enriched our understanding of how ideas were "sold" in the medium of the early modern printed book.[2] Scholarship has pointed out the various ways in which publishers, in particular, collaborated with authors and vice versa. A more original and intriguing perspective, however, is related to the way early book producers and sellers cooperated with each other, or, more broadly, were aware of their local and international competitors and adjusted their products accordingly (Hinks and Feely 2017). Here we get to observe the social and economic mechanisms within the business of printing and publishing, subscribing to the assumption that processes of the circulation of knowledge are determinant for the formation and development of scientific thinking. This specific viewpoint is relevant because it allows us to discover how this circulation of knowledge actually worked on a social and material basis. Eventually, it promises insights into the business model(s) that emerged in the aftermath of Gutenberg's enterprise of the second half of the fifteenth century.

There were many different kinds of relations between printers and publishers, and most of them are well known to scholars of book history: cooperation between publishers based on close family ties, on a wider group of relatives, or (in an extended sense) as the result of inheritance—what we might call family businesses. Against a broader social background transcending the boundaries of a single family, other types of cooperation between publishers/printers existed, too, and extended, for instance, to the lending, borrowing, and purchasing of woodblocks and types. Such forms of cooperation often resulted in the founding of printers' associations. These were sometimes established ad hoc for the production of particularly demanding individual editorial initiatives (Nuovo 2013) (Chap. 6). In other cases, publishers/printers cooperated in order to sell a particular text (produced in one print run) to a business partner, who might then have assembled it with other textual parts in a new edition, or might have merely replaced or adjusted the title page, leading to so-called reissues. A further form of relationship between book producers could be called "mutual awareness." This relation implies that two or more book producers did not actively cooperate on a social and economic basis but still knew about their competitors' businesses, and adjusted their own business accordingly, e.g., by specializing in a different field of publication or by actively competing with it through the practice of reprinting. Mutual awareness, in fact, means that book producers observed what other producers put on the academic book market and might have consequently decided to borrow ideas for the content of their own editions or taken aspects related to the production itself—such as format, visual apparatus, *mise-en-page*, or types. Mutual awareness therefore could turn into mutual imitation to an extent that two editions by different

---

[2] Historical research dedicated to individual early modern publishers and printers is very active and has produced innumerable great pieces of literature in the recent years. Concerning our perspective, we would like to mention just two of them as representative: (Lowry 1979; Gerritsen 1991).

book producers could look almost identical (reprint). In this respect, we can speak of relations among book producers, though only on an abstract level.

In spite of the fact that several relationships among book producers are well known and have been investigated in historical scholarship, so far there is no commonly acknowledged systematic method to analyze these relationships among the book producers and sellers of the period.

As a matter of fact, our knowledge of these sorts of relationships is almost entirely based on many discrete case studies, a generalization of which might not always be justifiable. In other words, scholarship lacks a concise research approach to investigating the emergence of economic relationships among the players involved in the production and distribution of early modern printed books. In the present work, we would therefore like to present and discuss an approach that might prove useful in identifying what we might call *potential* relationships among book producers. This approach, however, is not based on individual analyses and case studies—in fact, it does not even require comprehensive historical research on single publishers or printing houses. Instead, it is based on large-scale patterns emerging through network analysis. Based on a network constituted from bibliographic metadata of the publication of so-called *Sphaera* treatises—a genre to be introduced in the next section—and the circulation of so-called paratexts within these publications—a literary genre that will be explained in another section in greater detail as well—we hope to argue convincingly for such an approach that can serve as a blueprint or template for other bibliographical corpora and their underlying networks.

3 The Corpus

One condition for the realization of the aforementioned research approach is a well-defined corpus of printed editions. However, the definition of such a corpus can be based on different parameters and characteristics. One possibility would be a corpus based on a specific subject, a specific discipline, or even a specific genre. This would lead to a bibliographical record based on publications with similar content. Yet, due to the late emergence of specializations for printers and publishers who focused on books within one specific genre (Chap. 9) (Pantin 1998), the corpus could also be based on a multiplicity of subjects and genres, and be further defined by geographic limitations. For example, all of the books printed in Leipzig (Chap. 12) could be represented by such a corpus. With constraints placed on provenance, the corpus could be based on the books preserved in one specific library or archive. Our corpus, in any case, is structured around the content of the editions and is thus not based on geographical limitations. It is however limited to a specific time period: from the advent of print in the second half of the fifteenth century until 1650—on the assumption that after this period the rules and output rate of the book market changed considerably.

What follows is based on the "*Sphaera* corpus," a set of 359 printed editions defined on the basis of a specific subject or content, namely editions that contain, though in different forms, one specific work: Johannes de Sacrobosco's (1195–1256) *Tractatus de sphaera.*[3] This work was originally compiled in the thirteenth century in Paris, where Sacrobosco was appointed as a lecturer in the then recently founded university. The work is a qualitative introduction to geocentric cosmology and was used for teaching in the context of the quadrivial curriculum. The treatise met with tremendous success and became the most widely used textbook for the introduction of astronomy all over Europe up to the second half of the seventeenth century (Gingerich 1988; Valleriani 2017). The 359 editions collected in the corpus are all printed books that contain this particular treatise; the manuscript tradition is disregarded in this context for pragmatic reasons. Although a comprehensive description of the corpus has already been offered in another study (Valleriani 2020), it is perhaps useful to briefly summarize the main aspects of the corpus here.

The two first printed editions are dated 1472, while the last considered here was printed in 1650. As mentioned, the treatises of the corpus have been collected, generally speaking, because they contain Sacrobosco's treatise. They also might contain other texts. We distinguished between five different kinds of books: a) those that exclusively contain the treatise of Sacrobosco (sixteen editions); b) those that contain a commentary on Sacrobosco's text, namely a text printed on the same page in which portions of the original text are also printed (forty-seven editions) (Fig. 1); c) those we call "compilations," which contain Sacrobosco's original text and other texts that are related to the original one or to some of its subjects, so that the entire book can be considered an enlarged commentary (forty-five editions); d) those containing both commentaries of type b) and texts of type c) (125 editions); and e) adaptions of the treatise, namely works on the same general subject and with the same introductory character, following the same compositional order at least in their largest part and make at least a partial use of the same visual material while containing a different, new text instead of Sacrobosco's treatise (125 editions) (Fig. 2).

The great majority of these printed editions were printed in Latin (295 editions), while treatises, either translations or adaptations, were also produced in Italian (twenty-four), French (twelve), English (ten), Spanish (eight), German (seven), and Portuguese (four).

The treatises collected were printed in forty-one different European cities, with one exception of a treatise printed in what is now Mexico City (Chap. 7). Not surprisingly, Venice and Paris are the most relevant production centers from a quantitative point of view (seventy and sixty-nine editions, respectively). Wittenberg, in spite of the fact that its first *Sphaera* edition only appeared in 1531, is in the third position (fifty editions). Leipzig and Antwerp follow after (twenty-one and twenty editions), although the production in Leipzig came to a halt in 1520 and the publication of *Sphaera* editions in Antwerp only started in 1543.

---

[3] The database of the corpus is available through the project website: https://sphaera.mpiwg-berlin.mpg.de. Accessed 8 June 2021. For a critical edition and an English translation of Sacrobosco's treatise, see (Thorndike 1949).

facit diē naturalē: cōpofitū ex.24. horis: fcdo mō diftiguius diē nālē:i diē artificialē & noctē: & ad argu-
mentū dico φ eadē eclypfis ē i eodē iftāti tpis:mefurātis primū motū:nō aūt mefurātis motū folis: & ideo
aliqbus appet i tertia:aliqbus i pria hora noctis:uñ apud oēs ē idē tps mefurās primū motū:tñ nō eft idem
mefurās m otū folis:Adhuc aliqs poffet iftare:pria & tertia hora noctis funt q faciūt diuerfificari hmōi ap
parētias:fz hmōi horæ fūt ā nūero.24.horarū mefurātiū primū motū:ergo ēt ex mefura primi mot°acci-
dit diuerfitas:ergo ñ ē idē apd oēs: & primū argumētū adhuc ftat:Rñdeē φ hora pt accipi ul ut ē tale tps:
& tale iftas: & fic ē idē apd oēs:& tē i eodē iftati apd oēs appebit:ul'alr ut hora tale ul'talē fcipit denoiatio
nē:uidelz φ dicaē pria uel tertia:& ifto mō uariaē:qa uariā fcipit denoiationē:
qa oēs fere in italia icipiūt numerare horas ab occafu folis:& qa fol aliqbus citius aliquibus tardius occi-
dit:hic ē φ idē tps uarias fortiē denominationes:& fcdm ifta horologia cōputauit autor: græci aūt & ipi
uigīti quattuor horas habebāt:fed incipiebant numerare oriēte fole: hifpani aūt horologia dimidia hñt:
q numerāt.12.horas: & iterū recurrūt ad prinā: & hæc in meridie & media nocte initiū fummunt: & ħ de
di uerfitate cōtingēte in tpe qd mefuraē p diuerfa horologia.

Q uod aqua fit rotunda

Vod āt aq hēat tumorē & accedat ad rotūditatē fic patet,Ponaē fignū i littore maris:
& exeat nauis a portu : & itrm elogeē φ oculus exñs
iuxta pedē mali nō poffit uidere fignū,ftate uero na
ui oculus eiufdē exñtis i fumitate mali bñ uidebit fignū illd.Sed oculus exñtis iuxa pedē mali melius ui
beret uider fignū:q q ē i fumitate: cū fit ppiqor figno
ficut patet p lincas ductas ab utroqa ad fignū:& nul-
la alia huius rei cā ē q tumor aq. Excludaē .n. ōia
alia ipedimēta:ficut nebulæ & uapores afcēdétes.Itē

Radius uifualis   fit nū littorie

cū aq fit corp°hōgeneū totū cū ptib°eiufdē erit rōis:
fed ptes aq(ficut i guttulis & rorib°herbaz accidit)
rotūda nālr appetūt formā: ergo & totū cui°fūt ptes

Q VOD āt aq hēaē. Probat duplici rōe
aq tūorofitatē: feu rotūditatē qz pria
pz maxie ex figura textuali:quoz nauigā
tes dū fūt a terra diftātes:& qrūt uide lo
cū rēotū afcēdāt i altū fupra malū: & ui
dēt ide q uideri ñ poterāt ex loco ifimo
nauis:qd alia de cā accidere ñ pōt:nifi tu
more aq ipediēte uifione ei° q ē fub ma
lo:nō āt illius q ē i fumo loco:qa tumor
aq nō pueniat ad lineā ufqa eius uifualē.

Pro rōe fcda ē notādū φ corpoz nāliū
ueluti phyfi, dclarāt duplex ē maneries:
qdā nāqa funt fimplicia:& hōgenea dcā-
nām eādē pprietatē & dnoiatōnē i toto
hñtia:& ptib°qbuflibet fēfibilibz:& fūt
elta & multa alia mixta.eādē nāqa nā p-
prietas & denoiatio ē i tota aq:& carne:
& eaz fingul'ptibus:qlibet nanqa ps aq
qa dr:ficut & tota aq:& qlibz ps carnis:
caro noiaē: ceu & caro tota: q de cā fim
plicia & hōgenea noiaē: diuerfitate in
ptib° ñ hñtia. Alia uero dñr ætheroge-
nea & cōpofita: i toto & ptibus eādē na
turā & dnoiatōnē ñ retinētia:ut ē hō &
oēs eius inftfales ptes.ps nāqa hois ñ de
noiaē hō:ut manus:& caput:qa hō nō ē
eiufdē naturæ cū ptib° ei°: cū cōponaē
ex ptib°diuerfis i fpē:falē largo mō:ut
carne offe neruis &c.Et ē diuerfis fcdz

figurā:ut manu capite pede & hmōi:& iō tale cōpofitū ex ptibus diuerfaz rōnum: a phis & medicis ē de-
noiātū.In hoc igiē c orpora homogenea differūt ab ætherogeneis:qa hōgenea hñt i toto & ptibus eādē
naturā & rōnē.minie uero ætherogenea.Sed aq ē corpus homogeneū ut patet:qa eādē naturā & ppria ac
cidētia hēt tota aq:q ptes eiufdē.Sed ptes aq tēdūt naturalr ad figurā rotūdā:ceu demōftrāt guttæ aq ca-
dentis & rores i herbis iuēta:q qa nō cōtinent corpe terminato figura rotūda figuraē.Figura igiē naturalis totius elemēti aq ad rotūditatē tēdit:qd ē intētū.An āt rōes hæ demōftrēt ul'pbabilr cōcludāt nolo i
pñtiaz difcutere.Ptole.āt prio almag.cap.iiii.aq rotūditatē expimēto:qd cū pria rōne coicidere uidet: oñ
dit dices. Si fit nauis i mari tm a terra elogata φ nullo mō poffit uidere terrā & alia q fūt i ea:intm appro
piquet φ eā icipiaē uidere:pri°uidebunē fūmitates mōtiū & cacumia rez altaz:deinde qto magis dcā nauis terræ uicinaē:tāto mōtiū & hoz uidebit ptē maiorē:ac fi furgerēt ex aq: & qñ ppiqua erit: tot° mōs
uidebiē:quare cū prius apparuerit fūmitates q ptes ifimæ:arguiē aq rotūditas:q ipediebatur uidereē ps
ifima mōtis:cacumē uero minime:ut patet p figurā textuali. Vez qa hæ rōnes pcedūt potius ex qbufdā
apparentiis:& pbabilibus principiis fidē tm facientibus:quom ipertinens fit aftronomo declarare figuras
eltoz:fed pho ptinet hoc negotium:nā rotunditas terræ & aq:ē quoddā principiū in aftrologia:principia
autē cū fint nota:i fcia cuius funt pricipia rōne nō pbātur:fed leui iductōe & fuafiōē tm:tefte Auer.
ii.& iiii.phy.& prio de cælo.Ideo formo rōnē Arift.ii. de cælo tex.xxxi. figuram aquæ effe rotūdā neceffi.
tatem:pfupponendo primo φ aq nata ē fluere ad locū declutorē nō ipedita:uñ naturalr naturalitate cōi

             c iiii

Fig. 1 Composition of a typical early modern scientific commentary for quadrivial teaching. The text of reference is printed with bigger font size, the commentary text is positioned around it, a visual apparatus is added. From (Sacrobosco et al. 1508, 12r). Courtesy of the Library of the Max Planck Institute for the History of Science

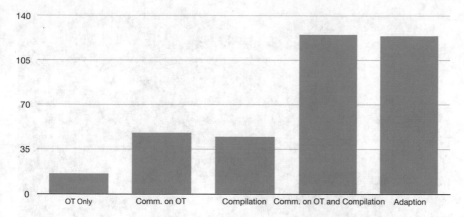

Fig. 2 A typology for the editions constituting the corpus of early modern printed commentaries on the *Sphaera* of Johannes de Sacrobosco: editions that contain the original medieval tract (OT) only; those that contain the original treatise with commentary; those that contain the original treatise and other treatises (compilations); those that contain the original treatise, commentary, and other texts; and adaptions. Authors' plot

Apart from two books in *sextodecimo* format, the dominant formats are folio, quarto, and octavo, though the last is the format that dominates this corpus (thirty-two, 118, and 206 editions, respectively).

The temporal distribution of the production of these treatises, moreover, was not constant, but notably increased around 1550 and maintained this peak until 1585 (Fig. 3).

The fact that these editions were mostly textbooks for use at universities or other educational institutions means that the corpus is not only defined on the basis of a specific scientific subject, but also on the basis of the specific institutional role which played in the context of teaching. In other terms, investigating the relationships among publishers and printers of these editions results in an investigation of their business model(s) in the framework of the academic book market. This also allows us to consider institutional and pedagogical developments of the period as well as the institutions' relationship to the book market. Viewed from this perspective, printing shops and individual publishers appear to be closely connected to the domains of learning and teaching cosmology. This, once more, proves their important role within the dynamic between the book market itself and the market's target customers: students and professors.

The textual content of the treatises has been analyzed through a process of atomization into text parts. A text part is a text portion that clearly has a beginning and an end, and which could be read independently from other text parts published in the same book. Such a text part could, for instance, be an epigram or an entire treatise on the orbits of the planets. We additionally distinguish between text parts that are original texts on one hand and those that are commentaries and translations on the other. Original text parts can be texts of reference, such as Sacrobosco's treatise itself, new texts written by contemporary authors, or older texts which were

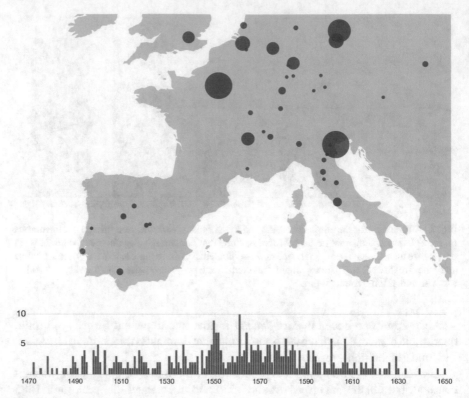

Fig. 3 Geo-temporal distribution of the production of the treatises belonging to the corpus considered here. Authors' plot

published *in the corpus* for the first time. In total, 447 original text parts and 119 commentaries and translations were identified. What is most relevant here is that text parts often reoccurred. It is through the analysis of their reoccurrences that we investigate how knowledge evolved over time.[4] Previous research (Valleriani 2020) has also sharpened our understanding of the role played by the authors of early modern commentaries within the dynamics of the *Sphaera* corpus, reflecting the low dominance of contemporaries in respect to the total number of credited authors (i.e., authors credited on the title pages of the books).

---

[4] The reoccurrence of text parts as a basis for investigating evolution of knowledge is now also used as a method in legal history. In this context, a text part is commonly identified with a paragraph or otherwise well-defined section of the text of a law or a legal corpus. For a pioneering implementation of this method, see (Funk and Mullen 2018). For a more comprehensive description and taxonomy of the text parts in our corpus, see (Valleriani et al. 2019). In the same work, on the basis of complex-network theory analytical tools, we were able to identify families of treatises (epistemic communities) whose contents influenced and shaped the contents of all the other treatises produced in the succeeding periods. The most dominant epistemic communities emerged in 1531 and 1538 and were initiated by Wittenberg's well-known printer Joseph Klug (Chaps. 4 and 5). See also (Zamani et al. 2020).

For this specific study, a systematic approach is proposed that seeks *potential* collaborations among authors when a) their texts were published in the same book and b) the authors were alive at the time of publication. Extensive research on the intellectual profiles of the authors of the commentaries revealed that they were all active in the area of university teaching and of the quadrivial disciplinary scheme. By considering "credited authors" (credited on the title pages), a total of 166 persons can be distinguished (among them twenty-two anonymous). Only fifty-eight of them were alive when their texts were published and only eighteen of them were involved in *potential* relationships among one another. Such relationships were identified by searching for pairwise authors who were alive at the time of publication and whose text parts were printed in the same edition so that a potential relationship, via book producer, among them could be established. This (thin) result suggests that the process of transformation of knowledge—as it can be historically reconstructed against the background of this corpus—was not driven by the authors themselves; against the background of the network analysis, the scientific debate was, in other terms, not primarily conducted by the scholars. In this respect, the hypothesis emerged that a leading role, in the case of textbooks, was taken over by the book producers—hence the necessity to investigate their mutual relationships (cooperation, competition, or mere business awareness) in order to understand whether there is a relationship between the transformation of knowledge and the formation of social communities.

4   Methodological Considerations

Book producers may have acted in more than one role within our corpus. For instance, one individual may have been the publisher of one book and the printer of another or even the author of a text. Only three people in the corpus were authors, printers, and publishers, three were both authors and printers, nine were both authors and publishers, seventy were only printers, sixteen were both printers and publishers, 102 were only publishers. Eighteen people were also identified as translators and seventeen with the roles of both translator and author.

The systematic approach we would like to suggest in identifying such potential relationships (of kinship and/or economic nature) among book producers considers paratexts and their circulation.[5] In linguistics, paratexts are texts that are, in its most generic definition, complementary to one or more main texts. Paratexts often frame the interpretation of main texts.[6] In our corpus of early modern printed editions, these paratexts are not always clearly distinguishable from main texts, but as a rough guide, we treat texts as paratexts if they introduce or conclude longer texts, and particularly—and more importantly for the argument of this paper—if they relate to

---

[5] A similar line of research concerning paratexts is followed by (Brown et al. 2017).

[6] For studies concerning the paratext as a genre and its function, see (Genette 2001; Töpfer 2007; Wagner 2008; Enenkel 2015; Smith and Wilson 2014; Batchelor 2018; Tweed and Scott 2018).

or qualify a social relation between, for example, the author of a text and another person, such as a colleague.[7] Typical examples of paratexts in our corpus include dedication letters and epigrams or other forms of poetry that are not primarily a means of communicating a cosmological idea. Thanks to a taxonomic analysis conducted by Irina Tautschnig, we are able to identify 251 different paratexts in the corpus, which are in turn a sub-group of the 447 original text parts mentioned above. These paratext text parts are organized according to a specific taxonomy: Poetry (ninety-seven), Dedication letter (ninety-one), Letter to the reader or preface (forty-three), and others (twenty).

The undergirding assumption of our approach is that paratexts are a good means by which to make qualified assumptions about potential social and economic relations, including simply "mutual awareness," between book producers. This genre of texts often established or rather represented, social relations, and as such, the occurrence of many paratexts is strongly tied to the concrete geographical and temporal context of a given edition—a context highly dependent on the work of the book producers.[8] It goes without saying that not all paratexts fulfill this criterion, yet, at a large scale, the 251 different paratexts are a promising basis for our analysis, and in many examples, it seems highly probable that the book producer deliberately chose which paratexts should accompany his edition.

However, it is not the single occurrence of a paratext that interests us, but rather its reoccurrence—its circulation within the corpus. Here, the assumption is that when a book producer B republished a paratext originating from the specific temporal and local context of a previous edition published by book producer A, producers A and B shared a social or economic context or were at least aware of each other's business, and thus were in a potential relationship with each other. Why is that? While main texts—such as a treatise on cosmology—widely circulated and could be the basis for many printing businesses completely unrelated to one another, a paratext frames this main text in a way that not only reflects a common interpretational framework but more importantly seems to suggest a more deliberate choice of B to follow the editorial agenda of A. Moreover, most paratexts were composed at the time of publication. Paratexts also mostly, in a strict sense, were not published alone, but were bound to one particular editorial project. By republishing a specific paratext, the previous printing project that embedded it is echoed in a way that renders some sort of relation between the book producers more plausible, as it requires at least an awareness of the edition that published the respective paratext before. Moreover, paratexts such as dedication letters are often testament to high authority and would probably not

---

[7] Paratexts such as titlepages, tables of contents, indices, and imprints or colophons are not considered here.

[8] "Die Paratexte der Drucke sind der Ort, an dem sich diese Transformationen des Produktions-, Distributions- und Rezeptionsprozesses am deutlichsten niederschlagen. In ihnen finden sich explizite Selbstaussagen der Produzenten, also der Drucker oder der Herausgeber des Buchs. Paratexte dienen daneben der Verwaltung des Buchs im Distributionsprozeß, indem sie seinen Inhalt identifizieren, aber auch der Rezeptionssteuerung, indem sie diesen Inhalt qualifizieren und die Attraktivität des Produkts betonen oder steigern." From (Wagner 2008, 135).

have been republished without previous agreement—a process that would necessarily involve the book producers as well. The very nature of the relations deduced from the reoccurrences of paratexts, however, needs to be investigated in a second step and will not be dealt with here, as this would require detailed historical research on the respective actors.

Technically speaking, we define the circulation of a paratext when this text is reprinted and republished at a later moment by a different printer and/or publisher. Specifically, we suggest identifying paratexts as mentioned and then grouping the editions by function of the reoccurrences of one particular paratext. If at this point the book producers responsible for the publication at hand were alive, we can deduce that they were mutual aware of each other or even in direct contact.

Our approach is defined more precisely by four conditions.[9] First of all, we distinguish whether the author of a paratext was alive at the time of publication or not, and we analyze the data according to both cases. For brevity's sake, we call the two resulting networks of book producers the "alive-network" and the "non-alive-network." These two perspectives allow us to take into consideration the real and strict social context on one hand (when the author was alive) and the role of the paratext in the design and conception of the new edition on the other hand. Certainly, there is considerable overlap between both networks, but the nuances do change considerably and will be sketched in a later section. In general, the author being alive means that we are dealing with a contemporary paratext; the non-alive-network would be built from paratexts that may come from much earlier times, and which therefore might have completely lost the social, local, or temporal context from which they originated by the time we encounter them.[10] This means, in the latter case, that the paratexts tend to become a main text. We have clearly identified cases in which some dedication letters, as time passes, became introductory texts, almost completely losing their interpersonal function.

The notorious dedicatory letter by Philipp Melanchthon (1497–1560) to Simon Grynaeus (1493–1541) is a good example, as the letter continued to be very important even after the Reformer's death. Rather than being printed as a letter, however, it was often simply used as an introduction to Christian astrology, even omitting mention of its author's or addressee's names altogether.[11] Here, we also faced a content-related

---

[9] To guarantee the reproducibility of our historical analysis, the dataset extracted from the database as well as the scripts that embed the following conditions and that are used to create the networks described and analyzed in the following sections are freely available at: https://gitlab.gwdg.de/MPIWG/Department-I/sphaera/sphaera-paratexts-data-prep. Accessed 8 June 2021.

[10] "Die Widmungsvorrede ist also schon früh mehr als eine persönliche Adressierung an einen individuellen Förderer, die sich zugleich an breite Leserkreise richtet, sondern löst sich vom ursprünglichen Entstehungskontext der Ausgabe, wird also nicht mehr mit einer spezifischen Ausgabe, sondern mit der in ihr erstmals vorgelegten Redaktion des Werkes selbst assoziiert." From (Wagner 2008, 152).

[11] Melanchthon's letter to Grynaeus was published and republished all over Europe and also beyond the prohibition in the Catholic countries. Often, we still find exemplars of texts in which the folios have been relocated, or the name of Melanchthon deleted or, later, the text published as if it were by an anonymous author. This text has been the object of numerous studies. We mention here (Pantin 1987; Lalla 2003; Reich and Knobloch 2004).

limit to our approach. The circulation of this text was, at least after a while, clearly due not to specific relationships among book producers but to the wish to publish a strong and authoritative text in defense of the study of astrology.[12]

The second condition, already mentioned, is that the book producers involved have been alive at the time of both publications linked by a common paratext. The time between both editions is called the "link age," which will be discussed below. One particular problem emerges in concomitance with this condition, namely the problem that birth and death dates of printers and publishers are not always known or not known with the necessary precision. In all uncertain cases, the active times of the book producers were specified—that is, the time period between the first and the last known edition produced by the respective actor. This is treated as equal to his career as an active book producer. However, even though these dates might bring some uncertainty into the equation, pushing both dates of birth and death by five years (i.e., extending the period by ten years) did not affect the results.

A further condition is that the link age must be at least one year. This means that we did not take into consideration the circulation of paratexts from one edition to the next when this process occurred within the same year. The simple reason for that is that it would require individual research to determine which edition preceded the other if both appeared in the very same year, as our network is oriented chronologically.

The fourth and last condition is what we call the "shortest temporal distance." If for instance a paratext was published once by publisher A and then many times by a second publisher B, we only consider the instance of nearest temporal proximity. We, therefore, prioritize the first republication of a paratext by a second book producer because we consider the time of the first republication to be the moment when the potential relationship was established.

Following these conditions, we created a network connecting all the editions with one another if, and only if, they both contain at least one identical paratext—what we called the circulation of a paratext above. If two editions share two or more paratexts, then the network contains the same number of links among each pair of editions as the number of paratexts they share.

5   The Network

The alive-network has 359 relations between the editions, while the non-alive-network has 622. This is to be expected—dropping a restrictive condition will increase the amount of links in a network. When ignoring the circulation of more than one paratext for each book, namely by deleting multiple links between the

---

[12] The approach delineated here may also be valid for disclosing social and economic relations as well as intellectual affinity. A systematic and formal taxonomy of paratexts that allows one to make such a distinction does not seem possible; the requirement therefore emerges of adding close-reading analyses of the detected paratexts in a second step in order to ultimately discover further relevant characteristics.

same couple of editions, the numbers decrease to 242 (alive) and 354 (non-alive) respectively.

Of the total 251 paratexts, only a small portion was republished in different editions of the treatises. In the alive-network this amounts to only fifty-six different paratexts, in the non-alive to seventy. This means that only fourteen paratexts were republished by different book producers after the death of the paratext's original author. It also means that the majority of paratexts were not republished, testifying to a possible singular and non-reproducible social context of the editions that contain them.

The five paratexts that circulated most according to these conditions are:

1.   the dedicatory letter by Philipp Melanchthon to Simon Grynaeus (8,6% for the alive-network and 26,2% for the non-alive-network);
2.   the epigram *De triplici ortu* by Philipp Melanchthon (9,1% for the alive- network and 14,4% for the non-alive-network);[13]
3.   the dedicatory letter by Élie Vinet (1509–1587) to Johannes Tacitus (?) (5,8% for the alive-network and 11,4% for the non-alive-network);[14]
4.   the carmen by Donato Villalta (1510–1560) dedicated to Pierius Valerianus (1477–1560) (0,14% for the alive-network and 9,6% for the non-alive-network);[15]
5.   the dedicatory letter by Christophorus Clavius (1538–1612) to the reader (5,4% for the alive-network and 7,4% for the non-alive-network).

The ten paratexts that circulated most are responsible for 38% of the links between editions in the alive-network and for 80% in the non-alive-network. This huge difference in percentage again points to the different status of a paratext in relation to the restrictive alive-condition. We could say emphatically that the older a paratext was, the more likely it was to be republished and thereby extracted from its original context. It also should be noted that paratexts one to four were very often published in the same editions; paratexts one and two were virtually always published in the same editions; paratexts three and four were virtually always published in the same editions and in the most editions that also included paratexts one and two. This means most links in the network are based on paratexts of one "tradition" or editorial context.

---

[13] The short poem of four verses—*De triplici ortu*—is taken to be a paratext although it does not appear at the beginning in most of the editions, but after the end of Sacrobosco's treatise. We count it as a paratext since it is supposed to conclude the treatise in a poetic manner, comparable to a doxology in religious texts. This paratext is closely related to the first paratext by Melanchthon, as they are virtually always published together (mutilated copies or other highly specific reasons explain the very few cases in which these text parts are not co-published).

[14] Élie Vinet's dedicatory letter is very closely related to the previous two paratexts by Melanchthon. Although this paratext was published in thirty-three editions, Melanchthon's dedication letter was not printed in only eight of them (or was removed due to censorship in the inspected copies).

[15] Donato Villalta's *carmen* is closely related to the paratext by Élie Vinet (no. 3), as they are always published together.

Table 1 List of the ten book producers with the most potential relationships

Alive-network			Non-alive-network		
Person	Links	Percentage (%)	Person	Links	Percentage (%)
Cavellat, Guillaume	20	4,44	Seitz, Peter I.	22	3,28
Richard, Jean	17	3,78	Krafft the Elder, Johann	20	2,99
Krafft the Elder, Johann	14	3,11	Cavellat, Guillaume	20	2,99
Seitz, Peter I.	14	3,11	Richard, Jean	19	2,84
Kreutzer, Veit	13	2,89	Heirs of Arnold Birckmann	18	2,69
Klug, Joseph	12	2,67	Kreutzer, Veit	17	2,54
Bindoni I., Francesco	11	2,44	Cholinus, Maternus	17	2,54
Crispin, Samuel	10	2,22	Richard, Thomas	15	2,24
Gabiano, Jean de	10	2,22	Bindoni I., Francesco	15	2,24
Ciotti, Giovanni Battista	10	2,22	Barbier, Symphorien	15	2,24

Looking at the temporal aspects of the results, the circulation of paratexts started as early as 1478 and ended in 1619 in the alive network, and in 1629 in the non-alive-network. The average age of the links is of five and seven years, respectively, while the oldest links are thirty-one and fourty-one years.

Coming finally to the potential relationships, we find that 102 book producers are involved in the circulation of paratexts in the alive-network and 118 in the non-alive-network. Their relations, if they are considered reciprocal, amount to 450 and 670. However, because of the need to order them chronologically, the network must be correspondingly oriented.[16] This in turn means that reciprocal relationships are represented by oriented links; therefore, they amount to 225 and 335 (i.e., the relations A-B and B-A are counted as one relation, not as two).[17]

Considering again only the absolute numbers, the ten book producers with the most potential relationships cover 29% and 26% (alive-network and non-alive-network) of the total amount of potential relationships detected (Table 1). Additionally, 7% and 5% of the book producers (in both cases exactly thirty-three persons) display only one potential relationship.

To take an example from a paratext, we can consider Jacques Lefèvre d'Étaples's (1450–1536) dedicatory letter to Carolus Borra (Charles Bourré) (d. 1498). In this

---

[16] Reciprocity here is an important feature, although the graph is directed chronologically. Yet, it must be assumed semantically that any cooperation is by definition reciprocal.

[17] Re-editions by the hand of the same printers and/or publisher have been excluded.

letter, d'Étaples formulates grandiloquent praise on behalf of the entire "academia" of Charles Bourré for his engagement in the teaching of mathematics.

This letter was printed and published first in 1494 in Paris by Wolfgang Hopyl (fl. 1489–1523) (Sacrobosco 1494). It was then reprinted and republished five years later in Venice by Simone Bevilacqua (1450–1518) (Sacrobosco et al. 1499b) and reissued by the same in the same year (Sacrobosco et al. 1499a). One year later it was republished by Hopyl again (Sacrobosco et al. 1500); then in 1507 by Henri d'Estienne I (1460–1520) in Paris (Sacrobosco et al. 1507); then by Giuntino Giunta (1477–1521) in Venice in 1508, but printed by the brothers Giovanni and Bernardino Rosso (fl. 1506–1512) (Sacrobosco et al. 1508); then in 1511 by Henri Estienne I again (Sacrobosco et al. 1511); then in 1521 by Simon de Colines (1480–1546) in Paris (Sacrobosco et al. 1521); then in 1531 by Lucantonio Giunta (1457–1538) in Venice (Sacrobosco et al. 1531); and finally in 1534 by Simon de Colines in Paris (Sacrobosco et al. 1534).

By applying the condition of the shortest temporal distance, these nine editions and one reissue (all containing the mentioned paratext by d'Étaples) result in a total of 17 potential relationships. It cannot be precisely identified from which edition a succeeding edition borrowed the text, or which book producer might have been in contact with which other book producers. Ordered according to the dates of publication of the editions from which the links originate toward other editions, the potential relationships, within the frame of the alive-network, are seen in Table 2.

6 Interpretation

Drawing definite conclusions from the data with regard to the social, economic, or intellectual relationships between book producers is hardly possible at this moment. Yet, some patterns and tendencies emerge that shall be sketched in what follows. In order to do so, we will mainly look at the geographical attributes of the network— i.e., the question of how centers of book production (cities) relate to one another. It turns out that the network indicates both transregional and local links between editions (and thereby between their producers) that were printed in cities of different regions, as well as those printed within one and the same city. A first step thus will be to draw a more precise picture of this situation, balanced against further data of the corpus. A second step will validate the data against other methods to trace possible relationships between editions, namely by analyzing the similarity of book layouts and typesetting based on fingerprints,[18] and by looking beyond the *Sphaera* corpus. Here we ask whether two book producers involved in a potential relationship

---

[18] The fingerprint of an edition is a unique identifier consisting of letters printed on specific pages of the respective edition. These fingerprints not only allow a more precise identification of a specific edition than a traditional bibliographical record, but also enable the detection of very similar prints or reissues of one print run. For the method of extraction of fingerprints in the *Sphaera* corpus, see (Beyer 2019). The fingerprints of the editions constituting the *Sphaera* corpus are available through the database mentioned above.

Table 2 Potential relationships resulting from the circulation of Jacques Lefèvre d'Étaples's dedicatory letter to Carolus Borra

Source year	Target year	Source publishers	Target publishers	Source printers	Target printers
1494	1499	Hopyl, Wolfgang	Bevilacqua, Simone	Hopyl, Wolfgang	Bevilacqua, Simone
1494	1499	Hopyl, Wolfgang	Bevilacqua, Simone	Hopyl, Wolfgang	Bevilacqua, Simone
1499	1508	Bevilacqua, Simone	Giunta, Giuntino	Bevilacqua, Simone	Giovanni & Bernardino Rosso (brothers)
1499	1508	Bevilacqua, Simone	Giunta, Giuntino	Bevilacqua, Simone	Giovanni & Bernardino Rosso (brothers)
1499	1507	Bevilacqua, Simone	Estienne I., Henri	Bevilacqua, Simone	Estienne I., Henri
1499	1507	Bevilacqua, Simone	Estienne I., Henri	Bevilacqua, Simone	Estienne I., Henri
1499	1500	Bevilacqua, Simone	Hopyl, Wolfgang	Bevilacqua, Simone	Hopyl, Wolfgang
1499	1500	Bevilacqua, Simone	Hopyl, Wolfgang	Bevilacqua, Simone	Hopyl, Wolfgang
1500	1531	Hopyl, Wolfgang	Giunta, Lucantonio	Hopyl, Wolfgang	Giunta, Lucantonio
1500	1521	Hopyl, Wolfgang	Colines, Simon de	Hopyl, Wolfgang	Colines, Simon de
1500	1508	Hopyl, Wolfgang	Giunta, Giuntino	Hopyl, Wolfgang	Giovanni & Bernardino Rosso (brothers)
1500	1507	Hopyl, Wolfgang	Estienne I., Henri	Hopyl, Wolfgang	Estienne I., Henri
1507	1508	Estienne I., Henri	Giunta, Giuntino	Estienne I., Henri	Giovanni & Bernardino Rosso (brothers)
1508	1521	Giunta, Giuntino	Colines, Simon de	Giovanni & Bernardino Rosso (brothers)	Colines, Simon de
1508	1511	Giunta, Giuntino	Estienne I., Henri	Giovanni & Bernardino Rosso (brothers)	Estienne I., Henri
1521	1531	Colines, Simon de	Giunta, Lucantonio	Colines, Simon de	Giunta, Lucantonio
1531	1534	Giunta, Lucantonio	Colines, Simon de	Giunta, Lucantonio	Colines, Simon de

also published/printed any non-*Sphaera* editions that display the same content. We take this to be a further indication of some sort of relation between the two book producers. Both complementary approaches ("fingerprints" and "similar editions") are taken to be spot-test validations of the data generated in the network on which our analysis is focused.

6.1   Geographical Distribution

As mentioned above, Paris, Venice, and Wittenberg are the places where most *Sphaera* editions were published. It is thus no wonder that in the present network (including data from both alive and non-alive networks) these three cities are most prominent with regard to the reoccurrences of paratexts.

Moreover, and probably in tight connection with this geographical observation, it must be taken into account that most links are based on a relatively small set of paratexts, mostly connected to the authors Melanchthon, Vinet, and Clavius. But these two presuppositions, however, do not necessarily warp the data and therefore present a problem for drawing valid conclusions. It rather qualifies the nature of the relationships between the book producers: those predominant paratexts, as has been underlined in a previous section, do not necessarily indicate the social and local context like other paratexts do. They rather suggest a broader awareness of the book producers toward certain intellectual trends in the cosmology (and wider academic) book market. The *Sphaera* editions introduced by Melanchthon's letter (editions often containing Vinet's dedication letter as well), and editions of Clavius's commentary were clearly disruptive developments in the publication of *Sphaera* editions and mark the emergence of new trends, materialized in the vast reprinting and republishing of these editions.

Bearing all of this in mind, interpreting the geographical distribution of the paratext-based links between *Sphaera* editions is greatly facilitated. A look at Table 3 immediately evidences the fact that most links between two *Sphaera* editions are created within one city, or more precisely, *within* those three cities that produced most of the *Sphaera* editions: Paris, Venice, and Wittenberg (marked by $\rightleftarrows$ preceding the city's name).

This data, on one hand, seems to suggest a rather local culture of relationships between book producers. This conclusion will be corroborated in the next section. On the other hand, the number of links among the three major centers of the production of *Sphaera* editions (and of editions in general) is not strikingly low. Does this indicate a more transregional aspect of the network and therefore contradict the apparent local nature of relationships? "Yes and no" might be the most precise answer. The transregional aspect follows from the immense number of reoccurrences of the prominent paratexts mentioned above. For very few links we can allege that book producers actively cooperated on a social, economic, or even contract-based level. We must keep in mind that those links, moreover, are links between editions, not between people. Although we, and we think with justification, regard links between editions to indicate potential relationships between their producers, those links are often links between one edition (A) and a plethora of other editions (B, C, D, etc.) featuring the same paratext on whose basis the links are established. It is, viewed from the perspective of economic history, rather unlikely that the producer of A was in a social relationship with all producers of B, C, D, etc. More likely, it would appear— and this is indeed what the sample tests sketched in the next section confirm—that

Table 3 Geographical distribution of the paratext-based links between *Sphaera* editions

Alive-network			Non-alive-network		
City/cities	Links	Percentage (%)	City/cities	Links	Percentage (%)
⇄Venice	37	7,57	⇄Venice	43	6,92
⇄Paris	28	5,73	⇄Wittenberg	42	6,76
⇄Wittenberg	26	5,32	⇄Paris	33	5,31
Venice → Wittenberg	22	4,50	Wittenberg → Antwerp	27	4,35
Venice → Paris	14	2,86	Antwerp → Wittenberg	24	3,86
Wittenberg → Venice	14	2,86	Venice → Wittenberg	22	3,54
Paris → Venice	13	2,66	Paris → Venice	20	3,22
Wittenberg → Paris	12	2,45	Venice → Paris	18	2,90
Paris → Antwerp	12	2,45	Wittenberg → Paris	17	2,74
⇄Lyon	12	2,45	⇄Antwerp	16	2,58
Cologne → Lyon	12	2,45	Paris → Wittenberg	14	2,25
Lyon → Venice	11	2,25	Wittenberg → Venice	14	2,25
Antwerp → Venice	11	2,25	Venice → Antwerp	14	2,25
Paris → Lyon	11	2,25	Lyon → Antwerp	14	2,25
Venice → Antwerp	10	2,04	Lyon → Paris	13	2,09
Antwerp → Wittenberg	10	2,04	Venice → Lyon	13	2,09
Cologne → Venice	10	2,04	Cologne → Paris	12	1,93
Wittenberg → Antwerp	9	1,84	Paris → Antwerp	12	1,93
Lyon → Cologne	9	1,84	⇄Lyon	12	1,93

A might have been in touch with only one book producer B, and B, in turn, might have had an impact on C or D, and so forth.

This interpretation does not disregard or violate the data at hand but rather tries to understand it in the context of an actual social situation. Additionally, if we think of the relationship between the book producers more in terms of their "awareness" of certain trends (e.g., the publication of Clavius's commentary or the text of Sacrobosco, always preceded by and therefore tied to Melanchthon's preface), the more transregional aspect of the network simply confirms that the reoccurrences of some paratexts (mirroring certain trends) are not mere chance but prove that book producers knew about those trends and adjusted their ventures accordingly. All of this could and often has happened without any social or economic relation between book producers who published the *Sphaera* editions that were used as templates, or manifestations of *Sphaera* editions that proved successful in other cities.

6.2   Validation and Corroboration for Local Cooperation

The fact that similar editions were printed within one city over a longer period of time is not a surprise. Many print shops were owned by families and dynasties, passing over the portfolio of the print shop or publishing house to the next generation. Within one city, mutual awareness of and occasional cooperation with local competitors, or a mere (and perhaps not always approved) imitation of their publishing program (or parts thereof) can more or less be taken for granted according to current scholarship and the many examples described in this volume. This holds true especially for places that did not control book production through privileges—like Venice and Paris did— for the printing of ancient and medieval authors and for the production of textbooks, such as Sacrobosco's *Tractatus*.[19]

A good example to illustrate the local dynamics of this network is Wittenberg, a small city, yet one giving home to many printers, and a highly productive center, especially with regard to religious books of the Reformation (Oehmig 2015). In-octavo *Sphaera* editions featuring Melanchthon's notorious letter to Grynaeus and presenting an amended text of Sacrobosco, probably edited under the auspices of Georg Joachim Rheticus (1514–1574) (Rosen 1974; Pantin 2020), must be seen as the vehicle of Wittenberg's success in the market for *Sphaera* editions. These many editions featuring Melanchthon's paratext also dominate the links generated in both the alive and the non-alive networks.[20]

By looking closer at editions from Wittenberg, we see that they not only feature the same paratexts but also resemble each other in ways that markedly underline how different print shops collaborated or influenced one another's businesses. By looking at so-called fingerprints (the multi-part code generated from the typesetting of several pages of an edition) some resemblances emerge so strongly that they cannot be ascribed to mere chance, but should be seen as some form of a relation between the printers or family-run print shops in early modern Wittenberg. For example, the treatise *Novae quaestiones Sphaerae* by Sebastian Dietrich (1521–1574)—a short reworking of Sacrobosco's treatise in the form of questions and answers, most likely written for university teaching of astronomy—was printed for the first time by Johann Krafft the Elder (1510–1578) in 1564 (Dietrich 1564). Fingerprints and a close inspection of this edition reveal that all seven subsequent editions produced in Wittenberg more or less have a very similar, almost identical *mise-en-page*. Krafft's second reprint of 1570 (Dietrich 1570) seems to have been the template for later editions printed by Anton Schöne (fl. 1569–1585) and Clemens Schleich (fl. 1569–1589) in 1573 (Dietrich 1573), by Matthaeus Welack (1540–1593) in 1583

---

[19] Many text parts in *Sphaera* editions however were written by contemporary authors and therefore publishers could be awarded with privileges for those editions, as is also proved by a considerable number of editions in our corpus. The role of privileges in the production of textbooks awaits further research. On privileges in general, see (Nuovo 2013, 195–257).

[20] Melanchthon died in 1560. Therefore, many links are disregarded after this year in the alive-network, while his preface did not cease to be an important supplement to many editions printed thereafter.

(Dietrich 1583), and by Lorenz Säuberlich (fl. 1597–1613) in 1605 (Dietrich 1605). Yet those editions were not reissues, and they differ in minor details: the woodblocks for some of the initials had already been replaced by Krafft himself in his later editions, and likewise in editions printed by some of his Wittenberg competitors and successors in the decades to follow. Another example of a similar kind is an edition of the *Libellus de Sphaera Iohannis de Sacro Busto* printed by Johann Krafft the Younger (fl. 1589–1614) and published by Zacharias Schürer & partners (fl. 1600–1626) in 1601 (Sacrobosco and Melanchthon 1601), which was then reprinted in 1629 by the widow and the heirs of Zacharias Schürer (fl. 1626–1640) (Sacrobosco and Melanchthon 1629). Those two editions were also not reissues, but new, yet strikingly similar editions—reprints using more or less the same typesetting but, for example, printing the initials from different woodblocks.

These spot tests in the Wittenberg market for *Sphaera* editions strongly suggest the existence of deep economic and social relationships among those book producers that also published *Sphaera* editions in this German city (Chaps. 4 and 5). Of course, these alleged relationships did not only extend to *Sphaera* editions, as can be confirmed by looking at books written by other authors, printed and published in early modern Wittenberg. For example, the print shop owned by Peter Seitz the Elder (d. 1548), who was later succeeded by his heirs, printed various, mostly religious treatises connected to the Reformation in the German language, most of them between 1550 and 1570. Those editions, authored by well-known theologians such as Urbanus Rhegius (1489–1541), David Chyträus (1530–1600), Peder Palladius (1503–1560), Martin Luther (1483–1546), and Johannes Garcaeus (1530–1574), had been published and printed earlier by Johann Krafft the Elder and Joseph Klug (1490–1552). Not only did the Seitz print shop produce treatises others had published before but also vice versa. For example, Ursula Seitz, widow of Peter Seitz the Elder, introduced Moritz Breunle's (b. 1500) *Ein kurtz formular und Cantzley buechlein* (Breunle 1548) to Wittenberg's print market in the year of her husband's death (1548). This successful so-called formulary was first printed in Leipzig and Augsburg in 1529 (Breunle 1529a, b), but, from 1552 onward, was also printed in Wittenberg at various times by Veit Kreutzer (fl. 1538–1563) (Breunle 1552, 1553, 1559, 1561) and the heirs of Peter Seitz the Elder (fl. 1548–1578) (Breunle 1554, 1556, 1557)—both also producers of *Sphaera* editions.[21] These examples, just as in the case of *Sphaera* editions, strongly suggest that Wittenberg's book producers were highly aware of their competitors' products and adjusted their book production accordingly, or even took over "rival assets."

---

[21] For Veit Kreutzer's and Peter Seitz I Heirs's production in the context of the *Sphaera* corpus, see respectively http://hdl.handle.net/21.11103/sphaera.100789 and http://hdl.handle.net/21.11103/sphaera.100789. Accessed 8 June 2021.

6.3   Validation and Corroboration for Transregional Awareness

Although local dynamics, as presented in the example of Wittenberg's production of *Sphaera* editions, show stronger support in the network and are also much easier to corroborate, some transregional or transnational aspects of the network also need to be addressed but await further confirmation through additional historical research. Much indeed could be said about specific transregional relations, and many of them can be, if not explained, at least interpreted against the background of historical and intellectual settings that are known to scholars of the field. For example, earlier research (Sander 2018) already shows that the *editio princeps* of the *Sphaera* including Melanchthon's letter, published by Joseph Klug in 1531 in Wittenberg (Sacrobosco and Melanchthon 1531), was not the only Wittenberg edition featuring Melanchthon that was republished shortly thereafter in Venice by Melchiorre Sessa I (1505–1565) (Sacrobosco and Melanchthon 1532). Obviously, Venice's book market demanded editions of scholarly texts that were somehow related to Melanchthon and his intellectual and humanist movement.

While this case was most likely not based on any economic relation between Klug and Sessa, other cases do in fact suggest such relations and cooperation. As Ian Maclean argues in this volume (Chap. 6), Francesco Zanetti (1530–1591) and Giovanni Battista Ciotti (1564–1635) might have collaborated in their undertaking to publish Clavius's commentary on Sacrobosco. That print shops that produced works by Jesuit authors (Chap. 11) might have benefitted from the order's transregional network goes without saying and yet awaits further in-depth research by book historians.

As for other *Sphaera* editions, once again a look at the editions' fingerprints is revealing. Although being rather the exception, one case, again related to Wittenberg and Melanchthon, is telling: It fell to Jean Loys (d. 1547), a Flemish printer who set up his business in Paris around 1535, to put Melanchthon's notorious letter as a preface on the French book market in 1542 (Sacrobosco and Melanchthon 1542). Fingerprints and a close inspection of the typesetting show that he did not use any of the four preceding editions from Wittenberg—Klug had published in 1531, 1534, 1536, and 1538—but the latter's edition of 1540 (Sacrobosco and Melanchthon 1540), at least to typeset Melanchthon's preface. Yet, as for the otherwise nearly identical typesetting of this paratext, Loys did not typeset the catchwords of Klug's print.[22] Moreover, the treatise by Sacrobosco, which was also newly edited (probably by Rheticus for Klug's edition of 1538), had been used by Loys as well. However, this text was not copied in terms of typesetting from any preceding edition produced in Wittenberg or anywhere else, and even new woodblocks seem to have been used. When Jean Richard (1516–1573) introduced Melanchthon's preface to Antwerp in

---

[22] A catchword is a word or syllable placed at the foot of a printed page that is meant to be bound along with other pages in a book. The word anticipates the first word of the following page. It helped the bookbinder to make sure that the leaves were bound in the correct order.

1543 (Sacrobosco and Melanchthon 1543b), he seems to have drawn on the typesetting of either Loys's edition of 1542 or of Klug's edition of 1540. As for Sacrobosco's text, Richard's typesetting differs in detail from both of those editions. Moreover, his edition also includes Sacrobosco's *Computus*, which was first published, together with his *Sphaera*, in Klug's edition of 1538 (Sacrobosco and Melanchthon 1538). But things get even more complicated: As a closer look reveals, Richard seems to have typeset the text of both Sacrobosco's *Sphaera* and *Computus* without a strict template. Albeit he took the typesetting of Melanchthon's letter to Grynaeus as his template from an edition of 1540 (Wittenberg) or 1542 (Paris), he compiled his edition by using textual parts (*Computus* and Melanchthon's dedication letter) as had only been done before in 1538 (Wittenberg) and 1543 (Wittenberg, printed by Peter Seitz the Elder) (Sacrobosco and Melanchthon 1543a). So, while the editions of 1540 and 1542 do not contain those additional textual parts, the editions of 1538 and 1543 contain Melanchthon's letter in different typesetting. This complex micro-analysis suggests that both Loys in Paris and Richard in Antwerp, in one way or another, were impacted by editions printed by Klug in Wittenberg. This impact, though not yet tangible through any documents providing an economic relationship, also indicates that printers used the typesetting of previous editions as templates and that Richard had clearly inspected more than one *Sphaera* edition to design his own publication.

Mutual awareness among book producers in different cities or even countries is by no means a phenomenon exclusive to *Sphaera* editions. As in the cases of local relationships, this can be further corroborated through spot tests of treatises published in various cities by different printers in our network. For example, Peter Seitz the Younger (d. 1577) published a commentary on Ovid in 1559 that originates in Georg Sabinus's (1508–1560) lectures at Kaliningrad (Sabinus 1559).[23] This work by Sabinus (a former student of Luther and Melanchthon in Wittenberg) was first published in Wittenberg in 1555 and 1556 (Sabinus 1555, 1556) by the print shop of the Heirs of Georg Rhau (fl. 1548–1566), who did not publish any *Sphaera* editions. After this edition was reprinted again in Wittenberg by Clemens Schleich and Anton Schöne—also printers of *Sphaera* editions[24]—in 1572 (Sabinus 1572), it found its way into the hands of Jérôme de Marnef (1515–1595) and the widow of Guillaume Cavellat, Denise de Marnef (fl. 1567–1616), two leading book producers of *Sphaera* editions in Paris.[25] Their edition of 1575 (Sabinus 1575) was reprinted twice in Paris (Sabinus 1579, 1580).

The Seitz print shop also published Gemma Frisius's (1508–1555) *Arithmeticae practicae methodus facilis* in 1550 (Gemma Frisius 1550). This extremely successful

---

[23] On Sabinus's commentary on Ovid and its early modern editions, see (Mundt 2019).

[24] For Clemens Schleich's and Anton Schöne's production of *Spheara* treatises, see respectively http://hdl.handle.net/21.11103/sphaera.100318 and http://hdl.handle.net/21.11103/sphaera.100317. Accessed 8 June 2021.

[25] For Jérôme de Marnef's and Guillaume Cavellat's production of *Sphaera* treatises, see respectively http://hdl.handle.net/21.11103/sphaera.100754 and http://hdl.handle.net/21.11103/sphaera.100726. The widow of Guillaume Cavellat, Denise de Marnef, also produced *Sphaera* treatises: http://hdl.handle.net/21.11103/sphaera.100281. Accessed 8 June 2021.

mathematical treatise was first published in 1540 in Antwerp by Gillis Coppens van Diest (1496–1572) (Gemma Frisius 1540). Reprinted at least sixty-five times thereafter, it was printed in Wittenberg several times by Georg Rhau (1488–1548) and by the heirs of Seitz the Elder, then several times in Paris, among other printings by Jean Loys, Thomas Richard (fl. 1547–1568), and Guillaume Cavellat (1500–1576)—all of them also producers of *Sphaera* editions[26]—in Lyon by the father of Jean de Tournes (1539–1615)—a printer of a *Sphaera* edition[27]—and in Leipzig and Strasbourg by printers with no business in *Sphaera* editions. Further overlaps with producers of *Sphaera* editions appear in reprints of Frisius's treatise, demonstrated through the prints of Maternus Cholinus (1525–1588) in Cologne (Gemma Frisius 1564, 1571, 1576), by Jean Bellère (1526–1595) in Antwerp (Gemma Frisius 1581), and in Wittenberg by the heirs of Krafft the Elder (Gemma Frisius 1579), Matthaeus Welack (Gemma Frisius 1583), Simon Gronenberg (fl. 1572–1602) (Gemma Frisius 1587, 1593), and Lorenz Säuberlich (Gemma Frisius 1604).[28] Although not all of these links are present in our network based on paratext recocurrences, most of them are, and the striking matches of the book producers in the cases of Sacrobosco and Frisius are certainly not to be taken as coincidences but can be interpreted as an indication of a shared market for books on astronomy and arithmetic. Both Sacrobosco and Frisius provided textbooks for two university-taught disciplines of the *quadrivium*, and there was certainly a market for these textbooks in university cities like Wittenberg, Antwerp, Cologne, and Paris.

7 Conclusions and Outlook

As mentioned at the beginning of this paper, in our quest for a more systematic approach, we used network analysis to detect potential relationships among book producers. Such relationships can be properly defined only by means of further historical research. They could be real relations of an economic nature, social relations on a broader level, or just "mutual awareness," indicating that the producers were observing and being influenced by one another's production. Taking into consideration the corpus of editions containing Sacrobosco's *Tractatus de sphaera*, we have

---

[26] For Jean Loys's and Thomas Richard's production of *Sphaera* treatises, see respectively http://hdl.handle.net/21.11103/sphaera.100816 and http://hdl.handle.net/21.11103/sphaera.100347. Accessed 8 June 2021.

[27] For Jean de Tournes's production of *Sphaera* treatises, see http://hdl.handle.net/21.11103/sphaera.100911. Accessed 8 June 2021.

[28] For Maternus Cholinus's, Jean Bellère's, Johann Krafft's I and his heirs' (Matthaeus Welack's, Simon Gronenberg's, and Lorenz Säuberlich's) production of *Sphaera* treatises, see respectively http://hdl.handle.net/21.11103/sphaera.100400, http://hdl.handle.net/21.11103/sphaera.100338, http://hdl.handle.net/21.11103/sphaera.100955, http://hdl.handle.net/21.11103/sphaera.100684, http://hdl.handle.net/21.11103/sphaera.100778, and http://hdl.handle.net/21.11103/sphaera.100294. Accessed 8 June 2021.

considered the circulation of paratexts to be an arguably dependable intimation of such relationships, at least as an impetus toward further historical research.

We admit that the absolute numbers of paratexts and publications constituting our networks might be too small for such a line of reasoning, but we are confident that, if a greater number of historical sources is considered, this method can become standard. The geographical network analysis in particular has shown that it is possible to draw inferences that at least sound plausible and can be corroborated by in-depth historical spot tests. As a preliminary result, we can state that the strongest and most frequent relationships between book producers in the context of the academic book market occurred within one and the same city, suggesting a few local centers of the network, particularly Venice, Wittenberg, and Paris. This is indeed in agreement with book historians' research on the production of school and textbooks (Gehl 2013). The analysis however also indicates transregional relationships between book producers. While economic relationships seem more likely in the local contexts, many of the transregional links seem to indicate a mutual (or occasionally one-sided) awareness of editions published by colleagues in other cities. Editions containing a similar set of text parts, especially the same paratexts, are arguably not coincidental or an effect of an unrelated yet similar demand for certain books in various cities. More likely, it seems that the transregional character of the early modern academic book market fostered a certain awareness for successful or highly demanded editions, later to be introduced into local markets with their own local dynamics.

Along with the first preliminary historical insights, our results allow historical researchers to prioritize close readings of the historical sources in order to find out what relationships really existed. Approaches along this line might include relating the *Sphaera* editions based on their fingerprints more completely and systematically than has been done here. Thereby possible reissues of the same text among different book producers can be identified, marking their collaboration as quite likely. Moreover, a comparison, by means of machine-learning technology, of the imagery within the treatises might indicate such collaborations even further, suggesting that printers exchanged, or at least reused, the same woodblocks for different editions.[29] These consecutive approaches may lead to further investigations regarding family or business relations between book producers. Interpreting the results of this study may also allow for a more political perspective. It is intriguing to read the results against the background of political alliances or relationships between cities. Finally, the relationships might reveal more about confessional boundaries (or their absence) as far as the book market was concerned. Here, Clavius's and Melanchthon's paratexts are obviously promising cases.

---

[29] A first step toward the completion of a machine-learning algorithm that allows for the discovery of similarities among early modern illustrations—a specific "Deep Similarity Model"—has already been achieved (Eberle et al. 2020).

Abbreviations

Digital Repositories

Sphaera Corpus*Tracer*   Max Planck Institute for the History of Science. https://db.
sphaera.mpiwg-berlin.mpg.de/resource/Start.   Accessed
07 June 2021.

References

Primary Literature

Breunle, Moritz. 1529a. *Ein kurtz formular unnd Cantzley buechlin, darynn begriffen wirdt wie man einem yegklichen wz standts wirde eheren und wesen er ist schreyben soll.* Augsburg: Heinrich von Augsburg Steiner.

Breunle, Moritz. 1529b. *Eyn kurtz formular und kantzley buechleyn daryn begriffen wird wie mann eynem yglichen was stands wirde eheren und wesen er ist schreyben sol vorhyn yn druck vorfasset und ytzunder auff das nawe mit fleiß ubersehen und an viel oertern mit nawen formularien Deutscher sendtbrieffe gebessert.* Leipzig: Jakob Thanner.

Breunle, Moritz. 1548. *Ein kurtz formular und Cantzley buechlein, darinn begriffen wird wie man einem jeglichen schreiben sol.* Wittenberg: Ursula Seitz.

Breunle, Moritz. 1552. *Ein kurtz formular undt cantzley büchlein, darinn begriffen wird wie man einem jeglichen was standes wirde ehren und wesen er sei schreiben sol.* Wittenberg: Veit Kreutzer.

Breunle, Moritz. 1553. *Ein kutz formular und Cantzley büchlein darinn begriffen wird wie man einem jeglichen was standes, wirde, ehren und wesens er ist schreiben sol.* Wittenberg: Veit Kreutzer.

Breunle, Moritz. 1554. *Ein kurtz formular und cantzeley buechlein, darin begriffen wirdt wie man einem jeglichen was standes wirde ehren und wesens er ist schreiben sol.* Wittenberg: Heirs of Peter Seitz I.

Breunle, Moritz. 1556. *Ein kurtz formular cantzeley büchlein.* Wittenberg: Heirs of Peter Seitz I.

Breunle, Moritz. 1557. *Ein kurtz formular und cantzeley buechlein, darin begriffen wird wie man einem jglichen was standes wirde ehren und wesens er ist schreiben sol.* Wittenberg: Heirs of Peter Seitz I.

Breunle, Moritz. 1559. *Ein kurtz formular, und cantzeley büchlein, darin begriffen wird, wie man einem jglichen, was standes, wirde, ehren und wesens er ist, schreiben sol.* Wittenberg: Veit Kreutzer.

Breunle, Moritz. 1561. *Ein kurtz formular und cantzeley buechlein, darin begriffen wird wie man einem jglichen was standes wirde ehren und wesens er ist schreiben sol.* Wittenberg: Veit Kreutzer.

Dietrich, Sebastian. 1564. *Novae quaestiones sphaerae, hoc est, de circulis coelestibus, et primo mobili, in gratiam studiosae iuventutis scriptae, a M. Sebastiano Theodorico Vuinshemio Mathematum Professore.* Wittenberg: Johann Krafft the Elder. https://hdl.handle.net/21.11103/sphaera.101072.

Dietrich, Sebastian. 1570. *Novae quaestiones sphaerae, hoc est, de circulis coelestibus, et primo mobili, in gratiam studiosae iuventutis scriptae, a M. Sebastiano Theodorico Vuinshemio Mathematum Professore.* Wittenberg: Johann Krafft the Elder. https://hdl.handle.net/21.11103/sphaera.101082.

Dietrich, Sebastian. 1573. *Novae quaestiones sphaerae, hoc est, de circulis coelestibus & primo mobili, in gratiam studiosae iuventutis scriptae, a M. Sebastiano Theodorico Vuinshemio, Mathematum Professore.* Wittenberg: Anton Schöne and Clemens Schleich. https://hdl.handle.net/21.11103/sphaera.101087.

Dietrich, Sebastian. 1583. *Novae quaestiones sphaericae, hoc est, circulis coelestibus et primo mobili, in gratiam studiosae iuventutis scriptae, A M. Sebastiano Theodorico Vuinshemio, Mathematum Professore.* Wittenberg: Matthaeus Welack. https://hdl.handle.net/21.11103/sphaera.101100.

Dietrich, Sebastian. 1605. *Novae quaestiones sphaericae, hoc est, de circulis coelestibus & primo mobili, in gratiam studiosae juventutis scriptae, a M. Sebastiano Theodorico Vuinshemio, Mathematum Professore.* Wittenberg: Lorenz Säuberlich for Samuel Seelfisch. https://hdl.handle.net/21.11103/sphaera.101053.

Gemma Frisius, Reiner. 1540. *Arithmeticae practicae methodus facilis.* Antwerp: Gillis Coppens van Diest for Gregorius de Bonte.

Gemma Frisius, Reiner. 1550. *Arithmeticae practicae methodus facilis, per gemmam Frisium medicum ac mathematicum.* Wittenberg: Heirs of Peter Seitz I.

Gemma Frisius, Reiner. 1564. *Arithmeticae practicae methodus facilis, per gemmam Frisium, medicum ac mathematicu, iam recens, ab ipso authore emendata, et multis in locis insigniter aucta. Hvc accesserunt Jacobi peletarii cenomani annotationes: eiusdem item de fractionibus astronomicis compendium: et de cognoscendis per memoriam calendis, idib. Nonis, festis mobilibus, et loco solis et lunae in zodiaco.* Cologne: Maternus Cholinus.

Gemma Frisius, Reiner. 1571. *Arithmeticae practicae methodus facilis, per gemmam Frisium, medicum ac mathematicum, iam recèns ab ipso authore emendata, et multis in locis insigniter aucta. Hvc accesserunt Jacobi peletarii cenomani annotationes: eiusdem item de fractionibus astronomicis compendium: et de cognoscendis per memoriam calendis, nunc verò à Joanne Stein recognita, et novis aucta additionibus.* Cologne: Maternus Cholinus.

Gemma Frisius, Reiner. 1576. *Arithmeticae practicae methodus facilis, per gemmam Frisium, medicum ac mathematicum, iam recens ab ipso authore emendata, et multis in locis insigniter aucta. Hvc accesserunt Jacobi peletarii cenomani annotationes: eiusdem item de fractionibus astronomicis compendium nunc verò a Joanne Stein recognita, et novis aucta additionibus.* Cologne: Maternus Cholinus.

Gemma Frisius, Reiner. 1579. *Arithmeticae practicae methodus facilis, per gemmam Frisium medicum ac mathematicum.* Wittenberg: Heirs of Johann Krafft I.

Gemma Frisius, Reiner. 1581. *Arithmeticae practicae methodus facilis, in eandem annotationes, de fractionibus astronomicis compendium & de cognoscendis per memoriam kalendis, idibus, nonis, festis mobilibus, locoque solis & lunae in zodiaco.* Antwerp: Jean Bellère.

Gemma Frisius, Reiner. 1583. *Arithmeticae practicae methodus facilis, per gemmam Frisium medicum ac mathematicum.* Wittenberg: Matthaeus Welack.

Gemma Frisius, Reiner. 1587. *Arithmeticae practicae methodus facilis, per gemmam Frisium medicum ac mathematicum.* Wittenberg: Simon Gronenberg.

Gemma Frisius, Reiner. 1593. *Arithmeticae practicae methodus facilis, per gemmam Frisium medicum ac mathematicum.* Wittenberg: Simon Gronenberg.

Gemma Frisius, Reiner. 1604. *Arithmeticae Practicae Methodus Facilis, Per Gemmam Frisium Medicum ac Mathematicum.* Wittenberg: Lorenz Säuberlich.

Sabinus, Georg. 1555. *Fabularum Ovidii interpretatio tradita in academia Regiomontana. A Georgio sabino.* Wittenberg: Heirs of Georg Rhau.

Sabinus, Georg. 1556. *Fabularum Ovidii interpretatio tradita in academia Regiomontana. A Georgio sabino.* Wittenberg: Heirs of Georg Rhau.

Sabinus, Georg. 1559. *Fabularum Ovidii interpretatio tradita in academia Regiomontana a Georgio sabino.* Wittenberg: Peter Seitz II.

Sabinus, Georg. 1572. *Fabularum Ovidii interpretatio, tradita in academia Regiomontana, a Georgio sabino.* Wittenberg: Clemens Schleich and Anton Schöne.

Sabinus, Georg. 1575. *Fabularum Ovidii interpretatio tradita in Academia Regiomontana*. Paris: Jérôme de Marnef and the widow of Guillaume Cavellat.

Sabinus, Georg. 1579. *Fabularum Ovidii interpretatio*. Paris: Charles Roger for Jérôme de Marnef and the widow of Guillaume Cavellat.

Sabinus, Georg. 1580. *Fabularum Ovidii interpretatio*. Paris: Charles Roger for Jérôme de Marnef and the widow of Guillaume Cavellat.

Sacrobosco, Johannes de. 1494. *Textus De Sphera Iohannis de Sacrobosco Cum Additione (quantum necessarium est) adiecta: Nouo commentario nuper editio Ad utilitatem studentium philosophice Parisien. Academie: illustratus*. Paris: Wolfgang Hopyl. https://hdl.handle.net/21.11103/sphaera. 101126.

Sacrobosco, Johannes de, Pierre d'Ailly, Francesco Capuano, Robert Grosseteste, Jacques Lefèvre d'Etaples, Joannes Regiomontanus, Georg von Peurbach, and Bartolomeo Vespucci. 1508. *Nota eorum quæ in hoc libro continentur. Oratio de laudibus astrologiae habita a Bartholomeo Vespucio florentino in almo Patavio Gymnasio anno.M.d.vi. TEXTUS SPHAERAE IOANNIS DE SACRO BUSTO. Expositio sphaerae Eximii artium & medicinae doctoris Domini Francisci Capuani de manfredonia. Annotationes nonnullae eiusdem Bartholomei Vespucii hic ide itersertae. Iacobi frabri stapulensis Commentarii in eandem sphaeram. Reverendissimi Domini Petri de aliaco Cardinalis & episcopi Cameracensis in eandem quaestiones subtilissimae numero xiiii. Reverendissimi episcopi Domini Roberti linconiensis sphaerae compendium. Disputationes Ioannis de regio monte contra cremonensia deliramenta. Theoricarum novarum textus cum expositione eiusdem Francisci Capuani omnia nuper diligentia summa emendata*. Venice: Giovanni and Bernardino Rosso for Giuntino Giunta. https://hdl.handle.net/21.11103/sphaera.100915.

Sacrobosco, Johannes de, Cecco d'Ascoli, Francesco Capuano, and Jacques Lefèvre d'Etaples. 1499a. *Sphera Mundi cum tribus Commentis nuper editis vz. Cicchi Esculani, Francisci Capuani de Manfredonia, Iacobi Fabri Stapulensis*. Venice: Simone Bevilacqua. https://hdl.handle.net/21. 11103/sphaera.100273.

Sacrobosco, Johannes de, Cecco d'Ascoli, Francesco Capuano, and Jacques Lefèvre d'Etaples. 1499b. *Sphera Mundi cum tribus Commentis nuper editus vz. Cicchi Esculani, Francisci Capuani de Manfredonia, Iacobi Fabri Stapulensis*. Venice: Simone Bevilacqua. https://hdl.handle.net/21. 11103/sphaera.100021.

Sacrobosco, Johannes de, Gerardus Cremonensis, Georg von Peurbach, Nūr al-Dīn Ishāq al-Bitrūgi, Prosdocimus de Beldemandis, Luca Gaurico, Francesco Capuano, Joannes Regiomontanus, Michael Scot, Jacques Lefèvre d'Etaples, Pierre d'Ailly, Robert Grosseteste, and Bartolomeo Vespucci. 1531. *Spherae tractatus Ioannis de Sacro Busto Anglici viri clarissimi. Gerardi Cremonensis theoricae planetarum veteres. Georgii Purbachii theoricae planetarum novae. Prosdocimi de beldomando patavini super tractatu sphaerico commentaria, nuper in lucem diducta per. L. GA. nunque amplius impressa. Ioannis baptistae capuani sipontini expositio in sphaera & theoricis. Ioannis de monteregio disputationes contra theoricas gerardi. Michaelis scoti expositio brevis & quaestiones in sphaera. Iacobi fabri stapulensis paraphrases & annotationes. Campani compendium super tractatu de sphera. Eiusdem tractatulus de modo fabricandi spheram solidam. Petri cardin. de aliaco epi cameracensis. 14. Quaestiones. Roberti linconiensis epi tractatulus de sphaera. Bartholomei vesputii glossulae in plerisque locis sphaerae. Eiusdem oratio. De laudibus astrologiae. Lucae Gaurici castigationes & figurae toto opere diligentissime reformatae. Eiusdem quaestio Nunquid sub aequatore sit habitatio. Eiusdem Oratio de inventoribus & laudibus Astrologiae. Reverendissimo cardin. epo. D. Bernardo Tridentinor principi dicata. Alpetragii Arabi theorica planetarum nuperrime Latinis mandata literis a calo calonymos hebreo neapolitano, ubi nititur salvare apparentias in motibus Planetarum absque eccentricis & epicyclis*. Venice: Lucantonio Giunta. https://hdl.handle.net/21.11103/sphaera.100999.

Sacrobosco, Johannes de, Bonet de Lattes, and Euclid. 1500. *Textus de Sphera Johannis de Sacro bosco Cum additione (quantum necessarium est) adiecta: Novo commentario nuper edito Ad utilitatem studentium Philosophice Parisien. Academie illustratus Cum Compositione Anuli Astronomici Boni Latensis. Et Geometria Euclidis Megarensis*. Paris: Wolfgang Hopyl. https://hdl.han dle.net/21.11103/sphaera.100889.

Sacrobosco, Johannes de, Bonet de Lattes, and Euclid. 1507. *Textus De Sphera Johannis de Sacrobosco Cum additione (quantum necessarium est) adiecta: Novo commentario nuper edito Ad vtilitatem studentium Philosophice Parisien. Academie illustratus Cum Compositione Anuli Astronomici Boni Latensis. Et Geometria Euclidis Megarensis*. Paris: Henri Estienne I. https://hdl.handle.net/21.11103/sphaera.100029.

Sacrobosco, Johannes de, Bonet de Lattes, and Euclid. 1511. *Textus De Sphera Johannis de Sacrobosco, Cum additione (quantum necessarium est) adiecta: Novo commentario nuper edito, Ad utilitatem studentium Philosophice Parisien[sis] Academie illustratus, cum Compositione Anuli Astronomici Boni Latensis. Et Geometria Euclidis Megarensis*. Paris: Henri Estienne I. https://hdl.handle.net/21.11103/sphaera.100919.

Sacrobosco, Johannes de, Bonet de Lattes, and Euclid. 1521. *Textus de sphaera Ioannis de Sacrobosco: introductoria additione (quantum necessarium est) commentarioque, Ad utilitatem studentium philosophiae Parisiensis Academiae illustratus. Cum compositione Annuli astronomici Boneti Latensis: Et Geometria Euclidis Megarensis*. Paris: Simon de Colines. https://hdl.handle.net/21.11103/sphaera.100995.

Sacrobosco, Johannes de, Bonet de Lattes, and Euclid. 1534. *Textus de sphaera Ioannis de Sacrobosco: Introductoria additione (quantum necessarium est) commentarióque, Ad utilitatem studentium Philosophiae Parisiensis Academiae illustratus. Cum compositione Annuli astronomici Boneti Latensis: Et Geometria Euclidis Megarensis*. Paris: Simon de Colines. https://hdl.handle.net/21.11103/sphaera.100099.

Sacrobosco, Johannes de, and Philipp Melanchthon. 1531. *Liber Iohannis de Sacro Busto, de Sphaera. Addita est praefatio in eundem librum Philippi Melanchthonis ad Simonem Gryneum*. Wittenberg: Joseph Klug. https://hdl.handle.net/21.11103/sphaera.100138.

Sacrobosco, Johannes de, and Philipp Melanchthon. 1532. *Liber Ioannis de Sacro Busto, de Sphaera. Addita est praefatio in eundem librum Philippi Melanchthonis ad Simonem Gryneum*. Venice: Giovanni Antonio Nicolini da Sabbio & Brothers for Melchiorre Sessa I. https://hdl.handle.net/21.11103/sphaera.100118.

Sacrobosco, Johannes de, and Philipp Melanchthon. 1538. *Ioannis de Sacro Busto libellus, De sphæra: Eiusdem autoris libellus, cuius titulus est Computus, eruditissimam anni & mensium descriptionem continens. Cum praefatione Philippi Melanth. & novis quibusdam typis, qui ortus indicant*. Wittenberg: Joseph Klug. https://hdl.handle.net/21.11103/sphaera.101106.

Sacrobosco, Johannes de, and Philipp Melanchthon. 1540. *Ioannis de Sacro-Busto libellus, de sphaera. Cum praefacione Philippi Melanth. & nout quibusdam typis, qui ortus indicant*. Wittenberg: Joseph Klug. https://hdl.handle.net/21.11103/sphaera.101025.

Sacrobosco, Johannes de, and Philipp Melanchthon. 1542. *Ioannis de Sacrobusto de sphaera liber. Plurimis novis typis auctus & illustratus. Praemissa Philippi Melanchthonis doctiss. praefatione, qua utilitatem sphaericae scientiae, & Christiano homini non negligendam probat*. Paris: Jean Loys. https://hdl.handle.net/21.11103/sphaera.101027.

Sacrobosco, Johannes de, and Philipp Melanchthon. 1543a. *Ioannis de Sacro Busto libellus de sphaera. Accessit eiusdem autoris Computus ecclesiasticus, & alia quaedam in studiosorum gratiam edita. Cum praefacione Philippi Melanthonis*. Wittenberg: Peter Seitz I. https://hdl.handle.net/21.11103/sphaera.101029.

Sacrobosco, Johannes de, and Philipp Melanchthon. 1543b. *Ioannis de sacro busto Libellus de Sphaera. Eiusdem authoris libellus, cuius titulus est Computus, eruditissimam anni & mensium de scriptionem continens. Cum praefacione Philippi Melanthonis et nouis quibusdam typis, qui ortus indicant*. Antwerp: Jean Richard. https://hdl.handle.net/21.11103/sphaera.100180.

Sacrobosco, Johannes de, and Philipp Melanchthon. 1601. *Libellus de Sphaera Iohannis de Sacro Busto. Accessit eiusdem Autoris Computus Ecclesiasticus, Et alia quaedam, in studiosorum gratiam edita. Cum Praefatione Philippi Melanchthonis*. Wittenberg: Johann Krafft for Zacharias Schürer & Partners. https://hdl.handle.net/21.11103/sphaera.100646.

Sacrobosco, Johannes de, and Philipp Melanchthon. 1629. *Libellus de Sphaera Iohannis de Sacro Busto. Accessit ejusdem Autoris Computus Ecclesiasticus, Et alia quaedam, in studiosorum gratiam edita. Cum Praefatione Philippi Melanchthonis*. Wittenberg: Widow & Heirs of Zacharias I. Schürer. https://hdl.handle.net/21.11103/sphaera.100303.

Secondary Literature

Batchelor, Kathryn. 2018. *Translation and paratexts: Translation theories explored*. London: Routledge.

Beyer, Victoria. 2019. *How to generate a fingerprint*, Preprint 499. Berlin: Max Planck Institute for the History of Science.

Brown, David M., Adriana Soto-Corominas, and Juan Luis Suárez. 2017. The preliminaries project: Geography, networks, and publication in the Spanish golden age. *Digital Scholarship in the Humanities* 32 (4): 709–732.

Eberle, Oliver, Jochen Büttner, Florian Kräutli, Klaus-Robert Müller, Matteo Valleriani, and Grégoire Montavon. 2020. Building and interpreting deep similarity models. *IEEE Transactions on Pattern Analysis and Machine Intelligence*. http://doi.org/10.1109/TPAMI.2020.3020738.

Enenkel, Karl Alfred Engelbert. 2015. *Die Stiftung von Autorschaft in der neulateinischen Literatur (ca. 1350-ca. 1650): zur autorisierenden und wissensvermittelnden Funktion von Widmungen, Vorworttexten, Autorporträts und Dedikationsbildern*. Leiden: Brill.

Funk, Kellen, and Lincoln A. Mullen. 2018. The spine of American law: Digital text analysis and U.S. legal practice. *The American Historical Review* 123 (1): 132–164. https://doi.org/10.1093/ahr/123.1.132.

Gehl, Paul F. 2013. Advertising or *fama*? Local markets for schoolbooks in sixteenth-century Italy. In *Print culture and peripheries in Early Modern Europe: A contribution to the history of printing and the book trade in small European and Spanish cities*, ed. Benito Rial Costas, 69–100. Leiden: Brill.

Genette, Gérard. 2001. *Paratexts: Thresholds of interpretation*. Cambridge: Cambridge University Press.

Gerritsen, Johan. 1991. Printing at Froben's: An eye-witness account. *Studies in Bibliography* 44: 144–163.

Gingerich, Owen. 1988. Sacrobosco as a textbook. *Journal for the History of Astronomy* 19 (4): 269–273. https://doi.org/10.1177/002182868801900404.

Hinks, John, and Catherine Feely (eds.). 2017. *Historical networks in the book trade*. London: Routledge.

Lalla, Sebastian. 2003. Über den Nutzen der Astrologie: Melanchthons Vorwort zum "Liber de sphaera". In *Gedenken und Rezeption: 100 Jahre Melanchthonhaus*, ed. Günther Frank, 147–160. Heidelberg: Verlag Regionalkultur.

Lowry, Martin. 1979. *The world of Aldus Manutius: Business and scholarship in Renaissance Venice*. Ithaca, NY: Cornell University Press.

Maclean, Ian. 2012. *Scholarship, commerce, religion: The learned book in the age of confessions, 1560–1630*. Cambridge, MA/London: Harvard University Press.

Mundt, Lothar (ed.). 2019. *Fabularum Ovidii interpretatio = Auslegung der Metamorphosen Ovids: Edition, Übersetzung, Kommentar*. Berlin: De Gruyter.

Nuovo, Angela. 2013. *The book trade in the Italian Renaissance*. Leiden: Brill.

Oehmig, Stefan (ed.). 2015. *Buchdruck und Buchkultur im Wittenberg der Reformationszeit*. Leipzig: Evangelische Verlagsanstalt GmbH.

Pantin, Isabelle. 1987. La lettre de Melanchthon à Simon Grynaeus: Avatars d'une défense de l'astrologie. In *Divination et controverse religieuse en France au XVIe siècle*, 85–101. Paris: Cahiers V. L. Saulnier.

Pantin, Isabelle. 1998. Les problèmes de l'édition des livres scientifiques: l'exemple de Guillaume Cavellat. In *Le livre dans l'Europe de la Renaissance: Actes du XXVIIIe Colloque international d'Etudes humanistes de Tours,* ed. Bibliothèque Nationale, 240–252. Paris: Promodis, Editions du Cercle de la Librairie.

Pantin, Isabelle. 2020. Borrowers and Innovators in the Printing History of Sacrobosco: The case of the in-octavo tradition. In *De sphaera of Johannes de Sacrobosco in the early modern period: The authors of the commentaries*, ed. Matteo Valleriani, 265–312. Cham: Springer Nature. https://doi.org/10.1007/978-3-030-30833-9_9

Reich, Karin, and Eberhard Knobloch. 2004. Melanchtohns Vorreden zu Sacroboscos *Sphaera* (1531) und zum *Computus ecclesiasticus*. *Beiträge zur Astronomiegeschichte* 7: 13–44.

Rosen, Edward. 1974. Rheticus as editor of Sacrobosco. In *Scientific, historical and political essays in honor of Dirk J. Struik*, ed. S. Robert Cohen, John Stachel, and W. Marx Wartofsky, 245–248. Dordrecht: Springer.

Sander, Christoph. 2018. Johannes de Sacrobosco und die Sphaera-Tradition in der katholischen Zensur der Frühen Neuzeit. *NTM Zeitschrift für Geschichte der Wissenschaften, Technik und Medizin* 26 (4): 437–474. https://doi.org/10.1007/s00048-018-0199-6

Smith, Helen, and Louise Wilson (eds.). 2014. *Renaissance paratexts*. Cambridge, UK: Cambridge University Press.

Thorndike, Lynn. 1949. *The Sphere of Sacrobosco and its commentators*. Chicago: The University of Chicago Press.

Töpfer, Regina. 2007. *Pädagogik, Polemik, Paränese: die deutsche Rezeption des Basilius Magnus im Humanismus und in der Reformationszeit*. Tübingen: Niemeyer.

Tweed, Hanah C., and Diane G. Scott (eds.). 2018. *Medical paratexts from Medieval to Modern: Dissecting the page*. Cham: Palgrave Macmillan.

Valleriani, Matteo. 2017. The tracts on the *Sphere*: Knowledge restructured over a network. In *Structures of practical knowledge*, ed. Matteo Valleriani, 421–473. Dordrecht: Springer.

Valleriani, Matteo. 2020. Prolegomena to the study of Early Modern commentators on Johannes de Sacrobosco's *Tractatus de sphaera*. In *De sphaera of Johannes de Sacrobosco in the early modern period: The authors of the commentaries*, ed. Matteo Valleriani, 1–23. Cham: Springer Nature. https://doi.org/10.1007/978-3-030-30833-9_1

Valleriani, Matteo, Florian Kräutli, Maryam Zamani, Alejandro Tejedor, Christoph Sander, Malte Vogl, Sabine Bertram, Gesa Funke, and Holger Kantz. 2019. The emergence of epistemic communities in the *Sphaera* corpus: Mechanisms of knowledge evolution. *Journal of Historical Network Research* 3: 50–91. https://doi.org/10.25517/jhnr.v3i1.63

Wagner, Bettina. 2008. An der Wiege des Paratexts: Formen der Kommunikation zwischen Druckern, Herausgebern und Lesern im 15. Jahrhundert. In *Die Pluralisierung des Paratextes in der Frühen Neuzeit: Theorie, Formen, Funktionen*, eds. Frieder von Ammon and Herfried Vögel, 133–156. Berlin: Lit.

Zamani, Maryam, Alejandro Tejedor, Malte Vogl, Florian Kräutli, Matteo Valleriani, and Holger Kantz. 2020. Evolution and transformation of early modern cosmological knowledge: A network study. *Scientific Reports–Nature*. https://doi.org/10.1038/s41598-020-76916-3

Matteo Valleriani is research group leader at the Department I at the Max Planck Institute for the History of Science in Berlin, Honorary Professor at the Technische Universität of Berlin, and Professor by Special Appointment at the University of Tel Aviv. He investigates the relation between diffusion processes of scientific, practical, and technological knowledge and their economic and political preconditions. His research focuses on the Hellenistic period, the late Middle Ages, and the early modern period. Among his principal research endeavors, he leads the project "The Sphere: Knowledge System Evolution and the Shared Scientific Identity of Europe" (https://sphaera.mpiwg-berlin.mpg.de), which investigates the formation and evolution of a shared scientific identity in Europe between the thirteenth and seventeenth centuries. In the context of this project, Matteo Valleriani implemented the development of machine learning technology and of the physics of complex system in the humanities. The project is also part of the investigations led by Matteo Valleriani in the context of BIFOLD (https://bifold.berlin). Among his publications, he has authored the book *Galileo Engineer* (Springer 2010), is editor of *The Structures of Practical Knowledge* (Springer Nature 2017), and published *De sphaera of Johannes de Sacrobosco in the Early Modern Period. The Authors of the Commentaries* (Springer Nature 2020).

Christoph Sander is a postdoctoral researcher at the Bibliotheca Hertziana, Max Planck Institute for Art History in Rome. Currently, he investigates the production, typology and use of diagrams in early modern science. In the past years he has collaborated with the research group "The Sphere" at the Max Planck Institute for the History of Science in Berlin. He has published widely on the history of early modern philosophy and its institutional embedding. His monograph study on magnetism—*Magnes: Der Magnetstein und der Magnetismus in den Wissenschaften der Frühen Neuzeit*—was published in 2020 by Brill Publishers.

Chapter 11
The *Sphaera* in Jesuit Education

Paul F. Grendler

Abstract When the Jesuits began to teach mathematics, they adopted the existing European curriculum which included Sacrobosco's *Sphaera*. Christoph Clavius, the most influential Jesuit mathematician, published a commentary on the *Sphaera* in 1570 which was widely used. Its publication also marked a change in publication policy by the Roman Jesuits. As the Jesuits prepared a uniform curriculum for the Society's schools in the 1580s and 1590s, Clavius offered a comprehensive mathematics curriculum and urged the Society to teach more mathematics and to train more Jesuit mathematicians. But some Jesuit philosophers rejected mathematics as unscientific. The Ratio Studiorum of 1599 included the *Sphaera* but did not expand Jesuit mathematical education. Jesuits continued to teach the *Sphaera* and to use Clavius' commentary until about 1650.

Keywords Sacrobosco · *Sphaera* · Christoph Clavius · Benet Perera · Jesuits · Society of Jesus · Mathematics · Astronomy · Natural philosophy · *Ratio studiorum* · Roman College · Vittorino Eliano · Giovanni Battista Eliano

1 Introduction

Johannes de Sacrobosco's (1195–1256) *Sphaera* was a major text in Jesuit mathematical education in the first century of the Society.[1] And Christoph Clavius (1538–1612), the leading Jesuit mathematician, published a commentary on it that one

---

[1] I am grateful to Mordechai Feingold for reading an early draft of this paper, and Nelson Minnich for help with a Latin passage. I am grateful to members of the seminar for their comments and especially to Christoph Sander and to Andrea Ottone, who also carefully edited my article. The remaining mistakes are mine.

[2] "il commento alla *Sphaera* di Sacrobosco fu la più ampia sintesi di astronomia elementare disponibile nei cinquanta anni dal 1570 al 1620" (Baldini 1992, 135).

P. F. Grendler (✉)
University of Toronto, Toronto, ON, Canada
e-mail: paulgrendler@gmail.com

© The Author(s) 2022
M. Valleriani and A. Ottone (eds.), *Publishing Sacrobosco's* De sphaera *in Early Modern Europe*, https://doi.org/10.1007/978-3-030-86600-6_11

scholar calls "the amplest synthesis of elementary astronomy available in the fifty years between 1570 and 1620."[2]

The Jesuits were not the first to teach Sacrobosco's *Sphaera*. Teachers and students across Europe taught or learned from it, because it was part of Renaissance mathematical and astronomical education, which were viewed as one whole discipline. Renaissance mathematics included astronomy and sometimes astrology. And the position of mathematics in the curriculum was different in the collegiate universities of northern Europe and Spain from its position in Italian law and medicine universities. A collegiate university was a university dominated by colleges, which were combination teaching and residence institutions that concentrated on teaching humanities and philosophy to young students pursuing Bachelor of Arts and Master of Arts degrees. Paris and Oxford were famous collegiate universities. Collegiate universities also taught a great deal of theology, but little law and medicine. Mathematics instruction in collegiate universities was part of the broad collection of studies leading to the Bachelor of Arts and Master of Arts degrees. Hence, many teachers taught, and practically all students learned, a little mathematics. But collegiate universities might or might not have had a professor of mathematics.

Italian universities concentrated on teaching law and medicine to older students pursuing doctorates in those disciplines. All larger Italian universities also had professors of mathematics who lectured on mathematics and astronomy at an advanced level. Galileo Galilei (1564–1642) at the University of Padua is the best known. But a limited number of students attended mathematics lectures, because they were not a prerequisite for law or medicine doctorates. Nevertheless, some medicine students attended mathematics lectures in order to learn about medical astronomy and astrology, on the belief that the movements of heavenly bodies influenced the progress of a disease and, properly understood, could tell the physician when to apply a therapy. And some students, whatever their major interest, were fascinated by the heavens. Although collegiate and Italian universities approached instruction in mathematics differently, one thing was the same. Practically every mathematics course included some use of Sacrobosco's *Sphaera* or a commentary on it with additional information and speculation on the movements of heavenly bodies.

2 The Early Years of Jesuit Education

In 1548, the Jesuits opened their first school in Messina, Sicily. The Jesuits had no master plan when they began and certainly not for mathematics. Nevertheless, they moved unevenly toward a tripartite school structure. The lower school taught Latin grammar, the humanities, a little Greek, and rhetoric, based on the ancient classics. This was the Renaissance *studia humanitatis* curriculum. Next came an upper school teaching Aristotelian philosophy by lecturing on individual texts of Aristotle (385–323 BCE) rather than by subject matter.[3] For the vast majority of students, philosophy ended their Jesuit schooling, as they left for employment or university study. The few remaining students—mostly Jesuits and other clergymen—studied Scholastic theology for three or four years and were ordained priests. Scholastic theology had a well-established tradition, which the Jesuits adopted and altered for their purposes.

The first Jesuits were uncertain about the content and organization of philosophical instruction. Because all ten founding Jesuits, plus other key early Jesuits, had studied at the University of Paris, they looked to it for guidance. The preparation for the Bachelor of Arts degree at Paris consisted of lectures on logic, the *Physics* of Aristotle, metaphysics, moral philosophy, and other topics, plus lectures "on some mathematical books, especially the *Sphaera* of Sacrobosco" (Schurhammer 1973, 144).[4] This was a broad but not sharply focused program that influenced Ignatius of Loyola (1491–1556), who had studied at Paris for seven years. In the section on universities in the Jesuit Constitutions adopted by the Society in 1558, Ignatius wrote: "Logic, physics, metaphysics, and moral philosophy should be treated, and also mathematics, with the moderation appropriate to the end which is being sought" (*Constitutions* 1996, pt. 4, ch. 12, 451, 180).[5] The view of Ignatius echoed the position of mathematics in collegiate universities. Ignatius endorsed mathematics, but provided no further guidance.

Jesuit schools in their first twenty years taught mathematics irregularly, but did teach Sacrobosco's *Sphaera* at least part of the time. For example, in 1555 the University of Perugia appointed a young Jesuit to be extraordinary professor of rhetoric and Greek without stipulating what he should teach. He surprisingly lectured on Sacrobosco's *Sphaera* for all or part of the academic year 1555–1556 (Springhetti 1961, 109–110).[6] The Jesuit school at Cologne taught the *Sphaera* in 1557 and 1561. So did Jesuit schools in Coimbra in 1562 and Évora in 1563 (MP, III, 530, 547, 319, 591). Of course, the Jesuits taught other texts as well. In 1548, the Messina school taught *De mundi sphaera* (1542) of Oronce Finé (1494–1555) (MP, I, 26).

In the 1560s and 1570s, Jesuit schools moved toward a more structured three-year cursus philosophicus of logic, natural philosophy, and metaphysics based on Aristotelian texts. But they continued to pay limited attention to mathematics. In the late 1570s, there were twenty-eight Jesuit schools in Italy, of which seven taught philosophy. Only two, the Collegio Romano (the leading Jesuit school founded in 1551, with accomplished Jesuit scholars as teachers, and the broadest curriculum) and the Milan school offered a mathematics course. The situation was similar in

---

[3] See (Grendler 2019b, 13–17), for the early development of philosophy teaching. For more information about the three levels of Jesuit education, see (Grendler 2019a, 15–19, 23–27).

[4] Only the necessary minimum of secondary sources are given in order to hold the paper to an appropriate length.

[5] Ignatius wrote this section of the Constitutions in late 1553 and early 1554, when there were still only a few Jesuit schools.

[6] See (Grendler 2017, 348–351), for the context.

northern Europe (Grendler 2004, 488–499).[7] This was the state of mathematics when Christoph Clavius began to teach mathematics and to train future mathematics teachers.

3   Clavius: The Academy of Mathematics and *Sphaera* Commentary

Clavius was the most influential Jesuit mathematician from 1570 until at least 1620. He was born in the city of Bamberg, capitol of an imperial prince-bishopric, or near it, in 1538. His name may have been a Latinized version of Clau or Schlüssel.[8] Nothing more is known about his early life until February 1, 1555, when he became a Jesuit novice in Rome. Ignatius of Loyola himself accepted him. The normal next step would have been to finish his novitiate and then study philosophy and theology at the Roman College. But in the autumn of 1555, Ignatius sent him to the University of Coimbra to boost the Jesuit presence there. In 1555, the Jesuits were given control of the Colégio des Artes, the most important part of the university; it was the first time that the Jesuits secured a strong institutional position in a university. Consequently, Ignatius sent some excellent Jesuit scholars to teach and some promising students to study at Coimbra.

However, the Colégio des Artes of Coimbra lacked a mathematics teacher; whether Clavius studied with the sole mathematics professor of the university, Pedro Nuñez (1502–1578), who was hostile to the Jesuits, is unknown. Clavius wrote that he was self-taught; if he meant that literally, he taught himself extremely well. He was well acquainted with the works of all the important ancient, medieval, and Renaissance mathematicians and astronomers.

Clavius was recalled to Rome in 1560 to study theology at the Roman College. He was appointed mathematics lecturer there in 1563 and spent the rest of his life in Rome except for short trips. Clavius lectured on mathematics from 1563 to 1571, possibly in the academic year 1575–1576, certainly in 1577–1578, and possibly other years in the 1580s for which records are missing. But he stopped by 1590 (Baldini 1992, 568).[9] In the academic year 1564–1565, he lectured on Sacrobosco's *Sphaera* at an advanced level and prepared a first draft of his commentary on the *Sphaera* (Baldini 2003, 51, 70). He became the key figure in the papal commission that prepared the reformed Gregorian calendar of 1582. His reputation soared.

---

[7] There may have been fewer mathematics courses in France and Germany in the 1570s but slightly more in the 1590s (Fischer 1978, 1983).

[8] This and the following paragraph are based on the good short biographies of (Baldini 1999a), and (Homann 2001).

[9] The information in (Villoslada 1954, 335) is incomplete.

In 1563, Clavius assumed leadership of an academy of advanced mathematical studies at the Roman College.[10] While it had begun in 1553 under a previous mathematics lecturer, Clavius made it into the most important center for mathematical studies in Europe. In Renaissance Italy, the word "academy" brings to mind informal associations in which working writers and dilettante nobles met periodically to discuss literature and enjoy each other's company. Clavius' academy was nothing of the sort. It was a group of men engaged in intense study of a range of advanced mathematical topics under Clavius' tuition and leadership. Numbers were small, ten or fewer, most often about five. A large majority were Jesuits; a few were laymen. The primary purpose of the academy was to train the members in a broad range of mathematical skills to serve their future needs. Typical members of the academy included future Jesuit mathematics teachers, who learned at the academy what they would teach, and future Jesuit missionaries who brought European mathematics to distant lands. The most famous of the latter was Matteo Ricci (1552–1610). While studying philosophy at the Roman College from 1572 to 1577, he attended the public mathematics lecture course in 1575–1576 taught by either Clavius or another Jesuit. He was also a member of Clavius' mathematical academy, possibly in the summer of 1576 and certainly in the academic year 1576–1577. In China, Ricci translated several of Clavius' mathematical works into Mandarin, including his commentary on Sacrobosco's *Sphaera* (Peking in 1607) (Baldini 2013, 138–147, 153–154).[11] Most Jesuit academy members studied for one or two years, then returned to their home provinces. However, Christoph Grienberger (Hall, Tyrol 1564–1636 Rome), who lived in Rome and often taught the public mathematics lectureship of the Roman College, was a member of the academy for about twenty years.

Clavius taught in the academy, assisted by other Jesuit academicians who served as research assistants in his later years. Clavius' major mathematical publications were largely the products of his academy teaching and research, rather than his public lectures in the Roman College. Young members of the academy occasionally taught in the academy in order to demonstrate their mastery of a body of material, or to present the results of their research. Much of the research by members began in the academy and was finished and published years later. The academy had its own library of astronomical and mathematical texts separate from the main library of the Roman College. The academy also served as a center for the exchange of scientific information, as the academy disseminated its results, received news from abroad, and welcomed visitors.

Clavius published numerous mathematical works, including extensive commentaries on Euclid's *Elementa*, and books on practical arithmetic, algebra, practical geometry, sun dials, the astrolabe, and cosmography.[12] The first and most often reprinted was *Christophori Clavii Bambergensis, ex Societate Iesu, in sphaeram*

---

[10] All studies of Clavius mention his academy. The best is (Baldini 2003).

[11] The Mandarin translation of Clavius' commentary on the *Sphaera* is number 31 in (Sommervogel 1960, VI, 1794).

[12] For the list, see (Baldini 2003, 74–76). Clavius also composed sacred music. Whether it was performed during his life is unknown. Amazon.com does not list any recordings.

Ioannis de Sacro Bosco commentarius (Sacrobosco and Clavius 1570). It was probably the most used post-medieval book on astronomy (Baldini 1999a, 18). The *Sphaera* database lists twenty-one subsequent European editions.[13]

Clavius' book is often called a textbook. If so, it was a textbook intended for teachers, practicing astronomers, and advanced students. Like many other Renaissance authors, Clavius used the commentary format because Sacrobosco's brief text offered a familiar Ptolomaic framework. It enabled Clavius to discuss issues that had risen over the centuries, and to add new material, including his own astronomical observations and insights. Like Sacrobosco, Clavius accepted a geocentric universe throughout his life. But so did practically everyone else at the time. And Ptolemaic astronomy supported by the latest astronomical observations provided much useful information. Clavius' work was comprehensive, learned, and up-to-date. It is not surprising that "contemporaries and modern historians have judged Clavius' *Sphaera* to be the greatest of all *Sphere* commentaries."[14]

4 Clavius' Publisher and Jesuit Printing Policy

The publisher was a surprise. Vittorio Eliano (1528–1581)—just arrived in Rome and little-known—published the first edition of Clavius' commentary. Behind this development was a change in the attitude of the Roman Jesuits toward printing and publishing.

In the last one to two years of his life, Ignatius of Loyola (died July 31, 1556) keenly wanted the Jesuits of the Collegio Romano to have their own press. He wanted them to be able to print the works of the teachers at the Collegio Romano, to produce inexpensive books for students, and to publish editions of ancient authors purged of morally objectionable passages for young readers. There were some misadventures. Ignatius ordered some typeface to be cast in Venice, but upon arrival in Rome, it proved to be defective. A German printer was hired, but it was discovered that he did not read Latin. Jesuit students were pressed into service as correctors (Villoslada 1954, 44–46; Garcia Villoslada 1990, 1001–1002).[15]

Despite the setbacks, beginning in 1555 the words "Romae: in aedibus Societatis Iesu," rendered into "Tipografia del Collegio Romano" by the *Universal Short Title Catalogue* (USTC), began to appear on the title pages of modest in-house publications. Many were sixteen-or-eighteen-page booklets of theses that a student at the Roman College intended to defend in a public disputation.[16] The Tipografia del Collegio Romano volumes often indicated that "Blado," that is, the firm of Antonio

---

[13] See Ian Maclean's article (Chap. 6) in this volume for comments on many of the editions.

[14] For a summary of Clavius' commentary and its relationship to Sacrobosco's text, see (Lattis 1994, 37–105), quotation on 37. See also (Baldini 2000, 15–48).

[15] See also (Sachet 2020, 200–206) for additional information. This book arrived too late to be fully used.

[16] For a typical example, see (*Assertiones* 1558). For several more see (Ascarelli 1972, 13–14).

Blado (1490–1567), Rome's most prolific and important publisher at that time, had done the printing.[17]

Despite the urging of Ignatius of Loyola, the Jesuits were not certain that printing and selling books were an appropriate ministry for a religious order. After Ignatius' death, the First General Congregation of the Society met from June 19 to September 10, 1558, with twenty-seven senior Jesuits, including the five remaining members of the ten founders, in attendance. Its first task was to elect a new superior general, which was dispatched by electing Diego Laínez (1512–1565), who had been vicar general since the death of Ignatius. It also debated and passed 168 decrees of great and small importance.

Decree 105 After the Election read "The printing and sale of books is left to Father General's judgment." The official summary of the congregation's deliberations stated the questions that were discussed, and the resolution.

> Is it tolerable or even praiseworthy in the Society to print and sell books for the sake of the general good that would ensue? Or rather should it be forbidden, lest we appear to be engaging in business? It was decreed that nothing should be decided in favor of either side of the question, but rather that it should be left to the discretion of the superior general. It seemed worthy of consideration, however, so that nothing may be done that would damage the [vow of] poverty or the institute of the Society (*For Matters of Greater Moment* 1994, 96).

In other words, the participants at the General Congregation meeting were divided. So, they left the decision to Laínez. In practice, the Jesuits carried on as before. The Tipografia del Collegio Romano published lists of theses to be disputed and house publications, including college rules, the Constitutions of the Society, the Spiritual Exercises of Ignatius, and the *Annuae litterae Societatis Iesu*, the annual reports of Jesuit activities around the world distributed to Jesuit colleges. The vast majority of these publications were works of fewer than one hundred pages in octavo format. According to USTC, the Tipografia del Collegio Romano published 136 editions between 1556 and 1635, but only four after 1617.[18] It did not publish significant scholarly and pedagogical works written by Jesuits at the Roman College or elsewhere.

Then in 1570, Vittorio Eliano published Clavius' commentary on the *Sphaera*. Eliano was a very surprising choice. A Jew born in Rome in 1528 with the first name of Yosef, he was the son of a merchant and the grandson on his mother's side of Rabbi Elia Levita (1469–1549), a famous Hebrew grammarian, scholar, and poet. Yosef learned the printer's trade at an early age. While living in Isny im Allgäu, a free imperial city in Württemberg, Yosef and his younger brother Elia (see below) assisted Paul Fagius (Büchelin, 1504–1549), a Protestant Hebraist and printer. Fagius collaborated with Elia Levita to print twelve editions of Latin and Hebrew works in 1541 and 1542, including texts and studies of the Old Testament in Hebrew and

---

[17] For Blado see (Barberi 1968; Menato 1997b).

[18] See also (Ascarelli 1972, 13, 354), for booklets of theses published in 1554 and 1555 by Blado and the Tipografia del Collegio Romano.

Latin and a Hebrew-German grammar (Clines 2020, 208 n. 31). Fagius later taught Hebrew and biblical studies at Strasbourg and the Universities of Heidelberg and Cambridge. Yosef next moved to Venice where he converted to Catholicism between 1544 and 1546 and changed his name to Vittorio Eliano. He then became the printer and collaborating publisher of some landmark Hebrew titles in Venice, Cremona, and elsewhere in northern Italy. He also censored Hebrew books in these cities (Casetti Brach 1993, 1997). He moved to Rome in 1568 or 1569 and began to publish under his own name in 1569. His first publication was an expurgated version of the poems of Horace for use in Jesuit schools (*Quinctus Horatius* 1569).

Eliano's publication of the Clavius commentary on the *Sphaera* in 1570 carried a comprehensive papal privilege dated January 12, 1569.[19] It granted to Vittorio Eliano and unnamed associates the exclusive right to publish a long list of potential major and lesser works by Jesuit authors. The papacy granted to Vittorio Eliano the right to print and publish the following works (here listed in the order found in the privilege and identified or explained in English).

Francisco de Toledo, commentary on the *Summa Theologica* of Thomas Aquinas
Francisco de Toledo, *Instructionem sacerdotis sive casus conscientiae*
Francisco de Toledo's commentary on Aristotle's *De anima*
Clavius' commentary on the *Sphaera*
Jesuit letters from the Indies
Selected epigrams
Selected orations
The same from Cicero, Terence, Plautus, and Horace "cleansed of all indecency"
A *catena* (chain or series) of (verses from the) Gospels by Doctor Emanuele of the Society of Jesus.[20]

The papacy granted to Eliano and his associates exclusive rights to many titles, some precisely indicated, others generic and vague. The extended list of titles was unprecedented. Otherwise, the terms were similar to those of a large majority of papal privileges issued in the sixteenth century. The privilege was for ten years. The texts were subject to religious censorship by the Inquisition or the Master of the Sacred Palace in Rome (the pope's theologian, usually a Dominican, whose duties included pre-publication book censorship in Rome). No one else was allowed to

---

[19] The privilege, "Pius Papa V. Motu Proprio &c" is found on the recto and verso of the second leaf of (Sacrobosco and Clavius 1570). It concludes "Datum Romae apud Sanctum Petrum Quarto Idus Ianuarij, Anno Quarto." The IV Ides was January 12. Because Pius V was elected on January 7, 1566, the fourth year of his pontificate began on January 7, 1569, and ended on January 6, 1570. Hence, the date of the privilege was January 12, 1569.

[20] "Nonnulla opera, scilicet, Commentaria doctoris Francisci Toledi Societatis IESU in summam Theologicam S. Thomae, eiusdem Instructionem Sacerdotis sive casus conscientiae, Eiusdem commentaria in libros Aristotelis de anima, Magistri Christophori Clavij in sphaeram Sacroboschi, literas Indicas patrum Societatis IESU, Epigrammatica selecta, orationes selectas, Synonyma exCicerone, Terentium, Plautum, Horatium ab omni obscoenitate purgatos, Catenam in Evangelia doctoris Emanuelis Societatis IESU…" (Recto of the second leaf of (Sacrobosco and Clavius 1570), no signature).

publish these works in Latin or Italian without the express permission of Eliano, nor could publication rights be transferred to others without his permission. The privilege was in force in all areas in Italy and beyond. Violators were subject to automatic excommunication and a fine of 500 gold ducats to be paid to the papal treasury. The justification for the privilege was that Eliano and his associates were printing these books "for the general and universal benefit of students" (*ad communem omnium studiosorum utilitatem*) and therefore deserved compensation for their expenses and labor. This was a variation on "the common good" justification frequently found in papal privileges.[21] In short, the terms of the privilege were unexceptionable except for the long list of potential titles.[22]

The expansive privilege argues that the Roman Jesuits had decided to move from a small in-house publishing operation to working with an external commercial press for the purpose of publishing major works of scholarship, the majority prepared by Jesuits teaching in the Roman College. The obvious choice for such a publisher would have been Antonio Blado, had he not died in 1567, and the Jesuits did not make an arrangement with his heirs. Instead, they chose a publisher who had just arrived in Rome and was known for publishing Hebrew books. Why Vittorio Eliano?

Giovanni Battista Eliano (1530–1589), Vittorio's younger brother, may have brought the Jesuits and Vittorio together. Giovanni Battista Eliano was born a Jew with the name of Elia in Rome in 1530. Elia lived in Italy, then in Isny im Allgäu where, like his brother, he assisted Paul Fagius (Clines 2020, 31–32). He next lived in Constantinople, Cairo, and elsewhere, and was intended to be a merchant like his father. Then his family sent him back to Venice to try to persuade Vittorio to return to Judaism and move to Cairo. Instead, Elia met André des Freux (ca. 1515–1556), an important early Jesuit, and also became a Catholic in 1551. Elia took the Christian name Giovanni Battista Eliano; he was also sometimes called Giovanni Battista Romano.[23] He immediately joined the Society of Jesus and followed des Freux to Rome, where he studied philosophy and theology at the Roman College (Ioly Zorattini 1993, 472–473; Libois 2001, 2:1233; Clines 2020, 25–42).

Literate in Hebrew, Arabic, Turkish, Latin, German, Italian, and Spanish, and intensely spiritual, Giovanni Battista Eliano became a valued member of the Jesuit community in Rome and useful to two popes. Giovanni Battista Eliano is best remembered for his years-long religious and diplomatic missions undertaken for the papacy to Egypt and Lebanon in an effort to bring Coptic Christians into union with Rome, which was unsuccessful, and to promote closer relations with the Maronite Church,

---

[21] The indispensable study of papal privileges is (Ginsburg 2019), especially pages 115–139.

[22] None of the approximately 430 papal privileges listed in (Ginsburg 2019, 141–284) gave permission to print so many potential titles.

[23] In 1586, he published under the name Giovanni Battista Romano a catechism with twenty-seven illustrations showing good and sinful actions; it had at least four editions. The 1591 edition and its title is cited: *Dottrina christiana nella quale si contengono i principali misteri della nostra fede rappresentati con figure per istruttione de gl'idioti & di quelli che non sanno leggere…Composta dal p. Giovanni Battista Romano della Compagnia di Giesu* (*Dottrina christiana* 1591). For the 1586 edition see (Eliano 1586). For editions of 1587 and 1608 and some illustrations from the work see (Grendler 1989, 354–356, 427).

where he had considerable success.[24] During these missions, he put his knowledge of printing to use by translating and printing books and by working to establish a press in Rome capable of producing books in Arabic and Syriac for Maronite Christians (Clines 2020, 31–32, 122–123, 142–145, 147, 157, 171).

When in Rome, Giovanni Battista Eliano taught at the Roman College. Although still studying philosophy and theology, and not ordained a priest until 1561, he began to teach Hebrew in the Roman College in 1552 or 1553. He continued to teach Hebrew to about 1561, from 1563 to 1570, and again in the academic year 1577–1578. Pius IV (1499–1565; ruled December 25, 1559–December 9, 1565) decided to promote the knowledge of Arabic for missionary reasons and to facilitate contacts with eastern Christians, and he recruited Eliano to this cause. He ordered the Roman College to teach Arabic. So Eliano added teaching Arabic to his duties in the academic year 1563–1564, and he continued to teach Arabic there at least through the academic year 1566–1567, and probably longer.[25] Pius IV also asked Eliano to write an Arabic version of a short Catholic profession of faith designed to persuade eastern Christians to join Rome, and this was published in 1566 by the Tipografia del Collegio Romano (*Fidei orthodoxae confessio* 1566).[26] And he asked Eliano to write an account of the Council of Trent in Arabic, which remains in manuscript.[27] Pius IV also wanted a press capable of printing an Arabic New Testament, grammars, and dictionaries. After his long trips to the Middle East, Giovanni Battista Eliano returned to Rome where he died in 1589 (Ioly Zorattini 1993; Libois 2001).

In short, Giovanni Battista Eliano was trusted by popes and had the attention of his Jesuit superiors; he had considerable experience with printing; and he was multi-lingual. He was in a position to bring his brother's printing expertise to the attention of the papacy and his Jesuit superiors. Finally, perhaps a common language helped Clavius and the Eliano brothers to bond. Giovanni Battista Eliano learned German as a child in Isny im Allgäu and Vittorio likely did the same, while Clavius came from Bamberg (Clines 2020, 31–32). Clavius and Giovanni Battista Eliano were probably the only German speakers teaching at the Roman College, which was filled with Italian, Spanish, and Portuguese Jesuits.

The comprehensive papal privilege granted to Vittorio Eliano meant that the Jesuits had decided to trust an external commercial press to publish scholarship emerging from the Roman College, plus some books for students. It was a delayed answer to the question that the General Congregation of 1558 left up to the general:

---

[24] Most of (Clines 2020) is a detailed study of those missions and how Eliano viewed his identity as a converted Jew.

[25] This incomplete summary of Eliano's teaching at the Roman College has been pieced together from various sources. The summary of his teaching in (Villoslada 1954, 326), omits his Arabic teaching and does not accurately indicate the years in which he taught Hebrew.

[26] According to (Somervogel 1960, III, 380), and (Ioly Zorattini 1993, 475), one printing was in Latin and Arabic, and the other in Latin only.

[27] Undated (but late 1565 or early 1566) annual letter of Juan Alfonso de Polanco S. J., Rome, in (*Polanci Complementa*, 1969, I, 560–561). For other references to Eliano see (*Polanci Complementa* 1969, I, 284, 422, 611; II, 628, 664). Pier Ioly Zorattini states that Eliano's Arabic manuscript on Trent is in the Biblioteca Nazionale Vittorio Emanuele II, Roma (Ioly Zorattini 1993, 475).

Should the Jesuits do their own publishing? However, the external press was not far outside the Society, because it was run by the brother of a Jesuit.[28] It is also likely that Vittorio Eliano was chosen for his ability to publish in Hebrew and to handle complex typographical matters, such as the planetary diagrams in Clavius' book. The decision was not a complete break with past policy, because the Tipografia del College Romano continued to publish books intended for a limited Jesuit readership.

However, Vittorio Eliano did not publish very many of the works for which he held a privilege. The 1569 privilege stated that Vittorio Eliano would publish three works of Francisco de Toledo (Córdoba 1532–1596 Rome, cardindal from 1593), a prominent Jesuit philosopher and theologian at the Roman College for many years. Eliano did not publish any of them. But it was not his fault. Toledo's commentary on Aristotle's *De anima* appeared in 1574/1575, published in Venice by Luc' Antonio Giunta II (1540–1602) (Toledo 1575; Camerini 1962–1963, II, nos. 769, 774) (Chap. 8).[29] Why this happened is unknown. Eliano often could not publish a work because the author did not finish it. Toledo's famous casuistry manual, *De instructione Sacerdotum*, was only published posthumously in 1599 in Milan, Cologne, and Lyon (Toledo 1599a, b, c; Sommervogel 1960, VIII, 70–71).[30] And his commentary on the *Summa Theologica* of Thomas Aquinas (1225–1274) was not published until 1869 in four volumes (Donnelly 2001, 4:3808).[31] Vittorio Eliano published only two of the many volumes that the papal privilege licensed him to publish.

Vittorio Eliano did publish other Jesuit scholarship, including Francisco de Toledo's work on Aristotelian logic in 1569 and 1572 (Toledo 1569, 1572), three complete and partial editions of the Jesuit Constitutions in 1570 (*Constitutiones* 1570a, b, c), and a Latin grammar manual of Manuel Álvarez (1525–1583) in 1572 (Alvarez 1572). Eliano's last publication of a Jesuit work was Juan de Polanco's (1517–1576) *Methodus ad eos aiuvandos, qui moriuntur*. Roma, apud Vittorio Eliano, 1577 (Polanco 1577), which offered advice and useful texts to those comforting the dying. Polanco was a very influential Jesuit who served as secretary to Ignatius of Loyola and two subsequent superior generals, and his book went into many editions plus translations. Overall, ten of the thirty works that Vittorio Eliano

---

[28] Vittorio Eliano's press was another link, not previously noticed, between *conversos* and the Society of Jesus in its first half-century. Scholars have noted that several prominent early Jesuits came from *converso* backgrounds, albeit sometimes several generations distant. The connection was brutally severed in 1593 when the Fifth General Congregation decreed that men of Jewish and "Saracen" background, meaning converts or descendants of converts from Judaism and Islam, would not be admitted into the Society of Jesus in the future. This rule was not completely abrogated until 1946. The reasons for this bitterly contested decision were complex; see (Maryks 2010) for the story.

[29] Some copies are dated 1574, others 1575. The edition carries the words *Cum Privilegio* but there is no further information. See also (Sommervogel 1960, VIII, 68–69).

[30] Plus many more editions.

[31] Another possible example of a work for which Eliano had a printing privilege that was not published until after the author had died was the *catena* of passages from the Gospels by "Doctor Emanuele." This might have been *Notationes in totam scripturam sacram…brevissime explicantur* of Manuel de Sá (ca. 1528–1596), a biblical scholar who taught theology at the Roman College. It was first printed in 1598 by the Plantin Press of Antwerp (Sa 1598). On Sà see (Leite 2001).

published in Rome between 1569 and 1577 were Jesuit texts. In 1578, he began to collaborate with Francesco Zanetti (1530–1591); they published three editions of the Hebrew Bible (1578, 1580, and 1581), and another Jesuit work, Robert Bellarmine's (1532–1621) Hebrew grammar. Nothing more is heard of Vittorio Eliano after 1581; he probably died in 1582 (Casetti Brach 1993, 1997).[32]

Domenico Basa (1500–1596), and Francesco and Luigi Zanetti (fl. 1590–1616) published the books of Clavius and other Jesuits for the rest of the century and beyond. In 1581, Domenico Basa and Francesco Zanetti published and printed a revised second edition of Clavius' commentary on the *Sphaera* (Sacrobosco and Clavius 1581). In 1585, Domenico Basa published the revised third edition (Sacrobosco and Clavius 1585). According to the USTC, Domenico Basa published and/or printed 125 Roman editions between 1579 and 1596 when he died, of which ten percent (thirteen) were books written by Jesuits. Again, according to the USTC, Francesco Zanetti published 219 Roman editions of which seventeen percent (thirty-eight) were authored by Jesuits.[33] Luigi Zanetti published 203 Roman editions between 1591 and 1606, of which twenty-five percent (fifty-one) were authored by Jesuits.

Most of Domenico Basa's Jesuit editions were works of Jesuit Latin scholarship including instructional manuals, which was the original purpose of publishing with Vittorio Eliano's press. But far less than half of the Jesuit editions published by Francesco Zanetti and Luigi Zanetti consisted of works in Latin. Their most frequent Jesuit publications were Italian translations of letters from the Oriental missions (ten by Francesco and thirteen by Luigi), which were probably read by many readers. At the same time, Francesco Zanetti collaborated in the publication of the first edition of Benet Perera's (also Benito Pereira, Pereyra, or Pererius; Valencia 1536–1610 Rome) important *De communibus omnium rerum naturalium principijs & affectionibus, libri quindecim* (1576) discussed below.

Overall, the initial alliance of the Jesuits with Vittorio Eliano led directly to three other Roman publishers who produced a substantial number of editions of Jesuit works. And other Roman presses also published Jesuit authors. The principle was established and put into practice that Jesuit authors would publish with external commercial presses. But after the privilege granted to Vittorio Eliano, no single publisher was favored. Hence, Roman Jesuit authors published with a variety of publishers in Rome and elsewhere. It is likely that the initiative of authors and printers, and the preferences of patrons who paid publication costs, played key roles in the choice of a publisher for the first editions of Jesuit works. The number of Jesuit books that publishers issued, and the expanding fraction of their total production

---

[32] For Bellarmine's Hebrew grammar see (Sommervogel 1960, I, 1151–1153).

[33] In these calculations, a small number of editions listed as published by the Tipografia Apostolica Vaticana and the Tipografia del Collegio Romano, but printed by Domenico Basa or Francesco Zanetti, are omitted. Further, the three editions of Clavius' commentary on the *Sphaera* are counted as a Basa publication, even though Basa and Francesco Zanetti collaborated on them. For Basa, see (Menato 1997a). The smaller number of editions listed by (Ascarelli 1972), included about the same percentage of Jesuit-authored editions. For example (Ascarelli 1972, 356–358), lists 163 editions published by Francesco Zanetti, of which nineteen percent (thirty-one) were Jesuit works.

that Jesuit books represented, demonstrated the commercial importance of Jesuit works to publishers.

5   Dedications and a Venetian Publisher

The 1570 edition of Clavius' commentary on the *Sphaera* carried a dedication letter dated March 20, 1570, Rome, from Clavius to Prince Wilhelm Wittelsbach of Bavaria (1548–1626). In 1579, he became Wilhelm V, called "the Pious," duke of Bavaria (abdicated in 1597). Hence, Clavius wrote a second dedication letter to him recognizing the new title dated September 18, 1581.[34] The 1581 dedicatory letter was more personal and informative. Clavius wrote that he had worked long into the night on the revisions, making changes and adding much new material. He stated that he was a German, and he recalled fondly the church and city of Bamberg "of which I am a foster-son." (*cuius ego alumnus sum*) (Sacrobosco and Clavius 1581, sig. ✤3r).[35] Clavius paid tribute to Duke Wilhelm V and his late father Duke Albrecht V (1528–1579) (Duke from 1550) for their firm support of Catholicism and the Jesuits. This was true. One example among many was their strong political support that enabled the Jesuits to take control of the University of Ingolstadt despite opposition from the rest of the faculty (*The Mercurian Project* 2004, 156–160, 169, 187, 189–193, 214, 216–17, 235–243). The 1581 dedicatory letter to Wilhelm V appeared in a large majority of the subsequent editions of Clavius' commentary on the *Sphaera*.

In 1611, Clavius dedicated volume three, comprising the *Sphaera* commentary and his *Astrolabium* (first published in 1593) of his *opera mathematica* to Johann Gottfried von Aschhausen (1575–1622), prince-bishop of Bamberg (ruled from 1609). Now ill and unable to teach, Clavius again mentioned his labors to revise the work, and he praised Gottfried as a good shepherd of his flock. Like Wilhelm V, Gottfried was a strong Counter Reformation ruler who helped the Jesuits found a college and school in Bamberg in 1611. As has been pointed out, Clavius' letters to northern European rulers as Wilhelm V and Prince-Bishop Gottfried helped create a stream of revenue to support his and other Jesuit publications (Baldwin 2003, 287–301).

The Venetian editions of Clavius' *Sphaera* commentary were the result of the initiative of Giovanni Battista Ciotti (ca. 1561–ca. 1629). A man of broad culture, Ciotti published a very large number of editions in Venice from 1590 through 1629 in many subjects, about seventy-five percent Italian editions and twenty-five percent Latin. He published the works of well-known Italian vernacular authors such as the poet Giambattista Marino (1569–1625) (Firpo 1981; Contò 1997). In 1591, he

---

[34] In the 1581 edition, the year of the letter is omitted. The letter is reprinted in (Sacrobosco and Clavius 1585), with the date 1581.

[35] Although the normal translation of *alumnus* is "foster-son," in Jesuit pedagogical documents it sometimes meant a student who was in some way unusual, such as a scholarship student in a school full of fee-paying students.

published Clavius' commentary on the *Sphaera,* his first Jesuit work. He wrote a dedication letter dated September 1, 1591, Venice, in which he praised Clavius as the prince of mathematicians of our time.[36] Three more editions followed: 1596 (published by Bernardo Basa, although it was Ciotti's edition), 1601, and 1603 (Sacrobosco and Clavius 1596, 1601, 1603) (Chap. 6).[37] All included the same dedicatory letter of Ciotti with a changed date of 1596.

After Clavius' *Sphaera* commentary, Ciotti published many other editions of Latin and Italian works by Jesuits. According to the USTC, Ciotti published 651 editions between 1590 and 1608, of which fifty-nine were written by Jesuit authors, that is, nine percent of the total.[38] Although Ciotti published such major works of Jesuit scholarship as Clavius' *Sphaera* commentary and theological works of Robert Bellarmine in Latin, he most often published the vernacular devotional works of Luca Pinelli (1543–1607), followed by "letters from the Indies," that is, vernacular accounts of Jesuit missionary labors in India, China, Japan, and the Philippines.[39] Ciotti created a Jesuit publication list that combined widely popular vernacular works, which sold well, with Latin scholarship important to university scholars. In 1606, the Jesuits were expelled from the Republic of Venice, which meant the closing of their churches and their five schools in the Republic. Ciotti continued to reprint Pinelli's vernacular devotional works but fewer other Jesuit titles, as the exile continued. The Jesuits were not permitted to return until 1657.

6 Clavius' Proposals for Jesuit Mathematical Education

Clavius wanted the Jesuits to teach more mathematics and astronomy. So, he tried very hard to persuade his own order to adopt a much-expanded mathematical curriculum. But he encountered opposition from Jesuit natural philosophers and some Jesuit provinces.

The opportunity to present the case for more mathematics and astronomy instruction arose in 1581. After General Everard Mercurian (b. 1514) died on August 1, 1580, fifty-nine leading Jesuits met in Rome from February 7 to April 22, 1581, for the Fourth General Congregation of the Society.[40] Although General Congregations normally occurred when a general died and a new one had to be elected, they

---

[36] "Christophorus Clavius nostrae tempestatis mathematicorum Princeps" (Sacrobosco and Clavius 1591, Sig. †2v).

[37] Bernardo Basa, nephew of Domenico Basa, was active as a publisher in Venice from 1582 to 1596 (Bruni 1997).

[38] This is based on a survey of Ciotti editions in USTC. Because the short titles in USTC often do not identify authors as Jesuits, a few Jesuit editions may have been missed.

[39] For Pinelli see (Sommervogel 1960, VI, 802–817; Regina 2001).

[40] All the provinces were European at this date, and they oversaw the overseas missionary outposts.

were also opportunities to chart the path forward.[41] After electing Claudio Acqua-viva (1543–1615) (General from 1581) the new superior general on February 19, the congregation passed sixty-seven decrees. The most important by far was decree thirty-one which appointed a commission of twelve Jesuits "to develop a plan of studies," that is, to draft an educational plan for the entire Society.[42] The decree was part of a general move in the Society toward more uniformity and tighter organization.

Jesuit schools and universities across Europe already had developed some common features. But the texts to be taught, pedagogical practices, and rules were not codified beyond individual schools or provinces and were subject to change. Moreover, like academics in all centuries, the Jesuits wanted to create the perfect curriculum in the ideal school. Indeed, beginning in 1551 Jesuit teachers had already written many lengthy treatises toward that goal. And also, like academics every-where, they disagreed about what the perfect curriculum should look like. Now the Society intended to create a uniform and mandatory plan of studies for all Jesuit schools.

Clavius seized the opportunity. He presented to the rector of the Roman College a multi-faceted proposal that would expand mathematical education and make perma-nent his academy. Titled "Ordo servandis in addiscendis disciplinis mathematicis" (The Order to Follow to Attain Proficiency in the Mathematical Disciplines), it was written in 1580 or 1581.[43] The curriculum that Clavius described was intended for the mathematical academy. But because of the action of the Fourth General Congrega-tion, his proposal became the focus of the discussion about mathematics in a universal plan of studies for the Society.[44]

Clavius' *Ordo* first described a twenty-two step program of comprehensive math-ematical and astronomical education to be accomplished over three years. For each step, Clavius indicated what should be taught, and the texts that teachers should use, including several of his own works. Step one in the three-year curriculum was to teach the first four books of Euclid's (323–285 BCE) *Elements* using Clavius' commentary on the first four books published in 1574. Step two consisted of basic arithmetical operations including addition, subtraction, multiplication, division of whole numbers and fractions, the use of proportions, and finding the square roots

---

[41] The Society continues to follow this practice today with a significant modification. The last two superior generals did not die in office. Instead, they announced their impending retirements about two years in advance, which allowed for lengthy preparation for the general congregations that followed.

[42] Quote in (*For Matters of Greater Moment* 1994, 176).

[43] The Latin text is found in (MP, VII, 109–115), and an English translation in (*Jesuit Pedagogy* 2016, 281–291). Although the Archivum Romanum Societatis Iesu (hereafter ARSI) manuscript lacks a date, Lukács, the editor, gives the date as not later than 1581 and ca. 1581 (MP, VII, 109), which (*Jesuit Pedagogy* 2016, 283) follows. Ugo Baldini believes that the date was "almost certainly" 1580 (Baldini 2000, 56); he also prints the first section with the three-year mathematical program (Baldini 1992, 172–175). All three editions include notes. Baldini's edition provides the most extensive notes and adds the most detailed identifications of the texts named on pp. 179–182.

[44] Several scholars have briefly analyzed or mentioned Clavius' plan for mathematical education. The following pages are a partial account that focuses on the position of his commentary on the *Sphaera*.

of numbers. He promised to compose a work on this material. In the meantime, he recommended two other works.[45]

Step three was to teach Sacrobosco's *Sphaera*: "The *Sphere* as briefly as possible or rather any other introduction to astronomy. The more important rules regarding ecclesiastical reckoning can be added to this. I will also put out a brief treatment. Meanwhile, my commentary on John of Holywood's *Sphere* will suffice, leaving out operations on curves, the treatise on isoperimetrics, and so forth since these will be treated below."[46] In later steps, Clavius included more instruction and texts concerning astronomy. Overall, Clavius' curriculum intended to teach a thorough and up-to-date Ptolemaic astronomy based on the best available works, including his own.

Step four was Euclid's *Elements*, books 5 and 6, again based on Clavius' commentary. Step five involved teaching the use of the geometer's square, the astronomer's quadrant, and other measuring instruments. Further steps involved teaching more Euclid, algebra, the use of the astrolabe, geography, the description and construction of sundials, some study of Archimedes (287–212 BCE), the study of some other ancient Greek mathematicians, and more.[47]

Clavius also included two less comprehensive alternate versions of his mathematics and astronomy curriculum. The "Ordo secundus brevior pro iis, qui non curant perfectissimam mathematicarum rerum cognitionem assequi" (A Second, Shorter Order for Those Who Are not Interested in Acquiring a Completely Thorough Understanding of Mathematics) consisted of nineteen steps over three years. It was basically the same as the first version, but with less detail. For example, Step three said simply "The *Sphere* and ecclesiastical calendrical reckoning in very brief fashion. John of Holywood with my commentaries." And he offered a third version, which was a two-year course: "Ordo tertius brevissimus et ad cursum mathematices, qui duobus annis absolvi debet, accommodatus" (A Third Order of Greatest Brevity and Adapted for a Mathematics Course that Ought to Be Finished in Two Years). The two-year course consisted of seven steps in the first year. Step two was "The *Sphere* and ecclesiastical calendrical reckoning, very concisely. These can be finished off by Easter." The second year included additional astronomical material. It is noteworthy that studying the *Sphaera* was a key part of all three versions. At the end of his document, Clavius made it clear that he much preferred the very comprehensive three-year mathematics and astronomy curriculum (MP, VII, 113–115; *Jesuit Pedagogy* 2016, 288–291).[48]

---

[45] In 1583, Clavius published his *Epitome arithmeticae practicae*, which included chapters on this material (Baldini 1992, 172; MP, VII, 110; *Jesuit Pedagogy* 2016, 283).

[46] See (MP, VII, 110; Baldini 1992, 172; *Jesuit Pedagogy* 2016, 283), for the translation.

[47] Because this study focuses on the *Sphaera*, the rest of the three-year mathematics curriculum is omitted.

[48] Quotes on (*Jesuit Pedagogy* 2016, 288, 289, and 290).

7   In Defense of Mathematics

Clavius next sought to raise the prestige of mathematics teachers within the Society, and he rebuked Jesuit philosophers who underestimated the value of mathematics.

In 1582, Clavius addressed to Superior General Acquaviva a second treatise entitled "Modus quo disciplinae mathematicae in scholis Societatis possent promoveri (A Method by Which Mathematical Disciplines Could be Promoted in the Schools of the Society).[49] Clavius began by stating that the mathematics teacher should possess "uncommon learning and authority." In order for the teacher "to have greater authority with the students, and for the mathematical disciplines to be more highly valued, and for the students to realize their usefulness and necessity," the teacher had to be invited to participate in solemn acts "at which doctors are created and public disputations are held" (MP, VII, 115; *Jesuit Pedagogy* 2016, 291).[50]

Clavius referred to solemn acts, which were important academic exercises in Jesuit schools and universities. A student who had completed three years of philosophical study and three or four years of theological study with great distinction was invited to be the defendant in a disputation lasting four or five hours in which he was questioned by his theology and philosophy teachers on any aspect of theology. Solemn acts were formal academic performances before an audience of teachers, students, and guests.[51] They were also festive celebratory occasions, because it was expected that the student would display much learning and be applauded. A non-Jesuit who did well in solemn acts might be awarded a doctorate of theology. Jesuits did not normally receive doctorates; successful completion of solemn acts was considered reward enough. Clavius wanted the Jesuit mathematics teacher to be a questioner of the candidate so as to be considered the equal of teachers of theology and philosophy.

Clavius then made a plea for the mathematical academy. "But so that the Society always has suitable professors for these sciences, twelve suitable people (i.e., Jesuits) would have to be chosen to take on this service and they would be instructed in a private academy in different mathematical matters." He warned that the mathematical sciences will not survive very long in the Society without the academy. On the other hand, having Jesuits learned in mathematics "would be a great enhancement for the Society, (mathematics) being very frequently brought up in conversation in talks and meetings of princes when those individuals realize that ours (Jesuits) are not ignorant of mathematics." But if Jesuits "fall silent in such meetings" there will be "considerable embarrassment and dishonor. Those to whom this has happened have often related it" (MP, VII, 115–116; *Jesuit Pedagogy* 2016, 291–292).[52]

Clavius next argued that mathematics was essential to the study of natural philosophy. Students must understand that mathematics was "useful and necessary for

---

[49] The text is found in (MP, VII, 115–117); English translation in (*Jesuit Pedagogy*, 2016, 291–294).

[50] I have slightly altered the translation of "actus solemniores" and "doctores creantur" to convey better the academic meaning.

[51] The *Ratio Studiorum* of 1599 called them "general acts" (actus generales) or "theological acts" (RS, 105–107).

[52] Again, I have very slightly modified the English translation.

correctly understanding the rest of philosophy." There was such a "mutual affinity" between mathematics and natural philosophy "that unless they aided one another, they could in no way maintain their own stature." It will be necessary "that students of natural philosophy study mathematics at the same time." He continued: "The experts agree that natural philosophy (*physica*) cannot be rightly perceived without mathematics, especially the parts that bear on the number and motion of celestial orbs; the number of intelligences; the effects of stars that depend on various conjunctions, oppositions, and the distances left among them; the infinite division of a continuous quantity; the ebb and flow of sea tides; the winds; the comets; the rainbow; the halo [around the sun and the moon] and other meteorological events; and the relationship of movements, qualities, actions, passions, and reactions, and so forth...I am leaving out innumerable examples in Aristotle, Plato (427–447 BCE), and their renowned interpreters, examples that can in no way be understood without an intermediate understanding of the mathematical sciences" (MP, VII, 116; *Jesuit Pedagogy* 2016, 292).[53] This was a sweeping claim that much of the astronomical part of mathematics instruction was absolutely necessary for the study and understanding of Aristotelian natural philosophy, which might also be called Aristotelian physical science.[54]

Clavius followed by criticizing Jesuit philosophy teachers for their ignorance of mathematics. "Indeed, on account of their ignorance, some professors of philosophy have very often committed many errors, and very bad ones at that; and what is worse, they have even committed them to writing. It would not be difficult to expose some of them. By the same token, instructors of philosophy ought to be proficient in the mathematical disciplines, at least to an intermediate degree, so that they do not founder on similar reefs to the great loss and dishonor of the reputation that the Society has in letters" (MP, VII, 116; *Jesuit Pedagogy* 2016, 292–293).[55]

Clavius then charged that some Jesuit philosophers ridiculed mathematics. "It will be a considerable contribution...if the instructors of philosophy stay away from those debates...[in which it is said] that the mathematical sciences are not sciences, that they do not have demonstrations, that they are distractions from metaphysics, and so forth. For experience teaches that this is a stumbling block for the students, one that is entirely unprofitable, especially because the instructors can hardly teach them without ridiculing these [mathematical] sciences (as has been gathered more than once from the reports of others)." Clavius wanted Jesuit philosophy teachers to encourage their students to study mathematics "and not...lead them away from

---

[53] The only change made to the English translation is translating *physica* as natural philosophy (*Jesuit Pedagogy* 2016) translates *physica* as "natural sciences" which is not precise enough in this case. In Jesuit curriculum discussions at this time *physica* meant natural philosophy, because the fundamental text studied in the natural philosophy course was Aristotle's *Physics*.

[54] For more comments on Clavius' arguments for the importance, usefulness, and certitude of mathematics, and as an essential part of Aristotelian science, see (Wallace 1984, 136–139; Lattis 1994, 35–36; Dear 1995, 37–40; Blum 1999, 246; Gilbert 2014, 3–6).

[55] In this passage, Clavius uses "praeceptores philosophorum," meaning all teachers of philosophy, not just teachers of natural philosophy.

studying them, *as many have in earlier years*" (author's emphasis) (MP, VII, 116–117; *Jesuit Pedagogy* 2016, 293). These were strong words directed against his colleagues.

Clavius concluded by proposing additional academic procedures by which Jesuit students would win support for mathematical studies and Jesuit philosophers would come to appreciate mathematics. Every month "all the philosophers" (meaning Jesuits studying philosophy and their teachers, who normally presided over or observed student presentations) should gather together. One student should "present a brief recommendation of the mathematical disciplines." Then with the help of another student or two "he should explain a problem in geometry or astronomy which is both interesting to the audience and useful in human affairs. "Or he should explain a mathematical passage from Aristotle or Plato, "(and passages of this kind are not scarce in their works)." Or he should introduce "new demonstrations of certain propositions of Euclid." Clavius believed that these exercises would stimulate "a burning desire" for mathematical studies. Finally, Jesuit students who wanted to obtain a master's degree or doctorate in philosophy or theology should be examined in mathematics as well, and a mathematician should be one of the examiners (MP, VII, 117; *Jesuit Pedagogy* 2016, 293–294).

8   Perera's Dismissal of Mathematics

Although Clavius did not name any Jesuit philosophy teachers who dismissed mathematics as unscientific, one of his very prominent colleagues at the Roman College did exactly that. This was Benet Perera. Perera entered the Society of Jesus in 1551 in Valencia. A brilliant student, he was sent to Rome where he studied at the Roman College and performed well in acts and public disputations. In 1558, at the age of twenty-three—quite young for a Jesuit—he began to teach the natural philosophy course at the Roman College. And he taught, wrote, and probably lived in the Roman College for the rest of his life. From 1558 through 1567, he thrice taught the three-year cycle of logic, natural philosophy, and metaphysics, always in that order (Baldini 1992, 569–570). He next taught Scholastic theology from 1567 to 1570. After a hiatus of six years, he taught Scholastic theology, positive theology, or scripture until 1597, although not all dates in which he taught each course are known, because of gaps in the records. It is possible that he was excused from teaching so that he might study and write between 1570 and 1576, again at one time or another between 1576 and 1597, and after 1597, because he published numerous long works, especially biblical commentaries (Villoslada 1954, 29, 41, 51–52, 59, 76, 78–79, 323, 324, 325, 327, 329, 331; Solà 2001). Some fellow Jesuits at the Collegio Romano viewed him as too much a follower of Averroes. Hence, in 1578 three fellow teachers at the Roman College plus another prominent Jesuit denounced Perera as an Averroist to Pope

Gregory XIII. But the pope referred the matter to Jesuit General Mercurian, who supported Perera, and exiled one denouncer to Turin.[56]

In 1547, Alessandro Piccolomini (1508–1579), a prominent Aristotelian natural philosopher, published *Commentarium de certitudine mathematicarum disciplinarum* (Piccolomini 1547). He argued that mathematical demonstrations lacked scientific certitude for two reasons. Mathematical axioms lacked the certainty of the syllogistic reasoning of logic which provided intellectually convincing proof, and mathematics presented abstractions. Hence, mathematics was not a real or universal science because it did not explain the nature of matter, as Aristotelian physics did.[57] This provoked a lively debate, as some strongly defended the certitude of mathematics, while Aristotelian philosophers often echoed Piccolomini's views.[58]

One of them was Perera. In his 1564 treatise on the best practices in humanistic studies, he wrote that "Mathematical sciences require one kind of proof; moral explanations and elaborations demand another; and theories and arguments involving nature demand yet another" (MP, II, 680; *Jesuit Pedagogy* 2016, 199).

Perera returned to this topic later. In 1576, he published his major philosophical work: *De communibus omnium rerum naturalium principiis & affectionibus libri quindecim* (Fifteen Books on the Principles and Properties Common to all Natural Things), a folio volume of nearly 600 pages (Perera 1576).[59] Fourteen confirmed, and two unconfirmed editions followed between 1579 and 1618, published in Rome, Venice, Paris, Lyon, Cologne, and possibly Strasbourg, in quarto or octavo, and the octavo editions were about 900 pages long.[60] The number and arc of editions of Parera's major philosophical work roughly paralleled the publication history of Clavius' commentary on the *Sphaera*.

Perera's book was a comprehensive study of the various branches of philosophy, their principles, and the relations between them. Its goal was to provide a unified approach to Aristotelian philosophy and the epistemology of scientific investigation.

---

[56] For a summary of the affair and bibliography, see (Grendler 2017, 405–407). For a good account of the philosophical and pedagogical issues involved, see (Sander 2014).

[57] See the brief summaries of (Rose 1976, 285–286); and (Lamanna 2014, 71–72).

[58] (Jardine 1988, 693–697; and Romano 1999, 153–162), each with more bibliography.

[59] There is a fuller bibliographical description in (Tinto 1968, item 244).

[60] The subsequent editions are Paris, apud Michel Sonnius, 1579 or Thomas Brumen, 1579 (Perera 1579a, b). It is the same edition, because the royal privilege was granted to them jointly. See the last page of the Michel Sonnius edition. Other editions are Rome 1585 (Perera 1585a); Paris 1585 (Perera 1585b); Lyon 1585 (Perera 1585c); Venezia 1586 (Perera 1586); Paris, Michel Sonnius, 1586 (Sommervogel 1960, VI, 499); Lyon 1588 (Perera 1588); Paris 1589 (Perera 1589); Venezia 1591 (Perera 1591); Köln 1595 (Perera 1595); Strasbourg, no further information, 1595 (Sommervogel 1960, VI 500); Köln 1603 (Perera 1603), Köln 1609 (Perera 1609a); Venezia 1609 (Perera 1609b), Venezia 1618 (Perera 1618), Köln 1618 (Perera 1618a, b). (Sommervogel 1960, VI, 499) also lists a first edition of Rome 1562. Almost all scholars dismiss this as a ghost, because no copy has been located. And Perera was only twenty-seven at the time and had been teaching for only four years. Jesuits did not publish large controversial works at such a young age.

The book was heavily based on his lectures on logic, natural philosophy, and meta-physics.[61] In the course of the book, Perera weighed the claim of mathematics to be a science against Aristotle's criteria for scientific truth in physics with its emphasis on causes and substance. He found mathematics wanting.

In book three, Chap. 4, Perera offered a particularly negative discussion about whether mathematical proof met the standard of demonstration as formulated by Aristotle in the *Posterior Analytics*. Perera began by posing the question, whether mathematical demonstrations can claim the first order of certitude. Perera wrote that although many accepted that they could, he completely disagreed. Although mathematics is called a science because it can produce connections and a most beautiful and admirable order, mathematical certitude does not go beyond mathematics itself. Mathematical demonstration does not address or determine causes. By contrast, a true demonstration clarifies the essence of something, how it depends on its cause, and is recognized as such. True demonstration is securely connected to matter, that is, natural things. Drawing mathematical abstractions from matter is not difficult, because they depend only on quantity; hence, the intellect can easily conceive them. They are not tied to a definite and certain material that depends on physical acci-dents (Perera 1579a, 118–121).[62] Mathematics does not require long experience and careful observation, as do the principles of physics (meaning natural philosophy) or medicine. For this reason, boys are able to escape the mathematicians, but not the natural philosophers.[63] He meant that students did not need to study mathematics, but must study natural philosophy in order to understand science. In other words, studying mathematics was not necessary. Whether or not Clavius had Perera in mind when he accused philosophers of ridiculing mathematics, the above comment fit his words.

It is likely that Perera made such comments in the classroom as well, because some of his book repeated word-for-word material from the manuscripts of his lectures (Blum 2012, 140–141). If mathematics could not offer real demonstration of cause and effect, it was not a true science. Because mathematics was not a true science, Perera saw no reason for young people to study it. Perera's position threatened the very existence of mathematical instruction in Jesuit schools (Romano 1999, 146).

The attacks on mathematics attracted enough attention that the Roman province discussed the matter in a meeting in 1575 or 1576. The Roman province wanted more attention to be given to mathematics lest it declines. And in the next sentence,

---

[61] Perera has attracted considerable scholarly attention in recent years. In addition to studies cited in the notes, see (*Benet Perera* 2014; Blum 2012, 138–182).

[62] There are many other passages rejecting the scientific validity of mathematics in the book. (Romano 1999, 142–145; Lamanna 2014, 69–80) quote and analyze some of them. The epistemic structure of Aristotelian physics made it very difficult for Aristotelians such as Perera to accept mathematical proof as scientifically valid (Baldini 1999b, 263–264).

[63] "Praeterea, principia Mathematicae non exigunt longam experientiam & diligentem observa-tionem, quaemadmodum principia Physicae vel Medicinae…propter quam pueri possunt evadere Mathematici, non autem Physici…" (Perera 1579a, 121).

it warned professors of philosophy not to speak publicly about trivialities.[64] Whether these two statements were linked, and the second was meant to rebuke Perera, is impossible to determine.

Perera was not the only philosopher at the Roman College who did not see mathematics as a true science in the years in which the place of mathematics in Jesuit education was hotly debated. Paolo Valla (1560–1622), who taught logic, natural philosophy, and metaphysics at the Roman College from 1584 to 1590, then theology from 1602 to 1605, held the same views concerning the superiority of Aristotelian physical proof over mathematical proof as Perera (Wallace 1984, 135–136).[65]

In short, at the very moment when Clavius was making a strong argument for much more mathematical instruction in the Jesuit curriculum, prominent Jesuit natural philosophers also teaching in the Roman College dismissed mathematics as not a true science and not worth studying. The battle was joined.

9 The *Ratio Studiorum* of 1599

As mentioned, the Fourth General Congregation in 1581 ordered the preparation of a Ratio Studiorum, an educational plan for the entire Society. The Jesuits labored on it for eighteen years and through several committees, and produced at least 1,800 pages of documents.[66] Clavius contributed his bit: Although he had already presented his views on the importance of mathematics at length, he added two more brief comments in the 1590s. In particular, he wanted his mathematical academy permanently incorporated into the Jesuit educational structure (MP, VII, 117–122; *Jesuit Pedagogy* 2016, 294–300). Other Jesuits argued just as strenuously for their disciplines. A committee produced a draft Ratio Studiorum of 1586 that was sent to all the twenty provinces for comments, which came in abundance. Another committee read the comments, made revisions, and drafted a second version in 1591. The Tipografia del Collegio Romano published both drafts, with Francesco Zanetti doing the printing

---

[64] "25. Curandum videretur ut maior adhibeatur diligentia circa disciplinas mathematicas, ne brevi contingat nullum reperiri qui eas praelegat. Simul et cavendum ne philosophiae professores eas publice coram auditoribus flocci faciant" (MP, IV, 254). This is a document that issued a number of admonitions about education in the province, most often concerning the Roman College and Roman Seminary. If the two sentences are not linked, it is possible that the warning to the philosophy teachers concerned the strife in the Roman College about whether Perera was an Averroist, which came to a head in 1578.

[65] Valla's major work on logic, which expressed his views, was published posthumously in 1622. For a translated quote from that work stating that mathematics did not deal with substances as Aristotelian physics did, see (Baldini 2000, 253 n. 42). For a brief biography of Valla, see (Pignatelli 2001). However, Valla began to teach natural philosophy at the Roman College in the academic year 1584–1585, rather than 1587 as Pignatelli states (Baldini 1992, 570).

[66] See (MP, 5–7). These are the documents, almost all of them from ARSI, that Lukács edited and printed. Other repositories may have more documents. There is no comprehensive study of the drafting of the Ratio Studiorum for the obvious reason that it would be an immense task. However, Ladislaus Lukács in (MP, V, 1*-34*; Lukács 1999, 2000) provides good short accounts.

for the 1586 draft (*Ratio studiorum* 1586, 1591).[67] The 1591 draft was also sent to the provinces to be tried out in the classroom for three years, with the provinces charged to report what needed improvement. Again, the provinces responded fully, as did many individual Jesuits. Arguments and differences of opinion were often sharp.

The provinces had mixed views about teaching mathematics and how much mathematics they should teach. The majority of provinces endorsed teaching mathematics but for only one year. The opinion of the Province of Milan was typical. Mathematics should be taught for one year to the natural philosophy students; the younger logic students should not be burdened further. And it was enough that the mathematics class should teach the first three books of Euclid, the *Sphaera*, the astrolabe, and arithmetic (MP, VI, 293). The Spanish provinces did not want to teach any mathematics at all. The Province of Aragon commented that some logic students were incapable of studying mathematics; hence, it did not recommend compelling any students to study mathematics. The Province of Toledo wrote that Jesuit students could hear lectures on mathematics in universities (MP, VI, 294).[68]

Although the provinces teaching mathematics in some of their schools were not required to report the texts studied, some did. The Province of Portugal reported that it offered one year of mathematics in which it taught the *Sphaera* to students who did the three-year philosophical cycle (MP, VI, 294). The Jesuit school in Louvain had a mathematics class that taught practical arithmetic and the *Sphaera* of Sacrobosco (MP, VI, 296). It is likely that teachers and/or students used Clavius' commentary in these classes.

The provinces responded negatively to the proposal of a mathematical academy in Rome to which they would send Jesuits with mathematical aptitude. They believed that young Jesuits should study mathematics and theology in their home provinces. The provinces of the Rhine and Upper Germany were more favorably inclined and praised Clavius by name. But they too were not convinced that it was necessary to send Jesuits to Rome in order to learn to be good mathematicians (MP, VI, 293–294). One reason for the reluctance to send young Jesuits out of the province was that many provinces were experiencing a teacher shortage, because the Jesuits were opening new schools as quickly as possible. In summary, the majority of the provinces approved of one year of mathematics instruction consisting of daily lectures for the natural philosophy students. They agreed that the lectures should include Euclid, the *Sphaera*, and arithmetic. But they opposed a Roman mathematical academy.

The Jesuits finally completed the *Ratio studiorum* in 1599. The first edition was published not in Rome but in Naples by the press of Tarquinio Longo (fl. 1598–1620), who published many other Jesuit works (*Ratio studiorum* 1599).[69] Other editions followed, including the first Roman edition in 1606 published by the Tipografia del Collegio Romano (*Ratio studiorum* 1606). Its full title was *Ratio atque institutio*

---

[67] For illustrations of the title pages see (MP, V, 41, 229).

[68] For more comments from the provinces, see (MP, VII, 136, 147, 165, 175, 235).

[69] For a reproduction of the title page, see (MP, V, 355). For the Jesuit editions published by Longo, see USTC. Why a Neapolitan publisher rather than a Roman publisher was chosen is unknown.

studiorum, which is universally shortened to Ratio Studiorum. It consisted of directions, practices, and rules to be followed by all teachers and superiors responsible for the schools. The Ratio Studiorum offered great clarity about Jesuit education, albeit in a dry and utilitarian presentation. It was mandated for use in all Jesuit schools, with some discretion in implementation permitted.

The 1599 Ratio Studiorum gave mathematics a limited place in Jesuit education. A student who undertook the entire Jesuit curriculum did not study mathematics until after four to six years of Latin grammar, humanities, and rhetoric instruction in the lower school and one year of logic in the upper school. The natural philosophy students then attended a forty-five minutes daily lecture on mathematics (RS, par. 38). That was all. And the mathematics lecture was in addition to twice daily hour-long classes in natural philosophy. After the natural philosophy year, the Ratio Studiorum did not require students to study any more mathematics. To be sure, it did not bar students not studying natural philosophy from attending mathematics lectures. How many did so is impossible to determine.

The 1599 Ratio Studiorum imposed a limited and flexible curriculum for the mathematics lecture. The teacher was to begin with Euclid's *Elements*. After about two months, he should teach "Geography or the Sphere," presumably Sacrobosco's text, or "about those things that are generally of interest." In addition, "Every month, or at least every other month, he [the mathematics teacher] should make sure that one of the students elucidates a famous mathematical problem at a large gathering of philosophers and theologians. Afterward, it ought to be submitted to disputation if it seems right to do so" (RS, par. 239, 240). This was something that Clavius wanted.

The Ratio Studiorum did not authorize an academy for advanced study in mathematics or in any other discipline.[70] Hence, it did not discriminate against mathematics. It did encourage students who attended the mathematics lecture course and had an interest in mathematics to study mathematics privately (RS, par. 38). There was one more reference to mathematics. The rules for professors of natural philosophy told them to devote the entire year to "topics in physics" followed by a list of topics that included "the different manners of proceeding in physics and in mathematics, about which Aristotle comments in *Physics*, book 2" (RS, par. 219).[71] Thus, the Ratio Studiorum pointed to passages about whose meaning Clavius and Perera sharply differed, but left it to the natural philosophy teacher—not the mathematician—to interpret them. Overall, Clavius must have been deeply disappointed with the Ratio Studiorum.

Why did the Ratio Studiorum fail to give mathematics a larger place in Jesuit education?

Perera's attack against the scientific value of mathematics probably played a role, because he was respected by other Jesuit philosophers and widely published. But

---

[70] The 1599 Ratio Studiorum did mandate the establishment of academies of Greek and Hebrew, meaning Jesuits who would meet for two or three times a week to practice these skills. It also mandated teacher-training academies in which an "expert teacher" would guide young Jesuits about to begin teaching Latin grammar and the humanities in "giving lessons, dictating, writing, correcting, and performing the other duties of a good teacher" (RS, par. 81, 83) (quote). These were not academies of advanced study.

[71] The relevant passages in the *Physics* are in II.ii.193b–194a.

the major reason was that the provinces did not want more mathematical education. The provinces usually mentioned practical reasons. There was also an unmentioned major reason: Members of the Society of Jesus did not see mathematics contributing greatly to their fundamental purpose, which was "to help souls" in this life and to help them reach salvation in the next.[72] The rest of the curriculum helped souls more. The lower-school curriculum of humanistic studies plus catechesis taught students to become morally good and eloquent individuals and citizens. Humanistic studies also gave students excellent Latin skills enabling them to earn a living or to pursue advanced education in law, medicine, or theology. Philosophy taught students how to find truth and separate it from error. Theology taught men about God and God's plan for human beings.

Overall, the Ratio Studiorum of 1599 mandated what the majority of Jesuit provinces were already doing in mathematics, but not more, as Clavius wanted. This was not surprising, because education does not change unless teachers, students, parents, and leaders of society with power over schools demand it. This was not the case for Jesuit schools in the 1590s. On the contrary, the requests from towns and rulers for more and more Jesuit schools teaching the humanities, Aristotelian philosophy, and Scholastic theology demonstrated high satisfaction with the status quo. But the Society missed an opportunity to build on the impressive mathematical accomplishments of Clavius and the academy of mathematics.

10 Mathematics Instruction after 1599

How many Jesuit schools offered the forty-five minutes daily lecture mandated for students enrolled in the natural philosophy course? What happened to Clavius' academy of advanced mathematical studies and with the *Sphaera*? What were the practical results for mathematics and astronomy?

The answer to the first question is that only a small minority of Jesuit schools in Europe offered mathematics lectures in the first half of the seventeenth century. They were usually the one or two most important Jesuit schools in a province. And the school was sometimes part of a Jesuit university or a civic-Jesuit university.

The Italian assistancy presented examples.[73] In 1600, it included five provinces of roughly equal geographical size and number of Jesuits: Milan, Venice, Rome, Naples, and Sicily (Table 1). Each province had seven to twelve schools open to Jesuits and external students (lay boys, youths, and men). However, the vast majority of Jesuit schools were lower schools that taught one to three classes in Latin grammar and humanities; a few added a logic class. Only schools that taught logic, natural philosophy, and metaphysics, plus some theology, were likely to offer the mathematics

---

[72] See the explanation of "to help souls" in (O'Malley 1993, 18–19).

[73] Assistancies were part of the structure of the Society. In 1609, there were five geographical assistancies: Italy, France, Germany, Portugal, and Spain. Each assistancy included several provinces. The Asia missions were part of the Portuguese assistancy.

lecture. In 1600, there were four mathematics lectureships in Italian Jesuit schools, all in schools that taught three years of philosophy plus theology.

As both the number of Jesuit schools and those that taught philosophy increased, so did the number of mathematics lectureships. There were seven mathematics lectureships in Italy in 1649 (Table 2).

In addition, the Jesuit university at Cagliari, Sardinia, had a lectureship of mathematics beginning in 1626 which continued until the suppression of 1773 (Fischer 1983, 85). However, the Sardinian Jesuits were part of the Province of Aragon and the Spanish assistancy, because Sardinia was ruled by Spain. Only in 1718 did Sardinia become part of the Duchy of Piedmont-Savoy.

This was the pattern across Europe. Only a minority of Jesuit schools taught mathematics, and these were schools that taught all three philosophical courses plus theology. In some assistancies and provinces, a larger minority of schools taught mathematics. For example, in 1616 the Province of Lyon had thirteen schools of

Table 1 Schools and Mathematics Lectureships in the Assistancy of Italy in 1600

Provinces	Number of Schools	Schools with Mathematics Lectureships
Province of Milan	7	1 (Milan)
Province of Venice	11	1 (Parma)
Province of Rome	8	1 (Roman College)
Province of Naples	12	1 (Naples)
Province of Sicily	12	0

Sources: *ARSI, Mediolanensis 47, Veneta 37, Roma 54, Neapolitano 80, and Sicula 60*

Table 2 Jesuit Schools and Mathematics Lectureships in the Italian Assistancy in 1649

Provinces	Number of Schools	Schools with Mathematics Lectureships
Province of Milan	20	1 (Milan)
Province of Venice	18	3 (Bologna, Parma, Mantua)
Province of Rome	22	1 (Roman College)
Province of Naples	26	1 (Naples)
Province of Sicily	21	1 (Palermo)

Sources: *ARSI, Mediolanensis 2, Veneta 40, Roma 59, Neapolitano 83, and Sicula 66. Because the Jesuits were barred from the Republic of Venice, the number of schools in the Province of Venice was lower than the other provinces*

which four taught mathematics (Romano 1999, 376).[74] Overall, the French assistancy and German assistancy (which included the Province of Austria) taught more mathematics, while the Italian assistancy and the Province of Portugal taught less mathematics. And some of the Jesuits who taught mathematics in Lisbon were German Jesuits. The Spanish assistancy mostly ignored mathematics.[75]

Jesuit mathematicians in Italy, especially former academy members, commonly taught Sacrobosco's *Sphaera* with the aid of the commentary of Clavius. Giovanni Giacomo Staserio (Bari 1565–1635 Naples) was a member of Clavius' academy for two years, 1595–1597, while studying theology at the Roman College (Baldini 2003, 73, 93 n. 93). After ordination, he returned to Naples, where he taught mathematics at the Jesuit school for eighteen years, with interruptions, between 1599 and 1624. He had considerable influence on mathematical instruction in Naples and exchanged numerous letters with Clavius (Gatto 1994, 75–89, 101–113, 150–160, 277, 308–325). Upon learning that Clavius had revised his *Sphaera* commentary for a new edition, he wrote to Clavius in 1606 asking him to print thirty copies for the Naples college (Brevaglieri 2008, 298; Giard and Romano 2008, 110–111).

At Parma, the Jesuit mathematics lectureship was part of the civic-Jesuit University of Parma founded in 1601. This was a university in which lay professors taught law and medicine, and Jesuits taught theology, metaphysics, natural philosophy, logic, mathematics, and rhetoric. Basically, the Jesuit upper school was incorporated into the university. Giuseppe Biancani (Bologna 1566–1624 Parma), a member of Clavius' academy of mathematics from 1598 to 1600 and probably the most accomplished mathematician in the Province of Venice, was the mathematics professor at the University of Parma from 1605 until his death.[76] The university *rotulus* (a list of the professors, the lecture schedule, and a brief summary of the contents of the lectures) for the academic year 1617–1618 indicated that he lectured on Euclid's *Elements*, and the "sphere" (Grendler 2017, 168).

It was the same in Mantua. In 1624, with the strong support of Duke Ferdinando Gonzaga (1587–1626; ruled 1613–1616), the Jesuits expanded their small school into a complete school teaching the humanities, philosophy, and theology. And in 1625, Duke Ferdinando created the civic-Jesuit University of Mantua, whose structure was the same as the civic-Jesuit University of Parma. The Jesuit who taught mathematics at the Jesuit upper school and then the university from 1624 to 1629 was Cesare Moscatelli (Bologna ca. 1585–1644 Modena). He studied mathematics at Parma, probably with Biancani. Four of the five *rotuli* for the years that he taught mathematics at Mantua survive. He always taught Euclid's *Elements* and the *Sphaera*, most likely on the basis of Clavius' commentary, plus another mathematical topic.

---

[74] (Romano 1999) provides a comprehensive study of mathematical education in Jesuit schools in France, which vastly expands the still useful data in (Fischer 1983).

[75] For the German assistancy, see (Fischer 1978); for the French and Italian assistancies, see (Fischer 1983). For the Spanish assistancy, see (Baldini 2000, 130 n. 4); for the Province of Portugal, see (Baldini 2000, 129–167).

[76] For his participation in Clavius' academy, see (Baldini 2003, 73). There is a large bibliography on Biancani.

Unfortunately, the University of Mantua only lasted four years. It closed as a consequence of the War of the Mantuan Succession, the plague which killed thousands in Mantua including several Jesuits, and the terrible Sack of Mantua in July 1630 (Grendler 2009, 81, 164, 169, 200–201, 252–253 *et passim*). Although the university did not return to its previous form, the Jesuits slowly rebuilt their school, which included a renewal of the mathematics lectureship in the middle of the 1640s.

The second question concerns mathematical academies. What happened to Clavius' academy? Were there any academies of advanced mathematics in Italian Jesuit schools after 1599? If not, how were Jesuit mathematicians trained?

Clavius' academy continued as before until he became ill and had to cease teaching in 1610 and died in 1612. At his death, his academy "was considered as second to no other European scientific institution" (Baldini 2003, 68). It did not formally dissolve but lived on for a while. Some Jesuits and a few non-Jesuits continued to receive advanced mathematical instruction in Rome from able Jesuit mathematicians such as Christoph Grienberger until about 1615. But by the 1630s there is little or no evidence of the existence of advanced mathematical instruction in Jesuit Rome. Outside of Rome, Italian mathematical academies have not been found in the rest of the seventeenth century.

Instead, advanced mathematics was taught one-on-one or in very small groups, as young Jesuits with mathematical aptitude connected with expert Jesuit mathematicians.[77] The best mathematician in the province usually taught in the most important Jesuit school where the majority of Jesuit scholastics in the province did their philosophical and theological studies. Hence, a young Jesuit with mathematical aptitude from somewhere else in the province went to Rome, Milan, Parma, Naples, or Palermo to study philosophy and theology, where he heard mathematics lectures as well. He might also ask his superiors to permit him to remain in Rome, Milan, et al., for another year or two in order to study with a master mathematician. Superiors were more likely to grant such requests than earlier, because the teacher shortage had eased, thanks to a wave of young men who joined the Society in the first quarter of the seventeenth century. Private instruction may or may not have been as productive as attending a mathematical academy, but it was a way to learn.

What happened at Bologna suggests one pathway to advanced mathematical study. The Society founded a day school in Bologna in 1551; it added a boarding school for noble boys in 1634, and a boarding school for citizen boys in 1645.[78] The noble and citizen boarders lived supervised lives in separate buildings, but attended classes together in the Jesuit day school. The University of Bologna was very concerned about competition from the Jesuit school, so it tried to prevent the Jesuits from teaching subjects, including mathematics, that the university taught. A compromise was reached: The Jesuit school was permitted to teach theology, philosophy, and

---

[77] The rest of this paragraph is speculative but likely. Much research needs to be done.

[78] In some Italian cities, "citizen" was a legal class lower than the nobility but above commoners. Certain governmental offices were reserved for citizens, while citizen merchants enjoyed tax concessions.

mathematics to fellow Jesuits and to boarding school students only. They were not permitted to teach these classes to day students, that is, lay boys and youths from the town, who comprised ninety-five percent or more of the student body in Jesuit schools. Hence, the Bologna Jesuit mathematics class had very few students; in 1646, it had only five students, all of them Jesuits.[79] Who the Jesuit mathematics students were is unknown; they might have been philosophy students, theology students, or older Jesuits with an interest in mathematics. In any case, there were five Jesuits interested in mathematics, ideal conditions for transforming the class into a de facto academy of advanced mathematics.

This may have happened. In 1672, students and professors from the University of Bologna complained that the Jesuit school was teaching the same advanced mathematics that university professors taught. Even worse, the mathematics students at the Jesuit school were posting their mathematical conclusions in the courtyard of the Archiginnasio, the university building in the center of the city (Grendler 2017, 309–310). That some advanced mathematics was taught at the Bologna Jesuit school is not surprising, because the Jesuit mathematics teacher at that time was Giuseppe Ferroni (Pistoia 1628–1709 Siena). He was a very able mathematician who had studied mathematics with one of the best Jesuit mathematicians in the Province of Venice and with two non-Jesuit followers of Galileo Galilei. Ferroni then taught mathematics in the Roman College from 1657 to 1660, at the Jesuit school in Mantua from 1660 to 1666, at Bologna from 1667 to 1686, and at the University of Siena and the Jesuit school in Siena from 1686 until his death. He had the knowledge to teach advanced mathematics in Bologna (Torrini 1973; Zanfredini 2001; Baldini 2002, 312; Grendler 2017, 310–311, 387, 389–390).

In their public instruction, Jesuits followed the papal decree of 1616 forbidding the teaching of heliocentrism as physical reality. As astronomers and mathematicians, they might—or might not—accept the concept of and evidence for heliocentrism as physically true. But as Jesuits, they were obliged to interpret the new data in light of the papal decree (Baldini 2003, 69). On the other hand, some Jesuits found ways to express their dissent. Again, Bologna and Ferroni offered an example.

Ferroni accepted the heliocentric position of Copernicus (1473–1543) and Galilei, but did not publish his views because of Jesuit censorship. But he did not completely hide them either. In 1680, an anonymous dialogue appeared in Bologna: *Dialogo fisico astronomico contro il sistema copernicano tenuto fra due interlocutori, Sig. Francesco Bianchini veronese sotto il nome di Adimanto, Sig. Ignatio Rocca piacentino sotto il nome di Silvio, convittori del Collegio del Beato Luigi Gonzaga della Compagnia di Gesù in Bologna* (A Physical and Astronomical Dialogue Against the Copernican System Between Two Interlocutors, Signore Francesco Bianchini From Verona Under the Name of Adimanto, Signore Ignatio Rocca of Piacenza

---

[79] For the enrollment in the mathematics class, see ARSI, Veneta 125, f. 3r, "Stato dei Studij del Colleggio di Bologna." For details about the Bologna Jesuit schools and the struggle between the Jesuits and the university, see (Grendler 2017, 297–316).

Under the Name of Silvio, Boarding Students of the College of Blessed Luigi Gonzaga of the Society of Jesus in Bologna), per Giuseppe Longhi, Bologna 1680.[80]

The title was inaccurate and Bianchini and Rocca were fictitious characters. The dialogue supported Copernicus, who was presented positively. The arguments supporting Copernicus were mathematical and physical, while the arguments favoring the Ptolemaic system were scriptural and miraculous. The dialogue also criticized the Holy Office prohibition against heliocentrism. The anonymous author was Ferroni. Setting a pro-Copernicus and anti-Ptolemy dialogue in a Jesuit boarding school (The College of Blessed Luigi Gonzaga was the Jesuit citizen school in Bologna) was a way to criticize Jesuit adherence to the papal prohibition barring Jesuit mathematicians from teaching heliocentrism as physical reality. The dialogue never mentioned the *Sphaera* of Sacrobosco, and it is unlikely that Ferroni taught it or Clavius' commentary in his classes.

By the late seventeenth century, the sharp separation between Jesuit Aristotelian philosophy and Jesuit mathematics had diminished greatly. Mathematicians developed an experimental methodology enabling them to describe physical reality in mathematical terms. And philosophers loosened or abandoned strict Aristotelian definitions of scientific proof. Mathematicians investigated philosophical issues and natural philosophers included mathematical evidence. They sought common ground. An institutional sign was that some Jesuits moved back and forth between the mathematics professorship and the natural philosophy professorship at the Roman College in the second half of the seventeenth century (Baldini 1999b, 269–270, 272–278; Udías 2015, 43–47; Raphael 2015). This would have been unimaginable in the time of Clavius and Perera.

Such developments meant the disappearance of the *Sphaera* from Jesuit mathematics instruction. Negative evidence from the Roman College supports this conclusion. The *rotulus* for the Roman College for the academic year 1696–1697 promised that the mathematics lectureship would teach geometry, the *Elements* (of Euclid), speculative and practical arithmetic, and general mathematical problems and theorems.[81] Noticeably absent was any reference to Sacrobosco or *sphaera*. The long pedagogical lives of Sacrobosco's *Sphaera* and Clavius' commentary had ended. Jesuit mathematics had entered a new era.

In 1730, the Society of Jesus finally decreed a formal cease fire between mathematics and Aristotelian natural philosophy, Clavius and Perera. The Sixteenth General Congregation meeting from November 15, 1730, to February 13, 1731, stated that they were both valid. It affirmed the Society's loyalty to Aristotelian

---

[80] The dialogue is printed with explanatory notes in (*La scienza dissimulata* 2005, 107–140).

[81] "EX MATHEMATICIS. Praeter Geometriae, atque Arithmeticae speculativae, & practicae Elementa, traduntur ex universa Mathesi, Problemata, & Theoremata." Archivio di Stato di Roma, Università 195, folio 23, which is the printed *rotulus* for the academic year 1696–1697. The descriptions of the mathematics course in the *rotuli* for the academic years 1697–1698, 1698–1699, and 1699–1700 are the same. ASR, Università 195, folios 23, 104, 185, 384. For photographic reproductions of those of 1696–1697 and 1699–1700, see (Grendler 2017, 328–329, 336–337). Only these four *rotuli* have been located.

philosophy including its philosophy of nature. It also endorsed mathematics and experimental physics.

> There is no opposition between the Aristotelian philosophy and the more attractive style of learning in physics, and especially in its more particularized branches, with which the more notable natural phenomena are explained and illustrated by mathematical principles and the experimentation of scholars; instead, there is complete agreement between them (*For Matters of Greater Moment*, 1994, 384, decree 36).

Although the contest had ended some time ago, the Congregation made it official. The eighteenth century witnessed a burst of Jesuit creativity in numerous scientific fields, including mathematics and astronomy.

11   Conclusion

The last phase of the pedagogical career of Sacrobosco's *Sphaera* was entwined with Jesuit mathematics instruction. Christoph Clavius, the most influential Jesuit mathematician, was the key figure. He wrote a commentary on the *Sphaera* that was widely reprinted and used. The first edition printed in Rome 1570 also marked a change in Jesuit policy concerning the publication of Jesuit works. With Clavius' book, the Roman Jesuits decided to entrust the printing and publication of major works of Jesuit scholarship to external commercial presses beginning with Vittorio Eliano, the brother of a Jesuit. Many other presses followed by publishing the books of Clavius and other Jesuits as part of their production.

Clavius believed that both mathematics and Aristotelian physical science offered scientific certitude, a view not shared by Benet Perera and some other Jesuit philosophers. Clavius fought to expand the mathematics curriculum of Jesuit schools and to make his mathematical academy a permanent part of Jesuit education. He was unsuccessful. The Ratio Studiorum of 1599 listed only a single class of mathematics in the standard Jesuit curriculum, and it did not make permanent his academy. Nevertheless, Jesuit schools did teach mathematics, and they used his commentary on the *Sphaera* until mathematical research and instruction changed in the middle of the seventeenth century. While it lasted, the alliance between a medieval astronomical text, its most important Renaissance commentary, and the preeminent Catholic teaching order was remarkable.

Abbreviations

Digital Repositories

Sphaera Corpus*Tracer*	Max Planck Institute for the History of Science. https://db. sphaera.mpiwg-berlin.mpg.de/resource/Start. Accessed 07 June 2021
USTC	Universal Short Title Catalogue. University of St. Andrews. https://www.ustc.ac.uk/. Accessed 07 June 2021

Archives and Special Collections

ARSI	Archivum Romanum Societatis Iesu
ASR	Archivio di Stato di Roma

References

Primary Literature

Alvarez, Manuel. 1572. *De constructione partium orationis liber. Emanuelis Aluani Lusitani ex Societate Iesu.* Roma: Vittorio Eliano ad instantiam Michele I Tramezzino. USTC 808937.

Assertiones theologicae disputandae in tempio Societatis Iesu, tempore electionis praepositi generalis. 1558. Roma: Tipografia del Collegio Romano. USTC 803886.

*Constitutiones Societatis Iesu. Cum earum declarationibus.*1570a. Roma Vittorio Eliano. USTC 763462.

Constitutiones Societatis Iesu. 1570b. Roma: Vittorio Eliano. USTC 763463.

Constitutiones et declarationes examinis generalis Societatis Iesu. 1570c. Roma: apud Vittorio Eliano. USTC 832415; USTC 832416.

The Constitutions of the Society of Jesus and Their Complementary Norms. A Complete English Translation of the Official Latin Texts. 1996. St. Louis: The Institute of Jesuit Sources.

Dottrina christiana nella quale si contengono i principali misteri della nostra fede rappresentati con figure per istruttione de gl'idioti & di quelli che non sanno leggere … Composta dal p. Giovanni Battista Romano della Compagnia di Giesu. 1591. Roma: presso Giacomo Ruffinelli ad istanza di Giorgio Dagano. USTC 828075.

Eliano, Giovanni Battista. 1586. *Dottrina christiana nella quale si contengono li principali misteri della nostra fede, rapresentati con figure.* Roma: nella stamperia de Vincenzo Accolti. USTC 828074.

*Fidei orthodoxae brevis et explicata confessio quam Sacrosanta Romana Ecclesia docet et iis maxime proponendam edit, quicunque ab orientalium, ad catholicae veritatis communionem accedere, et Romano pontifici praestare oboedientiam statuunt.*1566. Roma: Tipografia del Collegio Romano. USTC 804480.

For Matters of Greater Moment. The First Thirty Jesuit General Congregations. 1994. Ed. and trans. John W. Padberg, Martin D. O'Keefe, and John L. McCarthy. St. Louis: Institute of Jesuit Sources.

Jesuit Pedagogy, 1540–1616. 2016. Ed. and trans. Cristiano Casalini and Claude Pavur. Chestnut Hill, MA: Institute of Jesuit Sources.

MP. *Monumenta paedagogica Societatis Iesu*, ed. Ladislaus Lukács. 7 vols. Rome: Apud "Monumenta Historica Soc. Iesu" and Institutum Historicum Societatis Iesu, 1965–1992.

Perera, Benet. 1576. *De communibus omnium rerum naturalium principiis & affectionibus, libri quindecim. Qui plurimum conferunt, ad eos octo libros Aristotelis, qui De physico auditu inscribuntur, intelligendos. Adiecti sunt huic operi, tres indices; unus capitum singulorum librorum; alter quaestionum, tertius rerum.* Roma: impensis Venturino Tramezzino apud Francesco Zanetti. USTC 847492.

Perera, Benet. 1579a. *De communibus omnium rerum naturalium principiis & affectionibus, libri quindecim.* Parisiis: Apud Micaëlem Sonnium, via Iacobaea, sub scuto Basiliensi. USTC 170479.

Perera, Benet. 1579b. *De communibus omnium rerum naturalium principiis et affectionibus libri quindecim. Qui plurimum conferunt ad eos octo libros Aristotelis, qui de physico auditu inscribuntur intelligendos. Tres indices.* Paris: apud Thomas Brumen. USTC 138863.

Perera, Benet. 1585a. *De communibus omnium rerum naturalium principiis & affectionibus libri quindecim. Qui plurimum conferunt, ad eos octo libros Aristotelis, qui de physico auditu inscribuntur, intelligendos. Adiecti sunt huic operi tres indices, unus capitum singulorum librorum: alter quaestionum; tertius rerum.* Roma: excudebant Alessandro Gardane & Francesco Coattino ex typographia Giacomo Tornieri & Bernardino Donangeli. USTC 847493, USTC 847494.

Perera, Benet. 1585b. *De communibus omnium rerum naturalium principiis et affectionibus, libri quindecim. Qui plurium conferuntur ad eos octo libros Aristotelis, qui de physico auditu inscribitur, intelligendos. Tres indices.* Paris: Thomas Brumen. USTC 171908.

Perera, Benet. 1585c. *De communibus omnium rerum naturalium principiis et affectionibus, libri quindecim. Qui plurium conferuntur ad eos, octo libros Aristotelis, qui de physico auditu inscribitur intelligendos. Tres indices.* Lyon: sumptibus Sybile de La Porte. USTC 156545.

Perera, Benet. 1586. *De communibus omnium rerum naturalium principiis & affectionibus, libri XV. Qui plurimum conferunt ad eos octo libros Aristotelis, qui De physico audito inscribuntur, intelligendos. Adiecti sunt huic operi tres indices, unus capitum singulorum librorum, alter quaestionum, tertius rerum. Omnia verò in hac postrema editione denuò sunt diligentius recognita, & emendata.* Venezia: [ex Unitorum Societate] apud Andrea Muschio. USTC 847495.

Perera, Benet. 1588. *De communibus omnium rerum naturalium principiis et affectionibus, libri quindecim. Qui plurimum conferunt ad eos octo libros Aristotelis, qui de physico auditu inscribuntur, intelligendos.* Lyon: sumptibus Sybile de La Porte. USTC 142532.

Perera, Benet. 1589. *De communibus omnium rerum naturalium principiis et affectionibus libri quindecim qui plurimum conferunt ad eos octo libros Aristotelis qui de physico auditu inscribuntur, intelligendos.* Paris: apud Michel Sonnius. USTC 203448.

Perera, Benet. 1591. *De communibus omnium rerum naturalium principiis & affectionibus, libri XV. Qui plurimum conferunt ad eos octo libros Aristotelis, qui de physico auditu inscribuntur, intelligendos. Adiecti sunt huic operi tres indices, unus capitum singulorum librorum, alter quæstionum, tertium rerum. Omnia verò in hac postrema editione denuò sunt diligentius recognita, & emendata.* Venezia: apud Andrea Muschio. USTC 847499.

Perera, Benet. 1595. *De communibus omnium rerum naturalium principiis et affectionibus, libri quindecim. Qui plurimum conferunt, non tantum ad eos octo libros, qui de physico auditu inscribuntur: sed etiam multos alios difficillimos Aristotelis locos intelligendos.* Köln: Lazarus Zetzner. USTC 615757.

Perera, Benet. 1603. *Benedicti Pererii Societatis Jesu, De communibus omnium rerum naturalium principiis [et] affectionibus, Libri Quindecim: Qui plurimum conferunt, non tantum ad eos octo libros, qui de Physico auditu inscribuntur, sed etiam multos alios difficilimos Aristo.* Köln: sumptibus Lazarus Zetzner. USTC 2117770.

Percra, Benet. 1609a. *Benedicti Pererii Societatis Jesu. De communibus omnium rerum naturalium principiis [et] affectionibus, Libri Quindecim: Qui Plurimum Conferunt, non tantum ad eos octo libros, qui de Physico auditu inscribuntur; sed etiam multos alios difficillimos Arist.* Köln, Lazarus Zetzner USTC 2030945.

Perera, Benet. 1609b. *De communibus omnium rerum naturalium principiis & affectionibus, libri 15. Qui plurimum conferunt ad eos octo libros Aristotelis, qui de physico auditu inscribuntur, intelligendos. Adiecti sunt huic operi tres indices.* Venezia: apud Girolamo Polo. USTC 4034438.

Perera, Benet. 1618a. *De communibus omnium rerum naturalium principiis & affectionibus, libri 15. Qui plurimum conferunt ad eos Octo libros Aristotelis, qui de physico auditu inscribuntur, intelligendos. Adiecti sunt huic operi tres indices.* Venezia: apud Giovanni Antonio Giuliani. USTC 4027012.

Perera, Benet. 1618b. *De Communibus Omnium Rerum naturalium principiis [et] affectionibus, Libri Quindecim: Qui Plurimum Conferunt, Non tantum ad eos octo libros, qui de Physico auditu inscribuntur; sed etiam multos alios difficilimos Aristo.* Köln: Lazarus Zetzner (heirs of). USTC 2135370.

Perera, Benet. 1619. *Benedicti Pererii Societatis Jesu. De communibus omnium rerum naturalium principiis [et] affectionibus, Libri Quindecim: Qui Plurimum Conferunt, non tantum ad eos octo libros, qui de Physico auditu inscribuntur; sed etiam multos alios difficilimos Aristo.* Köln: Lazarus Zetzner (heirs of). USTC 2109282.

Piccolomini, Alessandro. 1547. *In mechanicas quaestiones Aristotelis, paraphrasis paulo quidem plenior Eiusdem commentarium de certitudine mathematicarum disciplinarum, in quo, de resolutione, diffinitione et demonstratione, necnon de materia, et fine logicae facultatis, quamplura concinentur ad rem ipsam, tum mathematicam tum logicam maxime pertinentia.* Roma: Antonio Blado, USTC 848295.

Polanco, Juan de. 1577. *Methodus ad eos adiuvandos, qui moriuntur: ex complurium doctorum, ac priorum scriptis, diuturnoque usu, et observatione collecta.* Roma: apud Vittorio Eliano. USTC 850080.

Polanci Complementa. Epistolae et Commentaria P. Ioannis Alphonsi de Polanco. 2 vols. 1969. Madrid: reprint Roma: Apud Monumenta Historica S. I.

Quinctus Horatius Flaccus ab omni obscoenitate purgatus. Ad usum Gymnasiorum Societatis Iesu. 1569. Roma: apud Vittorio Eliano. USTC 835875.

Ratio atque institutio studiorum per sex patres ad id iussu r.p. praepositi generalis deputatos conscripta. 1586. Roma: excudebat Francesco Zanetti: Tipografia del Collegio Romano. USTC 832489.

Ratio atque institutio studiorum. 1591. Roma, Tipografia del Collegio Romano, 1591. USTC 832504.

Ratio atque institutio studiorum Societatis Iesu. 1599. Napoli: ex typographia Tarquinio Longo. USTC 832515.

Ratio atque institutio studiorum Societatis Iesu. 1606. Roma, in Tipografia del Collegio Romano dei Gesuiti. USTC 4033963.

RS. *The Ratio Studiorum. The Official Plan for Jesuit Education.* Translation and Commentary by Claude Pavur. Saint Louis: The Institute of Jesuit Sources, 2005.

Sa, Manuel de. 1598. *Notationes in totam scripturam sacram, quibus omnia fere loca difficilia, brevissime explicantur. Tum variae ex Hebraeo, Chaldaeo, & Graeco lectiones indicantur. Opus omnibus scripturae studiosis utilissimum.* Antverpiae, ex officina Plantiniana apud Joannem Moretum. USTC 407048.

Sacrobosco, Johannes de and Christoph, Clavius. 1570. *Christophori Clavii Bambergensis, ex Societate Iesu, in Sphaeram Ioannis de Sacro Bosco commentarius.* Romae: Apud Victorium Helianum. https://hdl.handle.net/21.11103/sphaera.100365.

Sacrobosco, Johannes de and Christoph, Clavius. 1581. *Christophi Clavii Bambergensis ex Societate Iesu in Sphaeram Ioannis de Sacro Bosco commentarius Nunc iterum ab ipso Auctore recognitus, & multis ac varijs locis locupletatus.* Romae: Ex officina Dominici Basa. Colophon: Romae Apud Francescum Zanettum. https://hdl.handle.net/21.11103/sphaera.101117.

Sacrobosco, Johannes de and Christoph, Clavius. 1585. *Christophi Clavii Bambergensis ex Societate Iesu in Sphaeram Ioannis de Sacro Bosco commentarius Nunc tertio ab ipso Auctore recognitus, & plerisque in locis locupletatus. Permissu superiorem.* Romae: Ex officina Dominicio Basae. https://hdl.handle.net/21.11103/sphaera.101120.

Sacrobosco, Johannes de and Christoph, Clavius. 1591. *Christophi Clavii Bambergensis ex Societate Iesu in Sphaeram Ioannis de Sacro Bosco commentarius, Nunc tertio ab ipso Auctore recognitus, & plerisque in locis locupletatus. Permissu Superiorum.* Venetiis: Apud Ioan. Baptistam Ciotum Senesem, sub signo Minervae. https://hdl.handle.net/21.11103/sphaera.100360.

Sacrobosco, Johannes de and Christoph, Clavius. 1596. *Christophori Clavii Bambergensis ex Societate Iesu in sphaeram Ioannis de Sacro Bosco commentaius . Nunc tertio ab ipso Auctore recognitus, & plerisque in locis locupletatus. Maiori item cura correctus. Permissu superiorum.* Venetiis: Apud Bernardum Basam sub signo Solis.https://hdl.handle.net/21.11103/sphaera.100367.

Sacrobosco, Johannes de and Christoph, Clavius. 1601. *Christophori Clavii Bambergensis ex Societate Iesu In Sphaeram Ioannis de Sacro Bosco Commentarius. Nunc tertio ab ipso Auctore recognitus, & plerisque in locis locupletatus. Maiori item cura correctus. Permissu Superiorum.* Venetiis: Apud Io. Baptistam Ciottum, Sub signo Aurorae. https://hdl.handle.net/21.11103/sphaera.100673.

Sacrobosco, Johannes de and Christoph, Clavius. 1603. *Christophori Clavii Bambergensis ex Societate Iesu, In Sphaeram Ioannis de Sacro Bosco Commentarius. Nunc septimò ab ipso Auctore recognitus, & plerisque in locis locupletatus. Maiori item cura correctus. Permissu superiorum.* Venetiis: Apud Io. Baptistam Ciottum, Sub signo Aurorae 1603.https://hdl.handle.net/21.11103/sphaera.100370.

Sacrobosco, Johannes de and Christoph, Clavius. 1611. *Christophori Clavii Bambergensis e Societate Iesu Operum Mathematicorum Tomus Tertius Complectens Commentarium in Sphaeram Ioannis de Sacro Bosco, & Astrolabium.* Moguntiae: Sumptibus Antonii Hierat excudebat Reinhardus Eltz. https://hdl.handle.net/21.11103/sphaera.100383.

La scienza dissimulata nel Seicento. 2005. A cura di Emanuele Zinato, premesso di Paolo Rossi. Napoli: Liguori.

Toledo, Francisco de. 1569. *Introductio in dialecticam Aristotelis. Per r.p. Franciscum Toletum doct. theologum Societatis Iesu. Et ab eodem denuo recognita.* Roma: apud Vittorio Eliano. USTC 859187.

Toledo, Francisco de. 1572. *In universam Aristotelis Logicam commentaria, una cum quaestionibus, per doctorem Franciscum Toletum, societatis Iesu.* Roma: Vittorio Eliano ad instantiam Michele I Tramezzino. USTC 859183.

Toledo, Francisco de. 1575. *Commentaria una cum quaestionibus in tres libros Aristotelis De anima. Nunc primum in lucem edita.* Venezia: apud Lucantonio II Giunta. USTC 859194; USTC 859195.

Toledo, Francisco de. 1599a. *Summae de instruct. sacerdotum libri VII De peccatis liber unus: cum bullae coenae Domini dilucidatione. Quibus omnibus Christiani officii ratio explicatur.* Milano: ex offic. typog. Girolamo Bordone & C. USTC 859249; USTC 859250.

Toledo, Francisco de. 1599b. *Summa casvum conscientiae, sive de instructione sacerdotum, libri septem: item de peccatis liber unus, cum bullae coenae domini dilucidatione. Opus nunc primum in Germania excusum, et à mendis emendatum.* Köln/Münster: Lambert Raesfeldt. USTC 695127.

Toledo, Francisco de. 1599c. *Summae, de instructione sacerdotum libri septem: de peccatis liber unus: cum Bullae Coenae Domini dilucidatione.* Lyon, apud Horace Cardon. USTC 159102.

Secondary Literature

Ascarelli, Fernanda. 1972. *Le cinquecentine romane. "Censimento delle edizioni romane del XVI secolo possedute dalle biblioteche di Roma."* Milano: Editrice Etimar.

Baldini, Ugo. 1992. *Legem impone subactis. Studi su filosofia e scienza dei gesuiti in Italia 1540–1632.* Roma: Bulzoni.

Baldini, Ugo. 1999a. Clavius, Christopher. In *Encyclopedia of the Renaissance,* ed. Paul F. Grendler et al. 6 vols., 2, 17–18. New York: Charles Scribner's Sons.

Baldini, Ugo. 1999b. The development of Jesuit 'physics' in Italy, 1550–1700: a structural approach. In *philosophy in the sixteenth and seventeenth centuries. Conversations with Aristotle*, eds. Constance Blackwell and Sachiko Kusukawa, 248–279. Aldershot: Ashgate.

Baldini, Ugo. 2000. *Saggi sulla cultura della Compagnia di Gesù (secoli XVI–XVIII).* Padua: CLEUP.

Baldini, Ugo. 2002. S. Rocco e la scuola scientifica della provincia veneta: il quadro storico (1600–1773). In *Gesuiti e università in Europa (secoli XVI–XVIII).* Atti del Convegno di studi. Parma, 13–15 dicembre 2001. A cura di Gian Paolo Brizzi e Roberto Greci, 283–323. Bologna: CLUEB.

Baldini, Ugo. 2003. The academy of mathematics of the Collegio Romano from 1553 to 1612. In *Jesuit science and the republic of letters*, ed. Mordechai Feingold, 47–98. Cambridge, MA: MIT Press.

Baldini, Ugo. 2013. Matteo Ricci nel Collegio Romano (1572–1577): cronologia, maestri, studi. *Archivum Historicum Societatis Iesu* 82:115–164.

Baldwin, Martha. 2003. Pious Ambition: Natural philosophy and the Jesuit quest for the patronage of printed books in the seventeenth century. In *Jesuit Science and the Republic of Letters*, ed. Mordechai Feingold, 285–329. Cambridge, MA: MIT Press.

Barberi, Francesco. 1968. Blado, Antonio. *DBI* 10:753–757.

Benet Perera (Pererius, 1535–1610). 2014. *A Renaissance Jesuit at the crossroads of modernity.* Turnhout: Brepols; Bari: Edizioni di pagina.

Blum, Paul Richard. 1999. Aristotelianism *more geometrico:* Honoré Fabri. In *Philosophy in the sixteenth and seventeenth centuries. Conversations with Aristotle*, eds. Constance Blackwell and Sachiko Kusokawa, 234–247. Aldershot: Ashgate.

Blum, Paul Richard. 2012. *Studies on early modern Aristotelianism.* Leiden: Brill.

Brevaglieri, Sabina. 2008. Editoria e cultura a Roma nei primi tre decenni del Seicento: lo spazio della scienza. In *Rome et la science moderne entre Renaissance et Lumières,* ed. Antonella Romano, 257–319. Roma: École Française de Rome.

Bruni, Annalisa. 1997. Basa, Bernardo. *Dizionario dei tipografi* 1:81–82.

Camerini, Paolo. 1962–1963. *Annali dei Giunti.* Vol. 1 in 2 parts: Venezia. Firenze: Sansoni Antiquariato.

Casetti Brach, Carla. 1993. Eliano, Vittorio. *DBI* 42:475–77.

Casetti Brach, Carla. 1997. Eliano, Vittorio. In *Dizionario dei tipografi*, 404–405.

Cioni, Alfredo. 1970. Basa, Domenico. *DBI* 7:45–49.

Clines, Robert John. 2020. *A Jewish Jesuit in the eastern Mediterranean. Early modern conversion, mission, and the construction of identity.* Cambridge: Cambridge University Press.

Contò, Agostino. 1997. Ciotti, Giovanni Battista. In *Dizionario dei tipografi*, 292–295.

Dear, Peter. 1995. *Discipline & experience. The mathematical way in the scientific revolution.* Chicago: University of Chicago Press.

DBI. *Dizionario biografico degli italiani.* 1960–. Roma: Treccani.

DHCJ. *Diccionario Histórico de la Compañia de Jesús: Biográfico–temático*, ed. Charles E. O'Neill, SJ, and Joaquín M. Domíngues, SJ. 4 vols. Roma: Institutum Historicam S.I. and Madrid: Universidad Pontificia Comillas, 2001.

Dizionario dei tipografi e degli editori italiani. Il Cinquecento. 1997. 1 vol. Milano, Editrice Bibliografica.

Donnelly, John Patrick. 2001. Toledo, Francisco de. *DHCJ* 4:3807–08.

Escalera, José Martínez de la. 2001. Parra, Pedro de. *DHCJ* 3:3046.

Fischer, Karl Adolf Franz. 1978. Jesuiten-Mathematiker in der deutschen Assistenz bis 1773. *Archivum Historicum Societatis Iesu* 47:159–224.

Fischer, Karl Adolf Franz. 1983. Jesuiten-Mathematiker in der französischen und italienischen Assistenz bis 1762 bzw. 1773. *Archivum Historicum Societatis Iesu* 52:52–92.

Firpo, Massimo. 1981. Ciotti, Giovanni Battista. *DBI* 25:692–96.

García-Villoslada, Ricardo. 1990. *Sant'Ignazio di Loyola.* Traduzione dallo spagnolo di Anna Maria Ercoles. Milano: Edizioni Paolini.

Gatto, Romano. 1994. *Tra scienza e immaginazione. Le matematiche presso il collegio gesuitico napoletano (1552–1670 ca.).* Firenze: Leo S. Olschki.

Giard, Luce, and Antonella Romano. 2008. L'usage jésuite de la correspondance: sa mise en pratique par le mathématicien Christoph Clavius (1570–1611). In *Rome et la science moderne entre Renaissance et Lumières*, ed. Antonella Romano, 65–119. Roma: École Française de Rome.

Gilbert, Paul. . La preparazione della *Ratio Studiorum* e l'insegnamento di filosofia di Benet Perera. *Quaestio* 14: 3–30.

Ginsburg, Jane C. 2019. Proto-proprietà letteraria ed artistica: i privilegi di stampa papali nel XVI secolo. In *Privilegi librari nell'Italia del Rinascimento*, eds. Erika Squassina and Andrea Ottone, 103–287. Milano: FrancoAngeli.

Grendler, Paul F. 1989. *Schooling in Renaissance Italy. Literacy and learning, 1300–1600.* Baltimore: The Johns Hopkins University Press.

Grendler, Paul F. 2004. Italian schools and university dreams during Mercurian's Generalate. In *The Mercurian project. Forming Jesuit culture 1573–1580*, ed. Thomas M. McCoog, 483–522. Roma: Institutum Historicum Societatis Iesu and St. Louis: The Institute of Jesuit Sources.

Grendler, Paul F. 2009. *The University of Mantua, the Gonzaga, and the Jesuits, 1584–1630.* Baltimore: The Johns Hopkins University Press.

Grendler, Paul F. 2017. *The Jesuits and Italian universities 1548–1773.* Washington, DC: The Catholic University of America Press.

Grendler, Paul F. 2019a. *Jesuit schools and universities in Europe 1548–1773.* Leiden: Brill.

Grendler, Paul F. 2019b. Philosophy in Jesuit schools and universities. In *Jesuit philosophy on the eve of modernity*, ed. Cristiano Casalini, 13–33. Leiden: Brill.

Homann, Frederick A. 2001. Clavius (Klau), Christophorus (Christoph). *DHCJ* 1:825–26.

Ioly Zorattini, Pier C. 1993. Eliano, Giovanni Battista. *DBI* 42:472–75.

Jardine, Nicholas. 1988. Epistemology of the sciences. In *The Cambridge history of Renaissance philosophy*, eds. Charles B. Schmitt, Quentin Skinner, Eckhard Kessler, and Jill Kraye, 685–711. Cambridge: Cambridge University Press.

Lamanna, Marco. 2014. Mathematics, abstraction and ontology: Benet Perera and the impossibility of a neutral science of reality. *Quaestio* 14: 69–89.

Lamanna, Marco. 2019. Benet Perera: The epistemological question at the heart of early Jesuit philosophy. In *Jesuit philosophy on the eve of modernity*, ed. Cristiano Casalini, 270–294. Leiden: Brill.

Lattis, James M. 1994. *Between Copernicus and Galileo. Christoph Clavius and the collapse of Ptolemaic cosmology.* Chicago: The University of Chicago Press.

Leite, António. 2001. Sá, Manuel de (I). *DHCJ* 4:3454.

Libois, Charles. 2001. Eliano (Romano), Giovanni Battista. *DHCJ* 2:1233–34.

Lukács, Ladislaus. 1999. A History of the Jesuit *Ratio Studiorum*. In *Church, culture & curriculum. Theology and mathematics in the Jesuit Ratio Studiorum*, ed. Frederick A. Homann, 17–46. Philadelphia: Saint Joseph's University Press.

Maryks, Robert Aleksander. 2010. *The Jesuit Order as a synagogue of Jews. Jesuits of Jewish ancestry and purity-of-blood laws in the early Society of Jesus.* Leiden: Brill.

Menato, Marco. 1997a. Basa, Domenico. In *Dizionario dei tipografi*, 82–83.

Menato, Marco. 1997b. Blado, Antonio e Paolo. In *Dizionario dei tipografi,*147–149.

O'Malley, John W. 1993. *The first Jesuits*. Cambridge, MA: Harvard University Press.

Padberg, John W. 2000. Development of the *Ratio Studiorum*. In *The Jesuit Ratio Studiorum. 400th anniversary perspectives*, ed. Vincent J. Duminuco, 80–100. New York: Fordham University Press.

Pignatelli, Antonio. 2001. Valla (Valle), Paolo. *DHCJ* 4:3879–80.

Raphael, Renée J. 2015. Copernicanism in the classroom: Jesuit natural philosophy and mathematics after 1633. *Journal for the History of Astronomy* 46:419–440.

Regina, Francesco. 2001. Pinelli, Luca. *DHCJ* 4:3138–39.

Romano, Antonella. 1999. *La contre-réforme mathématique. Constitution et diffusion d'une culture mathématique jésuite à la Renaissance (1540–1640)*. Roma: École Française de Rome.

Rose, Paul Lawrence. 1976. *The Italian Renaissance of mathematics. Studies on humanists and mathematicians from Petrarch to Galileo*. Genève: Librairie Droz.

Sachet, Paolo. 2020. *Publishing for the popes. The Roman curia and the use of printing (1527–1555)*. Leiden: Brill.

Sander, Christoph. 2014. The War of the Roses. The debate between Diego de Ledesma and Benet Perera about the philosophy course at the Jesuit college in Roma. *Quaestio* 14: 31–50.

Schurhammer, Georg. 1973. *Francis Xavier. His life, His times*. Vol. 1: *Europe 1506–1541*. Trans. M. Joseph Costelloe. Roma: The Jesuit Historical Institute.

Solà, Francisco de Paula. 2001. Perera, Benito. *DHCJ* 3:3088–89.

Sommervogel, Carlos. 1960. *Bibliothèque de la Compagnie de Jèsus. Nouvelle Edition*. 11 vols. Bruxelles and Paris, 1890; reprint Louvain: Editions de la Bibliothèque S. J.

Springhetti, Emilio. 1961. Un grande umanista messinese, Giovanni Antonio Viperano (Cenni biografici). *Helikon: Rivista di Tradizione e Cultura Classica* 1: 94–117.

The Mercurian project. Forming Jesuit culture 1573–1580. 2004. Ed. Thomas M. McCoog, S.J. St. Louis: The Institute of Jesuit Sources and Roma; Institutum Historicum Societatis Iesu.

Tinto, Alberto. 1968. *Annali tipografici dei Tramezzino*. Venice-Rome: Istituto per la collaborazione culturale; reprint Firenze: Leo S. Olschki.

Tomasi, Franco. 2015. Piccolomini, Alessandro. *DBI* 83:203–08.

Torrini, Maurizio. 1973. Giuseppe Ferroni, gesuita e galileiano. *Physis* 15:411–423.

Udías, Agustín. 2015. *Jesuit Contribution to Science. A History*. Cham: Springer.

Villoslada, Riccardo G. 1954. *Storia del Collegio Romano dal suo inizio (1551) alla soppressione della Compagnia di Gesù (1773)*. Rome: Apud Universitatis Gregorianae.

Wallace, William A. 1984. *Galileo and his sources. The heritage of the Collegio Romano in Galileo's Science*. Princeton, N J: Princeton University Press.

Zanfredini, Mario, 2001. Ferroni, Giuseppe. *DHCJ* 2:1411.

Paul F. Grendler is Professor Emeritus of History of the University of Toronto. He has published eleven books and 140 articles. They include *The Roman Inquisition and the Venetian Press 1540–1605* (1977), *Schooling in Renaissance Italy* (1989), *The Universities of the Italian Renaissance* (2002), *The Jesuits and Italian Universities 1548–1773* (2017) and *Jesuit Schools and Universities in Europe 1548–1773* (2019). His books have been awarded six prizes. He was editor-in-chief of *The Encyclopedia of the Renaissance* (6 volumes, 1999). He has received fellowships from the Guggenheim Foundation and other organizations. The Renaissance Society of America awarded him the Paul Oskar Kristeller Lifetime Achievement Award in 2017 and the Society for Italian Historical Studies awarded him its Lifetime Achievement Award. In 2014 he received the Galileo Galilei International Prize presented annually to a non-Italian who has made major contributions to Italian scholarship. In 2018 he received the George E. Ganss S. J. Award for his scholarship in Jesuit Studies.

Chapter 12
Printing Sacrobosco in Leipzig, 1488–ca. 1521: Local Markets and University Publishing

Richard L. Kremer

Abstract During the first fifty years of its printing, the greatest number of Sacrobosco *Sphaera* editions appeared in Venice, Paris, and Leipzig. Indirect evidence suggests that the Venetian and Parisian copies circulated widely; Leipzig editions, however, primarily served the local market, i.e., university students in Leipzig, Erfurt, and Wittenberg. This paper analyzes fifteen editions of Sacrobosco's *Sphaera* issued by three Leipzig printers between 1488 and ca. 1521. These Leipzig Sacroboscos share a common text, set of woodcuts, and *mis-en-page* that were not much copied by printers beyond Leipzig. The paper also investigates the Leipzig masters who lectured on Sacrobosco during these decades; several became known locally as mathematicians, but most moved to other faculties or left the university. Three of the masters authored commentaries that printers added to their editions. One of these borrowed heavily from a lengthy commentary by the Paduan master, Francesco Capuano, printed 1499 in Venice. Another commentary would be reprinted several times in Cologne, but generally the Leipzig commentaries also remained local in their influence. After Wittenberg printers began issuing Sacroboscos in 1531, no further editions would be printed in Leipzig. The Leipzig Sacroboscos thus illustrate the dynamics of a local university market shaping the early printing history of this introductory textbook.

Keywords Johannes de Sacrobosco · *Tractatus de sphaera* · Leipzig · Local book market · Scientific commentaries

R. L. Kremer (✉)
Department of History, Dartmouth College, Hanover, New Hampshire, USA
e-mail: richard.kremer@dartmouth.edu

© The Author(s) 2022
M. Valleriani and A. Ottone (eds.), *Publishing Sacrobosco's* De sphaera *in Early Modern Europe*, https://doi.org/10.1007/978-3-030-86600-6_12

1 Introduction[1]

The early printing of Johannes de Sacrobosco's (d. ca. 1256) *Sphaera* appears to follow, geographically, the emergence of printing across Europe. Thirty-six editions of the *Sphaera* were printed during the incunabula period, from the mid-1450s to 1500. Most of these editions appeared in cities that were the leading centers for the production of incunabula editions (Table 1). Indeed, we notice a roughly linear relationship between the number of Sacrobosco editions printed in a place and the total number of incunabula editions produced there...with several exceptions. Despite robust early printing in Lyon and Augsburg, these places produced no Sacroboscos before 1501; the first Lyon edition appeared in 1564 and no Sacroboscos ever were printed in Augsburg. Inversely, Leipzig ranks second for printing incunabula Sacrobosco editions but only fifth in total incunabula editions. This pattern of early printing raises the question for this study: how did the presence of a local university (lacking in Lyon and Augsburg, but present in Leipzig) shape the production of the earliest printed editions of Sacrobosco's *Sphaera*. This elementary text, written early in the thirteenth century, emerged at the same time as the universities; it soon became a standard part of the quadrivial lectures in the arts faculties and has been preserved in hundreds of manuscript copies. Leipzig thus offers an ideal case study for exploring how university masters, early printers, and what we might call "student demand" interacted to create a local market for printed Sacroboscos.

Our analysis will proceed in three steps. We will first examine the fifteen editions printed in Leipzig and identify a set of common features that are found only in these editions, i.e., that did not circulate to editions printed elsewhere. Then, we will consider the Leipzig context in which these features emerged, including university regulations, the masters who lectured on the *Sphaera*, the three printing shops that issued editions, and the demand for *Sphaera* editions at the nearby universities in Erfurt and Wittenberg. Finally, we will analyze the three different commentaries on the *Sphaera* that were printed in Leipzig and authored by three local masters. And we will speculate on why, after 1521, no further editions of Sacrobosco were printed in Leipzig even as the university continued to grow during the sixteenth century, as would printing in Leipzig. We will conclude that Leipzig offered a local university market for *Sphaera* editions and that Leipzig's printers successfully filled that niche until more prominent masters and printers in Wittenberg, by 1531, would

---

[1] For their assistance in dating the Leipzig Sacrobosco editions I thank Falk Eisermann, Oliver Duntze and Matteo Valleriani. Michael H. Shank, Razieh Mousavi, Alena Hadravová, Petr Hadrava, Bettina Erlenkamp, Kristin Lippincott, Christoph Mackert, and the other authors in this volume who generously helped me with various bibliographical questions. For hosting me in Berlin as I began work on this paper, I am much obliged to Jürgen Renn, Matteo Valleriani, and Dept. I at the Max-Planck-Institut für Wissenschaftsgeschichte. And I am honored to acknowledge all the librarians who valiantly aided my research during a time of global pandemic.

Table 1 Printing Sacrobosco during the incunabula period[2]

Location	*Sphaera* editions	Total editions	Percent of *Sphaera* editions	Percent of 28,000 incunabula editions
Venice	12	4359	33	16
Paris	8	3872	22	14
Lyon	0	1635	0	6
Cologne	1	1753	3	6
Leipzig	9	1535	25	5
Strasbourg	2	1378	6	5
Augsburg	0	1367	0	5
Milan	1	1223	3	4
Bologna	2	637	6	2
Ferrara	1	130	3	<1

take over the local market with larger and more complex compilations of material on the *Sphaera*.

2   The "Leipzig Sacrobosco"

Although issued by three different printers and including commentaries by three different masters, the fifteen known *Sphaera* editions printed in Leipzig between 1488 and ca. 1521 (Table 2) feature a unique version of the text, a unique set of woodcut illustrations, and a nearly unique format that comprise what I will call the "Leipzig Sacrobosco." For his Sacrobosco edition, Lynn Thorndike in 1949 collated seventeen manuscripts and an unspecified group of early printed editions (the latter are noted only as "ed" in his apparatus and are not otherwise specified) (Thorndike 1949). The unique features of the Leipzig Sacroboscos do not appear in Thorndike's edition or in any of the pre-1488 printed editions that I have examined (Sacrobosco 1478, 1480; Sacrobosco et al. 1482, 1485). Apparently, the unique Leipzig elements, created by the initial Leipzig edition of 1488, were copied conscientiously by the subsequent Leipzig editions. Leipzig printers paid little attention to the editions issued by Venetian or other early printers. The Leipzig Sacroboscos also are unique in that they generally do not add other texts (e.g., the medieval *Theorica planetarum*, Georg von Peurbach's (1423–1461) *Theoricae nova planetarum* of 1456, Johannes Regiomontanus's (1436–1476) *Disputationes* of 1470) to their editions (Valleriani 2019, 5–7). Only two Leipzig editions, presenting a commentary by Conrad Tockler (1470–1530), include a short text ascribed to Thābit ibn Qurra (826–901) but, according to historian Barbara Obrist, probably compiled in the Latin West during the thirteenth century. Entitled *De recte imaginatione spere*

---

[2] Digital repositories: Sphaera Corpus*Tracer*; Gesamtkatalog der Wiegendrucke (henceforth GW).

Table 2 *Sphaera* editions printed in Leipzig[3]

Date	Printer	Commentary	Extant copies	Bibliography
1488	Martin Landsberg (d. 1523)		11	GW M14585 (Sacrobosco 1488)
1489	Konrad Kachelofen (1450–1528)		37	GW M14579 (Sacrobosco 1489a)
1489	Konrad Kachelofen		4	GW M14581 (Sacrobosco 1489b)
1494	Martin Landsberg		10	GW M14584 (Sacrobosco 1494)
1495	Martin Landsberg (A1)	Wenzel Faber (1455–1518)	26	GW M14582 (Sacrobosco 1495)
1498	Wolfgang Stöckel (1473–1539)		9	GW M14591 (Sacrobosco 1498)
1499	Wolfgang Stöckel	Wenzel Faber	12	GW M14592 (Sacrobosco 1499)
1503	Martin Landsberg (C)	Conrad Tockler (1470–1530)	7	VD16 J 711 (Sacrobosco et al. 1503)
1503	Wolfgang Stöckel	Casper Jacob (1480–1530)	1	VD16 ZV 29,655 (Sacrobosco 1503a)
1509	Martin Landsberg (C)	Conrad Tockler	4	VD16 J 714 (Sacrobosco et al. 1509)
1510	Martin Landsberg (D)		4	VD16 ZV 8725 (Sacrobosco 1510)
ca. 1515	Martin Landsberg		4	VD16 ZV 8724, GW M14589 (Sacrobosco ca. 1515)
ca. 1516	Martin Landsberg (D)	Wenzel Faber	3	VD16 ZV 20,627, GW M14583 (Sacrobosco and Faber ca. 1516)
ca. 1519	Martin Landsberg (D)	Wenzel Faber	12	VD J 716, GW M14587 (Sacrobosco and Faber ca. 1519)
ca. 1521	Martin Landsberg (E)	Wenzel Faber	2	VD16 J 715 (Sacrobosco and Faber ca. 1521)

et circulorum eius diuersorum, this text emphasizes imagined rather than drawn or physical representations of the world. *De recte imaginatione* had quickly become

---

[3] Martin Landsberg's printer marks in column 2 were taken from (Hamel 2014, 142–143), extant copies were identified from GW and "Verzeichnis der im deutschen Sprachbereich erschienenen Drucke des 16. Jahrhunderts" (henceforth VD16).

part of what historian Olaf Pedersen called the "corpus astronomicum," taught by faculties of arts in European universities.[4]

The unique Leipzig text is defined by particular additions and deletions not found in earlier Sacrobosco copies. For example, in chapter three, discussing the risings and settings of signs, Sacrobosco quoted Ovid's *Ex Pontus*, 1.8.28: "Quatuor autumpnos Pleias orta facit." All the Leipzig editions insert here another phrase from *Ex Pontus*, 1.2.25: "Cum sumus in ponto cum frigore cumque sagittis," expanding Ovid's complaint about his exile.[5] Later in chapter three, as Sacrobosco explained changing lengths of the days, he wrote: "… quando sol est in signis septentrionalibus; sed econverso, quando sol est in signis australibus." The Leipzig editions (inadvertently?) omit the second reference to the Sun, "sed est econverso quando est in signis australibus" (Thorndike 1949, 96, 103; Sacrobosco 1488, b8v, c5r). Such small variations presumably do not signal any editorial intention to revise Sacrobosco's text; but they do suggest that the Leipzig printers worked exclusively with earlier printed Leipzig editions and did not seek to "control" their text against non-Leipzig editions as they printed new editions through 1521.

The unique set of designs for the woodcuts in the Leipzig Sacrobosco is more significant. In her extensive study of the diagrams in thirteenth- and fourteenth-century manuscript copies of the *Sphaera*, art historian Kathrin Müller found considerable variety in the graphical additions to the text. Her first group of manuscripts has no diagrams. A second group (including the earliest known manuscript copy from c. 1240) offers one diagram, at the only point where Sacrobosco explicitly mentions a figure, "…in presenti figuratione continetur," depicting Aristotle's (384–322 BCE) cosmos of nine concentric spheres for the planets, fixed stars and *primum mobile* (Fig. 1). Dating from the second half of the thirteenth century, a third group inserts three more diagrams (climate zones, epicycle producing stationary points and direct/retrograde motion, and eclipses). Müller's fourth group adds another fifteen diagrams, with a Cambridge manuscript preserved at the Cambridge University Library (CUL), dated to 1276, providing the earliest copy of this material (CUL, shelfmark Li III 3). Finding the same colors and graphical designs in several late thirteenth-century copies of her fourth group, Müller concluded that an "early standardization of the diagram schema" [Gestaltung] had developed in the manuscripts (Müller 2008, 207–210).

For whatever reasons (cost, lack of knowledge of manuscripts from the fourth group, lack of technical ability to insert woodcuts into the type block?), the earliest printed editions of Sacrobosco include no woodcuts but rather offer large blank

---

[4] (Carmody 1956, 118–119; Thorndike and Kibre 1963, col. 924; Pedersen 1978, 319; Rashed 2009; Hasse 2016, 405–406). (Carmody 1960, 118–119, 140–144) describes this text, extant in nearly 100 manuscripts, as "Arabic unknown. Definitions and instructions in elementary astronomy, facts, and terminology." He judges Tockler's edition to have "no value" as a source for establishing the text, which he edits. The text was printed once more, in 1518, inserted at the end of the imprint as a lone quire and not found in all copies. See (Ps. Ptolemy et al. 1518, unfoliated sheet after colophon; Hamel 2014, 95–96).

[5] Ovid's full sentence reads: Hic me pugnantem cum frigore cumque sagittis/cumque meo fato quarta fatigat hiems (Tissol 2014, 48 and 34).

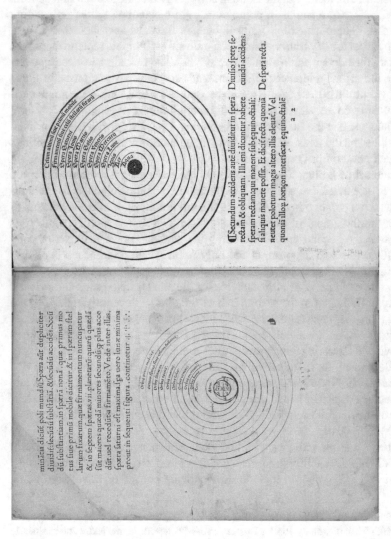

414   R. L. Kremer

Fig. 1 Concentric sphere diagrams in the early printed editions. Left: manuscript image. Venice, Florentinus de Argentina, 1472 (Sacrobosco 1472, 1v). Library of Congress, Rare Book and Special Collections Division. https://lccn.loc.gov/91127823; Right: woodcut image. Venice, Franz Renner, 1478 (Sacrobosco 1478, a2r). Courtesy of History of Science Collections, University of Oklahoma Libraries; copyright the Board of Regents of the University of Oklahoma

spaces in the page, affording readers a place to inscribe their own diagrams (Fig. 1, left). According to bibliographer Jürgen Hamel, woodcuts first appear in two 1478 Venice editions, printed by Franz Renner (d. 1494), who had come from Heilbronn and issued large, fine editions starting in 1471, and Adam von Rottweil (fl. 1470–1500), who had worked with Johannes Gutenberg (1397–1468) and in 1477 began printing in Venice (Hamel 2006, 114; 2014, 73–74). These two Venice editions each present three cuts (concentric spheres, climates, eclipses) of similar size and design; but the cuts differ, suggesting that each printer made his own blocks, one probably copying the other (Fig. 1, right).

As discussed by Catherine Rideau-Kikuchi in this volume, another German printer working in Venice, Erhard Ratdolt (1447–1527), would produce the first Sacrobosco edition with a pictorial apparatus approaching that found in Müller's fourth group of manuscript witnesses. Earlier a book-binder in Augsburg, Ratdolt specialized in the printing of astronomical, mathematical and "scientific" texts with fine initials and illustrations. The technical innovations in his 1482 edition of Euclid (4th–3rd cent. BCE) and 1483 edition of the Alfonsine Tables have attracted considerable attention from book historians (Baldasso 2009) (Chap. 3). Ratdolt's first Sacrobosco edition, printed in 1482, contains only four illustrations, featuring a new frontispiece showing an armillary sphere and designs that differ from those in the 1478 Venice editions (Sacrobosco et al. 1482). Ratdolt's next Sacrobosco edition, printed in 1485, repeats the 1482 frontispiece but massively expands the visual material from three to twenty-four individual woodcuts (Sacrobosco et al. 1485). The three earlier images (concentric spheres, climates, eclipses) are presented in full-page width (10×10 cm cuts); the remaining twenty-one cuts range in size from 3.5×3.5 cm to 10×3.75 cm. Ratdolt's twenty-four cuts differ, in content and design, from those found in Müller's fourth group of manuscripts (I use CUL, shelf mark Li III 3 to represent the "standard" manuscript set). Obviously, Ratdolt independently designed the cuts for his 1485 edition.[6]

Ratdolt's 1485 cuts provided models for the diagrams appearing in two 1488 editions, one printed in Venice by Johann Santritter (fl. 1480–1492) (Sacrobosco et al. 1488), the other by the first Leipzig Sacrobosco printer, Martin Landsberg (Sacrobosco 1488) (Fig. 2). Isabelle Pantin has examined the latter two editions, but she did not notice their reliance on Ratdolt's edition or the significance of the differences visible in the 1488 cuts.[7] Table 3 collates the diagrams in these three editions, based on Ratdolt's twenty-five cuts. Ratdolt's and Landsberg's cuts are numbered sequentially; Sanstritter's follow numeration provided by Pantin. Twenty-one of Ratdolt's cuts are copied, in content, in Santritter's edition; about half of these are nearly identical in graphical form. Twenty of Ratdolt's cuts are copied in

---

[6] Ratdolt's cuts often reappear in his editions of other texts. See, for example, (Leopold of Austria 1489).

[7] (Pantin 2020, 270–272): Pantin did note that Santritter made four large and significant additions of text to his edition; I shall here ignore the woodcuts Santritter created for these additions, material that he did not take from Ratdolt.

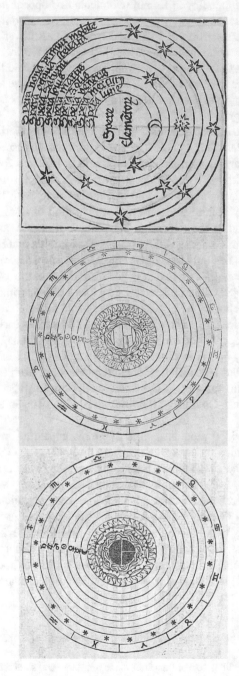

Fig. 2 Celestial and elemental spheres. Left: Ratdolt 1485 (Sacrobosco et al. 1485, 1–1v). Bayerische Stadtbibliothek. https://nbn-resolving.org/urn:nbn:de:bvb:12-bsb00036841-7; Center: Santritter 1488 (Sacrobosco et al. 1488, A9v). Courtesy of the Herzog August Bibliothek Wolfenbüttel: 16–1-astron-1. http://diglib.hab.de/inkunabeln/16-1-astron-1/start.htm; Right: Landsberg 1488 (Sacrobosco 1488, [a2]v). Courtesy of the Herzog August Bibliothek Wolfenbüttel: 16–1-astron-3. http://diglib.hab.de/inkunabeln/16-1-astron-3/start.htm

Table 3 Collation of woodcuts in Ratdolt 1485, Santritter 1488, and Landsberg 1488[8]

Content of the diagram	Ratdolt	Santritter	Landsberg
Frontispiece	1	1	1
Euclid's definition of the sphere	2	**2**	2
Theodosius's definition	3	**2**	2
Celestial and elemental spheres	4	**3**	3
Oblique and right spheres	5	**4**	4
Elemental spheres	6	**5**	5
Revolving heavens	7	**6**	6
Round heavens	8	**7**	7
Effects of refraction	9	**8**	
If the Earth is round	10	**9**	8 top
Visibility of stars at poles	11		8 bottom
Visual rays from a ship	12	10	9
Dividing the firmament in half	13		
Eye at the Earth's center	14		10
Zodiacal circle	15	12	11
Zodiacal signs	16	13	12
Colures through the poles	17	14	12
Right and oblique spheres			13
Five zones			14
Cosmic, chronic, heliacal risings	18	18	15 top
Astronomical risings and settings	19	19	15 bottom
Risings and setting of the signs			16
Circles of natural days	20	21	17
Dwellers at the equator			18
Six positions of the celestial sphere			19
Climates	21	**25**	20
Circles of the Sun	22	**26**	21
Circles of the Moon	23	**27**	22
Epicycle and retrograde motion	24	**28**	23
Eclipses	25	29, 30	24, 25

content but almost never in form by Landsberg. Landsberg's designs, we suggest, often sought to expand the explanatory content of the diagrams.

We begin our investigation of Landsberg's designs by considering the diagrams for the celestial and elemental spheres, the most commonly copied image in the

---

[8] Cuts denoted by bold font are very similar in form to Ratdolt's.

manuscript traditions. Ratdolt's and Santritter's cut prominently feature the astrological symbols for the planets and zodiacal signs and illustrate the four elemental spheres. Landsberg's cut is simpler, naming the planetary spheres in words and depicting neither signs nor elements. Landsberg also uses the same five-pointed mark for planets and fixed stars, depicting only the luminaries differently. In this diagram, placed early in the book, Landsberg appears to be reducing Ratdolt's content.

Elsewhere, however, Landsberg's cuts offer viewers more information. The next diagram in the three editions depicts the right and oblique spheres in an orthographic projection (Fig. 3). Ratdolt's diagram includes only two lines, confusingly marked "horizon obliqua" and "horizon rectus" (terms not found in the text). Santritter created two diagrams, naming the "equator" but otherwise retaining Ratdolt's design. Landsberg's cut provides two views of the celestial sphere, marked with the five parallel circles, one for the latitude of zero, the other for some latitude between the equator and pole, imaging the two situations explicitly discussed by Sacrobosco. Landsberg labels the cases "spera recta" and "spera obliqua," terms from the text. Compared with Ratdolt's, Landsberg's cut is less abstract and more elaborate with labels and additional lines.

A similar pattern appears in the most complex images to accompany Sacrobosco's text, the eclipse diagrams. The manuscripts had offered separate bird's eye views of solar and lunar eclipses, sizing the circles to show why a lunar eclipse is always visible "everywhere on earth" (actually a lunar eclipse is visible from half the earth's surface), unlike a solar eclipse visible only along a thin slice of that surface. These diagrams also mark the Moon's path around the Earth with two circles of similar diameter, intersecting to represent the lunar nodes and attempting to depict, we might say, the tilt of the lunar path out of the plane of the ecliptic (Fig. 4).

In his 1482 cuts, Ratdolt simplified these two images, confusingly placing the intersections of the two circles (lunar nodes?) not in line with the three bodies (Fig. 5).[9] In the 1485 edition, Ratdolt combined the two images into a single diagram, depicting the Sun twice and the Earth and Moon at roughly the same size. This cut does not label the nodes and presents visual rays from the Sun that do not illustrate the situation at eclipses; indeed, the two lines projecting from the Earth to the Moon on the right seem completely confused to this viewer. Santritter retained two separate diagrams with anthropomorphic images of the luminaries; but he did not depict the visual rays and, to my eye, did not illustrate Sacrobosco's point about the different visibilities of solar and lunar eclipses on the Earth's surface. Landsberg's simpler diagrams illustrate precisely that point. Going beyond the orthographic view, Landsberg uses the "figura" of the eclipse diagram, long described in canons to medieval astronomical tables and found in manuscript eclipse computations, to show what a terrestrial observer sees as the Moon moves through the Earth's shadow or across the Sun's surface (right, lower images in Fig. 5). A careful viewer, however, might note that the scales for the lower eclipse diagrams do not match the scales of the upper orthographic views in Landsberg's cuts. Unlike Ratdolt and Santritter, Landsberg

---

[9] No eclipse can occur when the Moon is 90° from its nodes, as depicted in Ratdolt's 1482 cut.

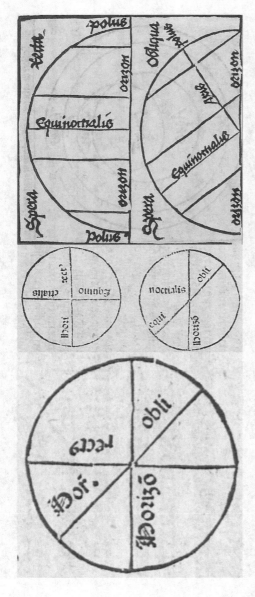

Fig. 3 Right and oblique spheres. Left: Ratdolt 1485 (Sacrobosco et al. 1485, 1-3r). Bayerische Stadtbibliothek. https://nbn-resolving.org/urn:nbn:de:bvb: 12-bsb00036841-7; Center: Santritter 1488 (Sacrobosco et al. 1488, [A10]r). Courtesy of the Herzog August Bibliothek Wolfenbüttel: 16–1-astron-1. http:// diglib.hab.de/inkunabeln/16-1-astron-1/start.htm; Right: Landsberg 1488 (Sacrobosco 1488, [a3]r). Courtesy of the Herzog August Bibliothek Wolfenbüttel: 16–1-astron-3. http://diglib.hab.de/inkunabeln/16-1-astron-3/start.htm

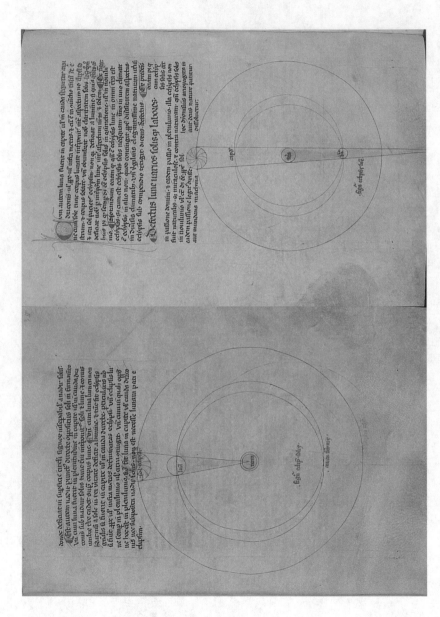

Fig. 4 Solar and lunar eclipses (CUL, shelfmark Li III 3, f. 35r–v). Reproduced by kind permission of the Syndics of Cambridge University Library

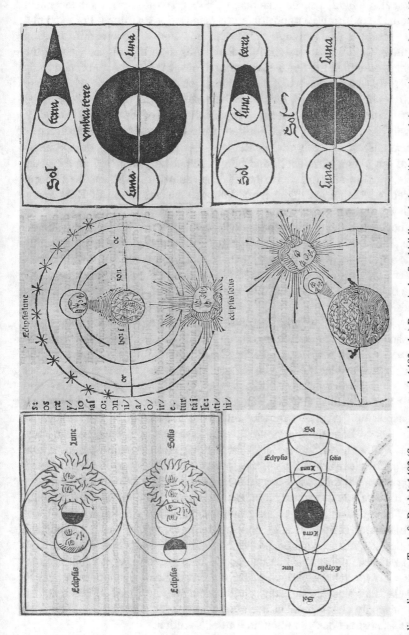

Fig. 5 Eclipse diagrams. Top left: Ratdolt 1482 (Sacrobosco et al. 1482, c1v). Bayerische Stadtbibliothek. https://nbn-resolving.org/urn:nbn:de:bvb:12-bsb000 54605-7; Bottom left: Ratdolt 1485 (Sacrobosco et al. 1485, 3–2r). Bayerische Stadtbibliothek. https://nbn-resolving.org/urn:nbn:de:bvb:12-bsb00036841-7; Center: Santritter 1488 (Sacrobosco et al. 1488, [BB12]v). Courtesy of the Herzog August Bibliothek Wolfenbüttel: 16–1-astron-1. http://diglib.hab.de/inkuna beln/16-1-astron-1/start.htm; Right: Landsberg 1488 (Sacrobosco 1488, [d7]r–v). Courtesy of the Herzog August Bibliothek Wolfenbüttel: 16–1-astron-3. http:// diglib.hab.de/inkunabeln/16-1-astron-3/start.htm

did not show the lunar nodes. Nonetheless, Landsberg's eclipse cuts add both clarity and detail to Ratdolt's abstract, flawed design.

Landsberg also included seven cuts that are either new to, or quite independent from, diagrams found in the earlier printed or manuscript traditions. For example, to illustrate Sacrobosco's discussion of the daily and annual movement of the Sun in chapter three, Ratdolt had designed a cut showing an orthographic view of the celestial sphere on which the Sun's daily motion, over the course of six months, is depicted (Fig. 6). Sacrobosco had called this motion "parallels, although they are not really circles but spirals," a description which Ratdolt's cut (as does Santritter's) seeks to show with looped lines extending beyond the edge of the sphere (Thorndike 1949, 101 and 133). Landsberg's cut, entitled "Centrum octoginta paralelli" (center of eighty parallels), completely revises Ratdolt's design. Employing stereographic projection of the planar astrolabe, Landsberg schematically depicted an astrolabe plate and rete, with the latter reduced to the off-center ecliptic divided into the zodiacal signs (Cancer and Sagittarius are labeled). Although the text speaks of 182 "circles of natural days" that are traced by the Sun's daily motion through the sky over the six months between summer and winter solstices, the diagram displays only eleven such circles. The pole, around which these circles are centered, is also prominently labeled. The astrolabe's horizon curve, however, is not shown; hence, Sacrobosco's discussion of how the Sun's annual motion gives rise to changing lengths of day and night is only implicit Landsberg's diagram. This topic, however, is ignored in Ratdolt's design.

Similarly, Ratdolt had illustrated Sacrobosco's discussion of the solar theory with a diagram showing an eccentric circle for the Sun. Two leaves later, now in Peurbach's *Theoricae novae planetarum* that he issued along with the *Sphaera*, Ratdolt illustrated Peurbach's solar theory with another cut showing the same theorica but embodied in physical realizations of orbs that Peurbach (following Ibn al-Haytham (965–1040)) had described (Fig. 7). Interestingly, Landsberg used the Peurbach design in chapter four of his Sacrobosco edition to illustrate the solar theory; but for the lunar theory, Landsberg presented Ratdolt's design of circles, not the Peurbachian orbs. Neither Landsberg nor any other Leipzig printer would issue an edition of Peurbach's *Theoricae novae planetarum*, first printed in 1474 by Regiomontanus.

In chapter three, Sacrobosco discussed the appearance of the Sun, over the course of the year, for dwellers in various climates (or geographical latitudes) of the Earth (Fig. 8). Landsberg added a unique illustration, showing the circles on the celestial sphere (pole, equator, ecliptic, tropics, and Arctic Circle) and the zenith and horizon for six latitudes between the equator and the pole. Each of these six panels is numbered but the text nowhere refers to these numbers. Likewise, the horizon (bottom line of each panel) is not labeled, which might make it difficult for untrained eyes to understand the diagrams. Presumably, the woodcut was designed to be explained orally in a university classroom. Once again, Landsberg's diagram is more detailed than the sparse, abstract designs found in Ratdolt's edition.

Leipzig's most prolific incunabula printer, Konrad Kachelofen, issued only two, nearly identical, Sacrobosco editions, both in 1489 (Sacrobosco 1489a, b). He used the same blocks in both editions and their designs consistently replicate those in

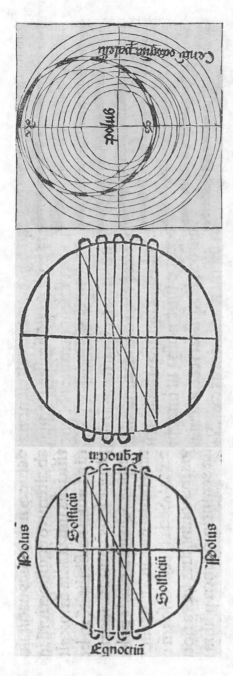

Fig. 6 Circles of natural days. Left: Ratdolt 1485 (Sacrobosco et al. 1485, 2–4r). Bayerische Stadtbibliothek. https://nbn-resolving.org/urn:nbn:de:bvb:12-bsb 00036841-7; Center: Santritter 1488 (Sacrobosco et al. 1488, BB5r). Courtesy of the Herzog August Bibliothek Wolfenbüttel: 16–1-astron-1. http://diglib.hab.de/ inkunabeln/16-1-astron-1/start.htm; Right: Landsberg 1488 (Sacrobosco 1488, [c4]r). Courtesy of the Herzog August Bibliothek Wolfenbüttel: 16–1-astron-3. http://diglib.hab.de/inkunabeln/16-1-astron-3/start.htm

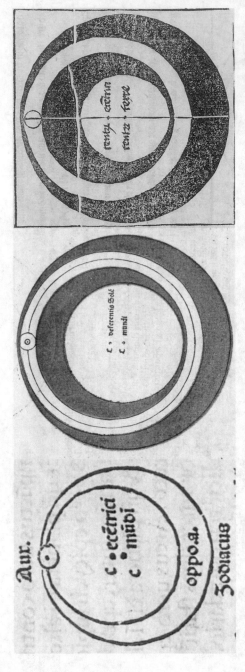

Fig. 7 Solar theory. Left and center: Ratdolt 1485 (Sacrobosco et al. 1485, 3r, 3–2v). Bayerische Stadtbibliothek. https://nbn-resolving.org/urn:nbn:de:bvb:12-bsb00036841-7; Right: Landsberg 1488 (Sacrobosco 1488, [d5]r). Courtesy of the Herzog August Bibliothek Wolfenbüttel: 16–1-astron-3. http://diglib.hab.de/inkunabeln/16–1-astron-3/start.htm

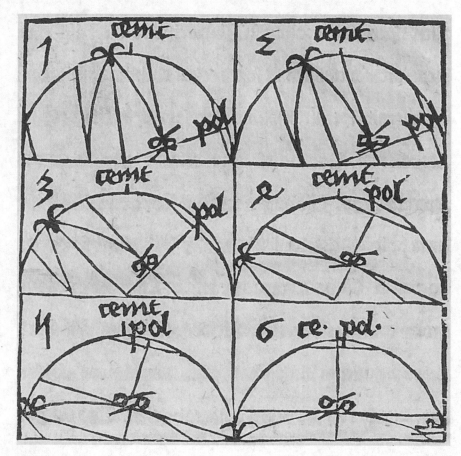

Fig. 8 Six positions of the celestial sphere. Landsberg 1488 (Sacrobosco 1488, [c7]v). Courtesy of the Herzog August Bibliothek Wolfenbüttel: 16–1-astron-3. http://diglib.hab.de/inkunabeln/16-1-astron-3/start.htm

Landsberg's 1488 edition. A "Leipzig look" for Sacrobosco's textbook had emerged already by 1489.

Landsberg's second and third editions (with Wenzel Faber's commentary) issued in 1494 and 1495, respectively, (Sacrobosco 1494, 1495), feature a new set of woodcuts, slightly enlarging the set of his first edition by expanding the six panels of cut 19 (Fig. 8) into three separate cuts, each with two panels, and adding one new diagram (Fig. 9). Not found in the standard manuscript images or in Ratdolt's early editions, this latter image illustrates Sacrobosco's very brief mention, in chapter one, of the motion of the eighth sphere backward against the sphere of the "last heaven" by "one degree in a hundred years" (Ptolemy's (b. 100) theory of precession). Sacrobosco offered no geometrical explanation for this motion; Landsberg's new woodcut, however, shows two small circles on the ecliptic, near the first points of Aries and Libra, a mechanical scheme usually attributed to Thābit and variously

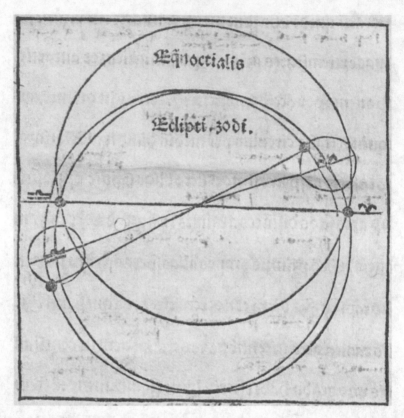

Fig. 9 Two contrary movements of the spheres plus precession. Landsberg 1494 (Sacrobosco 1494, [a5]v). Bayerische Stadtbibliothek. https://nbn-resolving.org/urn:nbn:de:bvb:12-bsb00029417-1

understood by medieval astronomers (Neugebauer 1962; Dobrzycki 2010). Presumably, the Leipzig lecturers had to explain this complex diagram to the students (see below) for Landsberg's 1494 edition offers no commentary.

Landsberg's 1494 edition also modified or added content to several of his earlier designs. First, he redesigned the frontispiece, removing the banderole ("Corpus spericum") and the text identifying the major circles on the celestial sphere and replacing the figure of Atlas, supporting the sphere on his back, with two elegantly draped angels embracing the sphere in their outstretched arms (Fig. 10). This "Christianized" design would appear in all subsequent Leipzig frontispieces (Hamel 2006, 119). Second, Landsberg returned to Ratdolt's design for Euclid's and Theodosius's (347–395) definitions of the sphere (Ratdolt 2 and 3 in Table 3). He removed the labels on the "zodiacal circle" diagram but added cartographic content to the "five zones" and the "climates" diagrams. Third, he reconfigured the "circles of natural days" (Fig. 6). Replacing the orthographic and stereographic projections, respectively, of Ratdolt's and Landsberg's earlier images, the revised design represents Sacrobosco's "spirals" with a bird's eye view of the sphere from a vantage point about halfway between the

Fig. 10 "Christianized" frontispiece in Landsberg 1494 (Sacrobosco 1494, A1v). Note that the zodiacal signs are incorrectly placed on this cut; given the slant of the ecliptic, what is shown as Aries should be Virgo, Taurus should be Leo, etc.[10] Bayerische Stadtbibliothek. https://nbn-resolv ing.org/urn:nbn:de:bvb:12-bsb00029417-1

---

[10] Did Leipzig students notice this fundamental error, rather ironic for a book on the *Sphaera*? Misplaced zodiacal signs on woodcuts of the *Sphaera* are not uncommon. See (Regiomontanus 1496: a3v).

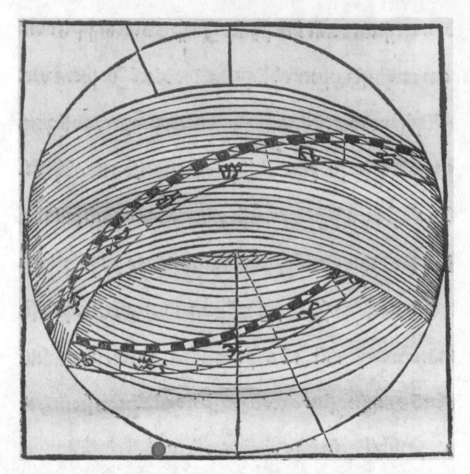

Fig. 11 Circles of natural days. Landsberg 1494 (Sacrobosco 1494, E1v). Bayerische Stadtbiblio-thek. https://nbn-resolving.org/urn:nbn:de:bvb:12-bsb00029417-1

lower pole and the plane of the equator (Fig. 11; see below for Faber's commentary on this diagram). I do not recall ever seeing such a design in medieval illustrations of the sphere!

The third Leipzig printer of Sacrobosco, Wolfgang Stöckel, issued his first edition in 1498 (Sacrobosco 1498). Its twenty-eight cuts carefully copy the designs of Landsberg's second set and are inserted at the same places in the text.[11] Stöckel's 1499 (Sacrobosco 1499) and 1503 (Sacrobosco 1503a) editions feature the same woodblocks.[12] Landsberg also used the same blocks for the eight editions he issued from 1495 to ca. 1521. Interestingly, for his first set of blocks, Landsberg had xylographically reproduced the text in the diagrams. However, for his second set,

---

[11] The collation of the editions differs, however. Landsberg 1494 is AG6, Stöckel 1498 is AE6FG4.

[12] Interestingly the 1499 and 1503 editions rotate the "circles of natural days" block by 180 degrees.

the text is typeset in the blocks. Abbreviations and placement of the typeset text in the diagrams vary among Landsberg's later editions, which suggests that he had to rework the blocks from time to time. But he managed to print nine editions with the second set of blocks. Incunabula woodblocks were known to be robust. For example, printing historian Christoph Reske has calculated that the half-page blocks in the *Nuremberg Chronicle* survived at least ten thousand impressions, the smaller cuts more than 25,000 impressions (Reske 2009, 76). We have no evidence concerning the size of Landsberg's Sacrobosco print runs. But even if he had printed as many as three-hundred copies per edition, the nine editions would have required his blocks to survive only 2,700 impressions.[13] Some breakage is noticeable in his later editions; but most of the blocks show little deterioration over the nine editions.

One final feature appears in the "Leipzig Sacrobosco," namely the leading of their forms or extra space set between the lines of text so that students could add interlinear glosses. Manuscript copies of university texts had long been prepared with wide margins and line spacings to accommodate student annotation. This format, by the 1470s, was taken up by early printers north of the Alps, especially in Leipzig. Printing historian Holgar Nickel has observed that small, leaded editions (twenty leaves or fewer) of classical Latin authors were a "hallmark" of early Leipzig printing.[14] All fifteen Leipzig Sacroboscos have leaded settings for the main text and compressed settings for the surrounding commentaries when present (Fig. 12). I have not found a similar concentration of this format in other cities where Sacrobosco editions were printed, especially not in Venice. Clearly, the Leipzig printers intended their leaded Sacrobosco treatises for the local university market.[15]

Hence, in text, diagrams, and *mis-en-page*, the "Leipzig Sacrobosco" remained quite stable over the fifteen editions. And they seem deliberately designed for the university's arts students, reading the *Sphaera* under the guidance of a lecturer.[16] The most significant change in the Leipzig editions, printed between 1488 and ca. 1521, would be the addition of commentaries by Wenzel Faber in 1495 and by Conrad Tockler and Caspar Jacob in 1503. Before analyzing these textual additions, we must consider the context in which the Leipzig *Sphaera* were produced, a setting quite different from the other leading centers of early printing such as Venice or Paris.

---

[13] Estimating print runs in the early decades of printing remains extremely difficult. See (Eisermann 2017, 178).

[14] Using GW's definition of editions "mit Durchschuss," I find that twenty percent of Leipzig's 1500 incunabula editions were leaded, far more than in other major printing centers such as Venice (0.3 percent of 4,300 editions), Paris (3 percent of 3,875 editions), Cologne (3 percent of 1,750 editions), Lyon (0.8 percent of 1,600 editions) or Augsburg (0.2 percent of 1,375 editions). Some surviving copies of these leaded textbooks are filled with student annotations. Historian Jürgen Leonhardt has even found examples of several copies of the same edition filled with identical marginalia, suggesting their original owners took dictation from the same lecturer (Nickel 1989, 135; Leonhardt 2004).

[15] For examples of classroom dictation, recorded in interlinear and marginal glosses in humanistic texts issued in leaded format by Leipzig printers, see (Bräuer et al. 2008).

[16] Note, however, that most of the surviving copies I have examined are clean, not annotated by early owners. This may suggest that sixteenth-century bibliophiles preferred unannotated books for their collections, or that Leipzig students were less than assiduous in reading Sacrobosco.

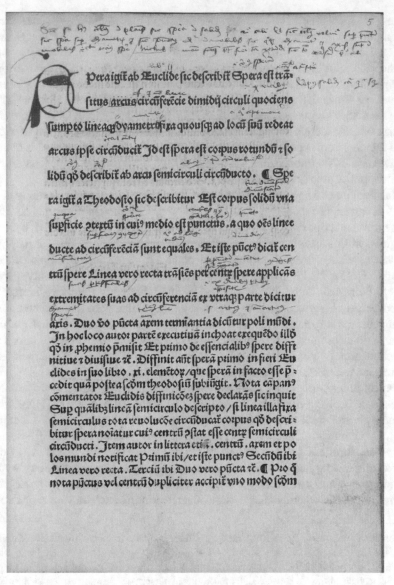

Fig. 12 Landsberg 1495 (Sacrobosco 1495, A5r), with leaded text, compressed commentary, and glosses and initial by an early reader. Courtesy of the Herzog August Bibliothek Wolfenbüttel: 171-quod-14. http://diglib.hab.de/inkunabeln/171-7-quod-14/start.htm

3 The *Alma Mater Lipsiensis* as a Local Market for the *Sphaera*

Founded in 1409 by two-thousand German students and masters who had withdrawn from the Bohemian university in Prague, Leipzig's university, the sixth to be chartered in the Holy Roman Empire, soon established itself among the Central

European universities.[17] The earliest extant statutes, from 1436 (unchanged in the 1471 reforms), show that the new university's curriculum and organization were quite typical for a fifteenth-century university.[18] Baccalaureate students were required to study three mathematical topics: the material sphere, arithmetic, and *computus*. Yet as is well known, university statutes do not always describe university practices. Not until 1502 (after another reform) do lectures on the material sphere, mathematics, arithmetic, music, and theory of the planets (*theorica planetarum*) regularly appear in the annual reports of Leipzig's deans (Table 4).

Despite gaps in the faculty records, Leipzig student notebooks provide scattered evidence that lectures on the *Sphaera* were held in the fifteenth century.[19] In the early 1440s a student's list of lectures he had attended to qualify for the baccalaureate examination (*cedulae actuum*) includes the *Sphaera* but no other mathematical topics. A similar document, prepared around 1451, lists Sacrobosco as well as John of Murs (1290–1351) on music and arithmetic, the *Theorica planetarum*, Euclid and Aristotle's *De caelo et mundo*. Another list from 1464 includes the *Sphaera*, arithmetic, and Euclid. Virgilius Wellendorffer's (1495–1534) lecture notes, copied in 1486, contain a richly illustrated and annotated Sacrobosco with an explicit indication that Master Wenzel Faber had, at the request of several students, decided to resume his lectures on the *Sphaera*. Although such documentation is sparse, we can assume that Leipzig masters in the fifteenth century had lectured, at least occasionally, on the *Sphaera*.

These lectures, however, were short and poorly remunerated. Scattered fifteenth-century evidence suggests that Leipzig's arts faculty would, during a semester, read on Aristotle's Ethics in 106 lectures, metaphysics and physics (100 lectures each), politics (86), *De caelo et mundo* (53), Euclid (74), and Johannes Peckham's (1230–1292) Optics (54). Mathematical subjects received less attention: twenty-two lectures for the *Sphaera*, twenty for theory of the planets, seventeen for music, twelve for arithmetic. With lecture fees pegged to the number of lectures, it is not surprising that the arts faculty had difficulty convincing its masters to lecture on Sacrobosco. For most of the fifteenth century, Leipzig's deans assigned elementary lectures by means of a lottery, forcing masters to cover the low-paid subjects like mathematics. Apparently, this practice proved unsatisfactory; in 1499 the faculty reformed its statutes, agreeing to remunerate mathematics lecturers with a fixed salary and room and board, a strategy that helped regularize lectures on these fields, as can be seen in

---

[17] The first wave of imperial universities included Prague (founded 1348), Cracow (1364), Vienna (1365), Heidelberg (1386), Cologne (1388), Erfurt (1392), Rostock (1419), and Leuven (1425). Leipzig did not acquire its law faculty until 1457 (Bünz 2008a, 13–15).

[18] Given Leipzig's tepid response to "Renaissance humanism" and the Reformation before 1542, historians have debated the extent to which continuities or disruptions characterize its history from the via *antiqua* of the early fifteenth century to the via *moderna* of later centuries. Cf. (Rothe 1961; Döring 1990; Häuser 2009–2010, Vol. 1). For a useful overview of university curricular statutes around 1400 (note that Ptolemy's *Almagest* appears only once, in the Prague statutes), see (Lorenz 1985, 223).

[19] For more information, see (Helssig 1909, 44–45, 89–92; Bodemann 2000; Pensel 1986, 182–183). See also UBL, Ms 1348, 86r–94v, Ms 1430, rear pastedown, Ms 1470, 402r–429v.

Table 4 Leipzig University, Arts Faculty, public lectures on the Sphere[20]

Years	Lecturer	Matriculation	Master
1502	Leonhard Baumgärtner de Gersbach	1494	1500
1504	Conrad Tockler de Nürnberg (1470–1530)	1493	1501
1503–1505, 1510, 1522	Sebastian Sibart de Mügeln (d. 1538)	1493	1502
1505–1506	Cristoph Schappeler de S. Gall (ca. 1472–1551)	1498	1501
1507	Sebastian Aurifaber de Weissenberg	1499	1504
1508, 1512, 1520	Gregor Breitkopf de Konitz (1472–1529)	1490	1497
1509	Peter Christanni de Freiberg	1499	1508
1509	Rempertus Giltzheim de Brunswick (d. 1532)	1499	1508
1511	Bartholomaeus Nägele de Lindau (ca. 1486–1520)	1500	1507
1513	Martin Schlautiz de Schleiz	1501	1510
1514–1515	Alexander Seckler (Birkhamer) de Esslingen	1494	1501
1515, 1518	Franciscus Richter de Hainichen (d. 1554)	1497	1508
1516, 1518	Henning Feuerhahn de Hildensemensis (d. 1546)	1507	1510
1517	Jakob Lincke de Glogovia (bacc. Cracow)	1510	1513
1518	Simon Eyssenman (fl. 1509–1519)	1505	1509
1523–1524	Caspar Börner (1492–1547)	1507	1518
1547–1553	Christophorus Kanisi (Montag) (d. 1554)	1521	1529
1555	Moritz Steinmetz (d. 1584)	1541	1550

Table 4. However, most of the masters in Table 4 lectured only once or twice on the *Sphaera*; they were not, it would appear, especially eager to lecture on Sacrobosco, despite the revised university statutes.[21]

---

[20] (Erler 1895–1902: passim). I find nothing on the *Sphaera* in the lists of lectures for 1527 and 1528 recorded in UBL, Ms 1470, 181r–184v.

[21] (Zarncke 1861, 455–456; Helssig 1909, 36–37, 53–61; Schmidt-Thieme 2002, 196; Bünz 2008a, 27). In 1521, the Wittenberg arts faculty had established two chairs for mathematics; not until 1542 would Leipzig, as part of its reform under Philipp Melanchthon's influence, create similar chairs, naming Georg Joachim Rheticus as *mathematice maiores* and Balthasar Klein as *mathematice minoris*. These professors, however, apparently did not lecture on the *Sphaera*. See (Erler 1895–1902, Vol. 2, 669; Stübel 1879, 548; Burmeister 2015, 292).

Who were these reluctant masters? Several would develop, at the least, local profiles in mathematics. Tockler, author of one of the Sacrobosco commentaries, we shall consider below. Simon Eyssenmann (fl. 1509–1519), who in 1518 served as dean of the arts faculty and rector of the university, authored an arithmetic textbook in 1511, a *computus* textbook in 1514, six annual German *Practica* and three Latin *Judica Lipsense* between 1514–1520. Although these treatises were mostly printed in Leipzig and were linked to the university, versions of his German Practica were also printed in Landshut, Nuremberg, Augsburg, and Lübeck. In 1522, Eyssenmann became Doctor of Medicine in Ingolstadt and eventually a city physician in Kaufbeuren (Swabia). Caspar Börner (1492–1547), who also had studied in Wittenberg with Martin Luther (1483–1546) and Philipp Melanchthon (1497–1560), would spend his life in Leipzig, first as a teacher in the Thomasschule and from 1539 in the arts faculty, energetically advancing its Protestant reform; collecting mathematical instruments, globes, and maps; and correcting Melanchthon's 1538 edition of Sacrobosco's *Computus* (Sacrobosco and Melanchthon 1538). Moritz Steinmetz (d. 1584) also remained in Leipzig until his death, teaching mathematics and botany in the arts faculty, publishing an arithmetic, a Greek and Latin edition of Euclid, a German *Practica* for 1565, and a historical survey of comets.[22] Nonetheless, not until Georg Joachim Rheticus (1514–1574) arrived in 1542 (after overseeing the printing of Nicolaus Copernicus's (1473–1543) *De revolutionibus* in Nuremberg) would Leipzig's faculty include a mathematician of wider reputation (Danielson 2006, 103–114; Burmeister 2015, 47–52).

Other Sacrobosco lecturers became known humanists or (anti) reformers. Bartholomaeus Nägele (ca. 1486–1520) was among that shadowy group of Leipzig masters who in 1515–1519 published anonymous, satirical attacks on scholastics and monks (*Epistolae obscurorum virorum*) that helped to pave the way for Luther's movement. In the early 1520s, Christoph Schappeler (ca. 1472–1551) preached radical reform in Swabia and co-authored a tract that rallied support for what eventually became the Peasant's War. Gregor Breitkopf (1472–1529), who lectured for twenty-five years in Leipzig's theology faculty, published editions of classical authors, but remained Catholic and authored a polemic against the Anabaptists. Henning Feuerhahn (d. 1546), who spent his life in the arts faculty, published new Latin poetry and co-edited anti-Lutheran polemics with Breitkopf.

Christophorus Kanisi (d. 1554) lectured seven times on the *Sphaera*, as a member of the theology faculty, but did nothing further to distinguish himself. Several Sacrobosco lecturers left the university to practice medicine (Giltzheim, Aurifaber). The remaining masters left Leipzig and disappeared from the historical record (Baumgärtner, Sekler, Sibart, Christanni, Schlautiz, Richter, Lincke). Lecturers on Sacrobosco, during the first half of the sixteenth century in Leipzig, did not tend to become leading stars in the university's firmament.

---

[22] (Zoepfl 1961; Kößling 2003, 45–74; Döring 2014, 191–212; Burmeister 2015, 112–114; Schwarzburger 1959, 355–356). For Börner's thirty-four canons on the celestial globe of Johann Schöner, see UBL, Ms 1489, 9r–33r (dated 1516).

A final bit of evidence about Sacrobosco at the university comes from public lecture notices. As is well known, medieval universities had, since their origins, publicly announced official news and lectures by hanging manuscript broadsides at designated locations ("das schwarze Brett" in German areas) in their towns. With the advent of printing, masters commissioned printed broadsides to announce their lectures. Survival rates for such ephemeral materials are, of course, quite random. For Leipzig, the few surviving announcements are mostly for humanistic lectures, starting in the 1460s for manuscript broadsides, in the 1490s for printed sheets. In 1506, the faculty formalized this practice in their statutes: "Notices of lectures by any faculty that are to be presented for fees in a public hall, should be printed and posted in many public places both inside and outside the city before the semester begins." Leipzig's only extant announcement for a mathematical lecture is for Sacrobosco (Fig. 13). The public lecture, in the winter semester of 1506–1507, would be held at the eleventh hour in a university hall. Sacrobosco's text was available from the Leipzig printer, Martin Landsberg, who also printed the announcement. The lecturer, Conrad Tockler, attached this printed notice to his personal copy of the 1499 Venice edition of Sacrobosco, noting below the printed text that he had offered the lectures

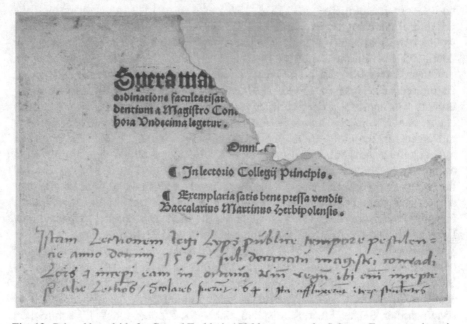

Fig. 13 Printed broadside for Conrad Tockler's 1506 lectures on the *Sphaera*. Front pastedown in an edition of texts on the *Sphaera*, printed by Simon Bevilacqua in Venice in 1499 (Sacrobosco et al. 1499) (Tockler's personal copy). UBL: Astron.15. https://nbn-resolving.org/urn:nbn:de:bsz: 15-0013-221052

to sixty-four auditors, despite an outbreak of plague in Leipzig (only 118 students had matriculated that semester at the university).[23]

Into this market of sporadic lectures, unenthusiastic lecturers, confessional strife, and even plague, the three Leipzig printers issued fifteen editions of Sacrobosco's *Sphaera*. Martin Landsberg, who printed the earliest edition in 1488, would become Leipzig's most prolific incunabula printer (about 480 editions); after 1500, he printed another 430 editions before his death in 1523. Originally from Würzburg, Landsberg had matriculated at the Leipzig University in 1472, earned a BA in 1475, and by 1485 began issuing imprints without dates or printer marks. His earliest extant edition is a German calendar (GW M15996). In the same year, he began printing annual calendars and astrological prognostications authored by the Leipzig master, Wenzel Faber, who would become the most widely published calendar maker of the incunabula period. Landsberg quickly established himself as the leading Leipzig printer of classical texts for the university lecturers (he lived and printed directly across the street from the university buildings); he also printed many contemporary authors and after 1518 became Leipzig's leading printer of both Reformed and anti-Reform titles (Geldner 1968–1970, Vol. 1, 245–246; Claus 1973, 108–109; Reske 2007, 515; Lehmstedt 2019, 96–105). The Sacrobosco thus represented a very small portion of Landsberg's massive production.

The next two Leipzig editions, both dated to 1489, were printed by Landsberg's rival, Konrad Kachelofen. As noted above, Kachelofen (who had not matriculated at a university) copied the number and design of Landsberg's woodcuts and placed them identically within the text (but he abbreviated the text and broke the pages differently). Kachelofen printed in Leipzig from 1484 until 1517. He used more than a dozen type sets and issued a wide range of titles, including humanistic university texts, but only about half as many editions of annual almanacs and *practica* as did Landsberg (Geldner 1968–1970, 241–244; Reske 2007, 514–515; Lehmstedt 2019, 85–94). In 1488, Wenzel Faber, together with Landsberg, provoked a public feud with another local calendar-maker, Paulus Eck (ca. 1440–ca. 1509) and his printer Kachelofen, resulting in dueling broadsides, accusations of failed astrological predictions, and indications of a rising (Bohemian) professor publicly ridiculing a (Bavarian) student who dared to publish annual almanacs. Perhaps this public attack deterred Kachelofen from printing further mathematical titles for the university; in any case, despite his major investment of twenty-four woodcuts for his 1489 Sacroboscos, Kachelofen apparently never again printed the *Sphaera*.

The third Leipzig printer to issue Sacroboscos was Wolfgang Stöckel, who had earned a BA in 1490 at the university in Erfurt. In 1495, he matriculated in Leipzig, and later that year married the widow of a Leipzig printer and took over the business. Stöckel printed in Leipzig from 1495 until 1526 when, as a result of financial and political problems, he moved to Dresden where he printed until his death in 1540. In Leipzig, he issued mostly university texts until 1518 when he abruptly shifted his

---

[23] (Sudhoff 1909, 87–88; Bertalot 1915; Jensen 2004; Eisermann 2004, A–513, H–536 to H–541, H–553, N–517, P–215, P–277). Interestingly, all of the ten known incunabula broadside lecture announcements are for humanistic lectures in Leipzig.

output to reformation tracts. In comparison with Landsberg and Kachelofen, Stöckel printed very few annual almanacs (2) and *practica* (18). As had Kachelofen, Stöckel copied Landsberg's woodcuts in number and design, but issued only three Sacrobosco editions, in 1498 without commentary (Sacrobosco 1498), in 1499 with Faber's commentary (Sacrobosco 1499), and in 1503 with a commentary by the Cistercian monk Caspar Jacob (Sacrobosco 1503a). The latter edition replaces the usual Leipzig title, *Opusculum...spericum*, with a title I can find in no other Sacrobosco edition: *Astronomice sciencie sphericum introductorium*.[24] Although university lectures on the *Sphaera* occurred nearly every year between 1503 and 1509 (Table 3), no further printed Leipzig editions appeared until Landsberg's in 1509. For whatever reason, Stöckel, like Kachelhofen, abandoned the *Sphaera*.[25] After 1503, Landsberg would remain the sole Leipzig printer of the *Sphaera*.

Like printers everywhere, Leipzig's sought to coordinate their selection of titles and print runs with their estimates of market demand. Printers who misjudged the market soon disappeared. Thus, we might speculate on how the fifteen known Leipzig editions of Sacrobosco correlated with the market.[26] Conceivably, Leipzig printers might have provided texts for not only local students but also for those in nearby Erfurt (roughly one-hundred kilometers from Leipzig) or Wittenberg (sixty kilometers distant), a new university founded in 1502. As can be seen in Fig. 14, Leipzig and Erfurt by 1450 had become relatively large universities, enrolling each year more than four-hundred new students (Fig. 14). Although Erfurt's matriculations declined by the 1470s, Leipzig's remained strong until 1520s when the outbreak of the Reformation and the Peasants' Wars emptied the universities of Central Europe. Interestingly, even though Erfurt printing began in the early 1480s, no edition of Sacrobosco is known from that town; and the first Sacrobosco to be printed in Wittenberg appeared in 1531. We might guess that Leipzig editions served students in all three universities until 1531 when Wittenberg's printers took over the market (forty-nine editions until 1629) (Chaps. 4 and 5). No Sacrobosco editions were printed in Leipzig after ca. 1521.[27]

As is well known, most students at late medieval universities studied only in the arts faculty; only a small portion completed a BA or MA degree. Summing the data depicted in Fig. 14, we find that about 31,000 students matriculated at the three universities between 1488 and 1531. If we assume that ten percent of these matriculants purchased a printed Sacrobosco, the fifteen Leipzig editions would have needed an average print run of about two-hundred copies to meet this demand. More research would be required to substantiate this estimate. But it does seem plausible that the Leipzig printers could have met the need at the three universities with only

---

[24] Only one of Landsberg's editions has a non-standard title; his 1510 edition with no commentary is entitled *Textus spere materialis*, echoing the title frequently announced for the university lectures (Sacrobosco 1510).

[25] Stöckel did print several other university mathematical texts, an arithmetic in 1505 and 1507, and a *computus* in 1515. See VD16 A 1875, A 1878, S 2366.

[26] I have not examined extant copies of Leipzig Sacroboscos for evidence of early ownership or provenance.

[27] For an overview of early astronomical imprints for university lecturers, see (Zinner 1964, 44–47).

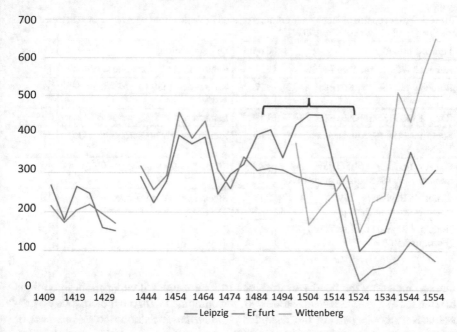

Fig. 14 Annual matriculation averaged by five-year intervals. Red bracket marks the period when Leipzig printers issued Sacrobosco editions. Author's plot[28]

fifteen editions over the forty-four years from 1488 until 1531. If fewer local students purchased Sacroboscos, we might guess that some portion of the Leipzig print runs were shipped to other university markets (e.g., Prague, Cracow, or Vienna) or were sold at Leipzig's annual book fairs. Interestingly, according to Matteo Valleriani's Sphaera Corpus*Tracer*, no edition is known that was printed in Prague, only one in Vienna (1518), and three in Cracow (1506, 1513, 1522). Not all cities with universities and print shops issued Sacrobosco editions.

The local market for Leipzig's Sacroboscos becomes more apparent when compared to the international market served by Venetian printers. A rough means to assess distribution of a printed edition is to survey surviving copies. Although sixteenth-century book databases do not yet provide reliable data on survival rates, GW does include numbers that enable comparative, if not definitive, conclusions for the fifteenth century. As can be seen in Table 5, many more Venice Sacroboscos are extant, in libraries scattered across the globe, than are Leipzig Sacroboscos (Table 5). Most of the Leipzig fewer editions are held in German, Austrian, or Polish libraries, with several in Uppsala (spoils from the Thirty Years War) and seven in London.

These details about the Leipzig arts faculty and its teaching of the quadrivium in the late fifteenth and early sixteenth centuries suggest that the Leipzig printers, in issuing fifteen Sacrobosco editions from 1488 until ca. 1521, primarily served the local

---

[28] Data from (Erler 1895–1902, Vol. 1, xc–xcvii).

Table 5 Extant copies of incunabula editions printed in Venice (four with fewer than fifteen copies not listed) and in Leipzig (From GW)

Venice editions	Copies	Leipzig editions	Copies
Renner 1478 (Sacrobosco 1478)	137	Landsberg 1488 (Sacrobosco 1488)	11
Ratdolt 1482 (Sacrobosco et al. 1482)	132	Kachelofen 1489 (Sacrobosco 1489a)	37
Ratdolt 1485 (Sacrobosco et al. 1485)	134	Landsberg 1494 (Sacrobosco 1494)	10
Santritter 1488 (Sacrobosco et al. 1488)	116	Landsberg 1495 (Sacrobosco 1495)	26
Bonetus Locatellus (for Octavianus Scotus) 1490 (Sacrobosco et al. 1490)	176	Stöckel 1498 (Sacrobosco 1498)	9
Simon de Bevilacqua 1499 (Sacrobosco et al. 1499)	243	Stöckel 1499 (Sacrobosco 1499)	12
Johannes Baptista de Sessi 1500 (Ferrariis 1500)	43		

market, including the nearby universities in Erfurt and Wittenberg. These editions offered the *Sphaera* as a stand-alone text, not combined with other elementary astronomical textbooks such as the *Theorica planetarum* or Sacrobosco's *Computus* or with modern works by Peurbach or Regiomontanus. It appears as if Leipzig's rather undistinguished lecturers desired low-cost imprints that offered their students Sacrobosco's text and little more. For further clues about how the text was used by Leipzig's masters, we turn finally to the commentaries.

4   The Leipzig Sacrobosco Commentaries

In the previous sections, we have suggested that the fifteen Leipzig Sacrobosco editions should be regarded as a local phenomenon, with Leipzig printers and masters with largely local reputations serving local students. However, might the three commentaries, printed in the Leipzig editions, have attracted wider attention and brought the Leipzig Sacrobosco into interplay with other printers and masters across Europe?

We first introduce our commentators. Wenzel Faber, from Budweis in Bohemia, would author the first Sacrobosco commentary to be published in Leipzig. He had matriculated at the university in 1475, earning a BA in 1477, an MA in 1479, and a Bachelor of Medicine in 1488. He was a member of the small Fürstenkolleg from 1483–1488, the large Fürstenkolleg from 1488 to 1508 (in these institutions, he received a regular salary); he served as rector in 1488, and as dean in 1489. The initial editions of Faber's multi-leaved annual prognostications (Latin) or *Practica* (German) were printed by Leipzig's first printer, Marcus Brandis (b. ca. 1455), from 1482–1485. But after Landsberg began printing, his shop issued essentially all of Faber's annual *Practica*, prognostications and broadside almanacs, more than 60 editions between 1488 and 1515.[29] If, as noted above, Faber had "resumed" lectures on the *Sphaera* in 1486, it seems quite likely that he had worked with the printer of his calendars, Landsberg, to produce the first Leipzig edition of Sacrobosco in 1488.

Our second commentator, Conrad Tockler—whose writings recently have been studied by Matteo Valleriani and Nana Citron—published a wider pallet of works than Faber, yet probably was not as well known. Born in Nuremberg, another major printing center, Tockler matriculated in Leipzig in 1493. In 1495, he received the BA, in 1502, the MA. From 1502 until 1510, Tockler lectured on the quadrivium (music, *Sphaera*, *Theorica planetarum*, optics, and Euclid) nearly every semester. In 1509, he became Bachelor of Medicine, in 1512 doctor of medicine. Apparently, he spent the remainder of his life in the medical faculty, but university records do not document his lectures there.[30]

Tockler worked closely with local printers to issue textbooks for Leipzig's arts students: Marsilius Ficinus's (1433–1499) *De sole* (Ficinus and Tockler 1502) with the printer Stockel; John of Murs's *Arithmeticae communis* (Muris and Tockler 1503), and Tockler's commentary thereupon as a separate imprint (Tockler 1503), both printed by Landsberg. Editions of Peter of Cracow's (1430–1474) *Computus novus* (Kremer 2007), printed in 1507 and 1511 by Landsberg, include a one-page astrological text attributed to al-Battani (858–929)—*De ortu quatuor triplicitatum secundum Conradum Noricum*—that Tockler presumably copied from the 1483 *editio princeps*.[31] In 1511, Landsberg printed two short texts authored by Tockler, canons for using tables and circular diagrams (neither given in the imprints) for simple calendrical and computistic operations. Tockler wrote commentaries on other quadrivial texts, including Peurbach's *Theorica novae planetarum* that remained in manuscript. Finally, Tockler, like Faber, published annual astrological calendars and almanacs for an extra-university audience (twenty editions from 1504 to 1515). Most were printed in Leipzig, but several Nuremberg printers issued editions, including several in Czech for distribution in Bohemia.[32] All of Tockler's annual calendars are traditional for the genre. Indeed, as Valleriani and Citron have emphasized, most of Tockler's editions are quite conventional for a late medieval lecturer on the quadrivium who was steeped in astrological medicine.

---

[29] For Wenzel Faber, see (Haebler 1915; Bruckner 1975; Eisermann 2008, 168–170; Skemer 2007; Kremer 2017, 355). For the editions here referred to, see GW and VD16.

[30] (Valleriani and Citron 2020): Three manuscript codices, written and collected by Tockler, are known. UBL, Ms 1605, ÖNB 5274 and ÖNB 5280 contain texts on quadrivial topics (including Peurbach's *Theorica nova planetarum*), astronomical instruments, sundials, and one astrological miscellany, but nothing on medicine.

[31] Leipzig printers Landsberg, Stöckel, and Johannes Thanner had printed editions of Peter's text in 1487, 1499, 1504 and 1506. Clearly the *Computus cracoviensis* accompanied university lectures on this topic. Cf. (Nallino 1977, Vol. 1, xxviii; Thorndike and Kibre 1963, col. 1449). For bibliographies of Tockler's printed works, see (Schmidt-Thieme 2002; Valleriani and Citron 2020, 135–136). *Editio princeps* of the Ficinus had appeared in Florence in 1493 (GW 9880), of the al-Battani in Venice in 1493 (GW M36394). Tockler's edition of John of Murs's arithmetic may be its *editio princeps*: see GW M2570020 (undated) and VD16 T157 (Vienna, 1515).

[32] For the broadside Czech almanacs, printed by Hieronymus Höltzel in Nuremberg, see NLCR, Teplá 503 (fragment, for 1507); Teplá 505s (fragment, for 1510) (Boldan 2008).

A final Leipzig Sacrobosco commentary, appearing in a single edition printed in 1503 by Stöckel, was authored by Caspar Jacob, a monk from the Cistercian monastery in Grünhain (about sixty kilometers south of Leipzig) (Sacrobosco 1503a). Shortly after the founding of Leipzig's university, the Cistercians had established a "Bernard College" there for their monks; similar colleges had been set up at universities across Europe, from Paris (1246), to Prague (1366), Cracow (1416), Oxford (1437), Erfurt (1443), Rostock (1444), and Greifswald (1487).[33] By providing living quarters supervised by the abbot of a nearby monastery, the Bernard colleges sought to enable monks to follow the Cistercian Rule in the non-cloistered setting of the university. Despite these colleges, and despite Pope Benedict XII's (1285–1342) bull of 1335 (*Fulgens sicut stella*) requiring houses of his order to send some brothers to universities, the Cistercians of the fourteenth and fifteenth centuries were not known for their embrace of academic study, especially of the arts curriculum. In Leipzig, about four-hundred Cistercians had matriculated before the closure of Bernard College in 1536, after the upheavals of the Reformation. Most of these monks attended during the years from 1480 to 1520. They heard lectures on the Bible and Peter Lombard's (1096–1160) *Sentences* but rarely finished a baccalaureate; only a very few became masters in the theology faculty. In 1488, however, the Cistercian masters did receive the privilege to join the arts faculty. However, despite such activity, Caspar Jacob's Sacrobosco commentary is the only example I have found of a Leipzig Cistercian interested in the quadrivium.[34] And it is one of the few known examples of a Cistercian monk, in the fifteenth or sixteenth centuries, authoring a non-theological university text.

Jacob dedicated his commentary to Father Peter Tumner (Tümpner), ordinary from the Cistercian Altzella house about eighty kilometers east of Leipzig and "provisor" (instructor) at the Bernard College. Although Jacob describes Tumner as a master of art and philosophy and a bachelor of theology, the latter does not appear in Leipzig's matriculation records and I have found no traces of Tumner in standard bio-bibliographical reference works.[35] Jacob likewise does not appear in the Leipzig matriculations; indeed, he appears only once in the university records, listed in 1501 as a "determinator" (bachelor) of the Bernard College now recognized as a master by the arts faculty.[36] He does not appear among the masters who lectured on the *Sphaera* (Table 4 above); however, the 1503 printing of his edition with commentary suggests that he was teaching the *Sphaera*, perhaps privately within Bernard's

---

[33] For more information, see (Schmidt 1899; Dietrich 1914; Bünz 2008b; Häuser 2009–2010, Vol. 1, 115–118; Erler 1895–1902, Vol. 2, 304–305).

[34] Of the 370 medieval English manuscripts of Cistercian provenance, only a handful contain content related to the trivium and quadrivium (including copies of Sacrobosco's *Sphaera*). See (Bell 1989, 83; Fitzpatrick 2010, 1 and 10). For the historiography of Cistercian ambivalence toward university study, see (Schneider 1999).

[35] (Sacrobosco 1503a, A2r; Dietrich 1914, 335). Tumner is not found in the Repertorium academicum germanicum (RAG).

[36] In 1472, a Caspar Carnificis de Grunenhayn did matriculate, whom Erler suggested may have been our Caspar Jacobi de Grunhayn (Erler 1895–1902, Vol. 1, 287 and 282, Vol. 3, 378–379).

College.[37] His commentary is preserved only in the 1503 Sacrobosco edition, itself extant in a single, lightly annotated copy.

Faber's commentary was the first to appear in Leipzig, issued by Landsberg in 1495. At several points, Faber referred to 1491 as "our time;" apparently, he wrote the commentary several years before Landsberg printed it (Sacrobosco 1495, B3v, H1v). Appropriate for the university context we described above, Faber's commentary is very "pedagogical" in form and content. Although occasionally his remarks extend over several pages, most of Faber's forty-seven interventions are short glosses, directly tied to a passage in Sacrobosco. Much of the commentary is a kind of reading guide, identifying larger topics in Sacrobosco's text, numbering points made by the earlier author, and explicating some technical terms.[38] Faber's commentary is much simpler and more straightforward than the larger, more expository commentaries of, for example, Jacques Lefèvre d'Étaples (1450–1536) of 1495 (Chap. 2) and Pedro Sánchez Ciruelo (1470–1554) of 1498 in Paris (Chap. 13), and Francesco Capuano (b. 1450) of 1499 in Venice—editions that comprise the earliest generation of printed Sacrobosco commentaries.[39]

Occasionally, however, Faber did offer additional materials or "suitably correct" (*bene correctum*) Sacrobosco (Sacrobosco 1495, H8r). Invoking Aristotle, *Physics* I—one should move from the general to the specific—Faber began with a lengthy introduction, defining astronomy and defending astrology by citing a host of medieval authors, going beyond Sacrobosco's rather sparing invocations of Ptolemy and Alfraganus (al-Farghānī) (d. 861). After Ptolemy, Faber mentioned Isidore of Seville (560–636), Albert the Great (d. 1280), Albumasar (Abu Ma'shar) (787–886), Pierre d'Ailly (1351–1420), Abraham ibn Ezra (1089–1187), Alcabitius (al-Qabisi) (d. 867), Haly Abenragel (d. 1040), and Leopold of Austria (fl. 1300). To undergird his distinction of theoretical and practical astronomy, Faber referred to Thābit, Peurbach, John Hispalensis (John of Seville) (1100–1180), Campanus of Novara (1220–1296), Almeon (al-Mamun)[40] and the Alfonsine Tables. By parading such a train of witnesses, Faber sought to elevate the status of the *Sphaera* in Leipzig's arts faculty. Seven benefits result, Faber concluded, from reading Sacrobosco's "short and easy" work: one learns about God by following the Creator's footsteps; one better understands other books on natural philosophy, such as *De caelo* or the *Meteorologica*; one becomes acquainted with the poets discussed by Sacrobosco (Virgil (70–19 BCE), Ovid (b. 43), Lucan (39–65)); one learns to understand the Alfonsine Tables, eclipse

---

[37] A 1514 inventory of the Altzella library, then called a *biblioteca publica*, includes—in addition to scholastic texts of theology, church history, and liturgy—significant numbers of humanistic and contemporary texts. Astronomical works include *De celo et mundo*, astrological classics (Albumasar, *Centiloquium*), the *Theoerica planetarum* and the Alfonsine Tables, both manuscript and the 1483 *editio princeps*. See ThULB, MNs. App. 22 A, esp. 22r–24r; (Palmer 1998; Mackert 2008).

[38] For example, Faber explained terms such as acute and obtuse angles, archetype, isoperimeter, cosimeter, homogeneity, diameter, pole, solstice, *cosmicus*, autonomastice, *chronicus*, chronicle, climate, and natural hours, often referring to Greek roots.

[39] See (Shank 2009) and the articles by Richard J. Oosterhoff, Tayra M.C. Lanuza Navarro and Elio Nenci in (Valleriani 2020, 25–110).

[40] (Pedersen 2002, 962). For the enigmatic "Almeon," see (North 1976, Vol. 3, 256–257).

predictions, and the motions of the planetary spheres; this prepares the mind to study other sciences; and one becomes worthy of being called a philosopher (citing Boethius (d. 524)) (Sacrobosco 1495, A2r–A4r).

Faber also enriched Sacrobosco by adding or expanding literary references. In considerable detail, he recited Greek myths on the origins of constellations such as Taurus, Sagittarius, the Pleiades, and the Corona borealis, referencing the classical sources. He summarized the astrological qualities of the twelve zodiacal signs and their triplicities. He explained the meaning of the term "pole" in the Christian liturgical chant for the Feast of the Ascension (*Iam Christus ascendit polum*). Citing Macrobius (370–430 BCE), he explained that the city of Syrene lies directly on the Tropic of Cancer because the Sun casts no shadows there at summer solstice. Glossing Sacrobosco's final sentence about Dionysius the Areopagite (5th–6th cent.) and the solar eclipse during Christ's Passion, Faber provided details about the topography of Athens, and confusingly conflated the early Christian saint and the sixth-century Neoplatonist philosopher (Corrigan and Harrington 2019). Faber's commentary took Sacrobosco far beyond the worlds of medieval mathematics or natural philosophy.

Faber also imported some astronomical and natural philosophical content to Sacrobosco's thin text, usually borrowing from other widely circulating late medieval textbooks. For example, at the woodcut depicting Aristotle's cosmos of concentric spheres (Fig. 1), Faber referred to unnamed "philosophers" who disagree on the relative roles of Aristotle's *primum mobile* and God in causing celestial motions. Likewise, philosophers demand uniform motions in the heavens, while astronomers "say that it is irregular, sometimes fast, sometimes slow," arguing that regular motion on an eccentric is seen, from the earth, as irregular against the concentric ecliptic circle. Faber inserted quantitative information on the relative sizes of stars in their six magnitudes, data he could have found in the recently printed edition of Leopold of Austria's *Compilatio de astrorum scientia*, a textbook composed roughly at the same time as Sacrobosco's. Several folios later, Faber listed the planetary distances in terrestrial radii, also compiled by Leopold.[41]

Faber took his readers fastidiously through Eratosthenes' (276–194 BCE) computation of the circumference of the Earth, giving all the intermediate values, and showing how to use Sacrobosco's rule for deriving the diameter from the circumference of a circle: subtract the twenty-second part from the circumference and divide the remainder by three (Sacrobosco 1495, C2r–C3r). After discussing Sacrobosco's summary of the seven climates on the earth's surface, Faber inserted a "Tabula climatum" that lists twenty-four climates, with the longest days ranging from twelve to twenty-four hours, at half-hour intervals. With its peculiar set of names for the climates and enigmatic quantitative values, the source of this table

---

[41] (Sacrobosco 1495, A7r, B2v; Leopold of Austria 1489, a8v). These cosmic dimensions, ultimately derived from Ptolemy's *Planetary hypotheses*, circulated widely in medieval Arabic and Latin texts, albeit marred by canonical sets of scribal errors. Faber's numbers follow exactly the peculiar values found in the 1489 *editio princeps* of Leopold, which makes it highly likely that our commentator used that edition (Goldstein 1967, 57; Swerdlow 1968, 130–131, 174–175; Helden 1985, 27, 34–35).

remains a mystery.[42] And at several points, Faber instructed his readers in practical astronomy. He described how to inscribe a local meridian line using a gnomon, how to measure the altitude of the pole with a quadrant or astrolabe, and how to find the length of a day with a vague reference to a globe or armillary sphere. Did he expect his Leipzig students to carry out hands-on exercises with brass instruments?[43]

Only a handful of times did Faber explicitly correct or update Sacrobosco. This author, Faber explained, had followed Ptolemy and al-Battani in giving the eighth sphere a single motion (precession). Thābit had added two small circles (Fig. 9) to the ecliptic to move the eighth sphere. More recently, King Alfonso X of Castile (1221–1284) had given the eighth sphere two motions, one of "access and recess" (the small circles rotating once in seven-thousand years), the other completing its course in 49,000 years, motions that in Ptolemy's time "had not yet been discovered" (Sacrobosco 1495, A6r–v, B3r–v, H2v–H3r). In other words, Faber here endorsed the scheme for precession plus trepidation of the Parisian Alfonsine Tables, developed in Paris around 1320 and used widely across Europe by the fifteenth century. Not surprisingly, Faber had used these tables (or the ephemerides of Regiomontanus based on those tables) for making all his annual almanacs and prognostications. Criticizing Sacrobosco's assertion that all the planets travel in a band within six degrees of the ecliptic, Faber obtusely described Ptolemy's geometrical models for planetary latitudes and concluded that Venus can reach a maximal latitude of $7;30°$, which requires Sacrobosco's band around the ecliptic to be widened. The value 7;30, not found in Ptolemaic latitude tables, is an erroneous value that appears in the widely copied Tables of 1322 by John of Lignères (14th cent.) and the 1483 *editio princeps* of the Parisian Alfonsine Tables (given as the maximal latitude for Mars, not Venus!).[44]

Faber criticized Sacrobosco on the philosophical status of geometrical models for the celestial bodies, a topic not centrally addressed, however, in the commentary. Concerning the daily motion of the Sun across the sky, Sacrobosco had written:

…the Sun, moving [over the course of the year] from the first point of Capricorn through Aries to the first point of Cancer with the [daily] sweep of the firmament, describes 182 parallels, to which parallels, although they are not really circles but spirals [spire], since there is no sensible error in this, no violence is done if they are called 'circles' (Thorndike 1949, 133).

---

[42] (Hadravová and Hadrava 2020, 283) indicate that Faber's table does not match any of the more than sixty tables of places they have found in manuscripts dating from the thirteenth to the sixteenth centuries. The computation of latitudes from these day lengths are very erratically computed for an obliquity of about $23;30°$. The column headed "Stadia latitutinis climatum," displaying very irregular first differences, I do not understand.

[43] A study of annotations in copies of editions of Faber's commentary might shed light on this question.

[44] For more information, see (Sacrobosco 1495, D2r; Alphonso X 1483, h1v; Neugebauer 1975, 226; Chabás and Saby Forthcoming, Table 8).

Most of the early printed editions, including all from Leipzig, render the term "spere" rather than "spire."[45] Presumably unaware of the typographic problem, Faber adds a term for "spiral," describing what we might call the continuous line visible in both Ratdolt's 1485 and Landsberg's 1494 diagrams (see Figs. 6 and 11):

> Note about the text: when the author says "circles," he speaks improperly, because a circle, when it is drawn, that is, proceeding from some point, the circumference is terminated at the same point from which the circumference of the same circle began. The circles in question here, however, are not terminated at the same point but are carried around in spirals [circumferuntur girative ita], so that all "circles," improperly so-called, endure as one line, etc. Therefore, he calls them much more appropriately "spheres" [referring to the misprinted 'spere'].[46]

Faber here seems to be instructing readers how to understand the diagrams, not the three-dimensionality of the cosmos or of philosophers' models. Later in the commentary, where Landsberg inserted a diagram for the Sun's motion that is very similar to one in Peurbach's *Theorica novae planetarum*, (Fig. 7), Faber again criticized Sacrobosco for referring only to "circles." Peurbach, emphasizing the three dimensionality of the *theorica*, had famously opened his tract with the phrase "sol habet tres orbes," a language that Faber did not accept, writing rather that "sol triplicem habet motum" (Peurbach 1474, [A1]r, Sacrobosco 1488, H2v). But Faber also commented that the Moon and planets move around the deferent, which is an "orb with a certain thickness (*in se spissitudinem*)" in whose "concavity" moves not a circle but a "small sphere" (*sperula*) called the epicycle. Faber's language here clearly mirrors Peurbach's. However, the equant, Faber continued, is "correctly called a circle or circumference" since it is "only an imaginary eccentric circle along which the center of the epicycle moves uniformly because it [moves] irregularly relative to the center of the deferent."[47] Faber's *theorica* were both physical and imaginary.

Finally, we note that Faber opened his remarks on the planetary *theorica* (chapter four) with an encomium to the Sun. All planets move irregularly with respect to the first movable (*primi mobilis*); hence, "the Sun is the king among the other planets because its motion determines the conditions that are necessary for the irregularities of the planets' motion." Faber did not specify here what "conditions" create which "irregularities," merely noting that all the planets' apogees process around the poles of the zodiac. He did not move to the language of heliocentrism, *De revolutionibus* Bk 1, 10: "...as though seated on a royal throne, the Sun governs (*gubernat*) the family of

---

[45] Ratdolt's 1482 edition gives "spere," his 1485 edition corrects to "spire." Most editions after 1500 give "spere." Note that an autograph by the Leipzig student, Virgil Wellendörffer, copied ca. 1486, gives "spere." UBL, Ms 1470, 420v.

[46] "Nota circa litteram: cum dicitur circuli, autor improprie loquitur, quia circulus, cum describitur procedendo videlicet ab aliquo puncto terminatur circumferencia in ipsum punctum, a quo circumferencia circuli eiusdem est inchoata. Circuli autem, de quibus hic est ad propositum, non terminantur in idem punctum, sed circumferuntur girative ita, ut omnes circuli improprie dicti manent una linea et cetera. Ideo vocat eas speras magis proprie loquendo" (Sacrobosco 1495, F2r).

[47] "Item alius ex circulis est equans qui proprie circulus vel circumferentia nominatur et est eccentricus circulus solum imaginatus super quo motus centri epicycli regulariter mouetur nam super centro defferentis irregulariter circumfertur" (Sacrobosco 1495, H4r).

planets revolving around it."[48] But his commentary did expand Sacrobosco's rather terse description of the planetary *theorica*.

The second Leipzig commentary on Sacrobosco was authored by Tockler. Since this work has been recently analyzed by Matteo Valleriani and Nana Citron, I need here only note several places where Tockler's treatment differs interestingly from Faber's (Valleriani and Citron 2020, 120–127). First and most obviously, Tockler provided much more astrological content, both medical and otherwise, for his readers. He often referred explicitly to Ptolemy's *Liber quadripartitus* and a group of practical astrological texts that had been printed together in Venice in 1493.[49] He quoted classic medieval astrological texts by Albumasar and Leopold of Austria, both of which had recently been printed by Ratdolt (Albumasar 1489; Leopold of Austria 1489). Tockler's students would have been introduced to the latest published works on practical astrology.

Unlike Faber, Tockler introduced his students to some of the mathematical details required to compute horoscopes and planetary positions with the Parisian Alfonsine Tables (Venetian editions had been printed in 1483 and 1492). The brief, incomplete samples of astronomical tables inserted into his commentary do not allow any actual computation, but they do introduce the genre.[50] Tockler's recomputation of Erastothenes' determination of the radius of the Earth, however, is filled with typographical and/or computational errors and would have confused his readers or forced them to fix the numbers in the margin, as was the case in a copy now in Munich.[51] Tockler indicated that the obliquity of the ecliptic is changing (from Ptolemy's 23; 51 to Almeon's 23;33 and now in "our time" to 23;28), but did not inform his readers that the latter value had just appeared in the *editio princeps* of Regiomontanus's *Epitome* (Sacrobosco et al. 1503, C6r; Regiomontanus 1496, b1r). He also nicely summarized Peurbach's physical description of the solar *theorica*, quoting (unlike Faber) Peurbach's phrase "sol habet tres orbes" but did not name the source or elaborate Peurbach's *theorica* for the other planets.[52] The second edition (Sacrobosco et al. 1509) of the commentary includes a lengthy text "recently added and diligently revised" by Tockler, giving instructions for constructing a "material sphere" of wood

---

[48] (Sacrobosco 1495, H4r). See (Copernicus 1973–1992, Vol. 2, 22).

[49] (Ptolemy 1493). Texts found in this edition, frequently quoted by Tockler, include aphorisms attributed to Hermes Trismegistes, Almansor (al–Mansūr), and Bethem (al-Battāni) as well as Zahel's (Sahl ibn-Bischr) *De significatore temporibus*. See (Hasse 2016, 406–407).

[50] Tockler's tables include a table of planetary periods and mean motions (B1v); of right and oblique ascensions of signs (D6v, E1v); of half-day lengths for the latitude of Leipzig (said to be 51°, but the data are for 50°, E2r); of day lengths for latitudes from the equator to the Tropic of Cancer, at intervals of 15 min (doubling the entries in Faber's similar table! F5r–v); and a table the positions of the solar apogee from 1500 to 1556 at four-year intervals (G1v) as well as another for the planetary apogees in 1503 and 1504 (G2v), both computed from the Parisian Alfonsine Tables.

[51] BSB, Res/2 A.lat.a. 199#Beibd. 5, D2v. The second edition of Tockler's commentary corrects some of these errors; see (Sacrobosco et al. 1509, D6v).

[52] (Sacrobosco et al. 1503, F6v). For Tockler's lengthy and apparently original commentary on Peurbach, see ÖNB Ms 5274, ff. 57r–120v.

and metal to represent the celestial bodies with ten concentric rings, a kind of armillary sphere.[53] Tockler did not, however, integrate his presentation of this material sphere with Peurbach's physical models.

The third Leipzig commentary, prepared by, and for, Cistercian monks, offers a quite different reading, emphasizing natural philosophy and contemplation of the "eternal governor" of the world. Sacrobosco's text becomes the site for discussion of natural causes, motion, the composition and disposition of celestial bodies, the elements, generation and corruption, the certitude of natural versus mathematical knowledge, and, as Jacob entitled his introduction, "the dignity of the science of astronomy." Readers might guess that Jacob enjoyed access to a well-stocked library as the commentary refers to many standard texts in the university arts curriculum. At one point, he cited one of the earliest published *Sphaera* commentaries by Francisco Capuano di Manfredonia, printed in 1499 by Simon Bevilacqua (1450–1518) in Venice.[54] A quick check reveals that Jacob took most of his commentary, verbatim, from that Paduan master and lecturer! The Cistercian commentary thus represents a transalpine circulation of knowledge, a dependency that, to the best of my knowledge, has not previously been recognized by bibliographers or historians of Sacrobosco.

A large part of Jacob's introduction, "De dignitate astronomice scientie," is lifted directly from Capuano's "Prologus." Three-quarters of the subsequent sixty glosses are copied, in full or part, from the opening sentences of the sections into which Capuano had divided his commentary. Often Jacob added his own comments to Capuano's. On nearly every folio, Capuano referenced the Ptolemaic textbook of thirty chapters by al-Farghānī (also frequently mentioned by Sacrobosco); he also quoted the standard university texts by Aristotle (*De caelo et mundo, Physica, De motu animalium, Metaphysica, De generatione et corruptione*), Averroes's (1126–1198) commentaries on these texts, Ptolemy's *Almagest* I and the *Centiloquium*, Albert the Great, the *Theorica planetarum*, and more sparingly, Euclid, Theodosius, and Haly Abenragel's *De judiciis astrorum*, an astrological compendium. These works hence appear in Jacob's commentary.

But Jacob also introduced new texts, most of which had been printed over the past two decades. Among classical authors, he cited Aristotle's *Meteorologica* and *Analytica posteriora*, Ptolemy's *Cosmographia*, Proclus's (412–485) *Sphaera*, Julius Firmicus's (4th cent.) *Mathesis*, Manilius's (1st cent. BCE) *Astronomica* (quoted in length several times). His medieval sources include Albumasar's *Introductorium in astronomiam* and *Flores astrologiae*, Avicenna's *Canon*, the Parisian Alfonsine Tables, Campanus's *Theorica planetarum* (a text that had not been printed),

---

[53] (Sacrobosco et al. 1509, C1r–C3v). I have searched, without success, for the source of this text, as I am not convinced that it was authored by Tockler. For a compilation of diverse texts on "the sphere" then circulating, see (Sacrobosco et al. 1499).

[54] (Sacrobosco et al. 1499) includes Sacrobosco's text with surrounding commentary by Cecco d'Ascoli, separate commentaries by Francisco Capuano (e1r–l5v or 41 leaves) and Jacques Lefèvre d'Étaples (l5v–o6r or 20 leaves), and Peurbach's *Theorica novae planetarum* surrounded by Capuano's commentary (p1r–9–3r or 62 leaves). According to (Shank 2009, 295) Capuano's was "perhaps the longest" commentary on Sacrobosco until Christopher Clavius's at the end of the sixteenth century.

and Regiomontanus.[55] And in his introduction, Jacob quoted from the patristics, Augustine's (354–430) *Soliloquia* and John of Damascus's (675–749) sermons.

Recently, Michael H. Shank has argued that Capuano's Sacrobosco commentary, with its extensive mingling of astronomical and physical themes and detailed arguments against a two-fold motion of the Earth (diurnal and annual), may have provided a foil against which Copernicus directed his heliocentric theory. Most of the provocative passages analyzed by Shank, quoting the final 1518 version of Capuano's commentary, do not appear in the 1499 version used by Jacob. Nonetheless, Jacob followed Capuano in taking a physical approach to astronomy. He emphasized that spheres must be imagined as "solid" and "dense" (A5r). He quoted Capuano on the terrestrial elements air and water participating in circular motions not proper to them, due to the drag (*raptus*) of the neighboring lunar orb.[56] Going beyond Capuano, he discussed Albert the Great's distinction, in *De meteoris*, between two types of elemental fire below the Moon, and quoted from Manilius's *Astronomica* I, 141–170, on the places of the four elements around the Earth (Sacrobosco 1503a, B4r–B5r; Magnus 1651, 2:5; Manilius and Manilius 1977).

Unlike Capuano, Jacob mentioned several times the "tabulas Alfonsi." In describing Alfonsine precession, Jacob suggested that one must "imagine" two motions, one of 1;28° in two-hundred years in the order of the signs (or a complete revolution in 49,000 years), the other of two "small circles" at the beginning of Aries and Libra (B1v–B2r) completing their revolution in seven-thousand years. Jacob contrasted these motions to Ptolemy's single motion of precession (1° in one-hundred years), urging readers to choose "the recent more thorough investigation of the motion of the spheres" (Sacrobosco 1503a, B1v–B2r, B6v–C1r). In a table, he lifted from the Sacrobosco commentary by Jacques Lefèvre d'Étaples (also in the 1499 *compendium*), Jacob listed some of the mean motions from the Alfonsine Tables, to a precision of sexagesimal fourths.[57] He did not, however, instruct his readers on how to compute quantitative planetary positions. Instead, he followed Capuano in describing the planetary *theorica*. Each planet has orbs for its deferent and epicycle; however, the equant is an "imaginary circle" so that the deferent is "not moved equally and uniformly around its center" (Sacrobosco 1499, l2v–l3r; 1503a, K4v–K5r). Despite their emphasis on a physical astronomy, neither Capuano nor Jacob offered their readers a consistent language for describing the Ptolemaic planetary *theorica*.[58]

---

[55] Jacob associated no specific text with Regiomontanus, referring to his and the Alfonsine Tables' presentation of a two-fold motion for the eighth sphere (precession plus trepidation). Perhaps Jacob meant here Peurbach's *Theorica novum planetarum*, printed by Regiomontanus in 1474, which does offer a complex *theorica* for the two-fold motion. See (Sacrobosco 1503a, B1v). Note that, in the final version of his commentary, Capuano also linked the Alfonsine Tables and Regiomontanus to the *theorica* for the eighth sphere. See (Ps. Ptolemy et al. 1518, f. 25va).

[56] (Shank 2009, 298–299). Sacrobosco also had used the physical term "raptus" to describe how the *primum mobile* "carries" all other spheres in daily motion. See (Thorndike 1949, 70, 119).

[57] (Sacrobosco 1503a, C1r; 1499, m3r). D'Étaples had listed the values to sexagesimal sevenths.

[58] For a recent study of medieval texts on "imagination" and the sphere, see (Obrist 2018).

Hence, Leipzig's masters had prepared three rather different Sacrobosco commentaries for their students. The earliest, by Faber, offers a "humanistic" reading with its attention to classical literary works and the semantics of technical terms. Tockler's is more "practical," emphasizing tools for the working astronomer/astrologer, albeit jumbled together without a clear focus and not providing enough details to, say, cast or interpret a horoscope. The Cistercian Jacob silently presented an abridged summary of the massive 1499 Paduan commentary by Francesco Capuano, citing many classic university texts on the natural philosophy of the heavens and earth but also introducing the recently printed Alfonsine Tables.

The commentaries by Tockler and Jacob were never reprinted. Faber's would appear, between 1501 and 1508, in seven editions printed in Cologne by Heinrich Quentell (d. 1501) and heirs (Sacrobosco 1500a, b, 1501, 1503b, c, 1505, 1508). Like Landsberg, Quentell printed primarily for the local university market, which may have prompted him to turn to an early university printing center like Leipzig for his material. As far as I know, Quentell and heirs were also the only printers to copy the Leipzig Sacrobosco in leading the text and reproducing the designs of all the woodcuts. Apart from this resonance in Cologne, the Leipzig Sacrobosco and its three commentaries did not circulate beyond Saxony.

5   Conclusion

Leipzig offers a case study of the early printing of Sacrobosco in a local university setting. As noted above, over the first fifty years (1472–1521) of printing the *Sphaera*, three cities had dominated: Venice with nineteen editions, Paris with seventeen and Leipzig with fifteen editions. Cologne follows with seven editions. Paris had a university, but its printers, like those in Venice, appear to have issued their editions for an international as well as local market. Although we have speculated that Leipzig editions may have been acquired by students in the nearby university towns of Erfurt and Wittenberg, it seems clear that the local market was the primary force shaping the production of Leipzig Sacroboscos from 1488 until ca. 1521. After that date, no further Leipzig editions would be printed; by 1531, the production of central European *Sphaera* editions moved north to printers in Wittenberg and to Melanchthon's powerful influence.

For their local market, Leipzig's printers and masters created what we have called the "Leipzig Sacrobosco," an edition comprised of idiosyncratic elements that do not appear in editions printed elsewhere. These include a particular version of Sacrobosco's text, a pedagogically inflected set of diagrams to accompany the text, leading of the print block so that readers could add interlinear annotations, and a reluctance to issue an edition containing works beyond the Sacrobosco. The Leipzig Sacrobosco was a slim *quarto* codex, apparently designed for elementary lectures on the *Sphaera* and little more.

We have, however, identified several examples where content from the Leipzig Sacrobosco did circulate beyond the local. Jacob structured his 1503 commentary on the more comprehensive commentary of the Paduan master, Francesco Capuano, printed in 1499 in Venice. Faber's commentary, printed five times in Leipzig by 1521, was also issued in Cologne. And the short text by ps.-Thābit that Tockler had added to his 1503 commentary, was issued again in a 1518 edition. But beyond these incidents, the Leipzig Sacrobosco did not become enmeshed in the network of texts, paratexts, authors, printers, and publishers that would characterize the later printing history of Sacrobosco's *Sphaera* stretching into the sixteenth and seventeenth centuries (Valleriani et al. 2019; Pantin 2020) (Chap. 10).

Abbreviations

Digital Repositories

GW	Gesamtkatalog der Wiegendrucke. Stiftung Preußischer Kulturbesitz. https://www.gesamtkatalogderwiegendruck e.de/. Accessed 07 June 2021
RAG	Repertorium Academicum Germanicum. Universität Bern. https://rag-online.org/. Accessed 07 June 2021
Sphaera Corpus*Tracer*	Max Planck Institute for the History of Science. http://db. sphaera.mpiwg-berlin.mpg.de/resource/Start. Accessed 07 June 2021
VD16	Verzeichnis der im deutschen Sprachbereich erschienenen Drucke des 16. Jahrhunderts. Bayerische Staatsbibliothek. https://www.bsb-muenchen.de/sammlungen/historische-drucke/recherche/vd-16/. Accessed 07 June 2021

Archives and Special Collections

BSB	Bayerische Staatsbibliothek, Munich
CUL	Cambridge University Library
NLCR	Národní knihovna České republiky, Prague
ÖNB	Österreichische Nationalbibliothek, Vienna
ThULB	Thüringer Universitäts- und Landesbibliothek, Jena
UBL	Universitätsbibliothek Leipzig

References

Primary Literature

Albertus, Magnus. 1651. *Opera*. Lyon: C. Rigaud.

Albumasar. 1489. *De magnis coniunctionibus, annorum revolutionibus ac eorum profectionibus, octo continens tractatus*. Augsburg: Ratdolt.

Alphonso, X. 1483. *Tabulae astronomicae*. Venice: Ratdolt.

Ferrariis, Georgius de. 1500. *Figura sphere: cum glosis Georgii de Monteferrato Artium et medicine Doctoris*. Venice: Giovanni Battista Sessa I. https://hdl.handle.net/21.11103/sphaera.100275.

Ficinus, Marsilius, and K. Tockler. 1502. *Preclari oratoris et philosophi Marsili Ficini Libellus de Sole*. Leipzig: Wolfgang Stöckel.

Leopold of Austria. 1489. *De astrorum scientia, decem continens tractatus*. Augsburg: Erhard Ratdolt.

Manilius, X.E., and M. Manilius. 1977. *Astronomica*. Cambridge: Harvard University Press.

Muris, Johannes de and Conrad Tockler. 1503. *Textus Arithmetice Communis: qui p[er] magisterio fere cunctis in Gymnasijs, ordinarie solet legi, correctus corrobatusque, perlucida quadam atque prius no[n] habita Commentatione*. Leipzig: Martin Lansberg.

Nallino, C.A. (ed.). 1977. *Al-Battani sive Albatenii opus astronomicum [1899–1907]*. Hildesheim: Georg Olms Verlag.

Peurbach, Georg. 1474. *Theoricae novae planetarum*. Nuremberg: Regiomontanus.

Ps. Ptolemy, Campanus da Novara, Pierre d'Ailly, Cecco d'Ascoli, Robert Grosseteste, Theodosius of Bithynia, Michael Scot, Joannes Regiomontanus, Jacques Lefèvre d'Étaples and Francesco Capuano. 1518. *Sphera cum commentis in hoc volumine contentis. videlicet. Cichi Esculani cum textu / Expositio Joannis Baptiste Capuani in eandem / Jacobi Fabri Stapulensis / Theodosij de Speris / Michaelis Scoti / Quaestiones Reverendissimi domini Petri de Aliaco etc. / Roberti Linchontensis Compendium / Tractatus de Sphera solida / Tractatus de Sphera Campani / Tractatus de computo maiori eiusdem / Disputatio Joannis de monte regio / Textus Theorice cum expositione Joannis Baptiste Capuani / Ptolemeus de Speculis*. Venice: Heirs of Ottaviano Scoto I. https://hdl.handle.net/21.11103/sphaera.101057.

Ptolemy. 1493. *Liber quadripartitus*. Venice: Bonetus Locatellus.

Regiomontanus. 1496. *Epytoma in Almagestum Ptolomei*. Venice: Johannes Hamann.

Sacrobosco, Johannes de. 1472. *Tractatum de Spera*. Venice: Florentinus de Argentina. https://hdl.handle.net/21.11103/sphaera.100685.

Sacrobosco, Johannes de. 1478. *Iohannis de Sacrobusto anglici viri clarissimi Spera mundi feliciter incipit*. Venice: Franz Renner. https://hdl.handle.net/21.11103/sphaera.100686.

Sacrobosco, Johannes de. 1480. *Ioannis de sacrobusto anglici viri clarissimi spera mundi*. Bologna: Dominicus Fuscus. https://hdl.handle.net/21.11103/sphaera.100264.

Sacrobosco, Johannes de. 1488. *Johannis de sacro busto spericum opusculum unacum utilissimis figuris textam declarantibus*. Leipzig: Martin Landsberg. https://hdl.handle.net/21.11103/sphaera.100265.

Sacrobosco, Johannes de. 1489a. *Iohannis de sacro busto spericum opusculum unacum utilissimis figuris textum declarantibus*. Leipzig: Konrad Kachelofen. https://hdl.handle.net/21.11103/sphaera.100267.

Sacrobosco, Johannes de. 1489b. *Iohannis de sacro busto spericum opusculum unacum utilissimis figuris textum declarantibus*. Leipzig: Konrad Kachelofen. https://hdl.handle.net/21.11103/sphaera.100266.

Sacrobosco, Johannes de. 1494. *Opusculum Iohannis de Sacro Busto spericum. cum figuris optimis et nouis textum in se. sine ambiguitate declarantibus*. Leipzig: Martin Landsberg. https://hdl.handle.net/21.11103/sphaera.100271.

Sacrobosco, Johannes de. 1495. *Opusculum Johannis de sacro busto spericum cum notabili commento atque figuris textum declarantibus utilissimis.* Leipzig: Martin Landsberg. https://hdl. handle.net/21.11103/sphaera.100886.

Sacrobosco, Johannes de. 1498. *Opusculum Johannis de sacro busto spericum. cum figuris optimis et novis textum in se sine ambiguitate declarantibus.* Leipzig: Wolfgang Stöckel. https://hdl.han dle.net/21.11103/sphaera.100887.

Sacrobosco, Johannes de. 1499. *Opusculum Johannis de sacro busto spericum cum notabili commento atque figuris textum declarantibus utilissimis.* Leipzig: Wolfgang Stöckel. https://hdl. handle.net/21.11103/sphaera.100888.

Sacrobosco, Johannes de. 1500a. *Opus sphericum Iohannis de sacro busto figuris et per utili commento illustratum.* Cologne: Heinrich Quentell. https://hdl.handle.net/21.11103/sphaera. 100891.

Sacrobosco, Johannes de. 1500b. *Opusculum Iohannis de sacro busto sphericum cum notabili commento atque figuris textum declarantibus utilissimis.* Cologne: Heinrich Quentell. https://hdl. handle.net/21.11103/sphaera.100890.

Sacrobosco, Johannes de. 1501. *Opus sphericum Johannis de sacro busto figuris et per utili commento illustratum.* Cologne: Heinrich Quentell. https://hdl.handle.net/21.11103/sphaera. 100023.

Sacrobosco, Johannes de. 1503a. *Astronomice sciencie sphaericum introductorium Ioannis de sacro busto, quibusdam notabilibus (ex eiusdem discipline diversorum autorum preclarissimis commentarijs iam noviter diligenter decerptis) dilucidatum.* Leipzig: Wolfgang Stöckel. https://hdl.han dle.net/21.11103/sphaera.100893.

Sacrobosco, Johannes de. 1503b. *Opus sphericum magistri Ioannis de Sacrobustho natione Anglici. figuris verissime exculptis et interpretatione familiari ad commoditatem desiderantium iucundissima Artis Astronomice callere principia pulcherrime et iterata recognitione illustratum.* Cologne: Heinrich Quentell. https://hdl.handle.net/21.11103/sphaera.100894.

Sacrobosco, Johannes de. 1503c. *Opus sphericum magistri Ioannis de Sacro busto natione angli figuris verissime exculptis et interpretatione familiari commoditatem desiderantium iucundissima Artis Astronomice callere principia pulcherrime et iterata recognitione illustratum.* Cologne: Heinrich Quentell. https://hdl.handle.net/21.11103/sphaera.100024.

Sacrobosco, Johannes de. 1505. *Opus sphericum magistri Ioannis de Sacro busto natione angli figuris verissime exculptis et interpretatione familiari ad commodita tem desiderantium iucundissima Artis Astronomice callere principia pulcherrime et iterata recognitione illustratum.* Cologne: Heinrich Quentell. https://hdl.handle.net/21.11103/sphaera.100028.

Sacrobosco, Johannes de. 1508. *Opus sphericum magistri Ioannis de Sacro busto natione angli magistrique Parrhisiensis figuris verissime exculptis et interpretatone familiari ad commoditatem desiderantium iucundissima Artis Astronomice callere principia pulcherrime et iterata recognitone illustratum.* Cologne: Heirs of Heinrich Quentell. https://hdl.handle.net/21.11103/ sphaera.100183.

Sacrobosco, Johannes de. 1510. *Textus spere materialis Ioannis de Sacrobusto.* Leipzig: Martin Landsberg. https://hdl.handle.net/21.11103/sphaera.100035.

Sacrobosco, Johannes de. ca. 1515. *Opusculum Johannis de sacrobusto spericum, cum figuris optimis et novis textum in se. sine ambiguitate declarantibus.* Leipzig: Martin Landsberg. https:// hdl.handle.net/21.11103/sphaera.100918.

Sacrobosco, Johannes de, Cecco d'Ascoli, Francesco Capuano, and Jacques Lefèvre d'Etaples. 1499. *Sphera Mundi cum tribus Commentis nuper editis vz. Cicchi Esculani, Francisci Capuani de Manfredonia, Iacobi Fabri Stapulensis.* Venice: Simone Bevilacqua. https://hdl.handle.net/21. 11103/sphaera.100273.

Sacrobosco, Johannes de, and Wenzel Faber. ca. 1516. *Opusculum Iohannis de sacro busto spericum cum notabili commento a Magnifico viro domino Wenceslao Fabri de Budrveysz medicine Doctore edito / cumque figuris textum declarantibus utilissimis.* Leipzig: Martin Landsberg. https://hdl. handle.net/21.11103/sphaera.100056.

Sacrobosco, Johannes de, and Wenzel Faber. ca. 1519. *Opusculum Iohannis de sacro busto spericum cum notabili commento a Magnifico viro domino Wenceslao Fabri de Budweysz medicine Doctore edito, cumque figuris textum declarantibus utilissimis.* Leipzig: Martin Landsberg. https://hdl.han dle.net/21.11103/sphaera.100994.

Sacrobosco, Johannes de, and Wenzel Faber. ca. 1521. *Opusculum Ioannis de Sacro Busto Spericum cum notabili commento a Magnifico viro domino Wenceslao Fabri de Budrveysz medicine Doctore edito / cumque figuris textum declarantibus utilissimis.* Leipzig: Martin Landsberg. https://hdl. handle.net/21.11103/sphaera.100993.

Sacrobosco, Johannes de, and Philipp Melanchthon. 1538. *Ioannis de Sacro Busto libellus, De sphæra: Eiusdem autoris libellus, cuius titulus est Computus, eruditissimam anni & mensium descriptionem continens. Cum praefatione Philippi Melanth. & novis quibusdam typis, qui ortus indicant.* Wittenberg: Joseph Klug. https://hdl.handle.net/21.11103/sphaera.101106.

Sacrobosco, Johannes de, Johannes Regiomontanus, and Georg Peuerbach. 1482. *Novicijs adolescentibus: ad astronomicam rempublicam capessendam aditum impetrantibus: pro brevi rectoque tramite a vulgari vestigio semoto: Ioannis de Sacro Busto Sphaericum opusculum. Contraque cremonensia in planetarum theoricas delyramenta Ioannis de Monte Regio disputationes tam acuratiss. quam utiliss. Necnon Georgij Purbachij in eorundem motum planetarum acuratiss. theoricae: dicatum opus utili serie et textum incobat.* Venice: Erhard Ratdolt. https://hdl.handle. net/21.11103/sphaera.100692.

Sacrobosco, Johannes de, Johannes Regiomontanus, and Georg Peuerbach. 1485. *Noviciis adolescentibus: ad astronomicam rempublicam capessendam aditum impetrantibus: pro brevi rectoque tramite a vulgari vestigio semoto: Ioannis de Sacro Busto sphaericum opusculum Georgiique Purbachii in motum planetarum accuratiss. theoricae Necnom contra Cremonensia in eorundem planetarum theoricas deliramenta Ioannis de Monte Regio disputationes tam accuratiss. que utiliss. dicatum opus utili serie contextum inchoat.* Venice: Erhard Ratdolt. https://hdl.handle.net/ 21.11103/sphaera.101123.

Sacrobosco, Johannes de, Johannes Regiomontanus, and Georg Peuerbach. 1488. *Spaerae Mundi Compendium foeliciter inchoat. Noviciis adolescentibus: ad astronomicam remp. capessendam aditum impetrantibus: pro brevi rectoque tramite a vulgari vestigio semoto: Iohanis de Sacro busto sphaericum opusculum una cum additionibus nonnullis littera A sparsim ubi intersertae sint signatis Contraque et cremonensia in planetarum theoricas delyramenta Iohannis de monte regio disputationes tam acuratiss. atque utills. Nec non Georgii purbachii in eorundem motus planetarum acuratiss. theoricae: dicatum opus: utili serie contextum fausto sidere inchoat.* Venice: Girolamo de Sanctis for Johannes Lucilius Santritter. https://hdl.handle.net/21.11103/sphaera. 100822.

Sacrobosco, Johannes de, Johannes Regiomontanus, and Georg Peuerbach. 1490. *Spaerae mundi compendium foeliciter inchoat. Noviciis adolescentibus: ad astronomicam rem publicam capessendam aditum impetrantibus: pro brevi rectoque tramite a vulgari vestigio semoto: Ioannis de Sacro busto sphaericum opusculum una cum additionibus nonnullis littera A sparsim ubi intersertae sint signatis: Contraque cremonensia in planetarum theoricas delyramenta Ioannis de monte regio disputationes tam acuratiss. atque utills. Nec non Georgii purbachii in erundem motus planetarum acuratiss. theoricae: dicatum opus: utili serie contextum: fausto sidere inchoat.* Venice: Bonetus Locatellus for Ottaviano Scoto I. https://hdl.handle.net/21.11103/sph aera.100885.

Sacrobosco, Johannes de, Konrad Tockler, and Thābit ibn-Qurra. 1503. *Textus Spere materialis Joannis de Sacrobusto cum lectura Magistri Conradi Norici in florentissmo Lipsensi gymnasio: nuper exarata. Verba Thebit acutissimi Astronomi de imagine tocius mundi atque corporis sperici compositione: introducendis ipsis incrementum magnum prebentia: per eundem: addita.* Leipzig: Martin Landsberg. https://hdl.handle.net/21.11103/sphaera.100931.

Sacrobosco, Johannes de, Konrad Tockler, and Thābit ibn-Qurra. 1509. *Textus Spere materialis Joannis de Sacrobusto cum lectura Magistri Conradi Norici in florentissimo Lipsensi gymnasio nuper exarata. Verba Thebit acutissimi Astronomi de imagine totius mundi atque corporis sperici compositione: introducendis ipsis incrementum magnum prebentia: per eundem ! addita. Ordi-natio spere materialis: et decem circulis ! huic operi inseriuens: per Magistrum Conradum Noricum novitem addita et diligenter revisa.* Leipzig: Martin Landsberg. https://hdl.handle.net/21.11103/sphaera.101042.

Tissol, Garth, (ed.). 2014. *Ovid, Epistulae ex Ponto, Book 1.* Cambridge: Cambridge University Press.

Tockler, Conrad. 1503. *Commentatio Arithmeticae communis.* Leipzig: Martin Landsberg.

Secondary Literature

Baldasso, Renzo. 2009. La stampa dell'editio princeps degli Elementi di Euclide (Venezia, Erhard Ratdolt, 1482). In *The books of Venice*, eds. Lisa Pon and Craig Kallendorf, 61–100. Venice: Biblioteca Nazionale.

Bell, David N. 1989. A Cistercian at Oxford: Richard Dove of Buckfast and London, B.L. Sloane 513. *Studia Monastica* 31: 69–87.

Bertalot, Ludwig. 1915. Humanistische Vorlesungsankündigungen in Deutschland im 15. Jahrhun-dert. *Zeitschrift für Geschichte der Erziehung und des Unterrichts* 5: 1–24.

Bodemann, Ulrike. 2000. Cedulae actuum: Zum Quellenwert studentischer Belegzettel des Spät-mittelalters. In *Schulliteratur im späten Mittelalter*, ed. Klaus Grubmüller, 435–503. Munich: W. Fink.

Boldan, Kamil. 2008. Sbírka minucí a pranostik z přelomu 15. a 16. století tepelského kláštera premonstátů. *Minulosté Západočeského Kraje* 43: 79–114.

Bräuer, Miriam, Jürgen Leonhardt, and Claudia Schindler. 2008. Zum humanistischen Vorlesungs-betrieb an der Universität Leipzig. *Pirckheimer-Jahrbuch für Renaissance- und Humanismus-forschung* 23: 201–216.

Bruckner, Ursula. 1975. Noch einmal: Wenzel Faber von Budweis oder Johannes Virdung?. *Beiträge zur Inkunabelkunde* N.F. 3 (6): 19–29.

Bünz, Enno. 2008a. Der Universität Leipzig um 1500. *Pirckheimer-Jahrbuch für Renaissance- und Humanismusforschung* 23: 9–39.

Bünz, Enno. 2008b. Kloster Altzelle und das Bernhardskolleg in Leipzig. In *Die Zisterzienser und ihre Bibliotheken: Buchbesitz und Schriftgebrauch im Kloster Altzelle*, ed. Tom Graber and Martina Schattkowsky, 247–288. Leipzig: Leipziger Universitätsverlag GmbH.

Burmeister, Karl Heinz. 2015. *Magister Rheticus und seine Schulgesellen: Das Ringen um Kenntnis und Durchsetzung des heliozentrischen Weltsystems des Kopernikus um 1540/50.* Constance: UVK Verlagsgesellschaft.

Carmody, Francis J. 1956. *Arabic astronomical and astrological sciences in Latin translation: A critical bibliography.* Berkeley: University of California Press.

Carmody, Francis J. 1960. *The astronomical works of Thabit b. Qurra.* Berkeley: University of California Press.

Corrigan, Kevin, and L. Michael Harrington. Pseudo-Dionysius the Areopagite. In *The Stanford Encyclopedia of Philosophy* (Winter 2019 Edition), ed. Edward N. Zalta, https://plato.stanford.edu/archives/win2019/entries/pseudo-dionysius-areopagite. Accessed 07 June 2021.

Chabás, José, and M.-M. Saby. Forthcoming. *The tables of 1322 by John of Lignères: An edition with commentary.* Turnhout: Brepols.

Claus, H. 1973. *Untersuchungen zur Geschichte des Leipziger Buchdrucks von Luthers Thesenan-schlag bis zur Einführung der Reformation im Hgt. Sachsen (1517–1539).* PhD diss., Humboldt Universität zu Berlin.

Copernicus, Nicolaus. 1973–1792. *Opera omnia.* 4 Vols. Warsaw: Officina Publica Libris Scien-tificis Edendis.

Danielson, Dennis. 2006. *The first Copernican: Georg Joachim Rheticus and the rise of the Copernican Revolution.* New York: Walker & Company.

Dietrich, Adolf. 1914. Studium und Studierende des Cistercienser Ordens in Leipzig. *Cistercienser-Chronik* 26: 289–301, 334–346, 360–366.

Dobrzycki, Jerzy. 2010. The theory of precession in medieval astronomy [1965]. In *Selected papers on medieval and Renaissance astronomy*, eds. Richard L. Kremer, 15–60. Warsaw: Instytut Historii Nauki PAN.

Döring, Detlef. 1990. *Die Bestandsentwicklung der Bibliothek der philosophischen Fakultät der Universität zu Leipzig von ihren Anfängen bis zur Mitte des 16. Jahrhunderts.* Leipzig: Bibliographisches Institut.

Döring, Thomas Thibault. 2014. Caspar Borner und seine Bibliothek. In *Buch und Reformation: Beiträge zur Buch- und Bibliotheksgeschichte Mitteldeutschlands im 16. Jahrhundert*, ed. Enno Bünz, 191–212. Leipzig: Evangelische Verlags-Anstalt.

Eisermann, Falk. 2004. *Verzeichnis der typographischen Einblattdrucke des 15. Jahrhunderts im Heiligen Römischen Reich Deutscher Nation.* 3 Vols. Wiesbaden: Reichert.

Eisermann, Falk. 2008. Die schwarze Gunst: Buckdruck und Humanismus in Leipzig um 1500. *Pirckheimer-Jahrbuch für Renaissance- und Humanismusforschung* 23: 149–179.

Eisermann, Falk. 2017. Fifty thousand veronicas: Print runs of broadsheets in the fifteenth and early sixteenth centuries. In *Broadsides: Single-sheet publishing in the first age of print*, ed. Andrew Pettegree, 76–113. Leiden: Brill.

Erler, Georg. 1895–1902. *Die Matrikel der Universität Leipzig.* 3 Vols. Leipzig: Giesecke & Devrient.

Fitzpatrick, Antonia. 2010. London, British Library Royal MS 8 A. XVIII: A unique insight into the career of a Cistercian monk at the University of Oxford in the early fifteenth century. *Electronic British Library Journal*: 1–35.

Geldner, Ferdinand. 1968–1970. *Die deutschen Inkunabeldrucker: Ein Handbuch der deutschen Buchdrucker des XV. Jahrhunderts nach Druckorten.* 2 Vols. Stuttgart: Hiersemann.

Goldstein, Bernard R. 1967. The Arabic version of Ptolemy's planetary hypotheses. *Transactions of the American Philosophical Society* 57: 3–55.

Hadravová, Alena, and Petr Hadrava, eds. 2020. *Sféra Iohanna de Sacrobosco: středověká ucebnice základu astronomie.* Prague: Akropolis.

Haebler, Konrad. 1915. Paulus Eck gegen Wenzel Faber. *Zeitschrift für Bücherfreunde* N.F. 6: 200–204.

Hamel, Jürgen. 2006. Johannes de Sacroboscos sphaera: Text- und frühe Druckgeschichte eines astronomischen Bestsellers. *Gutenberg-Jahrbuch* 81: 113–136.

Hamel, Jürgen. 2014. *Studien zur "Sphaera" des Johannes de Sacrobosco.* Leipzig: Akademische Verlagsanstalt.

Hasse, Dag Nikolaus. 2016. *Success and suppression: Arabic sciences and philosophy in the Renaissance.* Cambridge: Harvard University Press.

Häuser, Franz, (ed.). 2009–2010. *Geschichte der Universität Leipzig, 1409–2009.* 5 Vols. Leipzig: Leipziger Universitätsverlag.

Van Helden, Albert. 1985. *Measuring the universe.* Chicago: University of Chicago Press.

Helssig, Rudolf. 1909. Die wissenschaftlichen Vorbedingungen für Baccalaureat in Artibus und Magisterium im ersten Jahrhundert der Universität. In *Beiträge zur Geschichte der Universität Leipzig im fünfzehnten Jahrhundert: Zur Feier des 500 jährigen Jubiläums der Universität gewidmet von der Universitätsbibliothek*, ed. Karl Boysen, Vol. 2, 3–93. Leipzig: Harrassowitz.

Jensen, Kristian. 2004. Exporting and importing Italian humanism: The reception of Italian printed editions of classical authors and their commentators at the University of Leipzig. *Italia medioevale e umanistica* 45: 437–497.

Kößling, Rainer. 2003. Caspar Borner. In *Sächsische Lebensbilder*, ed. Gerald Weimers, Vol. 5, 45–74. Stuttgart: Franz Steiner Verlag.

Kremer, Richard L. 2007. "Abbreviating" the Alfonsine Tables in Cracow: The *Tabulae Aureae* of Petrus Gaszowiec (1448). *Journal for the History of Astronomy* 38: 283–304.

Kremer, Richard L. 2017. Incunable almanacs and practica as practical knowledge produced in trading zones. In *The structures of practical knowledge*, ed. Matteo Valleriani, 333–369. Dordrecht: Springer.

Lehmstedt, Mark. 2019. *Buchstadt Leipzig: Biografisches Lexikon des Leipziger Buchgewerbes, Bd. 1, 1420–1539*. Leipzig: Lehmstedt Verlag.

Leonhardt, Jürgen. 2004. Gedruckte humanistische Kolleghefte als Quelle für Buch- und Bildungsgeschichte. *Wolfenbütteler Notizen zur Buchgeschichte* 29: 21–34.

Lorenz, Sönke. 1985. Libri ordinarie legendi: Eine Skizze zum Lehrplan der mittelaltersuropäischen Aristenfakultät um die Wende vom 14. zum 15. Jahrhundert. In *Argumente und Zeugnisse*, ed. Wolfram Hogrebe, 204–258. Frankfurt a.M: Peter Lang.

Mackert, Christoph. 2008. Repositus ad bibliothecam publicam: Eine frühe öffentliche Bibliothek in Altzelle? In *Die Zisterzienser und ihre Bibliotheken: Buchbesitz und Schriftgebrauch des Klosters Altzelle im europäischen Vergleich*, eds. Tom Graber and Martina Schattkowsky, 85–170. Leipzig: Leipziger Universitätsverlag.

Müller, Kathrin. 2008. *Visuelle Weltaneigung: Astronomische und kosmologische Diagramme in Handschriften des Mittelalters*. Göttingen: Vandenhoeck & Ruprecht.

Neugebauer, Otto. 1962. Thabit ben Qurra, "On the solar year" and "On the motion of the eighth sphere:" Translation and commentary. *Proceedings of the American Philosophical Society* 106: 264–299.

Neugebauer, Otto. 1975. *A history of ancient mathematical astronomy*. Berlin: Springer-Verlag.

Nickel, Holger. 1989. Mit Durchschuß: Zu Preisrelationen im Buchwesen der Inkunabelzeit. In *Zur Arbeit mit dem Gesamtkatalog der Wiegendrucke*, ed. Ursula Altmann, 127–136. Berlin: Deutsche Staatsbibliothek.

North, J.D., ed. 1976. *Richard of Wallingford: An edition of his writings, with introductions, English translation and commentary. 3 Vols*. Oxford: Clarendon Press.

Obrist, Barbara. 2018. 'Imaginatio' and visual representation in twelfth-century cosmology and astronomy: Ibn al-Haytham, Stephen of Pisa (and Antioch), (Ps.) Masha-allah, and (Ps.) Thabit ibn Qurra. In *Image, imagination, and cognition: Medieval and early modern theory and practice*, eds. Christoph Lüthy, Claudia Swan, P.J.J.M. Bakker, and Claus Zittel, 32–60. Leiden: Brill.

Palmer, Nigel F. 1998. *Zisterzienser und ihre Bücher: Die mittelalterliche Bibliotheksgeschichte von Kloster Eberbach im Rheingau unter besonderer Berücksichtigung der in Oxford und London aufgewahrten Handschriften*. Regensburg: Schnell und Steiner.

Pantin, Isabelle. 2020. Borrowers and innovators in the history of printing Sacrobosco: The case of the in-octavo tradition. In *De sphaera of Johannes de Sacrobosco in the early modern period: The authors of the commentaries*, ed. Matteo Valleriani, 265–312. Cham: Springer. https://doi.org/10.1007/978-3-030-30833-9_9.

Pedersen, Fritz S. 2002. *The Toledan Tables: A review of the manuscripts and the textual versions with an edition*. Copenhagen: C. A. Reitzels Forlag.

Pedersen, Olaf. 1978. Astronomy. In *Science in the Middle Ages*, ed. David C. Lindberg, 303–337. Chicago: University of Chicago Press.

Pensel, Franzjosef. 1986. *Verzeichnis der altdeutschen und asugewählter neuerer deutscher Handschriften in der Universitätsbibliothek Jena*. Berlin: Akademie-Verlag.

Rashed, Roshdi, ed. 2009. *Thābit ibn Qurra: Science and philosophy in nineth-century Baghdad*. Berlin: de Gruyter.

Reske, Christoph. 2007. *Die Buchdrucker des 16. und 17. Jahrhunderts im deutschen Sprachgebiet*. Wiesbaden: Harrassowitz Verlag.

Reske, Christoph. 2009. Der Holzschnitt bzw. Holzstock am Ende des 15. Jahrhundert. *Gutenberg Jahrbuch*: 71–79.

Rothe, Edith. 1961. *Karl-Marx-Universität Leipzig: Bibliographie zur Universitätsgeschichte, 1409–1959*. Leipzig: Verlag für Buch- und Bibliothekswesen.

Schmidt, Ludwig. 1899. Beiträge zur Geschichte der wissenschaftlichen Studien in sächsischen Klöstern, 2: Grünhain, Buch, Pegau, Chemnitz, Thomaskloster in Leipzig. *Neues Archiv für sächsische Geschichte und Altertumskunde* 20: 1–32.

Schmidt-Thieme, Barbara. 2002. Konrad Tockler, genannt Noricus. In *Verfasser und Herausgeber mathematischer Texte der frühen Neuzeit*, ed. Rainer Gebhardt, 95–102. Annaberg-Buchholz: Adam-Ries-Bund.

Schneider, Reinhard. 1999. Wandlungen im Verständnis von Studium und Wissenschaft bei den Zisterziensern. In *Die rheinischen Zisterzienser: Neue Orientierungen in rheinischen Zisterzen des späten Mittelalters*, eds. Norbert Kühn and Karl Peter Weimer, 35–44. Cologne: Rheinischer Verein für Denkmalpflege und Landschaftsschutz.

Schwarzburger, Maria. 1959. Die Mathematikerpersönlichkeiten der Universität Leipzig, 1409–1945. In *Karl-Marx-Universität Leipzig, 1409–1959*, ed. Ernst Engelberg, Vol. 1, 350–373. Leipzig: Verlag Enzyklopädie.

Shank, Michael H. 2009. Setting up Copernicus? Astronomy and natural philosophy in Giambattista Capuano da Manfredonia's Expositio on the sphere. *Early Science and Medicine* 14: 290–315.

Skemer, Don O. 2007. Wenzel Faber von Budweis (ca. 1456/1460? –1518): An astrologer and his library in the early age of printing. *Gutenberg-Jahrbuch*: 241–277.

Stübel, Bruno. 1879. *Urkundenbuch der Universität Leipzig von 1409 bis 1555*. Leipzig: Giesecke & Devrient.

Sudhoff, Karl. 1909. *Die medizinische Fakultät zu Leipzig im ersten Jahrhundert der Universität*. Leipzig: Barth.

Swerdlow, N.M. 1968. *Ptolemy's theory of the distances and sizes of the planets*. PhD diss.: Yale University.

Thorndike, Lynn. 1949. *The Sphere of Sacrobosco and its commentators*. Chicago: University of Chicago Press.

Thorndike, Lynn, and Pearl Kibre. 1963. *Catalogue of incipits of mediaeval scientific writings*. Rev. and augmented ed. Mediaeval Academy of America.

Valleriani, Matteo. 2019. Prolegomena to the study of early modern commentators on Johannes de Sacrobosco's Tractatus de sphaera. In *De sphaera of Johannes de Sacrobosco in the early modern period*, ed. Matteo Valleriani, 1–23. Cham: Springer. https://doi.org/10.1007/978-3-030-30833-9_1.

Valleriani, Matteo, ed. 2020. *De sphaera of Johannes de Sacrobosco in the early modern period: The authors of the commentaries*. Cham: Springer. https://doi.org/10.1007/978-3-030-30833-9.

Valleriani, Matteo, and Nana Citron. 2020. Conrad Tockler's research agenda. In *The sphaera of Johannes de Sacrobosco in the early modern period: The authors of the commentaries*, ed. Matteo Valleriani, 111–136. Cham: Springer. https://doi.org/10.1007/978-3-030-30833-9_5.

Valleriani, Matteo, Florian Kräutli, Alejandro Tejador, Christoph Sander, Malte Vogl, Sabine Bertram, Gesa Funke, and Holger Kantz. 2019. The emergence of epistemic communities in the Sphaera corpus: Mechanisms of knowledge evolution. *Journal of Historical Network Research* 3: 50–91. https://doi.org/10.25517/jhnr.v3i1.63.

Zarncke, Friedrich. 1861. *Die Statutenbücher der Universität Leipzig aus den ersten 150 Jahren ihres Bestehens*. Leipzig: S. Hirzel.

Zinner, Ernst. 1964. *Geschichte und Bibliographie der astronomischen Literatur in Deutschland zur Zeit der Renaissance*, 2d ed. Stuttgart: Hiersemann.

Zoepfl, Friedrich. 1961. Der Mathematiker und Astrologe Simon Eyssenmann aus Dillingen. *Jahrbuch des Historischen Vereins Dillingen an der Donau* 61 (63): 86–88.

Richard L. Kremer is Professor Emeritus of history at Dartmouth College, where he taught history of science and curated that institution's collection of historic scientific instruments. Kremer's research focuses on the history of medieval and early modern European astronomy, especially on questions of practice, that is, computation, tables and instruments. He has edited or co-edited books on Johannes Kepler, Johannes Hevelius, Regiomontanus, and Alfonsine manuscripts and is completing a monographic study of medieval ephemerides. Kremer serves as Associate and Reviews Editor of the *Journal for the History of Astronomy*.

Chapter 13
Publishing Mathematical Books
of Parisian *Calculatores* (1508–1515)

Alissar Levy

Abstract Between 1508 and 1515, mathematical book production in Paris was particularly high and largely addressed to the Parisian calculatores: a group of masters and students mostly related to Parisian colleges of the Iberian tradition, such as Sainte-Barbe, Coqueret, and Montaigu. The mathematical publications of Pedro Sánchez Ciruelo, who taught in Paris during the last years of the fifteenth century, were a main inspiration for these Parisian *calculatores*; they also published several other books by themselves, mostly on arithmetic, proportions, and astronomy. This induced the development of a Parisian market for the publishing of the *calculatores*' mathematical books, mainly supported by printers and booksellers who did not published scholarly mathematical texts in other periods. This paper will question who the actors of this market were, why they published mathematical books, and how they contributed to the development of mathematical teaching in Paris.

Keywords Johannes de Sacrobosco · Mathematical books · Pedro Sánchez Ciruelo · Paris · Quadrivial education

1 Introduction

Johannes de Sacrobosco's (d. 1256) *Sphaera* took many forms during the Parisian Renaissance, from the first editions of this text in the late 1480s to the standard model printed by the middle of the sixteenth century (Table 1). These forms are associated with different ways of teaching mathematics, and they are particularly numerous by the 1510s, when the teaching of mathematics began to be more regular in Parisian colleges. This can be explained by the fact that there were two mathematical currents taught in Paris in this period: a traditional current, inspired in the work of ancient philosophers (Oosterhoff 2018), and the *calculatores* current, inspired in the work of fourteenth-century mathematicians (Biard and Rommevaux 2008; Calderon 1990;

A. Levy (✉)
École nationale des chartes, Paris, France
e-mail: alissar.levy@cnssib.fr

© The Author(s) 2022   459
M. Valleriani and A. Ottone (eds.), *Publishing Sacrobosco's* De sphaera *in Early Modern Europe*, https://doi.org/10.1007/978-3-030-86600-6_13

Table 1 Parisian models of Sacrobosco's *Sphaera*

Initiator of the *Sphaera* model	Publishers in Paris	Dates of publishing in Paris	Main characteristics of the Parisian editions
Erhard Ratdolt (Venice)	Wolfgang Hopyl, Antoine Caillaut, Georg Mittelhus, Félix Baligault, the Marnef brothers	1489–1494 ca. 1512	In-4. No commentaries. Few illustrations
Jacques Lefèvre d'Étaples (Paris)	Wolfgang Hopyl, Henri Estienne, Simon de Colines	1495–1538	In-folio. Commentaries. Several illustrations
Pedro Sánchez Ciruelo (Paris)	Guy Marchant, Jean Petit	1498–1515	In-folio. Commentaries. Several illustrations
Oronce Fine (Paris)	Regnault Chaudière	1516–1538	In-4. No commentaries. Several illustrations
Pierre Apian (Ingolstadt)/Philip Melanchthon (Wittenberg)	Jean Loys, Guillaume Richard, Guillaume Cavellat, Jérôme de Marnef, Denise Cavellat	1542–17th cent.	In-8. Commentaries. Several illustrations

Wallace 1969). Both were taught in Paris between the 1500s and the middle of the 1510s, after the development of two mathematical programs, respectively, prepared by Jacques Lefèvre d'Étaples (ca. 1450–1536) and Pedro Sánchez Ciruelo (1470–ca. 1560), at the end of the fifteenth century.

This paper will focus on mathematical book production in the *calculatores* current, between 1508 and 1515. We will mainly be interested on publishers (i.e., printers and booksellers) and their relationship with authors and the public. Although the *calculatores* were also interested in physics and philosophy, we will be mainly concentrated here on their mathematical works, namely works related to the *quadrivium* as well as proportions and movement. Through the study of mathematical books as material objects, we will be primarily attentive to how publishers influenced the development of mathematical teaching in Paris.

2   Pedro Sánchez Ciruelo and His Mathematical Program (1492–1500)

The development of mathematical teaching in Paris begins in the 1490s after the arrival of two scholars: Jacques Lefèvre d'Étaples (Chap. 2) and Pedro Sánchez Ciruelo (Chap. 7). At this time, there was no specific regulation on the teaching of mathematics at the University of Paris, even if Sacrobosco's *Sphaera* seems to have been regularly studied (Beaujouan 1997). Between 1489 and 1494, we know of at least four editions of this text (Sacrobosco 1489, 1493a, b, 1494) all printed in a small quarto format without commentaries, following Erhard Ratdolt's (fl. 1476–1528) publication (Sacrobosco 1485). However, unlike their model, the first Parisian editions of Sacrobosco's *Sphaera* are poorly illustrated and do not contain Peuerbach's complementary texts on the motion of the planets (Sacrobosco 1482, 1485).

Between 1495 and 1500, Lefèvre and Ciruelo improved mathematical teaching in Paris with two advanced mathematical programs. On the one hand, Lefèvre gave Parisian students seven mathematical texts organized in two books: the first on arithmetic and music, and the second on geometry and astronomy (Jordanus Nemorarius 1496; Sacrobosco 1500). Meanwhile, Ciruelo offered five mathematical texts, organized in four books, on theorical arithmetic, practical arithmetic, geometry, and astronomy (Ciruelo 1495; Bradwardine 1495, 1496; Sacrobosco 1498).[1] In both programs, astronomy is represented by an edition of Sacrobosco's *Sphaera*, with their respective commentaries. Most of these editions are *folios* and are widely illustrated. They were still regularly published in Paris in the middle of the 1510s.

2.1 Ciruelo's Arrival in Paris and His Mathematical Books

We do not seem to have administrative information about Ciruelo's stay in Paris, unlike his friend Pedro de Lerma (ca. 1461–1541) and other international students there, and we do not know much about the conditions of his Parisian studies. From his time in Spain, his Parisian publications, and some letters, we know that he stayed in Paris at least ten years, from 1492 to 1502, where he pursued his doctorate in theology and taught mathematics (Lorente y Péres 1921). Nevertheless, Ciruelo is quite unclear in his texts when he talks about his own activity. In a letter of 1526, published twenty-four years after his departure from Paris, he says that he went "through the most prestigious Parisian colleges of theology," but he does not specify the names of these colleges or his precise situation (Sacrobosco 1526). It is well known that the main Parisian colleges of theology were Sorbonne and Navarre, but students of these establishment generally appear in their administrative sources (Farge 1980).

Information about Ciruelo's teaching activity is also unclear. In this letter of 1526, Ciruelo says that "the profession of mathematician was necessary to buy food and clothes," but he does not describe the nature of this profession. The colophon of Thomas Bradwardine's (ca. 1290–1349) *Arithmetica speculativa* designates him as

---

[1] Dates given in bibliographical references are those which appear on the printed book. However, in France, until the middle of the 16th century, the year could begin on January 1 (new style) or on Easter (ancient style). There was no established dating system for printed books even at the level of a particular workshop (Veyrin-Forrer 2017).

a "lecturer of mathematics" (*mathematica legente*), but so far as we know, there was no position in Paris for the teaching of mathematics before the 1530s (Bradwardine 1495). In the other hand, in his letter of 1526, Ciruelo says that his edition of Sacrobosco's *Sphaera* was published "to be read in public" (*publice legeram*), which might imply that his teaching was recognized by the university. It is also common to find in secondary bibliography that Ciruelo taught at the college of Beauvais, but this information seems to come from a misunderstanding related to the interpretation of some colophons containing the mention *in Bellovisu* (Ciruelo 1505a); this actually means that the book was printed at Jean Marchant's (fl. 1504–ca. 1515) workshop, the *Beauregard*, named in Latin the *Bellovisu* (Renouard 1965, 294). The Latin translation of *Beauvais* would be *Belvacensis*, but it would be equally possible to designate the college by the adjoining street, *In clauso Brunelli*.

Despite this lack of administrative information, we have at least four mathematical books by Ciruelo published during his Parisian stay. Two of these books, the theorical arithmetic and the geometry, are editions of Bradwardine's texts, an Oxfordian *calculator* of the fourteenth century. The practical arithmetic, on the other hand, is an original publication. Finally, the astronomical book, as we said, is an edition of Sacrobosco's *Sphaera*, with Ciruelo's commentaries.

Nevertheless, unlike Lefèvre's publications—almost always justified and well-structured following a traditional organization of mathematical knowledge—Ciruelo's mathematical textbooks do not seem to be conceived as a planned program. For most of his books, Ciruelo does not explain the reasoning behind their publication. His edition of Sacrobosco's *Sphaera* is the only one to contain personal considerations, mostly on the importance of mathematical studies, but without references to his previous mathematical works. In addition, his books are printed in different formats, *quarto* and *folio*, which means that they most probably did not circulate together, as books of different formats were rarely bound together. This difference between Lefèvre's and Ciruelo's projects could be perhaps explained by the fact that Ciruelo was much younger than Lefèvre. In his following works, published after his departure to Spain, Ciruelo would be deeply marked by Lefèvre's Parisian program before taking some distance, probably for theological reasons (Ciruelo 1516; Sacrobosco 1526).

2.2 Guy Marchant, Publisher of Ciruelo's Mathematical Books

During his ten years at the University of Paris, Ciruelo always published his books at the same place: Guy Marchant's (fl. 1483–1505) workshop. Marchant was one of Paris' first printers: he began his activity in 1483, in a workshop named the Champ Gaillard, at Clopin Street, next to the college of Navarre (Renouard 1965, 293). He is mainly known for his role in the introduction of illustrated books in Paris, such the *Danse macabre* and the *Calendrier des bergers*, as well as the *Danse macabre des femmes* and the *Calendrier des bergères* (Hindman 1991). He seems to have obtained his master's degree by 1497, when he starts to sign colophons using the formula *a Magistro Guidone Mercatore*, instead of *per Guidonem Mercatoris*. His academic background as well as Marchant's interest in illustrated books could perhaps explain why he agreed to publish Ciruelo's mathematical texts. In fact, the publication of these books involved at least two technical problems: the acquisition of a specialized printing material and the reproduction of mathematical figures (Chap. 2). Printing mathematical texts involves mathematical figures, such geometrical and astronomical woodcuts, but also typographical characters representing Arabic numbers. In the beginning of the 1490s, these characters were still rare in Paris and often replaced by blank spaces, abbreviations, or Roman numbers (Levy 2020, 201–203).

The first three mathematical works prepared by Ciruelo, namely the two arithmetical texts and the geometry, were printed by Guy Marchant between 1495 and 1496. So far, we do not know if the arithmetical texts were printed before or after the geometry, as they are dated February 1495, which could also be, according to the Easter calendar, February 1496. The geometry, published in a folio format, contains multiple woodcuts representing mathematical diagrams. It was probably a costly book to print. The arithmetical texts do not contain so many mathematical figures, but the theorical arithmetic presents diagrams reproduced with typographical characters. They also include some historiated woodcuts, next to the title page and the colophon, coming from other books published in the same workshop. These illustrations do not have an explicit relationship with the content; they mainly represent persons talking or listening to others, which could evoke the scholarly dimension of these texts. In addition, both have a common illustration representing the nativity of the Christ, perhaps linking the two texts as separate volumes.

2.3   Jean Petit and Guy Marchant

From 1497, Guy Marchant began to regularly publish his works with another Parisian bookseller, Jean Petit I (ca. 1495–1540), who was active from 1495 at Saint-Jacques Street—the main street of the Parisian booksellers—and who quickly became one of the most influent publishers of the Parisian university neighborhood. However, unlike most of the publishers of his time, Petit was not a printer, which means that all the books that he published and/or sold were printed in workshops independent of his own bookshop (Renouard 1896). In the association between Jean Petit and Guy Marchant, Petit does not seem always to have had an editorial responsibility in the production of the books. In most cases, his name is mentioned on the title page but only Guy Marchant is credited in the colophon. This is the case of the last of Ciruelo's mathematical books published by Guy Marchant, Sacrobosco's *Sphaera*. The publication is dated February 1498, but according to Denise Hillard, Petit's device is from the beginning of 1499 (Hillard 1989).[2] The colophon only designates Marchant in the colophon, as the financier and the printer of the book. The name of Petit only appears in the copies sold by him, which were probably bought in advance to be printed with his personal device.

On technical matters, Sacrobosco's *Sphaera* presents more difficulties than the other mathematical books printed by Marchant, as it has not only text and images, but also extensive numerical tables, composed with metal strips. Pages of tables took longer to be produced than pages of text, as they required a lot of precision and a careful proofreading. Bibliographical analyses in Lefèvre's editions of Sacrobosco's *Sphaera* show that Wolfgang Hopyl (fl. 1489–1522) organized the typographical composition of the edition based on the difficulty of the sheets to compose (Levy 2020, 243–244). Therefore, compositors in charge of pages with a lot of tables and mathematical data worked on a smaller number of sheets than compositors in charge of pages mostly constituted of text. From this perspective, pages of tables seem to have required twice the composition time of a page of text. In addition, the printing process of these tables was a demanding operation: copies of Marchant's edition show how these tables could sometimes split out.[3]

Finally, this edition also contains a few decorative illustrations, again mainly from other publications of Guy Marchant. Two of them are used because of their astronomical or astrological subject: an astronomer holding an armillary sphere and a young man offering flowers to a lady under the astrological symbols of the month of April. The other illustration has a more hypothetical relation to the work, as it comes from a poem of the *Calendrier des bergers* entitled "The scream of death" and represents a young man blowing in a horn. In the original illustration, the letters "TO TO TO," cut as a part of the image, are coming out of the horn to express the sound of the instrument. In the *Sphaera* edition, the printer covers the two last "TOs" leaving only the first one. "TO" was the common representation of the Earth in medieval astronomy and a symbol generally used in early modern astronomical books. Therefore, it does not seem that this image was selected randomly: more likely, it was used because the detail echoed the astronomical nature of the work.

3   The Parisian Current of the *Calculatores* and Its Publishers (1508–1515)

In fact, Pedro Ciruelo did not publish mathematical books in Paris related to *calculatores* studies, such as texts on proportions or movement, but he was interested in the *calculatores'* works, notably in Thomas Bradwardine's texts. Mostly, he offered Parisian students an alternative to Jacques Lefèvre d'Étaples' mathematical program, still regularly published in the university neighborhood after his departure to Spain.

---

[2] The publisher's device is an image, generally printed in the title page, giving information about the publisher.

[3] Bibliothèque nationale de France, RES-V-203, f. H8.

3.1   Republishing Ciruelo's Arithmetical Texts Before *the* Calculatores

Even before Ciruelo's departure, his arithmetical texts seem to have been reprinted, but these imprints are so fragile and ephemeral that we are not sure we know them all.

To begin with, we know of an edition of Ciruelo's practical arithmetic also signed by Guy Marchant and presenting the same date as the first edition, February 1495, but containing a different composition of the text and the address of the *Beauregard*.[4] The *Catalogue of books printed in the fifteenth Century now in the British Museum [British Library]* (from now on referred to as BMC) points out that this workshop seems to have been acquired by Guy Marchant at the end of the 1490s and reports the date of this edition as 1498 (BMC 1963, 69). Philippe Renouard argues on the other hand that some books of Guy Marchant were already signed from the *Beauregard* before he officially acquired the place, just in front his original workshop, the *Champ Gaillard* (Renouard 1965, 293). In any case, the book was necessarily printed by 1505, before the end of Guy Marchant's activity (Renouard 1965, 293).

The subsequent editions of Ciruelo's arithmetical texts were published between 1502 and 1505 by Denis Roce (fl. 1490–1517) and Jean Lambert (fl. 1493–1514). Denis Roce began his publishing activity in 1490, but he did not print himself (Renouard 1965, 375–376). In March 1502, he published an edition of Bradwardine's arithmetical text following the 1495 publication (Bradwardine 1502): according to the colophon, the book was still printed by Guy Marchant, but all the title pages were to be distributed under Roce's device. In April 1505, Roce's also published an edition of Ciruelo's practical arithmetic, now printed by Jean Marchant, nephew and successor of Guy Marchant (Ciruelo 1505a). However, this edition presents a difference compared to previous publications: the addition of a mathematical problem at the end of the book. In this problem, students are invited to help Parisian scholars to administrate their finances, when, according to Ciruelo's own words, Parisian scholars were not particularly good in mathematics (Ciruelo 1521). Yet Ciruelo was no longer in Paris in 1505, which could mean that another edition of this book was published in the meantime.

Jean Lambert was active from 1503 as a printer and a bookseller. From at least the beginning of his career, he produced books in his own workshop (Renouard 1965, 133–234). Until the beginning of the 1510s, he was a neighbor of Denis Roce at Saint-Jacques Street, in front of the Saint-Benoît cloister (Renouard 1965, 133–234). In November 1505, he published two new editions of Ciruelo's and Bradwardine's arithmetical texts, also printing the mathematical problem published in Roce's 1505 edition in the practical arithmetic (Ciruelo 1505b; Bradwardine 1505). Finally, between Roce's and Lambert's publications, we know of at least three editions of Ciruelo's and Bradwardine's arithmetical texts, showing that these books were probably largely studied in Paris between the departure of Ciruelo and the main period of the *calculatores*. Nevertheless, Ciruelo's other mathematical books were mainly reprinted during the *calculatores* period.

---

[4] This information is given by the ISTC, ic00699600.

3.2   The Parisian Calculatores *and the Colleges of Iberian Tradition*

The Parisian *calculatores* current, as we said, followed the works of medieval theologians and mathematicians primarily active at Oxford and Paris in the four-teenth century. Unlike ancient philosophers, the *calculatores* argued that physics and mathematics were complementary disciplines, and they were particularly inter-ested in studies related to proportions and movement. These studies were quickly diffused in Europe, and by the fifteenth century, works on proportions also led to the improvement of algebraic principles (Veltman 2000, 401–404).

The *calculatores* works seem to have interested scholars from universities where these disciplines were taught at an advanced level at the end of the fifteenth century. In Paris, mathematical works from the *calculatores* began to be published by scholars who were primarily in Spain, such the Sicilian Renaud Montoro (15th cent–16th cent.), who did most of his theological studies at the University of Salamanca, but also at the University of Paris (Beltrán de Hereda 2001). By the beginning of the 1480s, during his Parisian stay, Montoro published an edition of Albertus de Saxonia's (ca. 1316–1390) *Proportiones*; Albertus Saxonia was a Parisian *calculator* of the four-teenth century (Saxonia ca. 1485). So too, as we said, Pedro Ciruelo also published mathematical works from Thomas Bradwardine, an Oxfordian *calculator* of the same period.

Even if the University of Paris was constantly attractive for international students, the end of the fifteenth century was an important moment for the constitution of an Iberian scholarly community. In fact, by the end of the 1490s, their arrival started to be recorded by three Parisian colleges, namely Coqueret, Sainte-Barbe, and Montaigu, neighbors at the Saint-Hilaire Mount, and progressively known for the attendance of Iberian scholars. These scholars were students of the liberal arts, as well as masters who were teaching these disciplines while finishing their doctorate, mostly on theology, but also on law or medicine (Quicherat 1860, 77).

The Iberian scholarly community at Paris developed primarily for diplomatic reasons. In 1498, John Standonck (1453–1504), master of the college of Montaigu, was reforming the establishment with the financial support of the admiral Louis Mallet de Granville (1438–1516), a former student of the college (De Matos 1950). Nevertheless, part of this financial support was taken from booty confiscated from a French privateer who attacked a Portuguese merchant ship; eventually, Emmanuel I, King of Portugal (1469–1521), took the money back (De Matos 1950). In conse-quence, John Standonck proposed to Emmanuel I to direct the money to the refor-mation of the college and promised in exchange to provide two grants for Portuguese students and to make the king himself a benefactor of the establishment (De Matos 1950). The king accepted, and soon Montaigu became the first destination for Portuguese students in Paris (De Matos 1950). By the beginning of the sixteenth century, the college also attracted Spanish students, and they were so numerous that the college of Sainte-Barbe, next to Montaigu, and the college of Coqueret, next to Sainte-Barbe, became two other main destinations for Iberian scholars.

The Parisian *calculatores* of the sixteenth century were not all Iberian scholars, but most of them were attached to colleges of the Iberian tradition. Thus, the *calculatores* current was first defined by a geographical and cultural space. Of the authors who published mathematical books, four of them were Iberian scholars—Juan Martinez Siliceo (ca. 1486–1557), Gaspar Lax (1487–1560), Juan Luis Vives (1492–1540), and Alvaro Thomás—but two of them were from other regions, namely Juan Dullaert (ca. 1480–1513), from The Netherlands and Jérôme de Hangest (d. 1538), from France. Extant scholarlship is not always clear about these authors' places of study and teaching, but according to Jules Quicherat, Lax and Vives studied together at Sainte-Barbe (Quicherat 1960, 88). In addition, it seems that Lax and Dullaert taught at the college of Montaigu, and Álvaro Tomás taught at the college of Coqueret (Farge 1980; Leitão 2000). Jérôme de Hangest, who was the first author to publish a mathematical book related to the *calculatores* current, was on the other hand not attached to a college of Iberian tradition but rather taught at the college of Reims, which was immediately next to Sainte-Barbe and Coqueret (Farge 1980).

3.3   Publishers of Mathematical Books During the Calculatores Current

Parisian mathematical production related to the *calculatores* current was mainly concentrated between 1508 and 1515. The *calculatores* themselves were principally interested in publications related to proportions, arithmetic, and astronomy, and their books were first published for their students. Because of this important Parisian demand for mathematical books, several Parisian printers and booksellers published mathematical books during this period. For the only time between the end of the fifteenth century and the middle of the sixteenth century, Parisian mathematical book production exceeded two percent of Parisian printing production in general.[5]

3.3.1   Jean Petit (1508)

The first mathematical book related to the Parisian *calculatores* current was Jérôme de Hangest's *Liber proportionum*, published by Jean Petit in June 1508 (Hangest 1508). Petit had been the main publisher of Hangest's works since the beginning of the sixteenth century. Two months later, however, Petit also published a second edition of Sacrobosco's *Sphaera* by himself, including Ciruelo's commentaries (Sacrobosco 1508), not reprinted in Paris since the first edition of 1498. Unlike the 1498 edition, shared with Guy Marchant, Petit alone seems to have been responsible for the 1508

---

[5] Numbers related to Parisian printed production are based on the Incunabula Short Title Catalogue (ISTC) and the Bibliographie des éditions parisiennes du 16ᵉ siècle (BP16).

publication: he is the only one explicitly mentioned in the colophon, and all the copies were distributed under his device. Jean Marchant, the printer, only appears trough the address of his workshop.

The 1508 edition of Sacrobosco's *Sphaera* presents two main differences compared to the 1498 publication. First, redrawing the historical woodcuts was not related to the subjects of the book: From the beginning of the 1500s, scholarly publication began to be formally different from books addressed to a larger audience. Then, the choice of a two-column layout, instead of a single line presentation of the text. This can perhaps be explained by way of financial reasons: as we said, Sacrobosco's *Sphaera* with Ciruelo's commentaries was particularly bulky, and therefore expensive. However, a two-column layout allowed a more readable text for smaller types. By choosing smaller characters, the publisher reduced the printing area of the text and saved fourteen sheets per copy (eighty folios instead of one hundred and eight folios), which is enough to print, for example, an equivalent run of fifty-folio *quarto* books.

3.3.2   The Marnef Brothers (1509)

In September 1509, the brothers Marnef (fl. 1485–1533) also started publishing mathematical books for the *calculatores*. They are known for their large network of bookshops around the country and their diversified production of printed books. The first mathematical text published by them was an edition of Ciruelo's *Algorismus* printed by Jean Marchant (Ciruelo 1509). So far, we do not know of any other edition of this text published between 1505 and 1513, and the Marnef brothers' edition is only known by one copy. Perhaps, other editions of Ciruelo's arithmetical books existed at the same time.

3.3.3   Poncet Le Preux and Guillaume Anabat (1509–1510)

By 1510, a neighbor of the Marnef brothers, Poncet Le Preux (fl. 1498–1559), published Álvaro Tomás' *Liber de triplici motu*, another book related to the *calculatores* current (Tomás ca. 1510). It is also one of the most advanced mathematical books published in Paris in the first half of the sixteenth century (Leitão 2000). The text is addressed to the students of the college of Coqueret, as is noted by a colleague of the author, Georges Bruneau de Vendôme, in a letter published at the end of the same edition. It presents all the formal characteristics of the *calculatores* books: gothic characters, two-column layout, red and black composition, and a heavy frame on the title page. The title page also explicitly refers to the original *calculatores* current as it mentions the fourteenth-century Oxfordian mathematician Richard Swineshead (d. 1354) and his *calculationes*.

The *Liber de triplici motu* does not contain a date of publication, but the *explicit* says that the author finished his work in February 1509, which could also be February 1510 according to the Easter calendar. The state of Le Preux's device is anterior to October 1510 (it does not contain a damage in the right side of the frame), so the book is probably published before this date (Valla 1510). The printing process is assured by Guillaume Anabat (fl. 1505–1510), who is mainly known for the publishing of religious texts. However, in the two last years of his career, from 1509 to 1510, he mostly printed scholarly books, especially for Iberian authors (Renouard 1964, 29–46). This could perhaps be explained by the fact that the printing of religious books required skills in red and black composition, a delicate process based on two press runs, also demanded for the publishing of Iberian scholarly books.

Individual copies of the *Liber de triplici motu* are particularly interesting because they present multiple states: some copies contain sheets entirely recomposed (i.e. four *folio* pages) but with no modification of the content. However, recomposed sheets differ from one copy to another and present red and black elements, while regular sheets are only printed in black. In addition, copies containing recomposed sheets, less numerous, were distributed under Guillaume Anabat's device, while regular copies were distributed under Poncet Le Preux's. The probable explanation here is that once all the sheets were printed, Le Preux allowed Anabat to keep the incomplete copies and to reprint the missing sheets so he could sell them under his own device. But Anabat went further, printing these sheets in red and black, and thereby highlighting his own run.

3.3.4 Thomas Kees (1510–1512)

For some printers, the publishing of mathematical books was indeed a way to stand out. Many examples can be found among the most important printers of the time, such as Erhard Ratdolt, Wolfgang Hopyl, or Henri Estienne I (fl. 1502–1520). But little-known printers also published mathematical books to highlight their technical skills and the quality of their work. Such is the case of Thomas Kees (fl. 1507–1515), a mysterious printer who had an important role in the publishing of mathematical books during the *calculatores* current. We do not know much about Kees, but according to Philippe Renouard, he was a modest printer, active from 1507 to 1515 (Renouard 1965, 223). He mainly printed for other publishers and never had his own device. By 1510, he had begun printing mathematical books. For the printing of these books, he acquired a large set of astronomical and astrological woodcuts often requested for the mathematical texts of the *calculatores* current.

The first part of this set was utilized in the publishing of an edition of Paulus Venetus' (ca. 1372–1429) *De compositione mundi* (Paulus Venetus, ca. 1510), an introduction to astronomy extracted from a *Philosophia naturalis* of the same author: an extensive commentary on Aristotle's (384–322 BCE) works. The Parisian edition was prepared by Juan Dullaert based on a Venetian edition published by Ottaviano Scotto (fl. 1490–1501) in 1498 (Paulus Venetus 1498). It was printed and distributed by Thomas Kees, who replaced the publisher's device with a historiated woodcut. The edition is not dated, but thanks to the second part of the set of illustrations, we know that it was not published after 1512.

The second part of the set was acquired for the publishing of an edition of Hyginus' (ca. 1st cent.) *Poeticon astronomicon* (Hyginus 1512), dated from May 1512. It was still printed by Thomas Kees, but shared with Olivier Senant (fl. 1505–1526), neighbor of Denis Roce and Jean Lambert. There is no secondary author mentioned in this edition, but a letter from Juan Luis Vives published in 1514 says that this book was also prepared by Juan Dullaert. For the printing of this text, Kees used woodcuts from his edition of the *De compositione mundi* and acquired new woodcuts based on a Venetian edition of the *Poeticon astronomicon* published by Giovanni Battista Sessa (fl. 1489–1505) in 1502 (Hyginus 1502). In other words, while the edition of the *De compositione mundi* was entirely realized on the basis of Scotto's edition, the *Poeticon astronomicon* was based on Scotto's *De compositione mundi*, as well as on Sessa's *Poeticon astronomicon*. The *Poeticon astronomicon* was therefore probably not published before the *De compositione mundi*.

3.3.5 Jean Petit (1511–1512)

Jean Petit sold and published mathematical books from the beginning of his career and especially during the *calculatores* current: after the two astronomical books published in 1508, he also published a philosophical and geometrical work with Henri Estienne in 1511, as well as a second edition of Bradwardine's *Geometria speculativa*. In fact, geometry was a discipline rarely taught in Paris compared to arithmetic and astronomy: Bradwardine's *Geometria*, first published in 1495 by Guy Marchant, was the only mathematical book from Ciruelo's bibliography not to be reprinted before the 1510s. So too, the beginning of Euclid's (323–285 BCE) geometry, regularly printed at the end of Lefèvre d'Étaples' editions of Sacrobosco's *Sphaera*, was entirely contained in three *folios* (Sacrobosco 1500). Finally, Charles de Bovelles' (1479–1567) introduction on geometry, published in a collective mathematical textbook of 1503 (Lefèvre d'Étaples 1503), when Bovelles was probably teaching in Paris, was not reprinted with the first part on the textbook in 1511 (Lefèvre d'Étaples 1511), after the departure of the author. On Bovelles' life, see (Klinger-Dollé 2016).

Because of the lack of teaching in this discipline, geometry was the first mathematical subject in Paris to be published by scholars for audiences other than students. In 1511, two geometrical texts were published addressed to other readers. Both were printed by Henri Estienne, who was the main publisher of Bovelles' works. The first was a collection of letters, mostly on philosophy, known by the title *Liber de intellectu* and addressed to an advanced scholarly audience (Bovelles 1511a). The four last letters, all on geometry, are regrouped under the name *Mathematicum opus quadripartitum*. The book is a bulky volume constituted by some two hundred *folios* richly illustrated with woodcuts especially realized for the publication. The second text, published in September of the same year, was the *Geometrie française*, a compendium of mathematical principles written and published in vernacular (Bovelles 1511b).

The *Liber de intellectu* was not only published by Henri Estienne. Even if all the title pages only mention his name, the colophon says that the book is printed by Henri Estienne and financed by Henri Estienne and Jean Petit I "associated in the art of copper." Estienne and Petit did not print many books together but their association with this publication might potentially be explained by the high costs of the printing of the book because of the amount of paper and the illustrations. On the other hand, the book was also too specialized to be sold by Jean Petit, who indeed does not seem to have sold any copies, at least under his own device. Instead, his name appears in the title pages of another book financed by Henri Estienne but sold by Estienne and Petit: Aristotle's (ca. 385–323 BCE) *Moralia*, published some weeks before Bovelles' *Liber de intellectu*, at the end of 1510 (Aristotle 1510).

Probably because of the publication of various geometrical books in Paris at the beginning of the 1510s, Jean Petit also financed the second edition of Bradwardine's *Geometria speculativa* (Bradwardine 1511) between 1511 and 1512. However, this edition does not appear to have been prepared by a professor of the university, as it contains many mistakes in Latin and mathematics, mainly related to abbreviations and the technical diagrams. In addition, this edition presents a more modern layout, with roman characters and long line disposition of the text (instead of two columns), far from the *calculatores* books of the beginning of the 1510s. Therefore, it does not seem that Petit was publishing this book for the *calculatores* and their students, but rather for Estienne's and Bovelles' audience: perhaps this seemed like a good time to publish a geometrical text.

3.3.6   The Marnef Brothers (1512–1515)

The Marnef brothers also saw an opportunity to publish mathematical books in the *calculatores* period. As early as 1509, an edition of Ciruelo's arithmetical text had already been financed by them. In the first half of the 1510s, two other mathematical books were published under their device: a compilation of three medieval texts on proportions, respectively from Albertus Saxonia (1316–1390), Thomas Bradwardine, and Nicole Oresme (ca. 1322–1382), and an edition of Sacrobosco's *Sphaera* following the first Parisian model (Saxonia et al. [1512–1515]; Sacrobosco [1512–1515]). These two books seem to have been published on the initiative of the booksellers, as they do not present the name of a scholar or any paratexts that could indicate that they were prepared for a specific course. The *Sphaera* book is very faulty in terms of typographical composition.

These two books published by the Marnef brothers are not dated, but they present the same publisher device (Renouard 1926, no. 718). The only dated occurrences of this device that we know of are from 1512 (Barletta 1512; Lemaire de Belges 1512). The two mathematical books published by the Marnef brothers were probably published by this date according to the state of their device. In addition, the text on proportions was probably published before 1515—before the end of the main period of the *calculatores* current—according to the subject of the book and the presentation of the text. The printer is not mentioned, but both editions include decorated letters used by Pierre Vidoue (fl. 1510–1543) in later publications (Erasmus 1520; Lefèvre d'Étaples 1533). We do not know of any book signed and dated by Vidoue before 1516, but according to Renouard and Jean de La Caille, he was active from 1510 (Renouard 1965, 428–429).

3.3.7 Thomas Kees, Jean Petit, and Jean Lambert (1513)

In 1513, Thomas Kees, Jean Petit, and Jean Lambert, all publishers of mathematical books, started a partnership together. They publish several books related to different subjects, among them three mathematical books: another edition of Ciruelo's *Algorismus* (Ciruelo 1513a), Juan Martinez Siliceo's *Liber arithmetica practice* (Siliceo 1513), and another edition of Paulus Venetus' *De compositione mundi* (Paulus Venetus 1513).

Unlike Ciruelo's *Algorismus*, published several times in Paris, Siliceo's *Liber arithmetice* practice was a previously unpublished text. Mostly known for his career in religion after his return to Spain (Quero 2014), Siliceo also published arithmetical texts reprinted several times in Paris until the middle of the sixteenth century. The first version of his arithmetic, on practical issues, was published in June 1513 by the three associated publishers. The text is explicitly addressed to the *calculatores*, as is mentioned in the complete title of the work and in the text itself. It is also the first time that special mathematical typographical characters, representing crossed numbers, were used in a Parisian printed book. The text is signed "Juan Martinez Blasius," a pseudonym of Juan Martinez Siliceo, who was named, in fact, Juan Martinez Guijarro (however, it does not seem that Juan Martinez Siliceo was the same person as Juan Martinez Población, who also published mathematical books in Paris in the 1510s (Levy 2020, 95–98).

Finally, Paulus Venetus' *De compositione mundi* was not only distributed by the three associated publishers, but by at least seven publishers, including Poncet Le Preux, Gilles de Gourmont (fl. 1499–1533), Claude Chevallon (fl. 1506–1537), and François Regnault (fl. 1501–1540). It was published three months after the premature death of John Dullaert (Elie 1951, 222), who prepared and taught the first Parisian edition. The title page announces the entire *Summa philosophia naturalis* of Paulus Venetus, specifying however that "the book starts with the *De compositione mundi*," which is the only text actually printed. In fact, the *Summa philosophia naturalis*, from which the *De compositione mundi* is extracted, was published in 1514 by most of the same associated printers and booksellers. The main party responsible for these publications seems to have been Thomas Kees, who is mentioned in the colophon as the printer and proofreader of the text. Jean Petit, on the other hand, was not among the publishers of this second part of the work.

3.3.8 Hémon Le Fèvre (1514–1515)

From 1514, Thomas Kees, Jean Petit, and Jean Lambert were no longer working as an association. However, another printer based in front of the Saint-Benoît cloister, next to Roce and Senant, started printing mathematical books: Hémon Le Fèvre (fl. 1509–1525). It is with Le Fèvre and Kees that Juan Martinez Siliceo published the second version of his arithmetical text, revised and extended, the *Arithmetica theoricen et praxim* (Siliceo 1514). This book presents a common second part to the *Liber de arithmetice practice*, but with many differences. First, it was printed in a smaller quarto format with roman characters, arringed in long lines, and without red. It also includes both theorical and practical issues and an additional introduction from the author, in which he states that Jacques Lefèvre d'Étaples and Josse Clichtove (ca. 1472–1543)—the greatest representants of the classical mathematical current—are among the most important mathematicians of the time.

Nevertheless, Kees and Le Fèvre also published many other books for the *calculatores*, before and after Siliceo's *Arithmetica theoricen et praxim*, much closer to the *calculatores* expectations. Therefore, the modern presentation of this book seems to have been realized on the initiative of the author himself, who was openly inspired by the authors and publishers of the classical current. In addition, Siliceo appears in this book as the main party responsible for its publication, as all the copies were distributed under his author's device and signed with his own initials "JMS" (Renouard 1926, no. 601). The address is however from Le Fèvre's workshop, and the colophon says that the book is published by Le Fèvre.

After the publication of Siliceo's text, Hémon Le Fèvre became a well-known Parisian publisher of mathematical books; he also published at least to other texts on these subjects: Tommaso Tedeschi's (1488–1527) *Sideralis abyssus* prepared by Nicolas Bérault (ca. 1473–ca. 1550) in 1514, and two books on arithmetic and proportions written in Paris by the Spanish *calculator* Gaspar Lax in 1515. The *Sideralis abyssus* is a text on the relation between astronomy and theology first published in Pavia in 1511 (Tavuzzi 1994). It is the first modern mathematical text published in another country before being reprinted in Paris (Radini Tedeschi 1514). It was also one of the first books published by Nicolas Bérault, who was preparing his doctorate on law and teaching liberal arts in Parisian colleges (Delaruelle 1902). The text was printed again by Thomas Kees with his mathematical illustrations. The content of the book is not precisely related to the *calculatores* intellectual current, but the book was printed following the conventions of the mathematical texts published for Iberian scholars: in the middle of the 1510s, Parisian scholarly mathematical book production was still divided in two main editorial currents.

Finally, between October and December 1515, Hémon Le Fèvre published Gaspar Lax's *Arithmetica speculativa* and *Proportiones* (Lax 1515a, b). Thomas Kees was not active anymore and was replaced by his neighbor, Nicolas de La Barre (fl. 1496–1528). The texts were published as two separate editions with their own title page and colophon but were probably meant to circulate together as they appear in Parisian copies of this book (Paris, Bibliothèque nationale de France, RES-R-141; Paris, Bibliothèque Mazarine, 2° 4621; Paris, Bibliothèque Mazarine, 2° 4621 bis). There is no Parisian copy of the *Proportiones* bound without the *Arithmetica speculativa*. The introduction of the *Proportiones* says that this text was published as a complement to the *Arithmetica speculativa*, but according to colophons, the *Proportiones* was printed two months before the arithmetical text: the formal distinction between these books was then premeditated and reinforced by a different arrangement of the text for each book. In addition, the red and black title page for the *Arithmetica speculativa* appears as an invitation to place this text before the *Proportiones*. Thereby, even if Lax's publications were close to the standard expectations of the *calculatores*, the book as an object is not less important, but considered part of the intellectual content.

3.3.9 Jean Lambert (1514)

Meantime, Jean Lambert kept publishing mathematical books, sometimes with other colleagues. By 1514, two astronomical texts, first prepared by John Dullaert, had been republished by Lambert: Hyginus' *Poeticon astronomicon* and Paulus Venetus' *De compositione mundi* (Hyginus 1514; Paulus Venetus 1514). Those were then revised by Juan Luis Vives, who was a student of John Dullaert and who writes about his project in a letter to his friend Johannes Fortius, published at the end of the *Poeticon astronomicon* (González 2015). The books are not explicitly dated but Lambert's *vignette* on the title page suggests a date before November 1513 and Vives' letter is signed from March 1514 (Paulus Venetus 1513). In addition, we do not know of any book published by Jean Lambert after 1514.

The form of these two books seems to have been influenced by Siliceo's *Arithmetica in theoricen et praxim*, as they were both printed in a small quarto with roman typographical characters. In his letter, Vives says that he himself asked the printer to publish the book in more readable types. Moreover, the two editions present Siliceo's author device, redesigned twice for their publication: in the *Poeticon astronomicon*, Siliceo's initials are replaced by a rooster and three faces, and in the *De compositione mundi*, the rooster and the three faces are replaced by the words "Spes mea Deus" printed in typographical characters. The woodcut was considerably damaged between these modifications, so we can situate the publications in this order chronologically. Lastly, both publications were printed with Thomas Kees material, but his name is not mentioned. We also do not know if he was still active at this moment, or if these two texts were printed by a successor. The *De compositione mundi* presents a messy application of the signature system (A^4 [A2 sig. A1] A^4 B-D^4 E^6 [E5 sig. F2] G-I^4); as this was an important element in the printing process of a book, it was perhaps produced by someone less experienced.

3.3.10   Following Editions of Ciruelo's Arithmetical Texts (1514–1515)

In the middle of the 1510s, Ciruelo's arithmetical texts were still regularly printed in Paris by several publishers, such as Denis Roce, Jean Lambert, and Olivier Senant, but also Jean de Gourmont (fl. 1506–1522) and Michel Lesclancher (fl. 1512–1520). From the end of the fifteenth century, Lambert and Roce offered many editions of Ciruelo's arithmetical texts, and between 1513 and 1514, they both published editions of Ciruelo's *Algorismus* (Ciruelo 1514a, b). Roce's edition, dated May 1514, does not present the name of a printer, but the material seems to come from Guillaume Des Plains (fl. 1512–1521), who only signed a few publications (Benoît 1521). Lambert's edition, on the other hand, presents the name of printer Antoine Aussourd (14th–15th cent.), but is dated from the beginning of the year, March 1513, which could also be March 1514 according to the Easter calendar.[6]

Meantime, Olivier Senant (probably influenced by the mathematical production of his neighbors) published an edition of Bradwardine's *Arithmetica speculativa*, following Ciruelo's editions of this text but in a *folio* format (Bradwardine 1514). This is an interesting choice, as all the other known Parisian editions of Ciruelo's and Bradwardine's arithmetical texts were published in a *quarto* format. For the most part, the format of an edition of a scholarly book was chosen according to the number of sheets used in the production of each copy. However, by choosing a *folio* format, Senant proposed something else: an edition of Bradwardine's *Arithmetica speculativa* that could be bound with other mathematical texts printed in the same format. The book is not explicitly dated, but the state of Senant's device is posterior to May 1514 (Jean de Jandun 1514). In addition, we do not know of any occurrence of this device after 1514, so the book was probably published that year.

Finally, the two editions published by Jean Gourmont and Michel Lesclancher, respectively on practical and theorical arithmetic, are different from the previous editions, as they are printed "with corrections and additions" (Ciruelo ca. 1515; Bradwardine ca. 1515). Neither are explicitly dated: Gourmont's edition was definitely published after 1513, as we find the same device in a much better state printed in a book containing a privilege from this year (Sabellico [1513]). In addition, the corrections to Gourmont's edition are not reported on Roce's editions of 1514, while they are present in a later edition (Ciruelo 1524) published in 1524 by Prigent Calvarin (fl. 1518–1566). Lesclancher's edition of Bradwardine's *Arithmetica speculativa* also contains modifications that are not reported in Senant's edition, which could perhaps indicate that his publication was printed after 1514. Between 1515 and the beginning of 1516, he also printed two mathematical books for other publishers, Jean Petit and Regnault Chaudière (fl. 1509–1554).

---

[6] The *Inventaire chronologique* dates the book from March 1314n.st. based on Lambert's address "next to the college of Coqueret" (Moreau 1977, 241), but according to Philippe Renouard, Lambert had been established at this address since 1511 (Renouard 1965, 234).

3.3.11 Jean Petit and Michel Lesclancher (1515–1516)

During his career, from 1512 to 1520, Michel Lesclancher mainly printed for other publishers. Therefore, he only signed a few books and he did not seem to have his own device. In 1515, he replaced Jean Marchant in printing Sacrobosco's *Sphaera* with Ciruelo's commentaries (Sacrobosco 1515), again under the responsibility of Jean Petit. However, in this new edition, Lesclancher's name and address are completely omitted, and his presence is only recognizable by the material employed for the printing of the book (Peuerbach 1515). In addition, for this publication, Jean Petit obtained the main historiated diagram used on Estienne's editions of Sacrobosco's *Sphaera*, representing Urania, Astronomia, and Ptolemy under an armillary sphere. This suggests that these two editions were not published concurrently, as they are indeed addressed to different audiences.

Moreover, in January 1515 or 1516 (probably the latter), Petit and Michel Lesclancher also published the first Parisian edition of Georg Peuerbach's (1423–1461) *Theoricae novae planetarum* (Peuerbach 1515), revised by Oronce Fine (1494–1555) and following the same disposition of text and paratexts used on the 1515 edition of Sacrobosco's *Sphaera*. Peuerbach's *Theoricae novae* was not related to the *calculatores* current, but it was probably designed to be gathered with the 1515 edition of Sacrobosco's *Sphaera*, as both were published by the same printer for the same bookseller, with the same general layout of the text. The colophon mentions Regnault Chaudière as co-publisher of this text, but the title pages seem to be distributed in the name of Jean Petit. In fact, it is possible that Chaudière's participation was mostly related to the presence of the technical diagrams, realized by Oronce Fine for his own edition of Sacrobosco's *Sphaera* (Sacrobosco 1516). Following editions were exclusively published by him.

4 The End of the *Calculatores* Current

Oronce Fine, Regnault Chaudière, and Simon de Colines (fl. 1520–1546) were among the main representatives of a new generation of actors in Parisian mathematical book production, which divided the first fifteen years of the sixteenth century between a classical teaching of the *quadrivium* and the *calculatores* current. The *calculatores* current, as we said, developed in a particular geographical and cultural space—the Parisian colleges of Iberian tradition—but it is also the product of a generation of scholars who taught liberal arts in these colleges. It spawned a local market for the publishing of mathematical scholarly books, notably around some personalities of the Parisian book trade, such as Jean Petit or Thomas Kees, and in some specific parts of the university neighborhood, such as in front the Saint-Benoît cloister.

The teaching of liberal arts was generally a provisory activity for graduated masters while preparing their doctorate (Quicherat 1860, 77). Between the middle of the 1510s and the beginning of the 1520s, most of these masters who were publishing mathematical books were no longer teaching liberal arts in these colleges: Vives moved to Flanders between 1512 and 1514 (González 2015, 45), Siliceo went back to Spain in 1517 (Villoslada 1938, 191), Hangest left Paris for Le Mans in 1519 (Bietenholz and Deutscher 2003, 409), Tomás had obtained his medical doctorate by 1520—after which we have no other information about him (Leitão 2000)—and Lax was in Spain in 1524 (Villoslada 1938, 406). From 1515, there are no more mathematical books published for the Parisian *calculatores* current, except a last reprint of Hyginus' *Poeticon astronomicon* in 1517 (Hyginus [1517]). By the middle of the 1520s, there were no more *calculatores* from the previous generation teaching mathematics in Paris.

In general, most of the books published for the *calculatores* current were not republished after the departure of the authors. This includes Ciruelo's books, even if the practical arithmetic was still republished in the 1520s by Prigent Calvarin (fl. 1523–1566). Bradwardine's *Geometria speculativa* was also republished in 1530 by Regnault Chaudière following the previous editions, but the name of Ciruelo is completely absent (Bradwardine 1530). The other books related to the *calculatores* were not reprinted after the end of the current, except Siliceo's *Arithmetica theoricen et praxim*, republished by Oronce Fine in 1519 and Thomas Rhaetus in 1526, both teachers of liberal arts in Paris (Siliceo 1519, 1526). In 1540, the Parisian publisher Jean Loys, personally interested in mathematics, gave a summary of this text, reprinted by Prigent Calvarin in 1542 (Siliceo 1540, 1542).

Most of the publishers who produced mathematical books for the *calculatores* current stopped publishing scholarly mathematical texts after this period. Some of them were not active anymore by the second half of the 1510s, such as Denis Roce, Jean Lambert, and Thomas Kees, but others were still publishing in the 1520s, for example, Olivier Senant, Hémon Le Fèvre, and Jean Petit. In the following years, mathematical texts were sometimes present in their publications addressed to a larger audience, like Anianus' (14th cent.) *Computus manualis*, the *Coeur de philosophie*, or the *Calendrier des bergers*, but from the second half of the 1510s mathematical scholarly books were mostly published by Henri Estienne's family network, including Regnault Chaudière from 1516, and Simon de Colines from 1520.

Finally, in the long term, mathematical publications related to the Parisian *calculatores* do not stand out as a reference in the teaching of these disciplines. However, they presented an alternative to the traditional conception of the *quadrivium* and had a great influence on the development of a properly modern conception of mathematics. Moreover, the *calculatores* current also led to an important increase in the production of mathematical textbooks in Paris, in the number of publishers interested in these publications, and in the presence of these disciplines in the Parisian market. In consequence, during the next fifteen years, Parisian professors and publishers would not look so much for the publication of new mathematical textbooks, but would invest their time and skills in the publication of more advanced mathematical books, primarily addressed to an international audience: from the 1520s, Paris would be a main pole of the European mathematical book trade.

Abbreviations

Digital Repositories

BP16	Bibliographie des éditions parisiennes du 16ᵉ siècle. Bibliothèque nationale de France. https://bp16.bnf.fr/. Accessed 07 June 2021
ISTC	Incunabula Short Title Catalogue. Consortium of European Research Libraries. https://data.cerl.org/istc/. Accessed 07 June 2021
Sphaera Corpus*Tracer*	Max Planck Institute for the History of Science. https://db.sphaera.mpiwg-berlin.mpg.de/resource/Start. Accessed 07 June 2021

References

Primary Literature

Aristotle. 1510. *Decem librorum moralium Aristotelis tres conversiones: prima Argyropili Byzantii, secunda Leonardi Aretini, tertia vero antiqua per capita et numeros conciliate, communi familiarique commentario ad Argyropilium adjecto*. Paris: Henri Estienne for Jean Petit and himself.

Barletta, Gabriele. 1512. *Sermones*. Paris: [Jean Barbier] for the Marnef brothers, Jean Petit and François Regnault.

Belges, Lemaire de. 1512. *Le Second Livre des Illustrations de Gaule et singularitez de Troye*. Paris: Geoffroy de Marnef; Blois: Hilaire Malican.

Benoît, saint. 1521. *Regula*. Paris: Guillaume Des Plains for the Marnef brothers.

Bovelles, Charles de. 1511a. *Liber de intellectu. Liber de sensu. Liber de nichilo. Ars oppositorum. Liber de generatione. Liber de sapiente. Liber de duodecim numeris. Epistole complures. Insuper mathematicum opus quadripartitum. De numeris perfectis. De mathematicis rosis. De geometricis corporibus. De geometricis supplementis*. Paris: Henri Estienne.

Bovelles, Charles de. 1511b. *Geometrie en francoys. Cy commence le Livre de lart et science de Geometrie: avecques les figures sur chascune rigle au long declarees par lesquelles on peult entendre et facillement compendre ledit art et science de Geometrie*. Paris: Henri Estienne.

Bradwardine, Thomas. 1495. *Arithmethica*, ed. Pedro Sánchez Ciruelo. Paris: Guy Marchant.

Bradwardine, Thomas. 1496. *Geometria speculativa recoligens omnes conclusiones geometricas studentibus artium et philosophie Aristotelis valde necessarias simul cum quodam tractatu de quadratura circuli noviter edito*, ed. Pedro Sánchez Ciruelo. Paris: Guy Marchant.

Bradwardine, Thomas. 1502. *Arithmethica*, ed. Pedro Sánchez Ciruelo. Paris: Guy Marchant for Denis Roce.

Bradwardine, Thomas. 1505. *Arithmethica*, ed. Pedro Sánchez Ciruelo. Paris: Jean Lambert.

Bradwardine, Thomas. 1511. *Geometria speculativa recoligens omnes conclusiones geometricas studentibus artium et philosophie aristotelis valde necessarias simul cum quodam tractatu de quadratura circuli noviter edito*, ed. Pedro Sánchez Ciruelo. Paris: Jean Marchant for Jean Petit.

Bradwardine, Thomas. 1514. *Arithmetica*, ed. Pedro Sánchez Ciruelo. Paris: Thomas Anguelart for Olivier Senant.

Bradwardine, Thomas. 1530. *Geometria speculativa Thome Bravardini recogliens omnes conclusiones geometricas studentibus artium et philosophie Aristotelis valde necessarias simul cum quodam tractatus de quadratura circuli noviter edito. Elementale geometricum ex euclidis geometria a Joanne Voegelin Haulpronnensi ad omnium mathematices studiosorum utilitatem decerptum*, [ed. Pedro Sánchez Ciruelo]. Paris: [Jean Bignon] for Regnault Chaudière.

Bradwardine, Thomas. ca. 1515. *Arithmetica speculativa nuper mendis plusculis tersa et diligenter Impressa*, ed. Pedro Sánchez Ciruelo. Paris: Michel Lesclancher.

Ciruelo, Pedro Sánchez. 1495. *Tractatus arithmethice pratice qui dicitur Algorismus*. Paris: Guy Marchant.

Ciruelo, Pedro Sánchez. 1505a. *Tractatus arithmethice pratice qui dicitur Algorismus*. Paris: Jean Marchant for Denis Roce.

Ciruelo, Pedro Sánchez. 1505b. *Tractatus arithmethice practice qui dicitur Algorismus*. Paris: Jean Lambert.

Ciruelo, Pedro Sánchez. 1509. *Tractatus arithmethice practice qui dicitur Algorismus*. Paris: Jean Marchant for the Marnef brothers.

Ciruelo, Pedro Sánchez. 1513b. *Tractatus arithmethice practice qui dicitur Algorismus*. Paris: Antoine Aussourd for Jean Lambert.

Ciruelo, Pedro Sánchez. 1513a. *Tractatus arithmethice practice qui dicitur Algorismus*. Paris: Thomas Kees for Jean Petit, Jean Lambert and himself.

Ciruelo, Pedro Sánchez. 1514. *Tractatus arithmethice practice qui dicitur Algorismus*, Paris: [Guillaume Des Plains] for Denis Roce.

Ciruelo, Pedro Sánchez. ca. 1515. *Tractatus arithmetice practice qui dicitur Algorismus. cum additionibus utiliter adjunctis*. Paris: Jean de Gourmont.

Ciruelo, Pedro Sánchez. 1516. *Cursus quattuor mathematicarum artium liberalium quas recollegit atque correxit magister Petrus Ciruelus Darocensis theologus simul et philosophus*. Alcalá de Henáres: Arnaldo Guillén de Brocar.

Ciruelo, Pedro Sánchez. 1521. *Apotelesmata astrologiae christianae*. Alcalá de Henáres: Arnaldo Guillén de Brocar.

Ciruelo, Pedro Sánchez. 1524. *Tractatus arithmetice practice qui dicitur Algorismus cum additionibus utiliter adjunctis*. Paris: Prigent Calvarin.

Erasmus. 1520. *De la declamation des louenges de follie stille facessieux et profitable pour congnoistre les erreurs et abuz du monde*. Paris: Pierre Vidoue for Galliot Du Pré.

Hangest, Jérôme de. 1508. *Liber proportionum*. Paris: Jean Barbier for Jean Petit.

Hyginus. 1514. *Aureum opus historias ad amussim pertractans una cum multis astronomice rationis ambagibus et signis poetarum locis prope infinitis exacte callendis, non mediocriter conducturis in lucem editum habes candidissime lector quod pauxilla a te pecunia comparari poterit*, ed. Juan Luis Vives. Paris: [Thomas Kees] for Jean Lambert.

Hyginus. 1502. *Clarissimi Hyginii astronomi de mundi et sphere ac utriusque patrium declaratione cum planetis et variis signis historiatis*. Venice: Giovanni Battista Sessa.

Hyginus. 1512. *Aureum opus historiasque ad amussim pertractans una periter cum multis astronomice rationis ambagibus et signis poetarum locis prope infinitis exacte callendis, non mediocriter conducturis in lucem editum habes candidissime lector quod pauxilla tibi pecunia compari poterit*, [ed. John Dullaert]. Paris: Thomas Kees for Olivier Senant and himself.

Hyginus. post 1517. *Higinii hystoriographi et phylosophi argumentissimi libri quattuor non solum poeticas et hystoricas verum et astronomicas permultas veritates acriori collectas lima et laconica brevitate enodantes studiose tibi (quicunque es) extant adjectis nuper multis in locis ex multorum poetarum hystoriographorum ac phylosophorum libris notis, cum recente et utili tabula*, ed. Juan Luis Vives. Paris: Pasquier Lambert.

Jandun, Jean de. 1514. *Habes humanissime lector librorum metheorum Aristotelis facilem expositionem et questiones super eosdem magistri Johannis Dullaert de Gandavo, in quibus diversae astrologicae veritates ab omni erroris vicio immunes et philosophicis pariter et medicis conformes esse probantur*. Paris: Thomas Kees for Gilles de Gourmont and Olivier Senant.

Lax, Gaspar. 1515a. *Arithmetica speculativa duodecim libris demonstrata*. Paris: Nicolas de La Barre for Hémon Le Fèvre.

Lax, Gaspar. 1515b. *Proportiones*. Paris: Nicolas de La Barre for Hémon Le Fèvre.

Lefèvre d'Étaples, Jacques. 1503. *Epitome compendiosaque introductio in libros arithmeticos divi Severini Boetii, adjecto familiari commentario dilucidata. Praxis numerandi certis quibusdam regulis constricta. Introductio in geometriam breviusculis annotationibus explanata sex libris distincta. Primus de magnitudinibus et earum circunstantiis. Secundus de consequentibus contiguis et continuis. Tertius de punctis. Quartus de lineis. Quintus de superficiebus. Sextus de corporibus. Liber de quadratura circuli. Liber de cubicatione sphere. Perspectiva introductio. Insuper Astronomicon*. Paris: Wolfgang Hopyl and Henri Estienne.

Lefèvre d'Étaples, Jacques. 1511. *Epitome compendiosaque introductio in libros arithmeticos divi Severini Boetii, adjecto familiari commentario dilucidata. Praxis numerandi certis quibusdam regulis constricta. Introductio in geometriam breviusculis annotationibus explanata sex libris distincta. Primus de magnitudinibus et earum circunstantiis. Secundus de consequentibus contiguis et continuis. Tertius de punctis. Quartus de lineis. Quintus de superficiebus. Sextus de corporibus. Liber de quadratura circuli. Liber de cubicatione sphere. Perspectiva introductio. Insuper Astronomicon*. Paris: Henri Estienne.

Lefèvre d'Étaples, Jacques. 1533. *In hoc opere continentur Totius Philosophiae naturalis Paraphrases*. Paris: Pierre Vidoue for Galliot Du Pré and Jean Petit.

Nemorarius, Jordanus. 1496. *Arithmetica decem librs demonstrata. Musica libris demonstrata quattuor. Epitome in libros arithmeticos divi Severini Boetii. Rithmimachie ludus qui et pugna numerorum appelatur,* ed. Jacques Lefèvre d'Étaples. Paris: Wolfgang Hopyl and Jean Higman.

Peuerbach, Georg. 1515. *Theoricarum novarum textus Georgii Purbachii cum utili ac preclarissima expositione domini Francisci Capuani de Manfredonia. Item in eadem reverendi patris fratris Sylvestri de Prierio perfamiliaris commentatio. Insuper Jacobi Fabri Stapulen. astrononomicum. Omnia nuper summa diligentia emendata cum figuris ac commodatissimis longe castigatius insculptis quam prius suis in locis adjectis*. Paris: Michel Lesclancher for Jean Petit and Regnault Chaudière.

Radini Tedeschi, Tommaso. 1514. *Sideralis Abyssus. Auctor ad lectorem. Argutum, sublime, novum, spectabile, tersum, depexum, varium, nobile, molle, teres, si nostrum non lector opus censebitur, arbor vere virens, laudem quo mereatur habet. Nos quinti necdum primum qui implevimus orbem lustri, alto tutos equore concha tulit. Dum legis abscedat livor, metabere dicens, gaudia sint qui mi gaudia tanta dedit,* ed. Nicolas Bérault. Paris: Thomas Kees for Hémon Le Fèvre.

Sabellico, Marco Antonio Coccio. [1513]. *Rerum veneratum panegyricus primus quod Genethliacon inscribitur. Ejusdem temporale Carmen de Italie tumultu. Carmen ad Cipicum Coriolanum de domus deflagratione*. Paris: Jean de Gourmont.

Sacrobosco, Johannes de. 1482. [Incipit] *Sphericum opusculum*. Venice: Erhard Ratdolt. https://hdl.handle.net/21.11103/sphaera.100692.

Sacrobosco, Johannes de. 1485. [Incipit] *Sphericum opusculum*. Venice: Erhard Ratdolt. https://hdl.handle.net/21.11103/sphaera.101123.

Sacrobosco, Johannes de. 1489. [Incipit] *Sphericum opusculum*. Paris: Wolfgang Hopyl. https://hdl.handle.net/21.11103/sphaera.100821.

Sacrobosco, Johannes de. [1493a]. *Tractatus de sphera*. Paris: Antoine Caillaut. https://hdl.handle.net/21.11103/sphaera.100825.

Sacrobosco, Johannes de. 1493b. *Tractatus de sphera*. Paris: Georg Mittelhus. https://hdl.handle.net/21.11103/sphaera.101125.

Sacrobosco, Johannes de. [1494]. *Tractatus de sphera*. Paris: Félix Baligault. https://hdl.handle.net/21.11103/sphaera.100270.

Sacrobosco, Johannes de. 1498. *Uberrimum sphere mundi commentum intersertis etiam questionibus domini Petri de Aliaco,* ed. Pedro Sánchez Ciruelo. Paris: Guy Marchant for himself and Jean Petit. https://hdl.handle.net/21.11103/sphaera.100274.

Sacrobosco, Johannes de. 1500. *Textus de sphera cum additione (quantum necessarium est) adjecta. Novo commentario nuper edito ad utilitatem studentium philosophice Parisiensis academie illustratus. Cum Compositione anuli astronomici Boni Latensis et Geometria Euclidis Megarensis*, ed. Jacques Lefèvre d'Étaples. Paris: Wolfgang Hopyl. https://hdl.handle.net/21.11103/sphaera. 100889.

Sacrobosco, Johannes de. 1508. *Uberrimum sphere mundi commentum intersertis etiam questionibus Petri de Aliaco nuper magna cum diligentia castigatum*, ed. Pedro Sánchez Ciruelo. Paris: Jean Marchant for Jean Petit. https://hdl.handle.net/21.11103/sphaera.100642.

Sacrobosco, Johannes de. [1512–1515]. *Tractatus de sphera*. Paris: [Pierre Vidoue?] for the Marnef brothers. https://hdl.handle.net/21.11103/sphaera.100989.

Sacrobosco, Johannes de. 1515. *Habes lector sphere textum una cum additionibus non aspernandis Petri Ciruello D. (a vero tamen textu apparenter distinctis) cum ipsiusmet sublimi et luculentissima expositione aliquot figuris noviter adjunctis decorata. Intersertis praeterea questionibus domini Petri de Aliaco. Omnia pervigili cura ad amussim castigata. Et rursus coimpressa*, ed. Pedro Sánchez Ciruelo. Paris: [Michel Lesclancher] for Jean Petit. https://hdl.handle.net/21.11103/sph aera.100988.

Sacrobosco, Johannes de. 1516. *Mundialis sphere opusculum nuper vigilantissime emendatum una cum figuris accommodatissimis cumque marginariis annotaciunculis recenter adjectis*, ed. Oronce Fine. Paris: Vincent Quignon for Regnault Chaudière. https://hdl.handle.net/21.11103/ sphaera.100991.

Sacrobosco, Johannes de. 1526. *Opusculum de sphera mundi, cum additionibus et familiarissimo commentario Petri Ciruelli Darocensis, nunc recenter correctis a suo autore intersertis etiam egregiis questionibus domini Petri de Aliaco*, ed. Pedro Sánchez Ciruelo. Alcalá de Henares: Miguel de Ehuía. https://hdl.handle.net/21.11103/sphaera.100884.

Saxonia, Albertus de. [1482–1485]. *Proportiones*, ed. Renaud Montoro. Paris: [Antoine Caillaut].

Saxonia, Albertus de, Thomas Bradwardine, and Nicole Oresme. [1512–1515]. *Tractatus proportionum*. Paris: [Pierre Vidoue?] for the Marnef brothers.

Siliceo, Juan Martinez [signed Blasius, Juan Martinez]. 1513. *Liber arithmetice practice, astrologis, phisicis et calculatoribus admodum utilis*. Paris: Thomas Kees for Jean Petit, Jean Lambert and himself.

Siliceo, Juan Martinez. 1514. *Ars arithmetica in theoricen et praxim scissa, omni hominum conditioni perquam utilis et necessaria*. Paris: Thomas Kees for Hémon Le Fèvre.

Siliceo, Juan Martinez. 1519. *Arithmetica Joannis Martini, Scilicei, in theoricen, et praxim scissa, nuper ab Orontio Fine, Delphinate, summa diligentia castigata, longeque castigatius quam prius, ipso curante impressa, omni hominum conditioni perquam utilis, et necessaria*, ed. Oronce Fine. Paris: Henri Estienne.

Siliceo, Juan Martinez. 1526. *Arithmetica joanis martini Silicei theoricen praxinque luculenter complexa, innumeris mendarum offuciis a Thoma Rhaeto, haud ita pridem, accuratissime vindicata quod te collatio huius aeditionis cum priore palam doctura est*, ed. Thomas Rhaetus. Paris: Simon de Colines.

Siliceo, Juan Martinez. 1540. *Arithmetica Silicei nuper permultis mendis vindicata, et commentariorum prolixitate*. Paris: Jean Loys for Jean de Roigny and himself.

Siliceo, Juan Martinez. 1542. *Arithmetica Silicei nuper permultis mendis vindicata, et commentariorum prolixitate*. Paris: Prigent Calvarin.

Thomas, Alvarus. [1509–1510]. *Liber de triplici motu proportionibus annexis philosophicas Suiseth calculationes ex parte declarans*. Paris: Guillaume Anabat for Poncet Le Preux and himself.

Valla, Lorenzo. 1510. *De lingua latina quam optime meriti, de ejusdem elegantia libri sex, deque reciprocatione libellus, cum Antonii Mancinelli lima suis locis apposita*. Paris: Jean Barbier for Poncet Le Preux and François Regnault.

Venetus, Paulus. 1498. *Expositio magistri Pauli Veneti super libros de generatione et corruptione Aristotelis. Ejusdem de compositione mundi cum figuris*. Venice: Ottaviano Scotto.

Venetus, Paulus. ca. 1510. *Librum maximum de compositione mundi, una cum figuris ad intelligen-tiam astronomie accommodatissimum pervigili cura Johannis Dullaert de Gandavo ad amussim castigatum*, ed. John Dullaert. Paris: Thomas Kees.

Venetus, Paulus. 1513. *Primus liber incipit de compositione mundi. Summa philosophie naturalis una cum libro de compositione mundi qui astronomie janua nuncupari potest notissime recognita et sine aliquo errore in lucem emissa*, ed. John Dullaert. Paris: Thomas Kees for Jean Petit, Jean Lambert and himself.

Venetus, Paulus. 1514. *Philosophia naturalis compendium una cum libro de compositione mundi qui astronomia janua inscribitur, opus egregium non solum phisices studiosis utile ac pernec-essarium sed intelligendis explicandisque veterum scriptorum ac poetarum scriptis locisque infinitis apprime conducibile nuper Lutecie emaculatissime impressum*, [ed. Juan Luis Vives]. Paris: [Thomas Kees] for Jean Lambert, Olivier Senant and Pierre Gaudoul.

Secondary Literature

Beaujouan, Guy. 1997. Le quadrivium et la Faculté des arts. In *L'enseignement des disciplines à la Faculté des arts (Paris et Oxford, XIIIe–XVe siècles)*, eds. Olga Weijers and Louis Holtz, 185–194. Turnhout: Brepols.

Beltrán de Hereda, Vincente. 2001. *Cartulario de la universidad de Salamanca (1218–1600)*. Vol. 3, 814–815. Salamanca: Universidad de Salamanca.

Biard, Joël, and Rommevaux, Sabine (eds.). 2008. *Mathématiques et théorie du mouvement, XIVe–XVIe siècles*. Villeneuve d'Ascq: Presses universitaires du septentrion.

Bietenholx, Peter G., and Thomas Brian Deutscher. 2003. *Contemporaries of Erasmus: a biographical register of the Renaissance and reformation*. Toronto: University of Toronto Press.

BMC. 1963. *Catalogue of books printed in the XVth century now in the British Museum [British Library]*. Part VIII. London: Trustees of the British Museum.

Calderon, Calixto P. 1990. The 16th century Iberian calculatores. *Revista de la Unión Matemática Argentina* 35: 245–258.

Delaruelle, Louis. 1902. Notes biographiques sur Nicole Bérault suivies d'une bibliographie de ses œuvres et de ses publications. *Revue des bibliothèques* 12: 420–445.

De Matos, Luís. 1950. *Les Portugais à l'Université de Paris entre 1500 et 1550*. Coïmbra: Universidade de Coïmbra.

Elie, Hubert. 1951. Quelques maîtres de l'Université de Paris vers l'an 1500. *Archives d'histoire doctrinale et littéraire du moyen âge* 18: 193–243.

Farge, James K. 1980. *Biographical register of Paris doctors of theology, 1500–1536*. Toronto: Pontifical Institute of Mediaeval studies.

González, Enrique González. 2015. Juan Luis Vives sur les presses parisiennes et le dialogue Sapiens (1514). *Réforme, Humanisme, Renaissance* 80: 39–67.

Hillard, Denise. 1989. *Catalogues régionaux des incunables des Bibliothèques publiques de France, vol. VI. Bibliothèque Mazarine*, n° 1151. Paris: Aux amateurs de livres.

Hindman, Sandra. 1991. The career of Guy Marchant (1483–1504): high culture and low culture in Paris. In *Printing the written word. The social history of books, circa 1450–1520*, ed. Sandra Hindman, 68–100. Ithaca, London: Cornell University Press.

Klinger-Dollé, Anne-Hélène. 2016. *Le De sensu de Charles de Bovelles (1511), Conception philosophique des sens et figurations de la pensée*. Geneva: Droz.

Leitão, Henrique. 2000. Notes on the life and work of Álvaro Tomás. *Bulletin do Centro Internacional de Matemática* 9: 10–15.

Levy, Alissar. 2020. *Du quadrivium aux disciplinae mathematicae: histoire éditoriale d'un champ disciplinaire en mutation (1480–1550). Une recherche de bibliographie matérielle et d'histoire sociale du livre à Paris au XVIe siècle*. Thèse pour le diplôme d'archiviste paléographe, dir. Christine Bénévent.

Lorente y Péres, José María. 1921. Biografía y análisis de las obras de matemática pura de Pedro Sánchez Ciruelo. *Publicaciones del Laboratorio y Seminario Matemático* 3: 264–349.

Moreau, Brigitte. 1972. *Inventaire chronologique des éditions parisiennes du XVIe siècle d'après les manuscrits de Philippe Renouard*, 5 vol. Paris: Imprimerie municipale.

Oosterhoff, Richard. 2018. *Making mathematical culture: university and print in the circle of Lefevre d'Étaples*. Oxford: University Press.

Quero, Fabrice. 2014. *Juan Martínez Silíceo, 1486?–1557, et la spiritualité de l'Espagne prétridentine*. Paris: H. Champion.

Quicherat, Jules. 1860. *Histoire de Sainte-Barbe: collège, communauté, institution*. Paris: L. Hachette.

Renouard, Philippe. 1896. Quelques documents sur les Petit, libraires parisiens, et leur famille. *Bulletin de la Société de l'histoire de Paris et de l'Ile-de-France*: 133–153.

Renouard, Philippe. 1926. *Les marques typographiques parisiennes des XVe et XVIe siècles*. Paris: H. Champion.

Renouard, Philippe. 1964. *Imprimeurs et libraires parisiens du XVIe siècle*, vol. 1. Paris: Services des travaux historiques de la ville de Paris.

Renouard, Philippe. 1965. *Imprimeurs parisiens, libraires, fondeurs de caractères et correcteurs d'imprimerie: depuis l'introduction de l'imprimerie à Paris (1470) jusqu'à la fin du XVIe*. Paris: J. Minard.

Tavuzzi, Michael. 1994. An unedited *Oratio* by Tommaso Radini Tedeschi (1488–1527). *Archivum Historiae Pontificiae* 32: 43–63.

Trzeciok, Stefan Paul. 2016. *Alvarus Thomas und sein Liber de triplici motu*, 2 vol. Berlin: Edition Open Access. https://edition-open-sources.org/sources/8.

Veltman, Kim. 2000. Mesures, quantification et science. L'Époque de la Renaissance. T.4. In *Crises et essors nouveaux (1560–1610)*, eds. Tibor Klaniczay, Eva Kushner, and Paul Chavy, 401–416. Amsterdam: John Benjamins.

Veyrin-Forrer, Jeanne. 2017. La question des styles en France pour les livres imprimés. In *Études bibliographiques à la mémoire de Jeanne Veyrin-Forrer*, ed. Wallace Kirsop, 18–28. Monash: Ancora Press.

Villoslada, Ricardo. 1938. *La universidad de Paris durante los estudios de Francisco de Vitoria*. Rome: Gregorian Biblical BookShop.

Wallace, William A. 1969. The *Calculatores* in early sixteenth-century physics. *The British Journal for the History of Science* 4 (3): 221–232.

Alissar Levy defended her École des chartes' thesis in 2020 on the Parisian mathematical book production between 1480 and 1550. She is now Research Assistant at the Rare books department of the Bibliothèque nationale de France where she works on the production of the Parisian bookseller Jean Petit (1495–1540).

Index

A

Abenragel, Haly, 441, 446
Abril, Simón, 240, 243
Abū Isḥāq, Ibrāhīm ibn Yaḥyā al-Naqqāsh al-Zarqālī, 297
Abū Maʿshar, Jaʿfar ibn Muḥammad ibn ʿUmar al-Balkhī, 449, 453
Acquaviva, Claudio, 383, 385
Acuña y Sarmiento, Diego, 242
Aguilar, Francisco de, 241
al-Battānī, Muḥammad ibn Jābir, 439, 443
Al-Ma'mun, 441, 455
al–Farghani, Aḥmad ibn Muḥammad ibn Kathīr, 441
al–Qabīsī, Abū al-Ṣaqr ʿAbd al-ʿAzīz ibn ʿUtmān, 441
Albert the Great, 441, 446, 447
Albrecht V, Duke of Bavaria, 381
Aleaume, Jean, 302
Aleaume, Jerôme, 302
Aleaume, Marie, 302
Aleaume, Pierrette, 302, 304
Alfonso X, King of Castile, 443
Alsted, Heinrich, 198
Álvarez, Antonio, 229, 239, 242, 243, 247
Álvarez, Manuel, 379
Amerbach, Johann, 29, 168
Ammon, Johann, 212, 213
Anabat, Guillaume, 468, 469
Anianus, 477
Anshelm, Thomas, 157, 162
Apian, Peter, 303, 460
Aquinas, Thomas, 66, ,=[
Archimedes, 384
Aristotle, 28, 34, 66, 156, 171, 228, 230, 232, 239, 248, 371, 376, 379,
386–389, 392, 413, 431, 441, 442, 446, 469, 471
Aschhausen, Johann Gottfried von, 381
Augustine St., 447
Augustus, Elector of Saxony, 119, 120
Ausonius, Decimus Magnus, 312
Aussourd, Antoine, 475

B

Badius, Conrad, 301, 302
Badius, Jodocus, 301, 302
Barbari, Jacopo de', 100
Barreira, Antonio de, 228, 234
Basa, Domenico, 201, 202, 380, 382
Basa, Isabetta, 203, 244
Basilea, Isabel de, 232
Bassantin, James, 299, 300
Battink, Rudolf, 299, 300
Bech, Philipp, 171, 172
Bellarmine, Robert, 380, 382
Bellay, Joachim du, 313
Bellère, Jean, 199, 359
Bellère, Pierre, 199
ben Ezra, Abraham, 441
Benedict XII, Pope, 440
Beraud, Symphorien, 205
Bérault, Nicolas, 473
Bernutz, Georg, 116, 139
Berthot, Claude, 312
Besold, Hieronymus, 152
Bessarion, Basilus, 69
Beurhaus, Friedrich, 197, 197
Bevilacqua, Simone, 81, 87, 90, 351, 352, 438, 446
Beyer, Christian, 107, 126
Biancani, Giuseppe, 395

Birckman, Arnold I, 281, 303, 350
Birckman, Arnold II, 303
Birckman, Franz, 303
Blado, Antonio, 374, 375, 377
Blume, Heinrich, 115, 137, 140
Boethius, Anicius Manlius Severinus, 27,
    442
Boissard, Jean-Jacques de, 212
Boissière, Claude de, 307–309, 312–315
Bordes, Guillaume des, 314, 315, 321
Börner, Caspar, 432, 433
Borro, Gasparino, 80, 81, 83, 277
Bouré, Charles, 350, 351
Bovelles, Charles de, 29, 36, 308, 310, 470
Brahe, Tycho, 43
Brandis, Marcus, 438
Breitkopf, Gregor, 432, 433
Breunle, Moritz, 356
Brocar, Arnao Guillén de, 231
Brucioli, Antonio, 282
Brucioli, Francesco, 282
Bry, Theodor de, 212
Bugenhagen, Johannes, 107, 111, 135, 165,
    167
Bullant, Jean, 309, 314
Burckhardt, Adam, 64
Buridan, Jean, 80

C
Cadamosto, Marcantonio, 296, 297
Caesarius, Johannes, 170
Caillaut, Antoine, 460
Campano da Novara, 82, 277
Cánova, Alejandro de, 232
Cantzler, Hans, 118, 119, 135, 142
Capuano da Manfredonia, Francesco, 446
Cardon, Horace, 209, 210
Castañeda, Rodrigo de, 228, 230, 232
Castro, León de, 240, 243
Cavellat, Guillaume, 17, 18, 199, 289, 290,
    300, 301, 303–305, 313, 358, 359,
    460
Celtis, Conrad, 71
Cerda, Luis de la, 242
Charles V, King of Spain, 244
Chaudière, Regnault, 293, 308, 460,
    475–477
Chaves, Jeónimo de, 229, 236, 242, 243
Chevallon, Claude, 472
Cholinus, Goswin, 192, 198
Cholinus, Maternus, 198, 359
Cholinus, Peter, 198

Chouet, Jacques, 194, 204, 207, 209, 210
Chyträus, David, 159, 160, 169, 356
Cicero, Marcus Tullius, 26, 71, 376
Ciotti, Giovanni Battista, 187, 190, 192,
    194–203, 207, 216, 244, 274, 276,
    282, 283, 285, 350, 357, 381, 382
Ciruelo, Pedro Sánchez, 441, 459–468,
    470–472, 475–477
Cisneroz, Jiménez de, 231
Citron, Nana, 429, 445
Claraevallensis, Bernardus, 271
Clavius, Christoph, 13, 14, 187–190, 192,
    196, 198–210, 215–217, 230,
    242–244, 276, 277, 349, 353, 354,
    357, 360, 369, 372–376, 378,
    381–396, 398, 399, 446
Clichtove, Josse, 29, 31, 32, 473
Colines, Simon de, 7, 25, 27, 32, 46, 50–52,
    293, 298, 301, 309, 316, 351, 360,
    460, 477
Commelin, Abraham, 212
Commelin, Isaac, 212
Continensis, Lucca Walter, 66
Copernicus, Nikolaus, 1, 164, 190, 314,
    397, 398, 433, 445, 447
Copius, Bernhard, 197
Copp, Johann, 297
Coppens van Diest, Gillis, 359
Cortés, Martín, 229, 239, 242, 243
Corvin, Matthias, 70
Corvinus, Antonius, 159, 160
Cranach, Christoph, 114, 143
Cranach, Lucas the Elder, 29, 100–102,
    106, 107, 111, 114, 115, 118, 120,
    124, 127, 138, 151, 155, 162
Cranach, Lucas the Younger, 115, 118
Crespin, Samuel, 187, 192, 200, 206, 207,
    210, 216
Creux, Jacques du, 206, 207
Cromberger, Juan, 229, 238

D
d'Ailly, Pierre, 82, 228, 277, 441
d'Ascoli, Cecco, 81, 82, 87, 277–279, 446
Danck, Johann, 70
Danfrie, Philippe, 296
Danti, Piervincenzo, 279, 282
David, Mathieu, 293
Dietrich, Sebastian, 111, 113, 129, 133,
    355, 356
Dietrich, Veit, 151, 152, 165
Dietz, Hans, 116, 125, 130, 139, 141

Dionysius the Areopagite, 442
Donà, Leonardo, 66
Döring, Christian, 111, 114, 115, 118, 124,
    136, 155, 431, 433
Draudius, George, 194, 195
Droscher, Paul, 115, 138
Dullaert, John, 467, 469, 470, 472, 474
Dürer, Albrecht, 36, 71

E
Eber, Paul, 128, 159, 165–167, 169
Eck, Paulus, 435
Edzard II, Count of East Frisia, 300
Eguía, Miguel de, 228, 230, 231, 247
Eleonora of Habsburg, Queen of France,
    299
Eliano, Giovanni Battista, 377, 378
Eliano, Vittorio, 201, 202, 374–380, 399
Erasmus of Rotterdam, 107, 135, 153, 195,
    231, 294, 472
Erastothenes, 445
Eschuid, Johannes, 80
Espinosa, Pedro de, 228, 232, 279, 280
Estienne the Elder, Henri, 7, 25, 27–32, 36,
    50, 51, 301, 351, 352, 460, 469–471
Estienne, Henri II, 309
Estienne, Robert, 28, 32, 34, 36, 293, 298
Euclid, 27, 34, 67, 73, 294, 307, 309, 311,
    312, 384, 387, 391, 398, 415, 431,
    433, 439, 446
Eyssenmann, Simon, 433

F
Faber, Wenzel, 412, 431, 435, 438, 439,
    441–445, 448
Fagius, Paul, 375–377
Faleiro, Francisco, 229, 238
Favre, Antoine, 192
Fernel, Jean, 215, 298, 299
Ferroni, Giuseppe, 397
Feuerhahn, Henning, 432, 433
Ficinus, Marsilius, 439
Finé, Claude, 308, 310
Finé, Jean, 308, 310
Finé, Oronce, 27, 36, 37, 43, 46–48, 51, 52,
    91, 173, 290, 293, 307, 308, 310,
    316, 460, 476, 477
Firmicus, Julius, 446
Florentinus, Antonius, 68
Florus, Lucius Annaeus, 312
Fontana, Christina, 66
Forcadel, Pierre, 307, 309, 311

Forfait, Benoist, 315
Fortius, Johannes, 474
Franceschi, Francesco de', 190, 202, 203,
    274
Francis I, King of France, 290, 296, 300,
    316, 321
Frederick III, Elector of Saxony, 120, 154
Freux, André des, 377
Froben, Johann, 29
Fuchs, Leonhart, 174, 175
Fusco, Domenico, 63, 66
Fusoris, Johannes, 296

G
Gabiano, David de, 205
Gabiano, Jean de, 192, 205–207
Galherde, Germão, 236, 246
Galilei, Galileo, 1, 190, 370, 398
Garamont, Claude, 316
Garcaeus, Johannes, 356
Gaurico, Luca, 280
Gazeau, Hugues, 210
Gelli, Giovanni Paolo, 192, 200, 201, 207,
    208, 210, 216
Gemma Frisius, Reiner, 153, 298, 300, 303,
    306–312, 358, 359
Gennep, Agnès van, 303
Gentil, Jean, 308, 314, 315
George, Duke of Saxony, 102, 155
Gerard of Cremona, 38, 53, 63–66, 82, 279
Gessner, Conrad, 158
Ghemart, Adrian, 229, 237, 243, 247
Girault, Ambroise, 304
Girault, Denise, 312, 313
Girón, Juan Téllez de, 236
Giunta, Bernardo (di Bernardo), 275,
    282–284
Giunta, Filippo, 256, 266, 274, 282
Giunta, Giovanni Maria, 263
Giunta, Giuntino, 82, 256, 279, 351
Giunta, Jacques de, 257, 258, 264, 280
Giunta, Lucantonio, junior, 52, 53, 82, 90,
    256–261, 263, 266–269, 271, 274,
    280, 281, 283, 351
Giunta, Lucantonio, senior, 259, 260, 279,
    281, 282, 283, 291
Giunta, Tommaso, 271
Giuntini, Francesco, 188, 190, 192, 204,
    205, 230, 243, 280, 281
Glarean, Heinrich, 304, 306
Goclenius, Rudolph the Elder, 198, 212,
    214, 215
Goldman, Andreas, 103, 104

Goltz, Moritz, 99, 110, 114–116, 118, 120,
    147, 151–154, 158, 164, 165–168,
    172, 175, 177, 178
Gómez Tejada de los Reyes, Cosme, 230,
    239
Gontier, Guillaume, 26
Gonzaga, Duke Ferdinando, 395
Gormann, Johann, 156
Goupyl, Jacques, 307, 312, 313
Gourmont, Gilles, 472
Gourmont, Jean de, 475
Gouvea, Diego de, 298, 299
Gracilis, Stephanus, 309, 312
Granville, Louis Mallet de, 466
Gregoras, Nicephorus, 297
Grienberger, Christoph, 373, 396
Griettan, Jean, 26
Griselle, Pierre, 26
Gronenberg, Simon, 359
Grosseteste, Robert, 82, 277
Grumme, Christoph, 119
Grünberger, Simon, 112, 113, 129, 134
Gruter, Janus, 212
Grynaeus, Simon, 347, 349, 355, 358
Guérende, Guillaume, 313
Gueullart, Jean, 311
Gutenberg, Johannes, 415

H
Hamel, Pascal du, 309, 310
Hangest, Jérôme de, 467, 477
Harambour, Augier d', 309, 311
Heffner, Claus, 100, 101, 120
Helt, Hugo, 300
Hempel, Frobenius, 119
Henrich, Henricus, 112, 134
Henrich, Merten, 112, 134
Henry II, King of France, 300
Heresbach, Konrad, 160
Hero, Albertus, 192
Herrera, Juan de, 230, 237, 244
Herwagen, Johann, 311
Hesse, Heinrich, 119
Hierat, Anton, 201, 207, 216
Higman, Jean, 25, 27–32, 50, 309
Hoffman, Andreas, 197
Hoffman, Daniel, 212
Hofman, Andreas, 115, 137
Hopyl, Wolfgang, 25–31, 49, 50, 80, 90,
    351, 352, 460, 464, 469

I
Ibn al-Haytham, Abu ʿAlī al-Ḥasan, 422
Ibn Rushd, Abū al-Walīd Muḥammad, 387
Italicus, Silius, 156

J
Jacob, Caspar, 429, 436, 440
Jacquinot, Dominique, 299, 308, 310, 321
Jenson, Nicholas, 68
João III, King of Portugal, 236, 299
John Frederick I, Elector of Saxony, 118,
    120
John of Damascus, 447
Jullieron, Guichard, 205
Junta, Juan de, 228, 230, 232, 234, 247,
    257, 259, 261, 264, 265, 266, 279,
    280

K
Kachelofen, Konrad, 412, 422, 435, 436,
    438
Kanisi, Christophorus, 432, 433
Keckerman, Bartholomäus, 198, 214
Kees, Thomas, 469, 470, 472–474, 476, 477
Kepler, Johannes, 1
Kersten, Wolle, 112
Ketton, Robert of, 296, 297
Klug, Joseph, 10, 108, 110, 111, 114, 151,
    153–155, 162, 165, 166, 177, 344,
    350, 356, 357
Klug, Thomas, 111, 112, 134
Köbel, Jacob, 297–299, 304, 306
Koberger, Anton, 291
Köln, Johann von, 388
Kopf, Peter, 194
Krafft the Elder, Johann, 99, 107, 113, 118,
    132, 138, 168, 199, 350, 355, 356,
    359
Krafft the Younger, Johann, 356
Krafft, Hans, 111, 118, 119, 132
Kreutzer, Veit, 107, 112, 134, 156, 168,
    175, 177, 185, 350, 356
Krüger, Thomas, 116, 140
Krumbfuß, Kilian, 111, 136

L
La Barre, Nicolas de, 473
Laínes, Diego, 375
Lambert, Jean, 465, 470, 472–475, 477
Lambertus Hersfeldensis, 157

Landsberg, Martin, 156, 412, 415–419,
        421–430, 434–439, 441, 444, 448
Lattes, Bonet de, 27, 297
Laux, David, 31
Lax, Gaspar, 467, 473, 477
Le Fèvre, Hémon, 473, 477
Le Preux, Poncet, 468, 469, 472
Lefèvre d'Étaples, Jacques, 17, 26, 29–36,
        41, 43, 45, 46, 49, 81, 82, 87, 90, 91,
        277, 286, 290, 298, 308, 310, 350,
        352, 441, 446, 447, 460–462, 464,
        470, 472, 473
Lehman, Jacob, 112
León, Juan de, 229, 236, 237, 246
Leopold of Austria, 415, 441, 445
Lesclancher, Michel, 475, 476
Levita, Elia, 375
Lignères, John of, 443
Lindau, Mark of, 70
Lob, Wenzel, 115, 137
Lombard, Peter, 311, 440
Longo, Tarquino, 391
López, Francisco, 240, 243
Lorraine, Christine de, 190, 309
Löslein, Peter, 67, 70, 73
Lotter, Melchior the Younger, 111, 114,
        115, 127, 138, 155
Loyola, Ignatius of, 371, 372, 374, 375, 379
Loys, Jean, 293, 299, 302–304, 312, 313,
        357–359, 460, 477
Lucanus, Marcus Anneus, 441
Lufft, Hans, 99, 110, 111, 113, 117, 118,
        120, 126, 132, 150, 163, 165, 167,
        168, 172
Luther, Martin, 29, 99, 102, 114, 115, 139,
        148–151, 153, 159, 162, 165, 167,
        197, 205, 356, 358, 433

M
Macrobius, Ambrosius Theodosius, 442
Madii, Francesco de, 66
Magellan, Ferdinand, 238
Magnien, Jean, 309, 311–313
Maius, Michael, 153
Maler, Bernard, 67, 68, 70, 73
Malines, Henry Bate of, 296, 297
Manilius, Marcus, 446, 447
Manutius, Aldus, 25, 28, 73, 291
Manuzio, Paolo, 202
Marcellus, Johannes, 126, 153
Marchant, Guy, 460, 462–465, 467, 470
Marchant, Jean, 462, 465, 468, 476

Marchetti of Siena, Salvestro, 190
Mareschal, Jean, 195, 210
Mareschal, Pierre, 210, 212
Marino, Giambattista, 381
Marnef, Brothers, 460, 468, 471
Marnef, Denise de, 304, 358
Marnef, Enguilbert II de, 312
Marnef, Jean III de, 312
Marnef, Jérôme de, 290, 304, 305, 311,
        358, 460
Marschalk, Nikolaus, 154
Martin, Jacques, 192
Martínez Siliceo, Juán, 304, 467, 472, 473
Māshāʾ Allāh, ibn Atharī al-Baṣrī, 296, 297
Mauro da Firenze, 282
Maurolico, Francesco, 245, 307
Maynemares, Jean de, 307, 314
Medici, Caterina de, 299, 307, 313
Medici, Cosimo I, 282
Medici, Cosimo II de, 190
Meietti, Roberto, 190, 195, 202
Mela, Pomponius, 67, 68
Melanchthon, Philipp, 9, 17, 99, 102, 110,
        115, 139, 147, 148, 150, 153,
        156–173, 176, 177, 191, 197, 303,
        304, 306, 310, 316, 319, 347, 349,
        353–358, 360, 432, 433, 448, 460
Mengewein, Hans, 119
Mercuriale, Girolamo, 158, 159
Mercurian, Everard, 382
Merian, Matthaeus, 212
Mesmes, Jean-Jacques de, 313
Mesmes, Jean-Pierre de, 307, 313, 314
Metko, Merten, 116, 141
Meyenburg, Michael, 158
Millanges, Simon, 312
Miranda, Luis de, 239
Mittelhus, Georg, 460
Mizauld, Antoine, 309, 312
Moibanus, Joannes, 158
Montmorency, Anne de, 314
Morel, Guillaume, 293, 312
Morel, Jean de, 309, 313
Morisanus, Bernardus, 187, 190, 192, 210,
        213–215, 217
Morpain, Francois, 312
Morrás, Domingo García, 230, 239
Moscatelli, Cesare, 395
Müller, Georg, 113, 135
Murs, John of, 431, 439

N
Nägele de Lindau, Bartholomaeus, 432

Navarre, Marguerite de, 299, 313, 314
Nebrija, Antonio, 227, 231, 235
Negro, Andalò del, 296, 297
Neumair, Christian, 164
Nuñes, Pedro, 192, 229, 236, 242, 243, 278

O
Offusius, Jofrancus, 314
Olschläger, Philipp, 115, 136
Oresme, Nicole, 38, 471
Osiander, Andreas, 165, 190
Osorio, Don Alonso de, 242
Ovidius Naso, Publius, 358, 413, 441

P
Pacioli, Luca, 36
Palladius, Peder, 356
Paraeus, David, 212
Paraeus, Philip, 212
Parix, Johannes, 225
Peckham, John, 156, 431
Peletier du Mans, Jacques, 303, 308, 309,
    312
Pena, Jean, 311
Penzio, Giacomo, 81, 83, 90
Perera, Benito, 214, 380, 387–390, 392,
    398, 399
Periegetes, Dionysius, 67, 68
Périer, Charles, 294
Perna, Pietro, 197
Peter of Cracow, 439
Petit, Jean I, 463, 471
Petit, Jean II, 291, 301, 305
Petit, Oudin, 291
Peurbach, Georg von, 2, 38–40, 43, 61, 63,
    64, 69–73, 75, 76, 80–83, 87, 90, 91,
    92, 166, 277, 293–295, 308, 411,
    422, 438, 439, 441, 444–447, 461,
    476
Philip II, King of Spain, 231
Philip IV, King of Spain, 242
Philip, Duke of Pomerania, 159
Philoponus, Johannes, 296, 297
Piccolomini, Francesco, 214, 307, 313, 388
Picquet, Pierre de, 304, 315
Pifferi, Francesco, 187, 189, 190, 191
Pinelli, Luca, 382
Pius IV, Pope, 378
Plains, Guillaume des, 475
Plato, 386, 387
Población, Juan Martinez, 298, 299, 306,
    308, 472

Polanco, Juan de, 378, 379
Possevino, Antonio, 267
Postel, Guillaume, 309, 311
Prigent, Calvarin, 475, 477
Proclus, 296, 297, 299, 304, 308, 313, 446
Psellos, Michael, 308, 313
Ptolemy, 27, 37, 38, 41, 43, 45, 164, 294,
    307, 309, 312, 398, 413, 425, 431,
    441–443, 445–447, 476
Puys, Jacques du, 291, 299

Q
Quentel, Arnold, 190, 204, 207, 208, 448

R
Ramírez de Prado, Lorenzo, 240, 243
Ramus, Petrus, 197, 198, 290, 311
Ratdolt, Erhard, 11, 34, 37–40, 61–63,
    67–70, 72–81, 83, 84, 86, 87, 89–92,
    415–419, 421–426, 438, 444, 445,
    460, 461, 469
Regiomontanus, Johannes, 38–40, 43–45,
    61, 63–73, 75, 76, 80–83, 86, 87, 91,
    92, 277, 299, 411, 422, 427, 438,
    443, 445, 447
Regnault, François, 293, 308, 309, 460,
    472, 475–477
Reinhold, Erasmus, 107, 135, 153, 154,
    294, 295, 307
Reinisch, Hans, 116
Renner, Franz, 61, 63–69, 72, 73, 80, 83,
    91, 414, 415, 438
Resch, Conrad, 291
Rhaetus, Thomas, 477
Rhau-Grunenberg, Johannes, 114, 115,
    140, 142, 155
Rhau, Georg, 99, 106, 107, 113, 117–119,
    125, 127, 151, 159, 167–169, 358
Rhegius, Urbanus, 153, 356
Rhenanus, Beatus, 29
Rheticus, Georg Joachim, 4, 128, 147, 148,
    151–153, 163–166, 173, 178, 355,
    357, 432, 433
Ricardo, Antonio, 245
Ricci, Matteo, 373
Richard, Guillaume, 302–304, 309, 460
Richard, Jean, 281, 302, 357
Richard, Thomas, 302–304, 359
Rioja, Francisco de, 242
Rocamorra y Torrano, Ginés, 230, 237,
    242–244
Roce, Denis, 465, 470, 473, 475

Rojas Sarmiento, Juan de, 299, 300, 321
Rolevinck, Werner, 70
Ronsard, Pierre de, 313, 314
Roomen, Fabiaan van, 202
Rörer, Georg, 151, 152
Rosso, Bernardino, 90, 351, 352
Rosso, Giovanni, 82
Rottweil, Adam of, 63, 64, 66, 73, 415
Rouillé, Guillaume, 240
Rüdem, Henning, 158
Ruehel, Conrad, 99, 108, 113, 116, 118–120, 140, 141
Ruehel, M. Johann, 118, 140
Ruehel, Maria, 116
Runge, Jakob, 159

S
Sabinus, Georg, 358
Sacrobosco, Johannes de, 1, 5, 13, 14, 16, 20, 25–29, 36, 41, 61–63, 66, 69–73, 75, 76, 79, 80, 87, 91, 92, 163, 164, 168, 187–191, 194, 200, 226–230, 232, 234, 236, 241–243, 246, 255, 277–280, 282, 293, 303, 304, 313, 337, 341, 355, 357, 359, 371, 374, 391, 398, 409–411, 413, 415, 418, 422, 425, 428, 429, 437, 431–449, 459
Salas de León, Christóbal, 240, 243
Sanctis, Girolamo de, 71, 72, 80, 81, 86, 206
Santritter, Johannes, 38, 40, 43, 61, 71, 72, 74, 80, 81, 83–87, 89, 90, 92, 415–419, 421–423, 438
Säuberlich, Lorenz, 113, 114, 117, 135, 138, 356, 359
Saup, Thomas, 116, 134, 140
Savetier, Nicolas, 298
Saxonia, Albertus, 466, 471
Schappeler de S. Gall, Christoph, 432, 433
Schedel, Hartmann, 71
Scheidewein, Heinrich, 172, 173
Schenck, Jakob, 159
Scheubel, Johann, 307, 311
Schierlentz, Nickel, 115, 139, 141
Schleich, Clemens, 112, 113, 132, 135, 355, 358
Schnellboltz, Gabriel, 110, 129
Schöne, Anton, 112, 355, 358
Schöner, Johannes, 297, 298, 307, 433
Schorer, Peter, 111, 114, 136

Schotte, Hans, 111, 114, 136
Schramm, Christoph, the Younger, 110, 115, 117, 118, 127
Schramm, Christoph. the Elder, 107, 108, 114, 115, 117, 118, 125, 127, 130
Schröter, Johann, 111, 112, 114, 115, 119, 126, 131, 136, 142
Schürer, Widow and Heirs of Zacharias, 356
Schürer, Zacharias, 114, 138, 197, 356
Schürer, Zacharias & Partners, 356
Schwertel, Johann, 116, 118, 141
Scot, Michael, 82, 277
Scoto, Girolamus, 18, 71, 270, 274, 276–278, 280
Scoto, Ottaviano, 38, 40, 43, 53, 81–83, 90, 277, 279, 280, 469
Scotto, Heirs of Ottaviano, 53, 82, 90, 279, 280
Scribonius, Wilhelm Adolf, 198
Scultetus, Abraham, 212, 213
Seitz the Elder, Heirs of Peter, 356
Seitz the Elder, Peter, 112, 113, 129, 167, 304, 356, 358, 364, 365
Seitz the Younger, Peter, 358
Seitz, Peter II, 112, 129
Selfisch, Samuel, 99, 106, 110, 111, 116–120, 127, 153
Senant, Olivier, 470, 473, 475, 477
Seneca, Lucius Annaeus, 28
Sertenas, Vincent, 314
Sessa, Giovanni Battista I, 81, 90, 291, 470
Sessa, Melchiore I, 82, 90, 199, 274, 357
Setzer, Johannes, 162, 167
Seville, Isidore of, 441
Seville, John, 441
Speyer, Johann von, 30, 71
Stabius, Johann, 297
Standonck, Jan, 466
Staserio, Giovanni Giacomo, 196, 217, 395
Stauffenbuel, Wolff, 113, 118, 119
Stöckel, Wolfgang, 155, 412, 428, 435, 436, 438–440
Stöffler, Johannes, 305, 307, 308, 315, 319, 322, 329, 331, 332
Suárez, Francisco, 214
Swineshead, Richard, 468

T
Taberniel, Jacinto, 230, 239
Tacitus, Johannes, 349

Tedeschi, Tommaso, 473
Thābit ibn Qurra, 411
Theodosius de Bithynia, 82, 90, 277, 417,
    426, 446
Thilo, Paul, 119, 131, 134
Thomas, Alvarus, 467, 468
Tinghi, Filippo, 205, 280, 281
Tockler, Conrad, 411–413, 429, 432–434,
    439, 445, 446, 448, 449
Toledo, Francisco de, 214, 231, 247, 376,
    379, 391
Torrecilla, Juan Flores, 240, 243
Tory, Geofroy, 36
Tossanus, Paul, 212
Tournes, Jean de, 312, 359
Tournon, Claude Michel de, 206, 207, 216
Trebelius, Hermann, 154, 155
Tremblay, Lucas, 308, 312
Trino, Gugliemo da, 81, 83, 87
Tumner, Peter, 440
Turon, Michael, 70

V
Valeriano, Pierio, 3, 192, 278, 349
Valla, Giorgio, 296, 297, 299
Valla, Lorenzo, 28, 469
Valleriani, Matteo, 2–4, 16–18, 20, 65, 75,
    90, 147, 148, 163, 228, 293, 338,
    341, 344, 410, 411, 437, 439, 441,
    445, 449
Vascosan, Michel de, 291, 293, 299, 316
Velcurio, Johannes, 170
Vendôme, Georges Bruneau de, 468
Venetus, Paulus, 469, 472, 474
Vérard, Antoine, 29, 150, 291
Verde, Simone de, 66
Vergilius Maro, Publius, 28, 71, 441
Viart, Guyone, 29, 32
Villalta, Donato, 3, 349
Vinet, Elié, 18, 192, 194, 200, 230, 278,
    306, 308, 312, 313, 349, 353

Vio, Tommaso de, 258
Vives, Juan Luis, 467, 470, 474, 476
Vogel, Bartholomäus, 99, 106, 107, 110,
    114–116, 118–120, 131, 151, 155
Vögelin, Gotthard, 166, 212
Volmar, Johannes, 164
Vostre, Simon, 232, 246

W
Wechel, André, 197, 311
Wechel, Christian, 291, 293
Weiß, Hans, 114, 137
Weiß, Severin, 128, 132
Welack, Matthaeus, 113, 117, 118, 142,
    355, 359
Wellendorffer, Virgilius, 431
Wettin, Ernest of, 100
Wilde, Simon, 169–171
Willer, Georg the Elder, 188, 189, 191
Willich, Jodocus, 170
Winner, Burchard, 134
Wittelsbach, Prince Wilhelm of Bavaria,
    212, 381
Wouwer, Gaspar van den, 202

Z
Zabarella, Jacopo, 192, 195, 198, 202, 210,
    214
Zamora, Alfonso de, 231
Zanetti, Bartolomeo, 282
Zanetti, Daniele, 203
Zanetti, Francesco, 202, 357, 380, 390
Zanetti, Luigi, 380
Zetter of Hanau, Paul de, 212
Zetter, Jacob, 212
Zetter, Peter, 212
Ziegler, Jacob, 297
Zwinger, Theodor, 158

Printed in the United States
by Baker & Taylor Publisher Services